TO RIDE THE STORM

The story of the British airship programme of 1924 to 1930, of HM Airship R.101 and the men who planned, designed and flew it – of a high endeavour to bring lighter-than-air travel to Empire air routes around the world – of Brigadier-General the Right Hon the Lord Thomson of Cardington, HM Secretary of State for Air – and of a princess and a carpet.

Outward Bound – HM Airship R.101C (five 650 bhp Beardmore Tornado IIIA compression-ignition engines) at the Cardington mooring tower on the afternoon of Saturday, 4th October 1930, before its departure for India, by way of Ismailia, Egypt, carrying 12 passengers, 42 crew and 25¼ tons of fuel.

To Ride the Storm

The Story of the Airship R.101

Sir Peter G. Masefield

WILLIAM KIMBER · LONDON

First published in 1982 by
WILLIAM KIMBER AND CO. LIMITED
Godolphin House, 22a Queen Anne's Gate,
London, SW1H 9AE

ISBN 0-7183-0068-8

Typeset by Grove Graphics
and printed and bound in Great Britain by
Robert MacLehose and Company Limited, Glasgow

Contents

PART TWO

The India Trail

*

EPILOGUE

How It Happened

*

List of Illustrations

List of Maps and Text Illustrations

Foreword

by

DR NOBLE FRANKLAND, OBE, DFC
Director of the Imperial War Museum, London

When she set course for India on the night of 4th October 1930, R.101 was the largest vessel in the world and the shed at Cardington in which she had been prepared was the largest building in the United Kingdom. The passenger accommodation in R.101 corresponded to what was offered by a substantial country house and the fabric of her outer-cover extended to more than five acres.

The voyage to India was the culmination of six years of pioneering endeavour and the brink of realisation for an Imperial dream; the age of the great airships would link the Empire and the Mother Country as never before; comfort, speed, regularity and safety would all be provided and Australia, India and Canada would only be half the journey the steamships of the surface knew. Perhaps even more than all these things, there would be a prestige of the air corresponding to that of the great ocean liners on the North Atlantic route; it would be world-wide and it would be British. It seemed to be a fairy tale about to come true. Yet within little more than two minutes at Beauvais in northern France only a few hours out from Cardington, R.101 crashed. She and nearly all those on board, including the Secretary of State for Air, Lord Thomson, were consumed in flame and the imperial idea of the great airship age perished with them.

The nation was stunned. The impact was not less and the results far more decisive than what followed the sinking of the *Titanic*. Those who remember the loss of R.101 will never forget it and those who now hear of it can scarcely not be profoundly moved. Yet curiously until now no one knew what happened, or to be more precise, exactly what happened.

The riddle of this paradox has set the bit between the teeth of Sir Peter Masefield and, over a period of nearly forty years, his remorseless but sympathetic curiosity coupled with his enduring patience and his profound experience and knowledge of aviation, have enabled him to produce the solution. If it seems strange that such a disaster should wait so long for its concluding chapter, there are stranger aspects than that of this extraordinary episode, which not only signalled the end of an era in aviation history but left a scar upon the very heart of the nation.

Sir Peter's book is important because it sets an historical record straight after it had wandered astray for so long and because it does so in a manner that is authoritative beyond doubt. It is fascinating in the same way as a well-founded detective story and it is moving because of the powerful elements of human endeavour and romantic involvement which perished with the ship and the men at Beauvais. Also, for me as I expect it will for countless others, it answers a need. Throughout all these years I have wanted to know; now I do.

TO RIDE THE STORM

'To ride the storm has
always been my ambition . . .'
LORD THOMSON OF CARDINGTON

INTRODUCTION

Imperial Dream

'R.100 gave me pleasure. R.101 will, I hope, give me joy. To ride the storm has always been my ambition and who knows but that we may realise it on the way to India – but not, I hope, with undue risk to human lives.'

LORD THOMSON in a letter to
the Princess Marthe Bibesco,
20th August 1930

'THERE ARE two tragedies in life,' wrote Bernard Shaw.* 'One is not to get your heart's desire. The other is – to get it.'

For Christopher Thomson, there came both.

Hence this book.

Its origins go back to wartime Whitehall. There, in 1943, I had been set to explore, once again, the merits of the rigid airship for postwar ocean travel. Those reassessments led back to R.100 and R.101, and from R.101 to His Majesty's Secretary of State for Air in 1924, and again, in 1929–30, to Brigadier-General the Right Honourable the Lord Thomson of Cardington, PC, CBE, DSO, psc – soldier, engineer, diplomat, writer, politician, traveller, the originator of the final British airship programme; a man of ability, humour, learning and of many depths – to whom, as yet, history has not accorded his due.

Down the trail of the airships and out of a dedication to a dream of Empire air routes, across a world now faded and gone, there comes a story of devotion to a cause and, with it, devotion to one of the loveliest and most gifted women of her day: to Marthe Lucie, the Princess Georges Bibesco.

In the pursuit of one heart's desire, the other was shattered, too – on a hillside, near Beauvais, in France.

*

* *Man and Superman*, Act IV, 1903.

Strangely, the story began with a carpet; not an ordinary carpet, but one of the most magnificent of ancient Persian carpets of Kashan, about whose existence – but not survival – I first learned on a winter's evening just before Christmas of 1943.

All that day, four of us had laboured at Gwydyr House, in Whitehall, among airship files from Air Ministry archives. Now, behind the blackout, in the warmth and comfort of the smoking room of The Royal Aero Club in Piccadilly, as we turned from our coffee to our port – a superb Martinez 1927 – up came the shadow of Lord Thomson's flying carpet, a shadow which gained in substance with the after-dinner talk.

Our minds went back to 1930, to the loss of the airship R.101, its dozen passengers and all but six of its 42-man crew.

'I've always felt that the carpet was – so to speak – the last straw,' said the veteran among us, Tom Cave-Browne-Cave,* he who had been one of the last at the mast-head on that wild October night.

'To the S of S, it was a matter of course that the carpet should go – as a symbol of something-or-other, so I was told. Anyway, up it went in the lift to the mast-head and it was stowed away in the forward gangway when the ship "slipped" and the voyage began.'

He paused.

'I've always thought that R.101 could have made it to India,' he went on thoughtfully. 'I believe that it would have got through all right, except for bad luck, bad weather, the fact that Scottie† was off-colour, that he determined to press on through wind and rain, which worsened every moment, and but for the fact that they had the Minister and all his entourage on board. Once started, then, of course, they couldn't turn back.

'Then, there was that extraordinary "miscalculation" – the only word to use for it, I suppose. The Secretary of State drove up to Cardington on that Saturday night with a van-load of baggage. I remember it well. There were several cabin trunks, a case of champagne for the State Dinner at Ismailia, and – all unheralded – that Persian carpet to go down on the floor. Magnificent, I'm sure. But, no light-weight when we needed to save every ounce.

* The late Wing-Commander Thomas Reginald Cave-Browne-Cave, CBE, FRAeS, MIMechE, formerly RN and RNAS and RAF (ret'd) (1895–1969).

† Major George Herbert Scott, CBE, AFC (1888–1930) who had commanded the British Airship R.34 on its two-way trans-Atlantic crossing in July 1919, and was, in 1930, Assistant Director of Airship Development (Flying and Training) at the Royal Airship Works, Cardington. He lost his life in R.101.

'Anyway, it was taken on board, done up in a roll, in the care of the Minister's young valet; the faithful James Buck.

'You know, I can't help feeling that the carpet loaded at the nose may have tipped the balance; not on its own, of course, but added to all the other things when the critical moment came.'

Airships were filling our minds that evening: their unfortunate history; whether they might have a future? Post-war planning was in the air. Every prospect was being dusted over and re-examined. I was specially involved through my lord and, temporary, master – Max Beaverbrook, Lord Privy Seal in Winston Churchill's War Cabinet, who had taken on, as a special responsibility, the planning of post-war civil air transport. From Lord Beaverbrook's office in Gwydyr House in Whitehall I had been appointed Secretary of the specific Cabinet Committee.

Odd as it may seem to-day, one of the items on the agenda for the after-Christmas meeting of the War Cabinet's Air Transport Committee in 1943 was to be a North Atlantic Airship Feasibility Study – a design assessment of a hypothetical helium-filled airship to be set alongside Leslie Frise's plans for the Bristol Type 167 land-plane. Seven-engine R.104,* or eight-engine 'Brabazon I'? Which might be the answer for post-war air services on the North-Atlantic route? Both had to meet Air Ministry Specification C.18/43 for a long-range transport capable of flying 5,000 miles non-stop carrying 50 passengers, their baggage and 7,500 lb of mail and cargo: 20,000 lb of payload in all.

Together with Jock M. Gray at the Ministry of Aircraft Production, I had roped-in two Lighter-than-Air stalwarts for the assessments. There was Wing Commander T. R. Cave-Browne-Cave, at that time the Director of Camouflage at the Ministry of Home Security but a great airshipman in his day. There was Lieutenant-Commander Nevil Shute Norway from the Admiralty, in earlier days one of Barnes Wallis's backroom boys on R.100 before he gave up aviation for literary stardom as 'Nevil Shute'.

R.104 is another story; as Lord Beaverbrook said, 'A pretty face, but no good in the kitchen'. But for the past few days at Gwydyr House we had lived with ghosts; ghosts who were manifest and articulate, as we pored over their day-to-day records, their careful

* R.102 and R.103 were design projects initiated by Colonel Vincent Richmond and Squadron Leader Michael Rope at the Royal Airship Works, Cardington, during 1929 and 1930.

plans, all neatly stored in the dusty files of the old Royal Airship Works, Cardington.

The whole sad saga was recorded there in the notes and minutes which had flowed back and forth between Cardington and Whitehall, some of them on notepaper from the same Gwydyr House, headed 'Secretary of State for Air'.

Through 1929 and 1930, there had been two central themes, R.100 and R.101 – but mostly R.101. How much enthusiasm, how much dedication, what thousands of man-hours, how many hopes – but so few fears – were inscribed on those faded pages.

Then, one dark and stormy night in 1930, they had all ascended in the Cardington lift, as though to the stars, and they had set out together, Minister and men, the flower of British airship knowledge, energy and experience, never to return. They left behind, in two neat files,* a monument to their aspirations for the future. They were the designs and estimates for R.102 and R.103, new airships to embody all they knew; all they had learned. These new airships would, they believed, make good the deficiencies of R.101; and R.100 besides. They would open new vistas for long-range travel; fast, comfortable – and safe.

Now, thirteen years later, as we settled into our Aero Club armchairs under the dim lights behind the blackout, the ghosts were still around us.

We remembered the energetic, debonair, fun-loving Sir Sefton Brancker, always ready with a joke but deadly serious about aviation and its future. As Chairman of The Royal Aero Club's Racing Committee he – with his chief – had done much to organise successive and successful British entries for the Schneider Trophy Contests of 1927 and 1929 which had finally brought the Trophy to a permanent site in the foyer of The Royal Aero Club below. And like his chief Brancker was an enthusiastic believer in Lighter-than-Air for long distance air services; a belief he had reiterated in numerous speeches and lectures.

We remembered, no less, his chief, Christopher Birdwood Thomson – always 'C.B.' to his friends – professional soldier turned politician; witty speaker, fluent writer, connoisseur of music and literature; tall, elegant, cosmopolitan; a man of character, of intelligence and charm; erudite and introspective with a sensitive nature and impeccable manners. A former Royal Engineer, Lord Thomson had become a

* Public Record Office (A/M) Files: AIR 5/987/3069 and AIR 5/988/3069.

devoted and informed convert to aviation, a convinced advocate of airships and Chairman of the Royal Aero Club from 1926 to 1929. Warm of heart, sometimes hot of head, he was an idealistic Socialist, who to-day would be termed a middle road Conservative. He held the post of Secretary of State for Air in each of the first two Labour Governments, for 865 days in all.

Thomson and Brancker had never known The Royal Aero Club at 119 Piccadilly, whither it had moved from 3 Clifford Street in June of 1931, eight months after the tragedy at Allonne. But their presence was still to be felt as they gazed out from among the portraits and photographs upon the walls. I recalled, too, that only a few mornings before, homebound from a day-bombing mission in a B-17F Flying Fortress of the 96th Bomb Group, US Eighth Air Force, I had myself looked down upon the fatal Beauvais Ridge – and remembered them.

Now, N. S. Norway chimed in on our thoughts.

'I was never a fan of R.101, as you know,' he said. 'For my money, R.100 was always the better ship by a long chalk. But R.100 was destroyed by that Beauvais wreck as completely as was R.101. I didn't know about the carpet, Cave. How odd. I can well imagine that – as you say – it might have been the final straw.'

'Who can tell?' said Cave-Browne-Cave. 'But R.101 was potentially much better than you think, Norway. There were a great many fine features about it. Its troubles – at least early on – were mostly to do with its outer cover and its gas bags; just the same sort of problems as you had with R.100. But, you know, there were other things as well. Scottie, in charge of flying, was a sick man. There were others, too. Even now, I won't say more. Of course, it all put an extra heavy strain on "Bird" Irwin, the Captain of the ship. With dear old Ralph "Mouldy" Booth, "Bird" – though pretty highly-strung – was one of the best airship captains we ever had. But, apart from "Grabby" Atherstone and "Sky" Hunt, the First Officer and Chief Coxswain of R.101 (and both very sound), Irwin wasn't well backed up on the flying side.

'There was a lady somewhere in the shadows, too, poor dear. There was even a rumour that there was a woman on board – though I know that wasn't true. But I never knew the full story and I doubt if anyone did. For all the good times of the British Airship Service – and they were good – it came as a tragic end. Unnecessary, too.'*

* Discussion recorded in notes made at the time; December 1943.

The conversation drifted on to other things. But Lord Thomson's carpet remained in my mind. Why had he taken it with him? What was its story? Who was the lady? Had she and the carpet really existed?

The quest to learn the truth was to lead on down a winding and misty trail over many years.

As for R.104 – that last official assessment gave, in 1944, the same answer as it would do to-day. The elegant and evocative shape of the rigid airship will never again be seen in the skies. For all the fervour of its advocates, advances in technology have passed it by.

The small, non-rigid, 'blimp' may still find a limited use. But the large rigid is too fragile, too ponderous, too slow and too unproductive to exist in the competitive world of to-day alongside the modern aeroplane, the heavy-lift helicopter and the container ship. The airship's buoyancy is too much affected by variations of outside temperature and in atmospheric pressure. The airship has perforce to fly too low to achieve either adequate safety or acceptance from those on the ground.

Most of all among these shortcomings, new large rigids would cost too much to develop, to build and to test, while, as in the past, their great bulk would be too vulnerable in wind and weather to survive.

*

The story of the rigid airship – which started with Count Ferdinand Zeppelin and the LZ-1, in July 1900, and ended with the breaking-up of the LZ-130, *Graf Zeppelin II*, in March 1940 – is an open-and-shut chapter in the forward path of aeronautics and of transport; but, nonetheless, a stage towards the reliability and the safety with which commercial transport aircraft girdle the world to-day.

Throughout the 1920's, while the fixed-wing aeroplane was still in its infancy, high hopes were held for the future of the airship as a serene and spacious means of long-range transport – hopes which came to be dashed, successively, in Britain, in France, in the United States and in Germany, when, one by one, between 1921 and 1937, the great rigids, R.38, *Dixmude*, the *Shenandoah*, R.101, the *Akron*, the *Macon* and the *Hindenburg*, all met with disaster. By the late 1930's, only the *Los Angeles* in the United States and the two *Graf Zeppelins* in Germany remained (the earlier two grounded) and by 1940 all of them had been broken up.

The seven rigid airship disasters, with the loss of 266 lives, brought to an end the aspirations of the devoted band of airshipmen from four nations whose vision, skill and dedication had remained steadfast towards Lighter-than-Air.

Central to the airship story is the British programme of 1924, which ended tragically with R.101, leaving behind a tangled web of myths, legends and distortions of fact, together with an unsolved mystery. It is a story of years of dedication to a dream of Imperial air routes in an age when journeys were still reckoned in weeks instead of hours; when leisurely travel by sea was accepted as the norm.

Most of all, it is the story of a drive to bring about a new era of long-range air transport some twenty years before that dream was achieved by another means.

From 1924 to 1930 – working on the, then, frontiers of knowledge – new ideas, new designs, new techniques of construction and two new airships were set in hand to produce what was, confidently, expected to be a leap forward in the transport of passengers, mail and goods across the world; in Lord Thomson's words:* 'to give a unity to widely scattered peoples, unattainable hitherto; to create a new spirit, or, maybe, to revive an old spirit which was drooping and to inculcate a conception of the common destiny and the mission of our race.'

A great ambition – to be attained, in large measure, a generation later when fixed-wing aircraft had grown in power and range sufficient to span the widest oceans.

There is an irony in the fact that – at a time when technological development inclines towards hydrogen fuels as the potential energy for air transport when, eventually, the oil runs out – a hydrogen conflagration ended British airship ambitions in 1930 and brought all civil rigid airship operations to an end with the *Hindenburg* in 1937.

But, there was much more to it than went up in those hydrogen-fuelled holocausts – and the helium-inflated *Akron* and *Macon* both came to no less tragic ends.

Over the years, the accident to R.101 has occupied much the same place in the history of air travel as the sinking of the *Titanic* has come to hold for the sea. The reasons for the loss of the *Titanic* – a fine ship sunk through a natural hazard compounded by over-confidence and human error – are clear and unclouded. The very similar reasons that led to the loss of R.101 have always been shrouded in doubt and confusion.

Bit by bit and piece by piece, new facts have come to light. One

* In 'Air and Empire', *The Observer,* London, Sunday, 10th October 1926.

by one, they have fitted together like parts of a jig-saw to reveal a pattern which, though simple enough, is fundamentally different from the suppositions and guesswork put forward by the Official Inquiry – and different, also, from the pattern of inaccuracies and myths which have been built up around the story from so many sources since then.

A proper understanding of the sequence of events and of the factors which led up to the R.101 disaster can be gained from an assembly of all the facts stored within Air Ministry files and Cabinet Minutes, (some only recently open to inspection) together with those from the diaries, the letters and the notes left behind by some of the chief actors in the tragedy, from discussions with survivors and some who were not on board, from ground and air inspections of the site of the disaster and from analysis of the store of weather knowledge which has been accumulated throughout the aviation scene in the years which have followed.

In addition to shedding new light on the background to the airship programme and to the disaster, this new evidence corrects four basic errors of fact from which a fabric of misconception had been woven around R.101.

First, it is now clear that there is no truth in the assertion – so often repeated that it became accepted lore – that the Secretary of State for Air brought personal and political pressure to cause R.101 to leave for India before the airship was ready and against the better judgment of those professionally concerned. That was not so. Indeed, on 2nd October 1930, at the final conference on arrangements for the flight, Lord Thomson emphasised that, 'There must be no rush on my account'. At dinner that night, he remarked that he was still uncertain whether the ship would be ready to leave for India during the next few days and that, if it did, whether he would – or could – fly with it as he so much wished.

There is, equally, no truth in the myth that Wing Commander Reginald Colmore, the Director of Airship Development, was told that, 'if the ship did not get the Secretary of State to India in time for him to arrive home for the Imperial Conference, no further money would be available for airship development – and none would be sought'.

On the contrary, before the end of September 1930, on the instructions of the Secretary of State, Sir Henry Self at the Air Ministry, had sought and gained approval in principle from the Treasury for an on-going airship programme and for plans for the start of regular airship services to India and Canada in 1931.

Nor is there truth in the statement by Major Oliver Villiers, (made under a misapprehension after the Public Inquiry) that Sir Sefton Brancker, the Director of Civil Aviation, had sought Lord Thomson's agreement to the abandonment of the India flight and that the Secretary of State had replied to him, 'If you are afraid to go, there are plenty who will be glad to have your place.'

Such a situation, and such a remark, would have been wholly out of character for both Brancker and Thomson, who were close friends, but the records show also that Brancker's name was not on the original passenger-list of R.101 for the India flight. He was included only after he had insisted that he be allocated the place of Wing Commander Tom Cave-Browne-Cave, who was to remain behind to press on with engine development for R.102.

Lord Thomson and Sir Sefton went together to see Mr J. H. Scullin, Prime Minister of Australia, on 3rd October 1930, the day before departure. Mr Scullin recalled: 'They were both full of enthusiasm on the subject of linking up Great Britain with Australia by airship. They undertook, on their return from their journey, to bring me their impressions so that we could discuss the matter further.'

Finally, the allegation is wholly false that those who designed, built and developed R.101 were incompetent and irresponsible. The Cardington design and construction teams included men of the highest ability and responsibility. They authorised the departure of R.101 to India only after they had assured themselves that – after an extensive refit – the ship was sound and airworthy and capable of undertaking the long voyage.

The Cardington team was a remarkable group. Some of those who were not aboard went on to achieve distinction in aeronautics, science and technology over a wide field; among them the Lords Baker and Kings Norton, Sir Alfred Pugsley and – at the Royal Aircraft Establishment – T. S. D. Collins and Miss Hilda Lyon.

Perhaps the most authoritative comment on R.101 and its builders was made by Dr Ludwig Dürr of the German Zeppelin Company; Chief Designer of the *Bodensee*, the *Los Angeles*, the *Graf Zeppelins* and the *Hindenburg*.

Following a lecture by Colonel Vincent Richmond, R.101's Chief Designer, at the annual meeting of the British Association on 8th September 1930, Dr Dürr said in the discussion: 'I regard R.101 as one of the best airships ever designed and constructed.'

Such an endorsement, from first-hand knowledge of R.101, on the part of the most experienced airship designer and constructor in the world, together with R.101's successful 16-hour trial flight on 1st and 2nd October 1930, reinforced the confidence of the Cardington team in their ship with its new cover and increased lift. They believed it to be airworthy and sound and there is good evidence now that, within the standards of the day, it was just that.

The departure for India – so long planned and so long delayed – was now looked upon as a matter of course. With very few exceptions, there was an atmosphere of assurance everywhere, free of thoughts about failure or concern about safety. Those who were to go, were the envy of all who were to be left behind.

So R.101 departed from Cardington for Egypt and India in the same aura of enthusiasm and confidence as attended the maiden voyage of the RMS *Titanic*, outward bound from Southampton to New York on 10th April 1912.

How, and why, R.101 crashed and how and why its story came to be so distorted afterwards is a tragic tale in itself – stuffed with long-dead prejudices, antagonisms and misconceptions, personal and political.

For all the deficiencies inherent in 'buoyant flight' in general, and in the inadequately-tried R.101 in particular, there is now no doubt that but for a combination of circumstances, against which the odds would have seemed very long indeed, and but for one last minute error in airmanship, the disaster to R.101 would not have happened. Among the contributions to catastrophe must be numbered an over-confident approach, some crew fatigue, a fatal malady, a – then unpredictable – combination of weather and geography, and a final snap decision to take exactly the opposite action from that which might have saved the ship and all on board.

With all of this, one must reflect that most of the critical factors which led to the disaster were common to both R.100 and R.101, as well as to other rigid airships of the time. In the same circumstances, on the same day, at the same place and with the same action, a similar fate would, almost certainly, have befallen R.100, or other airships, as came to R.101 at Allonne at 02.09 hrs on 5th October 1930.

With the understanding which is gained only with hindsight, we can see that R.101 ought not to have set out for India when it did. Against to-day's standards of airworthiness and of test-flying, R.101 should not have been scheduled to start on its first overseas flight so little proven – as did and were almost every airship before it. Of course passengers ought not to have been carried on an experimental flight. Of course in the face of the adverse weather forecast, the flight ought to have been postponed. But, at the time and on the day, in the atmosphere of confidence and assurance, all the essential requirements seemed to have been met. To have held back after so many years of preparation, would have been unthinkable. They believed themselves to be ready – and they went.

The subsequent Public Inquiry was, by modern standards of accident investigation, both superficial and inadequate. It did not establish the reasons for the disaster and it resorted to unsubstantiated guesswork: a gas-bag might have torn, the cover might have split; a sequence of events could be conjured up to support such suppositions. But, as Lord Brabazon, one of the assessors, candidly remarked to me, years later: 'We never did find out why R.101 crashed.'

More research, more consideration, later techniques of accident investigation and modern knowledge from wider experience of aeronautics, have made possible a more precise view of the whole affair.

In particular, during the almost half-century since R.101 crashed, there have been other accidents which have thrown new light on the circumstances of the night of 4th/5th October 1930. There have been occasions when bad weather has brought about conditions which were too much even for modern jet aircraft during approaches to land. Severe, low level turbulence in high winds, together with 'standing waves' over hilly ground have led to a number of fatal crashes of airworthy aircraft.

Analyses of the synoptic weather pattern around Beauvais on the night of 4th/5th October 1930, together with the evidence of French witnesses on the ground, all show that such adverse conditions did, indeed, exist. But there was much more to it; as is related in this book.

Although the Official Inquiry into the R.101 disaster discounted weather as a primary cause, insufficient evidence was taken on this subject and those concerned did not probe into the detail which, even then, was available.

However inadequate in such respects, the Official Inquiry did, however, underline some indisputable facts.

> What is so impressive [it records]* is the high courage and genuine confidence of all concerned. For four years, they had been preparing for this moment – and they believed themselves prepared for it.

* Report of the R.101 Inquiry: March 1931 (Cmnd 3825), pp. 67, 68 and 96.

> There was the fact, which never should be overlooked by those who now criticise, that more careful calculations had been made with regard to every part of R.101 than of any previous airship.
>
> And, there was the personality of the Secretary of State, whose enthusiastic backing of this part of British Air Policy must have been, from first to last, a most comforting support for the Cardington experts and whose resolution to travel himself in the ship . . . gave a cachet of Government support in the most striking way to the enterprise which completely filled the hearts and minds of the officers who perished.
>
> The event brought to a close, in a few moments of tragedy, the enterprise upon which years of concentrated effort had been directed on the part of these pioneers. The circumstances of their deaths can only add to the admiration evoked by their courage, skill and devotion.

Not for the first time – nor for the last – pioneers perished for their beliefs. 'The blood of the martyrs is the seed of the Church.'*

So, this book explores again the history of the British airship programme of 1924 to 1930, the story of those who created it, the last flight of R.101 and the complex series of incidents which led on to disaster.

Four central figures stand out.

First and foremost was the debonair and determined Christopher Birdwood Thomson – Brigadier-General the Lord Thomson of Cardington, Secretary of State for Air in 1924 and again from June 1929. Alongside him, during the critical launch-period in 1924, was William Geoffrey Hanson Salmond – Air Chief Marshal Sir Geoffrey Salmond, Air Member for Supply and Research at the Air Ministry from 1922 to 1927 – the man who conceived, organised and pressed forward the technical programme for the new airships. No officer in the Royal Air Force was more highly esteemed than Geoffrey Salmond and none inspired greater personal regard.

The third was Wing Commander Reginald Colmore, the Director of Airship Development from January 1930, a professional airshipman who had been with the airship programme since its start and had laid down the specifications to which both R.100 and R.101 had been built. Colmore was a man of skill, courage and integrity who – while devoted to airships – was well aware of their limitations.

The fourth principal was Lieutenant-Colonel Vincent Crane Richmond, Assistant Director of Airship Development (Technical) at the

* Tertullian (AD 160–225) Apol C. quoted by Sir Maurice Dean *The Royal Air Force and Two World Wars*, p. 86 (Cassell, 1979).

Royal Airship Works, Chief Designer of R.101 and leader of the research and construction team at Cardington. Magnificently supported by his chief assistant, Michael Rope, a design-engineer of genius, Richmond was a man of infinite pains and strong personality.

From the collaboration of these four dedicated men, with their teams, came R.101, and the plans and the designs for R.102 and R.103 to follow. Colmore and Richmond remained uninterruptedly with R.101 from start to finish. After the airship programme had been launched, Lord Thomson was out of office from November 1924 to June 1929 when he took up the reins again. In December 1926 Sir Geoffrey Salmond was appointed Air Officer Commanding Royal Air Force, India – to return in September 1931, as AOC-in-C Air Defence of Great Britain, and then, in April 1933, briefly to become Chief of the Air Staff a few weeks before his untimely death. Though in India in 1930, his interest in R.101 had not waned and he had arranged to meet the airship when it arrived at Karachi on 8th October, and to dine on board with Thomson, Brancker, Colmore, Richmond and Rope, thereby completing a reunion of the team which had been gathered together in 1924 to bring Lighter-than-Air travel to Imperial Air Routes.

The airship programme – an example of advanced technology of its day – thus provided an early example of the three-way collaboration (Government, administration and design and engineering teams) which has since become so familiar in, later, and much more sophisticated projects, from TSR–2 to Concorde, and Tornado and beyond.

At the centre, in Whitehall, was the small group who set the policy and held the purse-strings, headed by the Secretary of State for Air. At Cardington, forty-four miles north of Westminster, as the airship flies, the design and construction staff translated ideas into an elegant form of metal and fabric – a development team administered by the Director of Airship Development, Wing Commander Reginald Colmore (the link between the Air Ministry and Cardington) but dominated by Vincent Richmond.

Third in the triumvirate, also under DAD, but, as always in aviation, quite separate and independent in procedures and in outlook, was the flying staff; the officers and crew members of the two airships R.100 and R.101. They were headed by the Assistant Director of Airship Development (Flying and Training), Major George Herbert Scott, the man who had commanded R.34 on its pioneer double-crossing of the North Atlantic in July 1919. The two ships' captains, Ralph Booth for R.100 and Carmichael Irwin for R.101, reported to Herbert Scott.

The three project teams, Air Ministry, Royal Airship Works and flying staff, Cardington, all came together for the first time, fifty-four strong, on the evening of 4th October 1930 to board R.101 at the Cardington mast and to set out together to realise all that they had planned, over so long a time. At their head was Christopher Thomson, the originator of it all, now aged fifty-five, the full flowering of whose talents still seemed to lie ahead.

Christopher Birdwood Thomson

Christopher Thomson, the key figure in the story, was more than a soldier-turned-politician. He had entered the British Regular Army from Cheltenham College as a Gentleman Cadet at the Royal Military Academy, Woolwich, in January 1892 at the age of seventeen. He was commissioned a second-lieutenant in the Royal Engineers in March 1894, and subsequently gained long, practical, experience at home and abroad in the mechanical, the mining, the civil and the electrical engineering fields in both war and peace, including a term as Chief Royal Engineer of the 60th Division in Palestine in 1917/18. With a reputation as a man of clear views and the ability to express them, but always ready to listen to the views and experience of specialists and to give full weight to them, when he went to the Air Ministry in 1924, his attitude was notably different from the conventional run of ministers, who are primarily concerned with political affairs. Instead, while giving full attention to the political side and achieving, through the Cabinet, an immediate increase in the Air Estimates, he showed a studious enthusiasm for the advancement of both the military and the civil sides of his department and began to make the air his hobby as well as his work.

This interest in air affairs can be traced back to February 1900, when, as a first-lieutenant with the 11th Field Company, Royal Engineers, he was detailed to help with the RE Balloon Section on artillery observation duties outside Kimberley in the early stages of the South African War. It was his introduction to the wayward attractions of Lighter-than-Air.

Christopher Thomson – 'Kit' to his family – had been born at

Nasik* in the Bombay Presidency of India, on 15th April 1875. He came of a family distinguished in the Indian Army on both sides. His father was Major-General David Thomson, Royal Engineers; his mother the daughter of Major-General Christopher Birdwood, one-time commander of an Indian Army Division and grandfather not only of Christopher Birdwood Thomson, but, also, of the future Field-Marshal the Lord Birdwood of Anzac and Totnes (1865–1951) who commanded the Australian and New Zealand Army Corps in Gallipoli and France in 1915–16 and the 5th Army in France in 1918–19.

Kit and his younger brother, Roger, returned to England with their parents in 1879, and in due course they both went to Cheltenham College as day boys and on to the Royal Military Academy, Woolwich. On 28th March 1894 Christopher was at nineteen gazetted a second-lieutenant in the Corps of Royal Engineers, posted to the School of Military Engineering at Chatham for specialist training and then to Plymouth for a submarine mining course.

During the next twenty years, to the eve of the First World War, Christopher Thomson saw active service in the Mashonaland Campaign of 1896 (in what is now Mozambique and Zimbabwe-Rhodesia) throughout the South African War from 1900 to 1902, as a British Military representative with the Serbian forces in 1912, and again in the Second Balkan War of 1913. During this time, he served also two other tours of overseas duty in Mauritius and in Sierra Leone (where he commanded the 36th (Fortress) Company, RE), attended the Staff College at Camberley and served as a staff officer in the Operations Branch of the Chief of the Imperial General Staff at the War Office in Whitehall. At the age of twenty-nine, he was at once the youngest and – at 6 ft 5 ins – the tallest, major in the British Army; commended by Lord Kitchener and at the War Office highly regarded by the Director of Military Operations who was later to become Field-Marshal Sir Henry Wilson, Chief of the Imperial General Staff.

As a friend and colleague wrote of C.B. at this time:†

> The centre of all that was going on, delightfully cheerful at all times, a man of great energy, fine intellect and living sympathy, a master of foreign languages and of vivid and versatile conversation, a keen

* Nasik is now an aircraft manufacturing centre as a Division of Hindustan Aeronautics Ltd. (HAL). In 1981 the Nasik factory was producing under licence the Russian Mig-21 bis interceptor-fighter, at the rate of about thirty a year for the Indian Air Force as well as undertaking airframe overhauls. About 7,700 workers were employed at the Nasik aircraft factory in 1981.

† Lieutenant-Colonel Edward M. Mosley, DSO, RE, (1875–1950) in the Royal Engineers *Journal* for March 1931.

follower of the chase, throwing his heart over his fences, a good cricketer and a useful man at golf, he remained at heart a soldier of rare quality.

But, what made him loved and trusted, as he was by men of other ways of life and thinking, were his qualities of an idealism so unexpected that some of his friends could hardly believe it was there – a sympathy which led him to do all he could for anyone who asked. It is rare that a soldier – and a brilliant one at that – should possess such qualities in so much abundance as he did.

Major Thomson crossed to France on 14th August 1914, as a member of the Headquarters Staff of the British Expeditionary Force under Field-Marshal Sir John French. During the next six months, he took part, successively, in the Retreat from Mons, in the First Battle of Ypres and in a British Military Mission to the King of the Belgians, in company with HRH Prince Arthur of Connaught of the Royal Scots Greys. After some weeks on the staff of Sir Douglas Haig's 1st Army Corps at the First Battle of Ypres, Thomson was appointed Chief Military Interpreter between Sir John French and General Joffre.

By the end of February 1915, promoted temporary lieutenant-colonel, Christopher Thomson was on the move again, this time to Bucharest on Lord Kitchener's initiative, to use his Balkan experience as Military Attaché to Roumania with orders to bring that country into the war on the Allied side.

In Roumania, he had to play a lone hand against the heavy weight of German influence and intrigue. With diplomacy and tact – and in fluent French – he overcame the initial distrust harboured by the German-born King Ferdinand and he became on close terms with both the King and his consort, Queen Marie,* a grand-daughter of Queen Victoria and passionately proud of her British ancestry.

The situation in Roumania was, however, altogether different from the impression held of it in Whitehall. From Bucharest Thomson

* Queen Marie of Roumania (1875–1938) was the eldest daughter of HRH Alfred, Duke of Edinburgh and Saxe-Coburg-Gotha, second son of Queen Victoria. In 1892 Marie ('Missy') married Crown Prince Ferdinand of Hohenzollern-Sigmaringen (1856–1927), heir to the Roumanian throne, to the dismay of Prince George of Wales (later King George V), who had hoped to marry his cousin Marie himself. Ferdinand succeeded his uncle, King Carol of Roumania, in October 1914, and, with Marie, resisted German pressures to join in the war on the side of the Central Powers. On 27th August 1916 Roumania entered the war in the Allied cause; Bucharest fell to the Germans on 6th December 1916, and Roumania sued for peace on 7th May 1917. In the Treaty of Versailles, Roumania regained all her lost possessions together with additional territories formerly within the Austro-Hungarian Empire.

Twilight over Airships – Dusk falls on the evening of Saturday, 4th October 1930, as R.101 rides at the mooring tower at Cardington, a short while before its departure to inaugurate the first airship flight between England and India carrying the Secretary of State for Air and the Director of Civil Aviation on board.

[PLATE I]

The Secretary of State for Air – Brigadier-General the Right Honourable the Lord Thomson of Cardington, PC, CBE, DSO, psc, who was appointed Secretary of State for Air and a Member of the Cabinet from 23rd January to 7th November 1924, in the first Labour Government in Britain – and returned in the same office from June 1929, until his death in the R.101 disaster on 5th

The Princess Marthe Bibesco – Marthe Lucie Bibesco with her daughter, Princess Valentine, photographed in Paris in 1906 shortly after she had returned from a pioneering motor-car expedition from Bucharest to Teheran and back between April and June 1905. At the date of the photograph

quickly formed the view that the involvement of an unprepared and ill-armed country in a war on three fronts – against Austria-Hungary, Turkey and Bulgaria – would invite disaster and cause Roumania to become, not an asset but a heavy liability to the Allies.

These views, which were to be fully borne out, were brushed aside at home and on 13th August 1916, with dire forebodings, Thomson signed on behalf of the British Imperial Forces, a Military Convention in support of Roumanian arms. Roumania declared war on 27th August.

When, by the end of the year, disaster did indeed overtake the Roumanian armies, Thomson – now Head of the British Military Mission – strove hard to alleviate the consequences. At great personal risk he supervised the destruction of the Roumanian oil wells so as to deny this valuable source of fuel to Germany.

In May 1917, with Roumania about to capitulate, Colonel Thomson was recalled to England and then posted to Palestine as Commander Royal Engineers with the 60th (London) Division under General Sir Edmund Allenby. After taking part in the capture of Jerusalem on 9th December 1917, where he won the DSO, he took over command of the 180th Brigade in the advance on Jericho and beyond.

Thereafter, in February 1918, Thomson was called back to Paris by the CIGS, Sir Henry Wilson, promoted Brigadier-General and appointed a British Military Representative on the Supreme War Council at Versailles.

There, as a veteran of five campaigns, as an accomplished linguist, as an expert on Balkan affairs and as the draughtsman of clear and incisive military papers, Thomson was in his element and did brilliant work which in due course earned him the CBE. In October 1918 he had the satisfaction of bearing from Versailles to Mudros in the Aegean, the Allied terms for the surrender of Turkey and for the occupation of Constantinople. He was present on 30th October 1918 at the signing of the armistice with Turkey on board the Royal Navy battleship, HMS *Agamemnon*, anchored off Mudros. Thence he returned to the centre of affairs with the Supreme War Council until after the signing of the Treaty of Versailles on 28th June 1919; a treaty whose terms he roundly condemned as containing the seeds of another war.

Thomson's experiences in Roumania, at the Peace Conference at Versailles and of the world in general, now brought to him a growing disenchantment with the old, pre-war, social order together with a burning belief in a new justice which ought to be gained for a better post-war world.* In this frame of mind the simple and essentially moderate ideals of the emerging British Labour Party came to hold a magic appeal.

The Princess Marthe Bibesco

There was another deep and enduring influence as well. Bucharest in 1915 had brought to the still unattached Christopher Thomson the presence of one of the loveliest and most gifted women of her day; the Princess Marthe Bibesco, Roumanian by birth, French by upbringing, daughter of Jean Lahovary, leader of the Roumanian Conservative Party and one-time Roumanian Foreign Minister. At the age of sixteen, in 1902, Marthe Lucie Lahovary had married her cousin, the quixotic Prince Georges-Valentin Bibesco, eldest grandson of the last reigning Prince of Walachia. As a pioneer aviator Georges Bibesco rapidly became so pre-occupied with flying, motor-cars, sporting activities and other women that for the past ten years he had virtually deserted his wife and their young daughter, Valentine.

In these circumstances, Christopher Thomson and the princess had met for the first time on 24th March 1915, at a soirée given by King Ferdinand and Queen Marie at the Cotroceni Palace in Bucharest.

It had been love at first sight for the 40-year-old British colonel and for the beautiful and lonely 29-year-old Roumanian princess who, in having to make a life of her own while her wayward and neglectful husband occupied his time elsewhere, had perforce made herself equally at home in Bucharest, in Paris, in Madrid or in London, whence she had returned in March 1915 from a wartime visit to Bibesco cousins just in time to attend the royal soirée at the Cotroceni Palace.

During the next two years in which he was in Roumania the romance between Christopher Thomson and the enchanting Marthe† blossomed into a devotion which was to remain constant for the rest of his life. As a devout Roman Catholic the Princess Bibesco would not contemplate divorce from the absent Georges but throughout the

* Set out in Thomson's first book *Old Europe's Suicide; or the Building of a Pyramid of Errors, 1912–1919*, published in 1920 by George Allen and Unwin Ltd. and dedicated to: 'One I have always called "La Belle Sagesse", who greatly loves her garden by the "Sleeping Waters"; the subject of one chapter of the book.'

† Pronounced in the French manner: *'Mart'*.

fifteen years which followed C.B. never gave up hope that one day, somehow, he and Marthe might be able to marry. Georges was as ardent about flying as he was about other pursuits except, remarkably, his wife – and flying in 1915 was a dangerous occupation. As things turned out Christopher, not Georges, was to pay the price of pioneering in the air.

All this was, however, still far away during the spring and summer of 1915 in a Roumania not yet at war. The idyllic course of the romance in Ruritanian surroundings is recorded in the surviving diaries of both C.B. and Marthe.

On 5th May 1915, for instance, the Princess wrote cryptically:

> This evening Kit and I alone at Mogosoea.* Red roses. And one day perhaps I – the wife of an Englishman!

War immersed Roumania and took C.B. to Palestine and Marthe to Switzerland to escape from the German occupation. During the years which followed their devotion survived continual separations punctuated by rapturous meetings in Paris, in Brittany, in Paris, in London and again back in Roumania. It generated, too, a voluminous correspondence: 'Whenever I think of the future,' he wrote on 6th February 1918, 'you are the centre of the picture.'

On 29th August 1921, he was writing:

> You must remember that you are not only the person with whom I am just as much in love as ever but also a source of inspiration. There are few actions in my life of any importance which are performed without some thought of you.

Then, on 9th January 1928, he wrote from the Place Vendôme in Paris:

> I leave you with a sore heart but a grateful one. This visit to Paris has been perfect and it will leave a lasting fragrance. I will return as soon as may be. This time next year who knows under what conditions we may be living? But wherever we two may be I am and shall always be
>
> Your devoted
>
> Kit.

They were, indeed, an attractive and well matched pair who came to share a host of common interests and a wide circle of friends in France,

* The Palace of Mogosoea, the Princess's home, eleven miles north-west of Bucharest, on the banks of the Colentina.

in England and in Roumania. They were kept apart only by her deeply religious convictions which forbad the freeing of her bonds to the wandering Prince Georges Bibesco. Early on, as a diversion Marthe had turned to literature. In 1907 when she was only twenty-one, her first book *Les Huits Paradis*, an account of an adventurous motor expedition to Persia, was published in Paris. During the next sixty years she followed with forty-two more books: biographies, novels, historical romances, travelogues and memoirs, three of them made into films and four of them dedicated to C.B.T.; her *Grand Britannique*, who could match her, line-by-line with passages quoted from Gustave Flaubert, Anatole France, Marcel Proust and Joseph Renan.

Thomson's determination to carve out for himself a new life, eventually with Marthe, in a post-war world which, in 1918, seemed full of opportunities, led to his decision to resign his permanent commission in the Regular Army in July of 1919 and with great hopes of an impending General Election to seek adoption as the prospective parliamentary Labour candidate for Bristol Central.

Lean Years

There followed a time of hardship and privation for a lonely and sensitive man who, with only a slender Army pension* and a prospect of casual earnings as a Military Correspondent of the *Daily Herald*, had plunged himself into a way of life different from anything which his twenty-five years in the Army had prepared him.

The future first British Labour Prime Minister, James Ramsay MacDonald, wrote of him at this time: †

> It needed a rare courage to enter a sphere of activity for which he had no experience and in which he had no influence – especially as he was to enlist in the ranks of a Party whose creed and personnel were anathema in those circles in which he had hitherto moved. If, at first, he was regarded with rather surprised eyes, his warm, human, sympathies and his facility for comradeship, his instinctive urge to champion the cause of the weak and the oppressed – whether individuals or nations – and his simple sincerity were soon appreciated at their true worth. He did not speak their familiar language, but his

* £174 per annum after commuting half his pension for 25 years 5 months service for a lump sum of £2,347-4s-od from 19th November 1919.

† Introduction to *Smaranda*, published by Jonathan Cape, London, 1931.

> personality soon made that all right. His politics were unclouded by philosophic doubts and unalloyed by personal push or ambition. He believed and he served. He had built up no elaborate theories of Socialism; he was a practical man with an eye to see, a heart to feel and a head to be satisfied only by hard, practical, work.

When the long-delayed General Election eventually came on 15th November 1922, Thomson succeeded in almost halving Sir Thomas Inskip's previous majority of 5,095 votes over Ernest Bevin in the strong Tory stronghold of Bristol Central.* He was bitterly disappointed but not dismayed and poured his heart out in a long letter to Marthe in Roumania. Always fascinated with politics, and seeing a parallel, she encouraged him to try again.

> Hope and perseverance are watchwords for both of us [she wrote]. I like the saying of William The Silent, even if it must have been sotto voce – One does not need to be hopeful in order to undertake nor need one be successful in order to persevere.

Though he had failed to win the Bristol seat, and though he failed again in another Tory stronghold at St Albans in December 1923, Thomson's increasing skill as a speaker, his membership of the Fabian Society, his knowledge of the world and his ability to get on with people in all strata of society, brought him to the attention of those foremost in the Labour Party of that time – and, in particular, to a steadily closer friendship with its leader, Ramsay MacDonald.

> We shook hands [wrote MacDonald†] and, like David and Jonathan, that was enough. For ever after, we understood each other. It was one of those affections that require no growth. Perhaps, in some other life, we had met; perhaps our harmonies were so perfect that they required no tuning and no testing. We met for the first time as old familiars.

So it was that hard work, sincerity, and a gift of friendship, brought Christopher Thomson to a post for which another strange twist of fate had already prepared him.

* The result of the 1922 Election at Bristol Central was: Inskip (C) 15,568, Thomas (L) 12,303. Conservative majority 3,265; but a swing to Labour of 15.6 per cent compared with 1918.

† In a letter to the Princess Marthe Bibesco; 8th October 1930.

Eight years before, in 1915 in Bucharest, two of the visitors from the British parliamentary scene had been Sir Samuel Hoare, MP, later to become Air Minister in the Conservative Government of 1922/23, and Leo Amery, MP, later to be First Lord of the Admiralty in the same administration.

Both these leading Tories of their day formed a lasting friendship with the British Military Attaché, whose views on the potential for the winning of the war from the East were closely in line with their own. Even Thomson's decision to stand for Parliament as a Labour candidate after the war did not impair their regard and accord. This mutual friendship and their appreciation of C.B.'s intellectual qualities against his military-engineering background, led Hoare and Amery to agree, in 1923, to invite him to serve on an Air Ministry advisory panel as a counter-balance to the rumbustious, right-wing, Commander Dennistoun Burney who was at that time a passionate advocate for airships to be built under Vickers auspices.

Airships

The panel under the Chairmanship of the then Air Vice-Marshal Sir Geoffrey Salmond, had been set up to assist a Cabinet Sub-Committee on Airship Development (chaired first by Amery and then by Hoare) to examine the possibilities of airships for future long-range Imperial air routes.

Work on this panel in May and June of 1923,* first took Thomson to Cardington and brought him into close contact, and to a developing friendship, with Geoffrey Salmond and Sir Sefton Brancker, Director of Civil Aviation. Both Salmond and Brancker, like Thomson, had been through the Royal Military Academy and the South African War, while Salmond had been a contemporary of Thomson at the Staff College at Camberley from January 1909 to December 1910.

Geoffrey Salmond's advisory panel recommended to the Cabinet Sub-Committee that a new British airship programme should be set in hand, although Salmond, Brancker and Thomson all expressed reservations about Burney's optimistic proposals for a private operating company for airships, set up by Vickers but subsidised and guaranteed by the State – a company whose equipment would be handed over to Burney's old Service, the Royal Navy, in time of war.

* See Appendix 1, 'Airship Development'.

To the three former Regular Army officers, the Air Ministry not the Admiralty, and Royal Air Force not the Royal Navy, seemed the logical Service to foster airship work in times of peace and war.

Action on a potential airship programme was overtaken by political events in the autumn of 1923, leading to a General Election in December. Nevertheless, his brief encounter with Lighter-than-Air and its protagonists, served to engender in Thomson an informed interest in the possibilities for long-range Empire air services and to fire him with an enthusiasm for the development of a new generation of airships. It equipped him, also with some knowledge of air affairs generally, a concern for the future of the Royal Air Force and a high regard for two of the leading personalities in British aviation of their day – Geoffrey Salmond and Sefton Brancker.

Geoffrey Salmond's committee had been a stimulating interlude, but Thomson turned from it back to his political endeavours and to his literary work, necessary to keep the wolf from the door. He had just arranged to embark upon a well-rewarded lecture tour in the United States – encouraged by introductions from his new-found friend, Sefton Brancker – when an insistent invitation came from Ramsay MacDonald to come to see in the New Year of 1924 with him at the Chief's cottage, The Hillocks, at Lossiemouth in Scotland.

Labour might shortly come to power for the first time and MacDonald was engaging in the pleasant pursuit of Cabinet making.

No matter that Thomson had no parliamentary seat; that could be found. What did matter was that Thomson was something different in the Labour Party – an asset not to be wasted.

So it came about that on New Year's Day of 1924, at Lossiemouth, Thomson was first offered (and turned down) the Foreign Office should Labour take office. He refused it because he knew that Ramsay MacDonald wanted to add the Foreign Office to the Premiership; a double load which, in due course, he shouldered with relish. C.B. refused from his friend the War Office too, lest he should appear to lord it over his former Army chiefs. Instead, with his interest in air affairs so spurred by his association with Geoffrey Salmond and Sefton Brancker he embraced with enthusiasm the third offer now pressed upon him : that of becoming Secretary of State for Air.

So, with the resignation of Stanley Baldwin and his government in January 1924, Thomson found himself at the Air Ministry as its political chief, a member of the Cabinet with a seat in the House of Lords. It was one of the turning points in his life.

*

'A chorus of praise has a monotony which is tolerable only in Heaven' – and, with all his qualities, Christopher Thomson had his foibles like everyone else.

For example, though the main principles by which he directed his life were sound and clear enough, the fact that Thomson frequently enjoyed an argument just for argument's sake made him enemies as well as friends. His fondness for exploring and raising for discussion all aspects bearing upon any problem was one of the characteristics that won him the warm regard of the first Commandant at the Staff College during Thomson's time there and, likewise, the hostility of Wilson's successor at the College, Sir William Robertson. In the same way the unorthodox initiative which brought him the commendation of Lord Kitchener in South Africa brought about, also, a quarrel with Brigadier-General John Gough in front of Ypres.

Thomson's warm heart was matched by a hot temper. As time went on he mastered the latter while yielding even more to the first. While he never suffered fools gladly and was quick to show his dislike of humbug, he had little regard for money and plausible rogues rarely left his company without some quixotic act of generosity.

C.B. was constant in his devotion to Marthe Bibesco from March 1915, until the end of his life but he remained susceptible to feminine charms. But from 1923, second only to Marthe was his devotion to aviation in all its aspects – and to Lighter-than-Air most of all.

The Air Ministry

During his nine months at Adastral House – from 23rd January to 7th November 1924 – Christopher Thomson showed himself to be a man of vigour, judgment, imagination and firm decisions; an ardent and determined advocate of a strong, independent, Royal Air Force and of a progressive civil aviation. He got on well with his Service chiefs and his leading civil servants alike, while his speeches went down well in the House of Lords.

The position at the Prime Minister's right hand was made the more attractive by the presence at the Air Ministry of so many old friends and former colleagues from the Royal Military Academy at Woolwich, from service in the South African War and from the pre-1914 War Office. The Royal Air Force – now his Service – had at its head erstwhile professional soldiers like himself who had earlier than their new chief set their eyes and their ambitions upon the wide horizons of the air. Those links were to Thomson a supreme advantage denied to his predecessors and successor, Sir Samuel Hoare.

First among the former Army officers at the Air Ministry in 1924 was the Chief of the Air Staff, Air Chief Marshal Sir Hugh Trenchard, a powerful, inarticulate figure of fifty-one; tall, broad-shouldered, with a deep and booming voice, a gruff manner and a commanding presence. Originally convinced that the Royal Flying Corps would best remain linked to the Army, Trenchard was now the leading protagonist of a separate and independent air force, primarily of bombers, and of the need to build it to a position of strength and security as the third service. His office, on the sixth floor of Adastral House on the corner of Kingsway and the Aldwych, was an eyrie from which he commanded the Royal Air Force at home and overseas with absolute authority and a conviction about the future dominance of air power.

At Trenchard's right-hand in Adastral House was the Air Member for Supply and Research, Christopher Thomson's old friend of Army days and leading advocate of Lighter-than-Air craft, Air Vice-Marshal Sir W. G. H. Salmond – Geoffrey Salmond – the elder of the two Salmond brothers and forty-six years old in 1924*. Sir Geoffrey Salmond – small, precise, fertile of ideas and quick in the uptake – was another veteran of the Royal Military Academy, of the Staff College at Camberley (1911–12) and the South African War (1899 to 1902). His career had thus closely paralleled that of the new Secretary of State for Air up to the time when Salmond transferred to the Royal Flying Corps.

Alongside Salmond was Air Vice-Marshal Sir William Sefton Brancker, Director of Civil Aviation at the Air Ministry since 1922. Brancker had likewise been through the Royal Military Academy at Woolwich and had served in the Horse and Field Artillery in the South African War. At forty-seven in 1924 Brancker was a burning enthusiast for civil aviation in all its aspects; a man who got things done.

In the Secretary of State's private office, as the fourth member of this inner circle, was the 33-year-old Major Christopher Llewellyn

* Marshal of the Royal Air Force Sir John M. Salmond (1881–1968), the younger of the brothers, was in 1924 Air Officer Commanding British Forces in Iraq. On 1st January 1930, he succeeded Sir Hugh Trenchard as Chief of the Air Staff in advance of his elder brother, then – and since 1927 – Air Officer Commanding Royal Air Force, India. Sir Geoffrey Salmond was appointed to succeed Sir John as Chief of the Air Staff from 1st April 1933, but died on 27th April at the age of fifty-five before he could take up the post.

Bullock, the Minister's Principal Private Secretary; a brilliant, hard-working, combative and ambitious civil servant, devoted to the Royal Air Force, a former pilot and Air Staff Officer in the RFC and RAF during the war. Bullock had returned from France to serve Winston Churchill as his Private Secretary at the Air Ministry in 1919, and then had moved on as Personal Assistant to Hugh Trenchard. Thence, he became Private Secretary to Sir Samuel Hoare and then to Christopher Thomson.

In Trenchard, Salmond, Brancker and Bullock, the new Air Minister found that he had four powerful and single-minded allies. All four, with the Secretary to the Air Ministry, Sir Walter Nicholson, had been much concerned about the sort of chief they might have foisted upon them by the first Labour Government. Their relief was immense when they learned that the chief was to be Thomson. It was a relief which turned even to enthusiasm as C.B. established himself ready to listen, but clear in his ideas and firm in his decisions to turn them into fact.

In nine months, between the end of January and the beginning of November 1924, they accomplished together more of lasting significance than did any other Government Department of that first Labour administration.

Thomson's association with Salmond and Brancker in 1923 on airship policy had brought the whole subject of Britain's future place in the air to the forefront of his mind. By the beginning of 1924 he had developed his ideas into six clear objectives which, as the new Air Minister, he determined should form the framework of his policy. They contained not only ideas but also ideals on which to concentrate drawn against the background of precepts learned at the Staff College, twelve years before.

Thomson touched upon them in his maiden speech in the House of Lords on 4th March 1924:

'The flower of idealism has its roots in common sense. In my own mind I compare common sense and idealism with two things we used to study as Staff Officers; with strategy and policy. We were taught that they should always be in harmony and go hand in hand.'

And hand-in-hand they were made to go at the Air Ministry; some of them carefully considered continuations of the policies of his predecessor, Sir Samuel Hoare, who had been dominated by Hugh Trenchard, the CAS; others quite new. All of them were in harmony with the views of Trenchard for whom C.B. formed a high regard but from whom he allowed no similar domination.

Thomson stated his six objectives concisely:

First and foremost, a strong, independent, Royal Air Force; Britain's first line of defence.

Second: a government-supported commercial air transport company to pioneer air routes into Europe and beyond, inevitably confined to short-haul daylight services because of the limitations of the aircraft of the day.

Third: sufficient orders to maintain in being the manufacturing industry for British airframes and aeroengines; an industry capable of expansion in time of war.

Fourth: plans for new – or resurrected – military and civil air bases for land-planes, seaplanes and airships.

Fifth: means of enlisting increased public interest and support for aviation; the encouragement of 'an air-faring outlook' on the part of the British people to match the traditional British sea-faring outlook of days gone by.

And finally, his most ardent hope, a new airship programme designed to link the distant centres of the British Commonwealth and Empire by means of long-range air services flying day and night while providing space, comfort and serenity of progress not matched even by ocean liners.

Such were the ambitious concepts he outlined at a Press Conference on 7th February 1924, and in his maiden speech in the House of Lords on 4th March. They reflected his whole-hearted conversion to aviation against his background as a military engineer combined with a wide experience in world affairs.

Thomson's first and major battle and his first victory for the Ministry was the upholding of the independence of the Royal Air Force, then under powerful assault from the Admiralty, aided somewhat less enthusiastically by the War Office.

There can be no doubt that, but for Thomson's spirited and skilful defence in the Cabinet and the support he received from the Prime Minister, the small and financially-starved Royal Air Force (a shadow of its wartime strength in 1918) would have been split into a Royal Naval Air Service and an Army Air Corps. Then there might have been no Fighter Command to win the Battle of Britain in 1940 and no Bomber Command to carry the war to Germany thereafter.

As it was, against the bitter opposition of Earl Beatty, the First Sea Lord, the new Secretary of State for Air won the day and followed up by persuading the inherently-pacifist Labour Cabinet to approve a stepping up of the Air Vote for 1924/25 by some £2,840,000 – a

17 per cent increase over the previous year; and this at a time then the Votes for both the Army and the Royal Navy were cut.

On the civil side, on 1st April 1924 the Secretary of State for Air launched his new 'single chosen-instrument' to develop air transport services with landplanes and flying boats. Though it was a project initiated by his predecessor Thomson himself decided upon the euphoric name *'Imperial Airways Ltd'* in place of the ponderous 'Imperial Air Transport Company' of the earlier proposals.

For his third objective, the manufacturing field, Thomson pressed forward and extended the existing policy of building upon the small, struggling, British aircraft industry which, like the Royal Air Force, had been reduced to a shadow of the size to which it had grown at the end of the First World War. The result was that, even amid the financial stringencies of 1924, a series of new Air Ministry specifications were drafted and issued to the industry. From them came seven successful new types of aircraft, all of which went into production.* There could be no more practical example of Ministerial determination and skilled negotiations with a reluctant Treasury.

Aircraft manufacture, military operations and commercial air services are all dependent upon adequate bases from which to fly and Thomson's fourth objective was to ensure that more aerodromes should be provided where they were most needed. As a start in 1924, funds were allocated for the acquisition for the Royal Air Force of land at North Weald, Tangmere, Turnhouse and Waddington;† for the expansion of Croydon Airport for Imperial Airways and for the design and construction of mooring masts at Cardington, Ismailia and Karachi for a new generation of long-range airships.

As a back-up to all of this, the Air Ministry saw clearly that public support for the adolescent military and civil aviation of 1924 would be crucial to the success of its forward plans and therefore deserving of studious encouragement. As a move towards this objective, Thomson and Trenchard went together to enlist the support of Lord Haldane, Lord Chancellor, for the recruiting and equipping of thirteen new, non-regular, air squadrons of which six would be for

* The seven aircraft types were: The Blackburn Iris long-range flying-boat; The Boulton and Paul Sidestrand twin-engine day-bomber; the Fairey IIIF fleet-reconnaissance aircraft; the Handley-Page Hyderabad night-bomber; the Supermarine Southampton reconnaissance flying-boat and the two new rigid airships, R.100 and R.101.

† North Weald and Tangmere both became leading Fighter Command stations in the Battle of Britain, Waddington a Bomber Command station and Turnhouse the Edinburgh Airport.

Home Defence. Together they would form a new Auxiliary Air Force of 'week-end fliers' on the lines of Haldane's pre-war Territorial Army of 'week-end soldiers' – which had given such gallant and valuable service during the Great War of 1914–18.

As a civil counterpart to the Auxiliary Air Force and in order to popularise civil flying and to bring it within reach of many more people, Thomson and Brancker discussed together the idea of forming ten, regionally-based, flying clubs to be subsidised by the Air Ministry and to receive a grant of £10 for each new flying licence awarded through The Royal Aero Club. Thus, there would be gained a nucleus of trained pilots for commercial airlines in peace-time and for the Royal Air Force in time of war. This imaginative idea was announced by Thomson on 1st September 1924. In due course, it achieved much success once a suitable light aeroplane, the de Havilland Moth, arrived on the scene in 1925.

All these five separate but interlinked objectives were set in train with vigour and determination. They were to be the foundations for a new national endeavour to be formed around the development of British aviation. Concerned as he was to advance all five projects, Christopher Thomson's most fervent hopes of all were directed towards his sixth objective: his Imperial Airship programme.

Following consultations with Hugh Trenchard, Geoffrey Salmond and Sefton Brancker in February 1924, Christopher Thomson presented to the Cabinet a proposal* for the design and construction by the Air Ministry of an advanced new airship of 5,000,000 cubic feet capacity, twice the size of any previous lighter-than-air craft. Airship bases would be constructed in Egypt and India for overseas trial flights. He proposed that thereafter, in 1927, further airships should be constructed for both commercial and defence purposes on the basis of experience gained with the first ship.

The stated requirement was that the new airship should be designed to carry 100 passengers with their baggage (14 tons) in comfortable living and sleeping quarters and, in addition, 16 tons of goods (or the equivalent military load) for non-stop distances of up to 3,500 miles, cruising at 63 miles-an-hour, and that the airship should be capable of flying day and night in all weathers between England and the major centres of the British Empire.

This was an extraordinarily advanced concept for its day when the largest heavier-than-air craft carried no more than a dozen pas-

* AD (24)2 and CP (104)24.

sengers and had a range of little more than 250 miles.

Once again, this plan was strenuously contested by their Lordships of the Admiralty, who were in favour of airship development but only under Royal Navy control for use as 'the eyes of the fleet' in time of war.

After long and often bitter arguments, on 1st May 1924 Christopher Thomson gained Cabinet approval to a revised proposal for the design and construction of two airships; one to be built by the Royal Airship Works at Cardington and the other to be offered, under Air Ministry contract, to a subsidiary of Vickers Ltd., The Airship Guarantee Company Ltd, headed by Sir Dennistoun Burney.

A week later, on 7th May, the Cabinet, with Treasury endorsement, formally approved the recommendations of 'The Cabinet Committee on Airship Development' for the programme proposed by the Secretary of State for Air and agreed to seek Parliamentary sanction for a total expenditure of £1,400,000 on airships and air bases over the next three years. This expenditure was to be in addition to the normal Air Votes. The specific allocation for each year would be added to the Air Estimates, starting with approval for an expenditure of £400,000 in the financial year 1924/25.*

Here was a resolution of the arguments and vacillations which had continued in and out of government circles ever since 1918. The new Labour Government set its hand to the most forward-looking programme of aeronautical development so far projected by any country in the world. The inspirer and driving force was Christopher Thomson, with the active support and encouragement of Geoffrey Salmond as Air Member for Supply and Research and Sefton Brancker, the Director of Civil Aviation.

The 1924 Airship Programme

The proposals for the new airship programme were laid before the House of Comomns by the Prime Minister, on 14th May, debated and approved. They were presented in more detail to the House of Lords by the Secretary of State for Air a week later. He explained the plan as the most advanced airship programme in the world. It would start with a period of research and investigation in preparation for two practical tasks:

* Paper CP–285(24). (AIR 8″21/9848).

1. *The construction of* two airships of 5,000,000 cubic feet capacity each of such speed, range and load-carrying ability as would make possible long overseas voyages with reliability and safety against a substantial margin of lifting power; and,
2. *The demonstration of* the suitability of the airships for Empire air routes, by the successful completion of an initial series of flights to and from a new terminal in India.

As a new and forward-looking method of achieving these objectives, the design and construction of the two ships would be divided between the Air Ministry, at the Royal Airship Works at Cardington, and the Vickers-owned Airship Guarantee Company Ltd, subject to a satisfactory contract with the private company.

The new programme would be, at the same time, a rejection of the proposals put forward to successive governments by Commander Dennistoun Burney, MP – late of the Royal Navy and now a consultant to Vickers Ltd. The Burney Scheme had been an optimistic and grandiose plan for the construction and operation of six new airships by Vickers Ltd, at a cost of some £4,800,000, most of which should be provided by the Government which, said the Secretary of State, would be asked to take all the financial risks.

'The great objection to the Burney Scheme,' said Lord Thomson,* 'is that at the end of fifteen golden years the one company would have a complete monopoly. No other airship firm would be able to compete with it because of the immense sums of public money with which it would have been endowed as the one and only airship provider in the country.'

By contrast, under the new proposals, two airships would be ready for service in a shorter time, and at a much lower cost, and instead of a government-financed private monopoly, there would be two separate design and manufacturing plants. Research and development by the Air Ministry would be made freely available to the private company.

'My immediate object,' said Lord Thomson,* 'is to enlist the co-operation of private enterprise in these early stages, not so much in competition with the Air Ministry as, rather, in a spirit of emulation. The history of aviation is largely based on combinations of this sort as the efforts of the early pioneers proved. They wanted to make money, but they also made sacrifices. They tried and failed and tried again, while the world looked on. The pioneers not only proved their

* House of Lords *Hansard*: 21st May 1924.

case, but many of them made their fortunes.

'I may be mistaken, but I will take the risk of assuming that even the keenest business men are not inspired wholly by sordid motives when they co-operate with the Government in the Conquest of the Air.

'As in other branches of aviation, a great amount of patience and imagination are required. We need the power to discriminate and wait and not let ourselves be rushed.

'Finally', he said, 'under the Burney Scheme, if one airship crashed, there would be £400,000 gone. Now, if one of the first two ships crashed, we would still have the experts in Government employ, we would still have the air highway and we would still have the power of employing people and giving out contracts. We would still have the determination to go on with the resources of a government behind the plans and the endeavours.'*

But, would there be a Lord Thomson? Would there still be a drive and a determination to succeed?

In this way, the 1924 airship programme was launched. The Royal Airship Works at Cardington was brought back on to the active list on 1st July 1924, and a contract signed on 22nd October 1924 with the Airship Guarantee Company, for the design and construction of R.100, for the sum of £350,000, of which £150,000 was paid on signature.

So Christopher Thomson's six objectives were all set in train.

When the first British Labour Government fell in November 1924, Christopher Thomson left an Air Ministry that was more firmly established than it had been at any time in the past. A series of important programmes had been agreed and finance for them provided.

On 8th November 1924 Thomson addressed a 'thank-you' letter to all Air Ministry staff and, at the same time, and – characteristically – a welcoming note to his old friend, Sir Samuel Hoare, coming in again to succeed him on the fourth floor of Adastral House in Kingsway:

> I like to feel that you will find the Air Ministry and its affairs in better shape than when you left, not least in the security of the Royal Air Force and its steady expansion – and on airships too. I am sure that we share the same determination to keep them where they properly belong – in the forefront of progress.

* Lord Thomson of Cardington in the House of Lords on 21st May 1924.

Thomson had been a popular Minister; popular with the Service, popular with Parliament, popular with the Press. He had impressed his Labour colleagues. He was respected by his Tory and Liberal opponents. *The Times* commented:

> Lord Thomson has been loyal to Ramsay MacDonald through difficult days; and this has established his reputation. He is one of the best brains in the Labour Party – a thinker, a wit and, at the same time, one of the most charming of men.

Charles Grey, Editor of *The Aeroplane*, added:

> The only real regret at the fall of the Labour Government is that the nation will lose the services of Lord Thomson as Air Minister. His extreme keenness on his job has been one of the few bright spots in the Labour administration. It would not be a bad plan to adopt the French system of retaining good Air Ministers whatever government is in power.

And so, out of office from November 1924 to June 1929, Thomson set himself to widen and develop the close interest he had formed in aviation; especially in air power as Britain's first line of defence and in the large rigid airship as the most promising means of strengthening the links of Empire through long-distance air services. His belief in airships for ocean routes was shared in both the United States and Germany. In the ten years up to 1929 every attempted airship crossing of the North Atlantic had been successful and 239 airship passengers and crew had flown the Atlantic without accident or injury. By contrast, out of 27 North Atlantic attempts by aeroplane, 16 had failed with the loss of 21 lives.

Large rigid airships, with long endurance in the air and comfortable passenger accommodation, seemed to give promise of becoming the successors to the great ocean liners of the day. For the present, however, the only way across the ocean was by sea and Thomson took ship for extensive visits to the United States in 1925 and in 1926. In the process, he further widened his horizons, made many new friends and a useful amount of money from lecture tours from coast to coast. In 1928, he spent six weeks on a Trans-Canada Tour with a 53-man Empire Parliamentary Association delegation which travelled, mostly by special train, from Quebec to British Columbia and back to Nova Scotia – 'speaking, speaking all the way'.

At Westminster, during these years, Thomson took an increasing share of the work of the Opposition in the House of Lords. His interest in aviation was intense, further stimulated by his American visits. He began, also, to take an active part on committees of The Royal Aero Club at 3 Clifford Street. He enjoyed the society of the people he found there; a cross-section of pioneer aviators and aircraft designers, manufacturers and air transport operators, Royal Air Force Officers and Air Ministry officials, as well as the growing numbers of private owners of light aircraft. They welcomed his obvious enthusiasm and courtly manners. He had found a developing interest and a new circle of friends.

On 19th July 1926 this association was further cemented when Christopher Thomson was unanimously elected to succeed the Duke of Sutherland as Chairman of The Royal Aero Club of the United Kingdom, the eighth Chairman since its formation by Charlie Rolls and Frank Hedges Butler in a balloon over Sidcup in 1901 – as C.B. remarked, 'the Sidcup that cheers'. The chairmanship was an honour which he cherished greatly, a position into which he threw himself with his usual vigour and good humour. In addition to its social position as a meeting place for everyone in the, still small, circle of British aviation, The Royal Aero Club was the controlling body for all British sporting flying; it was still the official source of flying licences in the form of Royal Aero Club Aviators' Certificates and it was the national representative body on the Federation Aeronautique Internationale.

At the same time, Thomson became a patron of the Air League of the British Empire and, on 11th April 1927, a Member of the Royal Aeronautical Society. He was right in the aviation swim. For the next four years, in congenial company, it gave him growing delight.

Clearly he revelled in the way in which the Royal Aero Club continued to bring him into close contact with the progress of air affairs and with their leading personalities not only in government but also in the manufacturing and operating industries. It enabled him to champion their causes in the House of Lords, and, most significantly, to press upon Sir Samuel Hoare and Hugh Trenchard the importance of setting up a High Speed Flight of the Royal Air Force and to enter it, on behalf of the Royal Aero Club, as the official British team for the Schneider Trophy Contest of September 1927 at Venice.

In the event, the contest was won by Flight Lieutenant S. N. Webster, RAF, flying a Supermarine S.5 seaplane at 281 miles an hour over the 217 mile course; it was a great boost to British aeronautical prestige and, at the same time, the means of developing a new breed of fast monoplanes and high-powered aero-engines which were to lead directly to the Supermarine Spitfire and Hawker Hurricane of ten years ahead.

After the 1927 Schneider Contest, as the re-elected Chairman of the Royal Aero Club, Thomson, with Brancker as Chairman of his Racing Committee, pressed for renewed British participation in the race of 1929 and for the allocation of funds for a new series of aircraft and engines to defend the trophy against Italian, French and American competition at Cowes. In the middle of the political campaign of May 1929, Thomson headed a Royal Aero Club team to Portsmouth to negotiate detailed arrangements for the race with the Mayor and Corporation.

Ahead of this, in December 1928, as Chairman of The Royal Aero Club, Thomson had led a British delegation on behalf of the Conservative Government to an international aviation conference in Washington DC. There, as before, he made a notable impression for his knowledge, courtesy and wit.

So came the General Election of 30th May 1929, the first election at which the ranks of the voters were swelled by the attractive forms of more than five million young women between twenty-one and thirty who had now been enfranchised under the much publicised 'Flapper Vote' Act of April 1928. Labour emerged with 287 House of Commons seats against 261 for the Conservatives and the Liberals' 59. Although the Liberals thereby held the balance of power, Lloyd George quickly made clear that the support of the Liberal members would be thrown behind Ramsay MacDonald and denied to Stanley Baldwin and the Conservative Party.

Baldwin resigned the premiership on 4th June. King George V invited MacDonald to form a new government. On Saturday, 8th June, MacDonald accompanied by Thomson and other leading members of the new Cabinet went by train to Windsor and received their seals of office from the King.

Telegrams of congratulation came pouring in. Among them, to Christopher Thomson was a specially cherished cable from the Princess Marthe Bibesco in Roumania.

It was natural, therefore, that on 9th June 1929, the Sunday following the announcement of the new government, Christopher Thomson should sit down at his favourite corner-table in the library of the United Service Club in Pall Mall * (one of the most elegant rooms in London) and write to the source of so much affection :

* Now the Institute of Directors.

Thank you so much for your two letters and the telegram, just received, and, as ever, my constant devotion.

We 'kissed hands yesterday' and had a most charming interview with His Majesty who looked extremely well considering his recent illness – and spoke without notes.

Though the general situation has been so confused my own was clear so far as being in the Cabinet and, as you know, I wanted to get back to the Air Ministry for at least one more year.

We start under better conditions than last time. Then we took over our departments in a hurry and met Parliament almost at once. Now, we have until the 25th to get into our affairs. Then will follow a short Parliamentary session and the real work will begin in October, culminating in the Budget for 1930/31.

Now you will be arriving in London on 4th July and we shall be lunching together with Lady Colefax* on the 5th. I am so sorry that I cannot meet your train on the 4th but I shall have to be in the House for the debate which follows the King's Speech.

I want to take you to the British Museum after lunch on the 5th to see the new discoveries at Ur of the Chaldees – and then to Wimbledon to see the tennis. I so look forward to being together again.

God bless you – and as ever and for ever, with very best love – Kit.

*

Thus Thomson came back to his old department, acclaimed on all sides as the right man for the job. The Air Ministry was indeed fortunate, as few departments are fortunate, in acquiring again a political chief who understood its problems and was eager to press hard for their solution on practical lines. His interest in every branch of aeronautical progress had further increased after the fall of the first Labour government and with his Chairmanship of the Royal Aero Club and his active participation in the work of the Royal Aeronautical Society and of the Air League he now knew more about the aviation scene, and the leading figures in it, than anyone else in politics.

He was also the first trained engineer to reach Cabinet rank. But he was much more than that. How much more was cogently stated

* Lady Sybil Colefax, wife of Sir Arthur Colefax (1873–1936), barrister-at-law whom Lord Thomson invited to preside over two aircraft accident inquiries in 1924 and 1929.

by his long-time friend and Army colleague, Captain Sir Basil Liddell-Hart, who wrote of C.B. in 1930*:

> The tragedy of Beauvais has brought to an early close the career of one of the most remarkable of modern soldiers and the most remarkable career that any modern soldier has enjoyed.
>
> That term 'enjoyed' is apt. If the variety of his career enhanced his range of appreciation of life, so did that variety also provide him scope for the full enjoyment of it.
>
> For me at least, Lord Thomson gave the impression of one who savoured life with a connoisseur's delicate and cultivated palate, tasting the full quality of its variety.
>
> That connoisseur-like attitude to life was suggested in his very appearance. In his finely-chiselled features, his essentially aristocratic bearing and, above all, his expressive hands, the hands of an artist, there was a blended sensitiveness and strength enhanced by his air of natural dignity with sometimes a trace of hauteur that would, gradually or suddenly, melt into an irresistible charm. That charm often served him well in discussion and controversy – and its power lay in the fact that it was not an artifice, but essentially the bouquet of an inward mellowness.
>
> In conversing with him, there would be moments when one wondered how, as a Socialist, he could hold such conservative views, or how he could reconcile himself to some of the platform sentiments of his colleagues. Then, in a flash, would come some penetrating thrust at hoary conservatisms of belief or tradition. They showed how impossible would have been his presence on a Conservative platform and how disconcerting a fellow he must have been in a regimental mess.
>
> His adhesion to Labour was not only by conviction and of long-standing, but had been proclaimed at a time and in an atmosphere that hall-marked his sincerity and his moral courage.
>
> Even at the Staff College, Thomson made no secret of his leanings towards the moderate Socialist theories of the day and was ever ready to argue the case for the miners or the railwaymen. He was ready, also, to prick what he considered the bubbles of vacuous custom. If he indulged in such discussions, partly through a fondness for argument, this had the effect of enlarging his own outlook and it ended by convincing him of the rightness of the case he argued. Moreover, he had an abiding sympathy for the under-dog.
>
> His close knowledge of France and of the French character had a considerable influence in later years on the attitude adopted by the Labour Government in its relations and negotiations with France. No

* In *The Fortnightly Review* for 1st November 1930.

> other Minister in the 'inner circle' both spoke French so fluently or knew the French so intimately.
>
> In government, his influence on the general work of the Cabinet was far greater than was publicly realised. If this influence was a tribute to his knowledge of the world and his matured judgment, it probably owed much to his close friendship with the Prime Minister and, also, the Chancellor of the Exchequer.
>
> Such friendships play a larger part in the government of the country than academic students are apt to appreciate.
>
> If Thomson bore the strain better than most of his colleagues, it was partly because his economy of energy and variety of interest. He was an ardent lover of music and he would – when possible – slip away from the affairs of State to attend a concert or recital; returning refreshed and freshly poised. His love of literature was equally marked and, in particular, his capacity to quote French classics evoked admiring wonder.
>
> If these be trifles, they, nevertheless, serve to round off the portrait of a man who lent grace and distinction to everything he touched; who was always an exception in every circle and yet gave others the feeling that they were divergencies from the natural level. With his passing, a subtle flavour passes from the palate of the world who knew him.

Thus, Sir Basil Liddell-Hart – himself as good a judge of men as he was of military affairs.

Christopher Thomson was, thus, from a different mould compared with all his Cabinet colleagues, yet, for all the breadth of his background and for all his wide experience, he was still a lonely man. There was only one cure for that; fulfilment of the long cherished dream of a future with the lovely and beloved Marthe. One day, perhaps. . . .

Meanwhile in the high summer of 1929 that one other part of his heart's desire seemed at last to be coming close to fulfilment; the completion of the new airship R.101, a first flight to India and then the start of regular Empire air services with a new generation of airships. Britain would be in the lead.

'. . . R.101 will, I hope, give me joy. To ride the storm has always been my ambition . . '.

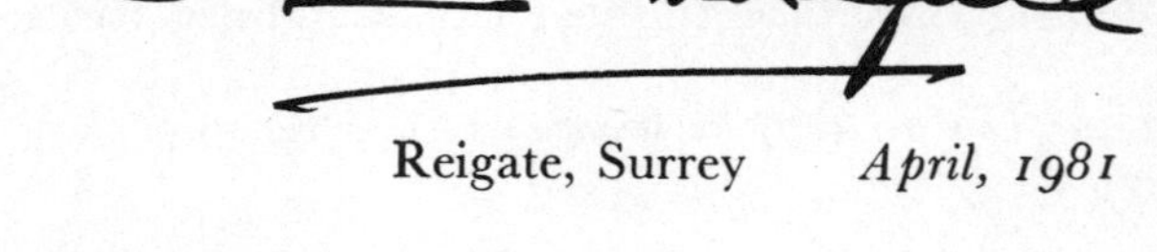

Reigate, Surrey *April, 1981*

PROLOGUE

CARDINGTON

Saturday, 4th October 1930

DEPARTURE

PROLOGUE

Cardington

Saturday, 4th October, 1930

Departure

DAYLIGHT was fading fast. In the gathering dusk, the massive form of the airship, the largest vessel in the world, swung gently at the masthead in the light, south-westerly wind; poised for flight. Fuelling was complete with 25-tons of diesel oil on board. The last stores were going up the lift. Everywhere, there was an air of eager expectation and happy belief in a resounding success.

A little crowd stood around the foot of the Cardington mooring mast – relatives and friends of the fifty-four who were going on board for the great flight to India: four officers, thirty-eight crew members and a dozen passengers in all. As they went up the mast in one's and two's, the evening was still clear, though there was a scent of rain upon the wind. Tomorrow should be fine and, by then, R.101 should be well on its way.

Now, 170 feet above the ground, the lights were going on at the top of the mast and in the airship's passenger accommodation. A golden glow had begun to shine behind the mid-ship strip of cabin windows. One by one, lights began to twinkle in each of the five power-cars slung beneath the great silvery hull.

At long last, the promised voyage to India was about to start. Planned for more than six years now, it was to be the first step towards a regular weekly Imperial Airship Service through Ismailia to Karachi and beyond. Soon Cardington and its Royal Airship Works would be on the map as a major port for Lighter-than-Air craft on Empire and Atlantic routes; employment would expand and be secure – a new era in air transport would begin.

The man who had planned it all would be on board. By a happy chance Lord Thomson of Cardington had been returned to office in time for the first flight of R.101 almost a year before. The airship had been designed and built at Cardington – the first fruit of Lord Thomson's airship programme of 1924. Now, extended in length and lifting capacity and proven to Air Ministry satisfaction by a 16-

hour test flight three days before (a flight which in the words of the Officer-in-Charge of Flying, Major Scott, had gone 'really well'), and with a total of almost 120 flying hours in eleven flights behind it, R.101 was ready at last. There were few who had the slightest doubt that this long-prepared flight to India would be anything but a resounding success even though, as at the start of every pioneering long-range flight by a prototype aircraft, those who knew most about it understood the manifold unknowns and trusted that they would enjoy good luck.

At 6.15 there was a stir among those at the foot of the mast. A large blue Daimler drove across the aerodrome. Out stepped the tall, elegant figure of the Secretary of State for Air, Brigadier-General the Right Honourable the Lord Thomson of Cardington; debonair, smiling, spruce in his dark overcoat, a white handkerchief in his breast-pocket, a black Homburg on his head. After him came his Private Secretary, Major Louis Reynolds, bald and bowler-hatted – even taller – red official bag in hand.

Lord Thomson shook hands warmly with the little group around the car. There was short, stocky, cheerful Sir Sefton Brancker, Director of Civil Aviation, monocle in eye – a man dedicated to aviation; like Thomson an enthusiast for airships; the life and soul of any party. There was beside him another passenger on the flight, thick-set, solid, reliable, Colonel Vincent Richmond, R.101's chief designer and a happy man this day. Fourth of the group was Squadron Leader Ernest Johnston, R.101's navigating officer, a Master Navigator;* smart in the new blue uniform of the Airship Service.

'Well, Brancks, so we're off at last,' said the Secretary of State as they shook hands.† 'The ship looks fine – and the weather's pretty good, isn't it?' and he turned to Johnston.

'Not at all bad, Sir,' replied the Chief Navigator. 'It looks as though it may blow up a bit with a spot of rain for a short while. But we'll get off ahead of that and we'll be in fine weather across France by morning. It's good to have you with us, Sir.'

By now the Ministerial baggage was being loaded into the lift to go to the mast-head. A long roll in a mackintosh cover went up last of all as Lord Thomson stood with Sefton Brancker and watched it go.

'Take care of my precious carpet,' he called to Herbert Mann, the

* And a founder of the Guild of Air Pilots and Air Navigators of the British Empire (G.A.P.A.N.).

† Recalled by Major Louis Reynolds at Gwydyr House, January 1944.

Cardington lift attendant. He turned to Brancker with a smile. 'It's almost by way of a charm,' he said. 'Came from Sulaimaniya in Kurdistan. You've seen it at home. We'll get it laid out for the State Banquet at the Ismailia Mast. Should go down well in every sense.'

The four of them, with Reynolds in the middle, posed for a final photograph; the great streamlined hull of the airship floating serenely above.

A few more words and then: 'Well, Louis,' said the Secretary of State, 'it's time we said good-bye and went up on board. See you in about sixteen days. Thank you for coming out with me. And by the way,' he added, 'I know they want to save weight. So, after all, I won't take this hefty pack of manuscripts with me,' – he waved to a heavy bag. 'Would you mind taking them back? Put the lot on my desk at Gwydyr House. I was going to work on them on the voyage but I've got those speeches to draft, letters to write and that Conference Paper to put together for when we get back. So I'll leave this lot behind.'

They shook hands. 'Good luck, Minister,' said Reynolds.

'Thank you,' said Thomson. 'As you know, Louis, this is a voyage I've been awaiting all my life – and more than ever since we started along this airship trail in 1924. I've never looked forward to anything so much. Well hardly ever,' he added with a laugh.

Not long afterwards watchers on the ground saw Thomson and Brancker, side by side, leaning on the rail behind the now brightly-lit windows of the saloon, looking down on the scene below as, in the falling dusk, suddenly the floodlights came on.

Now, as R.101's Captain, Carmichael Irwin, balanced up his ship, a sparkling mist of water-ballast deluged from the nose to add a fairy-like touch to the departure scene; disguising too the first raindrops as they began to fall on the rising gusts of wind.

It was almost dark and the underside of the airship's envelope shone silver in the lights as one by one the engines were started in their single cars. For a time the starting engine in the forward starboard car itself would not start but by just after 7.30 British Summer Time all five of the airship's Tornado heavy-oil engines were turning over with a low rumble.

At 7.36 precisely there was a cheer from the crowd which had now grown to substantial size. The mooring cone slipped from the mast-head and the nose of R.101 lifted clean and clear. It was a good start. The airship drifted backwards on the wind. As the ship rose and slipped into the darkness, the red port lights shone out while the control car and cabin lights were dimmed so that their occupants could better see the ground.

On the ground, amid the cheers, there were low sobs from some of the wives who were left behind; desolate at the thought of partings which might be for weeks . . . or more.

A few minutes later, as the rain began to fall in earnest, the watchers on the ground glimpsed again the lights of the airship as R.101 moved forward over Shortstown for a farewell circuit of its native Bedford as a salute to the families of its crew.

For minutes more the deep droning of the engines told of the ships voyage towards India and into history, bearing with her fifty-four souls and the hopes of a nation.

Having done with all things of England, R.101 advanced to her fateful appointment at Allonne.

PART ONE

SECOND TIME AROUND

June 1929 to September 1930

CHAPTER ONE

Fulfilment

Blue skies, smiling at me
Nothing but blue skies do I see.
IRVING BERLIN

Wednesday, 19th June 1929

ON THIS LOVELY summer's morning, after yesterday's rain, even the usually dim interior of London's St Pancras station was light with scattered sunshine. As the 9.25 train to Bedford and the North, pulled out behind its dark red Midland Compound, puffing fiercely in the struggle to gather speed up the bank towards Mill Hill, the brightness of the June day matched the sunny mood of the Secretary of State for Air. He settled his tall, sparse frame into his first class corner seat and smiled across at his companion opposite.

'A wonderful day, Bullock and, with any luck, a splendid summer ahead.' As he spoke, he opened the official red box on the seat beside him and took out a bulging file.

'You know, Bullock, whatever the sweets of independence in opposition, it's great to be back in office; to be on the inside again – and it's just at the time when we ought to be able to get things really moving ahead once more.'

'I'm sure that's right, Minister; and it's certainly good to have you back,' replied his Principal Private Secretary, Major Christopher Llewellyn Bullock, old friend and ally from five years before. 'I don't doubt that we'll be as busy this summer as we were last time; here, and overseas too.'

Christopher Thomson smiled.

'Another trip to the Middle East, you mean? Perhaps we'll come back with yet another flying carpet. But this time, maybe the flying carpet is R.101, and the airship will waft us on to India at last – I hope before Christmas. Then, to Canada in the New Year and, all being well, regular services after that. It seems that Fate's been kind. All we set in hand, with so much toil, in 1924 seems to be coming together now, at a happy moment.'*

* Conversation noted on 19th June 1929, by Sir Christopher Bullock (1891–1972) and related to the author on 14th May 1970.

On this brilliant June morning with its cloudless sky, and its promise of midday heat, Christopher Thomson had every reason for delight. Nine days before, on 10th June 1929, he had come back to the Air Ministry to take charge for the second time; borne in on the crest of victory at the Polls on – for the Labour Party – a Glorious First of June.

*

To outward appearances, Brigadier-General the Lord Thomson of Cardington might have seemed the least probable of Labour Ministers. If his suits were old, they were of perfect cut, carefully pressed to go with his general air of tall distinction. He was 6 feet 5 inches in height, slim with a slight stoop and a manner which extended an old-world courtesy to everyone he met. His ready smile gave no hint either of his inward loneliness or of his firmness in command. Yet, ever since he had resigned his commission in the Regular Army at the age of forty-four in 1919, and disappeared from the convivial company of Army life to carve a new career for himself in the Labour Party, he had devoted to that unfamiliar scene a loyalty born of 'an idealism so unexpected that some of his friends could hardly believe that it was there.' *

It was an idealism which went back to lessons learned at his mother's knee from which came a compassion and a concern for the less fortunate in life. Over the years, and especially after the insight he gained into the hardships of the Roumanian peasantry during 1916 and 1917, this side of his nature, and his sympathy for 'the poor under dogs', developed into a deepening disenchantment with the social order of the day. At the same time none of this lessened his appreciation of the good things of life even though they were largely denied to him during the lean years of 1920 to 1923.

That was the time in which, on a foundation of hard work, sincerity, good-humour, a gift for friendship and increasing skill as a speaker, he made his way through hardship and privation in the often hostile atmosphere of political aspirations. In this way, too, he moved from being an unknown outsider to a respected member of the Labour Party and to becoming a close friend of its leader. It was a friendship in which each man found in the other qualities he admired and did not possess himself.

* Lieutenant-Colonel E. M. Mozeley (1875–1950) writing in the *Royal Engineers Journal* of March 1931.

Fly on British Coal – Lord Thomson of Cardington as Chairman of the Royal Aero Club of the United Kingdom at Croydon Airport on 23rd February 1929, with Miss Winifred Spooner and a de Havilland Gipsy Moth flying on petrol distilled from coal.

To Windsor – The new Prime Minister, the Right Hon James Ramsay MacDonald, accompanied by the Secretary of State for Air, Brigadier General the Lord Thomson of Cardington and by the Secretary of State for Foreign Affairs, the Right Hon Arthur Henderson, board their train at Paddington to journey to Windsor Castle to receive their seals of office from HM King George V, following the formation of the second Labour administration on 8th June 1929.

[PLATE III]

The Cabinet – The Prime Minister, Mr James Ramsay MacDonald with his second Labour Cabinet at No 10 Downing Street. Mr MacDonald is flanked, on his right by Mr Philip Snowden, Chancellor of the Exchequer and on his left by Mr Arthur Henderson, Secretary of State for Foreign Affairs. The Right Hon William Wedgwood-Benn, Secretary of State for India is 'outside left' (seated on the right of the photograph) and Lord Thomson of Cardington behind Mr Snowden with the first woman Member of the Cabinet, Miss Margaret Bondfield, Minister of Labour, on his right.

[PLATE IV]

Back as a member of the Cabinet and installed as Secretary of State for Air for the second time around, after ten days in office his first tour of inspection outside the Air Ministry was to travel by train with his PS to see R.101 in its construction shed at Cardington. There, the new airship was nearing completion at last; evocative and tangible evidence of his first term of office five years before.

As he turned over the file of papers from his Red Box, settling himself back into his corner seat, Thomson spoke reflectively to Bullock of that May Day in 1924 when the first MacDonald Cabinet had set the seal of approval on his plan, so diligently and hopefully drawn up with Geoffrey Salmond, Sefton Brancker and Reginald Colmore, for an Imperial Airship Service to fly to the furthest corners of the Earth. Now, as if by Providence, he was back in office at the Air Ministry just in time to reap the fruits of the seed sown in 1924.

More than ever now, he saw aviation as a symbol of new opportunities for Britain in the years ahead. Here, better than in any other field of modern technology, the traditional British skill and enterprise in design, in manufacture and in long-range navigation could be used to bring the peoples of the Empire closer together than they had ever been before.

In his grandfather's day the passage to India had taken sixteen weeks by way of Cape Town and Durban. Even today the fastest liners required four weeks for the voyage through the Mediterranean, the Suez Canal and the Red Sea.

When R.100 and R.101 were in service, the time for the voyage to India could be cut drastically to five days or less – and flown in much greater comfort than could be provided by the weekly service by landplane, train and flying boat started by Imperial Airways three months before – on 30th March 1929 – scheduled to take eight days to Karachi with twenty-one stops on the way.

Of course the airships had taken much longer to produce than they had hoped and planned. Perhaps there was something to be said for that, however, not only because he had always advocated a policy of 'safety first' but also because, if the original schedule had been met, his friend and rival, Sam Hoare, would have been able to boast the first airship flight to India as he had already achieved a passage on the first commercial landplane flight early in 1927.

Thomson had set his heart on that airship voyage to the land of his birth and, from a recent talk with the Prime Minister, it would have a special significance for him too, even beyond showing for all to see the success of his airship policy of 1924.

Meanwhile he had been welcomed back into his old job at the Air Ministry. *Flight* wrote:

> Few inside British aviation circles will fail to applaud the appointment of Lord Thomson of Cardington to the post of Secretary of State for Air. When the first Labour Government was in office he proved himself to be not only an enthusiastic but also a very able Air Minister. That he quickly became exceedingly popular was due, partly to his obvious keenness but largely to his considerable personal charm.

The Aeroplane was no less positive.

> Lord Thomson has been an excellent Chairman of the Royal Aero Club [wrote C.G.G.]. As an after-dinner speaker at aviation banquets he has been an outstanding success. As the official British representative at the International Conference at Washington he did valuable work and won personal popularity and respect to an unusual degree. He has nursed British aviation assiduously because of a genuine interest. His re-appointment will be a valuable national as well as an international asset.

Now, as the train sped on its way to Bedford – past Hendon Aerodrome, where preparations were already in hand for the RAF Display of a month hence, and past the open grass fields beside the railway at Radlett where Handley-Page had told him, the week before, that he had bought land for a new factory and aerodrome – the Secretary of State opened the Air Ministry file and turned to the original papers of 1924. There was his Memorandum of 7th March which he had presented to the Cabinet's Airship Development Committee* only six weeks after taking office.

He looked up from those early pages with a quizzical smile at Bullock. An appendix was headed: 'Estimated rate of progress'. It detailed a step-by-step programme of research, design and construction over the course of two-and-a-half years to arrive at 'Experimental flight or flights to India' with the projected new airship.

All along, that flight to India and back had been the focus of the endeavour. It would mark the completion of the first phase. He had written: 'There can be little doubt but that once the new ship has been tried out successfully the future development of airships will

* Memorandum AD (24)2, 7th March 1924, by the Secretary of State for Air: 'Airship Development by the Air Ministry' (AIR 8/79/06990).

be rapid and to a certain extent automatic and it is for this reason that all other considerations should be subordinated to what is the main issue at the present stage, i.e. the construction and trying out of the new ship as proposed.'*

It had all been detailed with, perhaps undue, precision. After preliminary arrangements the programme had been due to start on 1st August 1924. The Royal Airship Works at Cardington had in fact been re-opened a month before that.

The original programme of 7th March 1924, had then been scheduled to run as follows:

1st August 1924	– Model experiments to begin at the National Physical Laboratory, Teddington.
1st February 1925	– Placing of contract for the Egyptian base.
30th April 1925	– Completion of research and testing of experimental girders.
31st May 1925	– Completion of enlargement of the airship shed at Cardington.
1st June 1925	– Placing of contract for the Indian base at Karachi.
1st July 1925	– Start of construction of the new airship.
1st August 1925	– Completion of the new mooring mast at Cardington.
1st September 1925	– Completion of the Egyptian mast and base.
1st October 1925	– Departure of R.33 or R.36 on a trial flight to Egypt to gather data.
31st December 1925	– Return to Cardington of R.33 or R.36 on completion of Egyptian tests.
31st July 1926	– New ship is floated and ready for flight.
31st January 1927	– New ship's first India flight completed and airship back at Cardington.

The Memorandum of 7th March 1924, which was there on the file, had gone on to say that:

> The practical steps which should follow immediately after the successful completion of the first flights to India are:
>
> 1. The building of a new shed at Cardington;

* 'Programme for Government Research, Experiment and Development of Airships'. Memorandum by the Secretary of State for Air. (7th March 1924) – (AD 24(2) – (AIR 8/74/06990).

2. A programme for the construction of additional airships;
3. The establishment with them of regular services between England and India and between England and Canada;
4. The adaptation of the design of the airships to meet defence requirements.

To these ends a sum of £500,000 has been provided in the estimates for the commencement of this work so as to avoid any delay in the development of airships and any break up of the organisation pending the settlement of policy.

A successful flight to India and back had thus been seen from the start as the key to the future development of airships for both commercial and military purposes. That first India flight had been scheduled for the end of January 1927. Here they were in June of 1929 and neither of the two new airships had yet flown. The whole programme was running about three years late – nor had either R.33 or R.36 ever flown to Egypt as intended to explore so many of the unknowns.

'I'm not complaining – but it's all pretty late, Bullock, isn't it?' observed the Secretary of State, looking up from the file. 'I see a note here from Geoffrey Salmond which gives some of the reasons early on.'*

Christopher Bullock prided himself on his knowledge of all the facts. Here was an opportunity to put the Secretary of State fully in the picture of what had happened in the five years since he had been closely in touch with progress. He reminded Lord Thomson that his original Cabinet Paper had specifically made the point that the outline programme for R.101 was very tight and made no allowances for possible delays.

But R.100 had lagged behind even more than had R.101 even though the Vickers' airship was deliberately intended to be much less experimental than R.101 which incorporated a great many new and advanced developments – not least the safety feature of the heavy-oil engines upon which the Air Ministry had insisted from the start.

The note from Geoffrey Salmond, dated 24th July 1925,† did indeed give some of the background to the initial delays. Written nearly four years before, it was, as the Secretary of State remarked, 'Now somewhat ancient history'.

* Conversation recalled by Sir Christopher Bullock to the author, 14th May 1970.

† Note by Air Member for Supply and Research to the Secretary of State (Sir Samuel Hoare) through Secretary (Sir W. E. Nicholson) on 'Airship Development Programme'. 24th July 1925, (AM5/1006/1M. 08109).

In July 1925 Sir Geoffrey had explained that, during the first twelve months of the airship programme, the design and drawing office staff at Cardington had been massively diverted from work on R.101; first on to the reconditioning of R.33 and R.36, then to the re-design of the nose of R.33 after its break-away from the Pulham mast in the great gale of April 1925. After that much time had been taken on the complexities of the new features embodied in R.101's test frames and on stressing the Cardington mooring mast. All in all it meant that extraneous circumstances had put back construction of R.101 by about nine months during the first year of the programme.

It had all drifted, explained Bullock, because Sir Geoffrey Salmond – the main force behind the airship programme – had not received the same support from Sam Hoare as he had from Christopher Thomson. Nor had Sam Hoare had the ear of his Prime Minister to anything like the same extent as Lord Thomson of Ramsay MacDonald. An example on the file was a further note from Salmond about the way in which the projected trial flights to Egypt by either R.33 or R.36 had been allowed to lapse, in spite of the importance which both Thomson and Salmond had placed upon them as collectors of vital – even essential – information on route requirements.

Thomson turned over the pages and came upon the note from AMSR dated 16th February 1926*. He read it out:

> I am afraid that I do not quite agree with the sentence which deals with the cancellation of the Egyptian flight – in its present form.
>
> I wish to emphasise the point that it was only the extreme financial pressure which induced me to agree with the cancellation of the flight and that to do so does involve a certain amount of risk in so much as this flight would have given the airship staff first-hand experience of the difficulties of airship operation and, more particularly, of mast mooring under tropical conditions. It is not only that greater precautions will now have to be taken on experimental flights of R.100 and R.101 than would otherwise have been the case, but we are adding to the number of unknown factors and thus increasing the difficulties of the experimental flights. Moreover, the absence of these trials was used by us an an argument against Commander Burney's proposals.
>
> I have therefore suggested that we should bring out the point that I would never have contemplated cancelling these experiments had it not been for extreme financial stringency.
>
> It is true that, owing to certain later developments I was not anxious

* Note on Memorandum to accompany the Air Estimates 1926/27 signed by W. G. H. Salmond (AMSR) 16th February 1926 (AIR 5/1006/1M/08109).

> to carry out the flight with R.33* but had funds permitted I would have completed the reconditioning of R.36 and used her for this purpose.

Thomson looked up. 'Penny-pinching, Bullock,' he observed. 'We may live to regret the cancellation of those trials, I shouldn't wonder.'†

Bullock agreed. He was sure that Geoffrey Salmond had been right. Certainly, it would have been most valuable to have learned with R.36 how airships behave in at least semi-tropical conditions.

R.36 had been built by Beardmore's at Inchinnanin, Glasgow, and completed in 1921. It had been the only British airship with an external passenger-car slung up against the hull underneath. R.36 – at a volumetric gas capacity of 2,101,000 cubic feet in twenty bags – was some seven per cent bigger than the 1,960,000 cubic feet of R.33 in nineteen bags. Moreover, R.36 had 23 per cent more power from its five engines, totalling 1,540 bhp, compared with R.33's 1,250 bhp.

According to Cardington's calculations, R.36 ought to have been capable of flying non-stop to Ismailia, carrying a 32-man crew, 4,540 gallons (15 tons) of petrol and a small number of scientists and observers on board.

Such a flight would have yielded valuable lessons if it had not been cancelled on those economic grounds.

'Ah, well, it can't be helped,' said Thomson. 'But I must admit that I wish I'd been there at the Air Ministry to push it along.'

In the short time since he'd taken office again he had not been able to gain more than a general briefing on the airship programme and now he asked Bullock to run over with him from the file how the whole project had progressed in performance and cost since 1924. Very little detail had been published; the Air Ministry had played its cards close to its chest.

Bullock opened his folder.

'Well, as you know, Minister,' he said, 'the estimated performance has gone down a bit and, with the delays, the cost has gone up.

He pulled out a note ‡ from Lieutenant-Colonel Vincent Richmond, Chief Designer of R.101 at the Royal Airship Works. It was dated only eleven days before: 8th June 1929.

Richmond had said that inflation of R.101 was now due to begin

* Severe corrosion had been found in R.33's structure, built in 1917.

† Recalled by Sir Christopher Bullock, 14th May 1970.

‡ RAW Note 'R.101 Lift Available', 8th June 1929 (AIR 11/155/M.8109).

on or about 1st July. That would mean the first flight sometime in August, unless it should be delayed by Works' holidays.

On the latest information at the Royal Airship Works the airship looked like coming out a good deal heavier than he had hoped. With the 25 tons of fuel and oil estimated to be needed for the Egypt–India route, the lift remaining for passengers and freight was now expected to be 16¾ tons – or 37,520 lb. That compared with the original, optimistic, estimate of 30 tons. The missing 13¼ tons had been swallowed up by the unforeseen additional weights of engines, structure and passenger accommodation. Even so, taking 350 lb for each passenger, his baggage, his food and water and cabin equipment, a payload of 16¾ tons would still make possible the carrying to Egypt and India of 100 passengers and about a ton of mail and cargo; or fifty passengers and about nine tons of dead load. That wouldn't be bad for what was, after all, still an experimental ship.

Bullock added that he thought that everyone at Cardington felt fairly cheerful about the ship's likely performance.

The subject of costs was rather less satisfactory.

Bullock reminded the Secretary of State that the cost of the original programme had been estimated in 1924 at £1,400,000, including contingencies, to be spent between 1st August 1924 and 31st January 1927. That figure had been reckoned to include research, the recommissioning of either R.33 or R.36, design and construction of the mooring masts at Cardington and Karachi, the enlargement of the Cardington shed, the design and construction of R.101 and a sum of £350,000 for a fixed price contract with the Airship Guarantee Company for the design and construction of R.100 at Howden in Yorkshire.

Looking through his papers, Bullock pulled out a sheet which showed that in 1924 the original estimates of costs had been set out under eleven heads:

1.	Research	£ 50,000
2.	Reconditioning of R.33 or R.36	£ 75,000
3.	Cardington shed and mooring mast	£103,000
4.	Cardington Staff and office expenses	£162,000
5.	New ship (R.101) materials and engines	£ 91,000
6.	Flying staff wages and salaries	£ 28,000
7.	Trials (R.101; 200 flying hours)	£ 30,000
8.	R.100 contract price	£350,000
9.	Egypt base and staff	£ 90,000
10.	India base and staff	£251,000
11.	Contingencies (at 12 per cent)	£170,000
	Estimated total (1924)	£1,400,000

Last February the Air Council had given the Treasury a revised estimate of the costs of the airship programme up to the completion of an out-and-back flight to India by R.101. It had come out at £2,051,000 to the end of January 1930. That was an increase of £651,000, or 46 per cent, on the original estimates in spite of the savings made by cutting out the flight of R.33 or R.36 to Egypt and back.*

It represented the cost of stretching out the programme by an additional three years compared with the original forecast which had set January 1927 as the date for the flight to India and back. The original expectation was that the cost of £1.4 million would be incurred over a 31 month period; that is at an average of some £45,000 a month. Now the expected cost of £2.051 million would be spread over 67 months which would work out at a little less than £31,000 a month. So, though the total was up, the annual rate of expenditure would be some 30 per cent less.

What this represented, Bullock observed, was the cost of delays and of quite a number of unforeseen happenings. The Air Ministry had certainly not been extravagant in any way and the policy laid down by Lord Thomson in 1924 of going ahead cautiously with the maximum of emphasis on safety, had been faithfully followed.

'Well; you've certainly got the situation at your finger-tips, Bullock,' remarked the Secretary of State with a smile. 'You almost persuade me that the delays have been worth while though I'm not sure that the Lords Commissioners of His Majesty's Treasury would wholly agree with that view.

'What you are really saying, I think, is that we're running about three years late with both ships; we've taken rather more than twice the time we estimated in 1924 to reach the present stage. On the other hand, you tell me that we'll spend only half as much again as we expected – and, I suppose it can be said that we've given more employment to more people for a longer time.

'But the real question is whether we're going to get value for money. I hope so. Even if the performance – at least at the start – does turn out to be a bit down on the original forecasts, it does look as if R.101 will do the job we hoped for – and India by Christmas.' †

He looked up. The dark shapes of the two great airship sheds at Cardington were there in view from the railway carriage window,

* No provision was necessary for money inflation in 1920–30.

† Conversation noted by Sir Christopher Bullock and recalled to the author, 14th May 1970.

with the tripod of the still-empty mooring mast half-a-mile to the south.

The Secretary of State looked at his watch. The time was 10.35.

'Here we are, Bullock,' he said. 'I must say I can't wait to see R.101 in the flesh at last. I'm sure the car from Cardington will be waiting for us at Bedford Station.'

It was.

CHAPTER TWO

The Royal Airship Works

O all ye works of the Lord, bless ye the Lord.
The Book of Common Prayer

19th June 1929

ON THE STEPS of the administration block at Cardington, waiting to greet the Secretary of State for Air, a welcoming party of three was headed by the Air Ministry's Deputy Director of Airship Development, Wing Commander Reginald B. B. Colmore, 'a quiet, studious, inherently sound man; the best type of Naval officer'* who had been for the past ten years one of the key figures in British airship development 'devoted to airships but clear about their limitations.' *

From July 1924 until only four days previously the Director of Airship Development at the Royal Airship Works had been Group Captain Peregrine Fellowes, DSO, like most British airshipmen a product of the Royal Navy and the Royal Naval Air Service before moving over, on 1st April 1918, to the newly formed Royal Air Force. In 1921 he surveyed and opened the Air Mail route between Cairo and Baghdad.

Fellowes had been selected by Geoffrey Salmond to return from the Middle East to take charge of the new Department at Cardington partly because of his knowledge of the projected airship route to India and partly because of his previous background in the Royal Navy which made him suitable to head up a largely ex-Naval Airship Service.

Fellowes had shown himself to be a skilful administrator as well as a fine aeroplane pilot and always a cheerful and attractive personality. His knowledge of airships was, however, nil although in 1927 he had conducted a meticulous survey of the potential airship routes to India, Australia and South Africa. Now, in 1929, with Cardington well established and practical airship flying about to begin, there had been a feeling at the Air Ministry that there was a need for a practical airship man to head up the Royal Airship Works and that Fellowes' qualities ought to be used elsewhere. Accordingly, on the previous

* Sir Harry Self in a letter to the author, May 1968.

Saturday, 15th June, Fellowes had left Cardington to take up duty at the Air Ministry from 1st July 1929 as Director of Personnel Services under Air Chief Marshal Sir John Salmond, Air Member for Personnel.*

That left Reginald Colmore in charge at Cardington; for all Fellowes' attributes, a popular move.

If Fellowes had been an airship amateur temporarily in command of the Royal Airship Works, his former subordinates were very much professionals in Lighter-than-Air affairs with which they had spent most of their working lives. Indeed, the RAW's 'Big Three' – Colmore, Richmond and Scott – had begun to learn their trade among non-rigid airships not long after the Admiralty had taken over responsibility for the 'Battle Bags' from the Royal Flying Corps on 1st January 1914. United in their devotion to lighter-than-air craft, the three were very different in most other characteristics.

All three were of such importance to the airship programme of 1924/30 that a brief note of their backgrounds and early service will serve to set the Airship Service in perspective.

Wing Commander R. B. B. Colmore, OBE, like his former chief, had started his Service life as a Royal Naval Cadet on board HMS *Britannia*. He was commissioned as a sub-lieutenant in the Royal Navy at the age of twenty in 1907, and was promoted lieutenant in 1909. Five years later, as a lieutenant-commander, RN, he first saw active service with the RN Armoured Car Division at Antwerp in 1914 and commanded the Armoured Car section in Gallipoli in 1915; later in the same year, he was Commanding Officer in the first campaign against the Senussi tribes on the Egyptian border.

Colmore transferred to the Airship Section of the Royal Naval Air Service in 1916 and, after serving at the Royal Naval Airship Station at Barrow-in-Furness, he was appointed to command the most active of the non-rigid airship stations at Mullion near the Lizard, in Cornwall, from which in 1917 and 1918 in every tolerable weather, patrols were flown over the Western approaches.

While he was at Mullion, Colmore evolved a new form of anti-submarine patrol which combined the use of airships and seaplanes in conjunction with surface craft to such effect that, in 1918, he was appointed (now as a lieutenant-colonel, Royal Air Force) as a Staff

* Air Commodore P. F. M. Fellowes, DSO, was appointed in 1933 to lead the Westland-Houston Expedition which, in April of that year, was the first to fly over Mount Everest. Fellowes had retired from the Royal Air Force in 1933 and died in South Africa in 1955.

Officer, 1st Class, Aircraft Operations under the Commander-in-Chief, Grand Fleet, Dundee, to set up his system around the whole British coastline.

In September 1916 Colmore had taken part in the detailed assessment of the wreck of the German Zeppelin, L.33, which had been forced down at Little Wigborough, near Mersea Island in Essex, and set on fire there by its crew. The structure of this new Zeppelin was, however, almost undamaged and drawings made from it were used as the basis for the Admiralty designs of R.33 and R.34 ordered from Armstrong-Whitworth and Beardmore respectively.

In 1919 Colmore was granted a permanent commission in the Royal Air Force, with the rank of squadron leader, and attached to the Air Ministry as Staff Officer (Airships) in the Department of the Controller General of Civil Aviation. There, in 1924, under Sir Sefton Brancker, he took a lead in the drawing up of the specifications for the new airships R.100 and R.101,* and in July 1924 went to Cardington as Deputy Director of Airship Development (DDAD).

Colmore was a dedicated airshipman of courage and judgement; good on both the technical and administrative sides of his responsibilities.

Second of the 'Big Three' at Cardington from July 1924 was Lieutenant-Colonel Vincent C. Richmond, Head of Design and Research at the Royal Airship Works – in practice, chief designer of R.101.

Richmond has been described by one who knew him well† as 'a strong but kindly leader of courage and ability, wholly dedicated to airships with a penetrating mind and a breadth of view which impressed us all. He was always ready to listen to others but never hesitant to take decisions.'

Modest and of sound judgment, he was without doubt the dominant personality at Cardington; a benevolent autocrat in the Drawing Office, supremely well versed in the theoretical basis of rigid airship design and construction yet always open to new ideas and prepared to discuss the merits of a case. These two sides of his character were seen in the complete reliance he placed, on the one hand, in his brilliant technical assistant, Squadron Leader Michael Rope, RAF, and, on the other, in the complementary practical knowledge and experi-

* Appendix 2—Airship Specifications, 1924 (AM Specification: A22/24 and A24/24.

† Professor Sir Alfred Pugsley, OBE, FRS, DSc, Hon FRAeS, Technical Officer at the Royal Airship Works, Cardington, 1926–31, and from 1968 Emeritus Pro-Vice Chancellor of Bristol University.

ence of John North, Chief Engineer of Boulton and Paul Aircraft Limited and leading authority of the day on metal aircraft construction.

So it was that Michael Rope's inventive mind produced a series of ingenious and effective advances in design for the new airship while the detailed design and manufacture of R.101's structural components was sub-contracted to the skilled Norwich company. And so it came about that a pillar of private enterprise, John North, worked happily in the closest collaboration with the leading State-designer, Vincent Richmond.

Third of Cardington's 'Big Three' was the Officer in Charge of Flying and Training (always known as 'FT'), Major George Herbert Scott, CBE, AFC, a veteran of the practical flying of British airships; the holder of 'British Airship Pilot Licence No A.1 (First Class)'.*

Herbert Scott had joined the Royal Naval Engineering College at Keynsham at the age of twenty-four in 1912. From there he had gone to the Airship Station at South Farnborough on the outbreak of war and then, early in 1915, to the Kingsnorth airship station near Hoo in Kent on the north bank of the Medway Estuary, north-east of Rochester. Scott's first RNAS command was British Naval Airship No 4; the *Parseval* PL-18 non-rigid imported from Germany in 1913 and, by 1915, based at Barrow-in-Furness. In 1916 Scott went on to command the Royal Naval Airship Station for non-rigids at Anglesea. Early in 1917 he was promoted to command the first British rigid airship to take the air: HM Airship No 9 built by Vickers at Walney Island, Barrow, and flown for the first time on 27th November 1916. That ship lacked lift and effective control but taught many lessons. It was employed for a year and 198 flying hours on experimental flying and training of crews. From No 9, Scott graduated to command its more airworthy development, Naval Airship No 23, also built by Vickers and first flown on 19th September 1917. Then in July 1919 Scott established his lasting reputation in command of the Beardmore airship R.34 on its historic return flight across the North Atlantic Ocean.†

Subsequently, Scott was on the technical staff at Cardington and,

* Issued after the war from the Air Ministry and dated 4th February 1921.

† R.34 flew from East Fortune in Scotland to Hazelhurst Field (later Roosevelt Field), Mineola, Long Island, between 2nd and 6th July 1919 in 108 hrs 12 mins and back to Pulham in Norfolk between 9th and 13th July 1919 in 75 hrs 3 mins. This was the first east to west flight across the Atlantic, the first return crossing by air and the only Atlantic flight with more than thirty people on board until the LZ.126 (ZR-3) airship was flown from Friedrichshafen to Lakehurst, New Jersey, in October 1924.

in 1921, Officer-in-Charge of the Royal Air Force Airship Station at Pulham St Mary in Norfolk. During that time, he developed his design for high mooring-masts for airships, examples of which were eventually built at Cardington, Ismailia, Karachi and Montreal, Canada.

With the start of the 1924 airship programme, Scott was appointed to be in charge of flying and training at the Royal Airship Works at Cardington. In 1927, he visited Canada to advise the Canadian Government on the location of an airship base. As a result, St Hubert, Montreal, was chosen and equipped.

A tough, calm, 'press-on' character, who was awarded an AFC in 1918 and CBE in 1919, Herbert Scott had a reputation in the Airship Service for determination, and also for 'ham-handedness'. He was commanding the *Parseval* PL-18 (No 4) when it collided with its shed at Barrow, in a fog, during November 1915. He was in command of R.36 when it was badly damaged during mooring to the Pulham mast in June 1921, and he was in charge of R.33 when it hit the shed doors at Pulham on its last flight in November 1925.

Airships are notoriously difficult to handle near the ground. Scott had demonstrated this fact dramatically with HMA No 4, and in bringing to an end the service lives of R.33 and R.36. By 1929 Scott had, sadly, lost some of his force and drive and, in the opinion of his colleagues, his judgment left a good deal to be desired. His reputation was such, however, as the most experienced British airship commander, that his role as one of the 'Big Three' at Cardington was unchallenged.

So, by the middle of 1929 Colmore, Richmond and Scott had, together, completed more than forty-three years of Lighter-than-Air experience interrupted on the practical side by a gap of three years in British airship flying between 1921 and 1925 after the loss of R.38 and now for the past two and a half years since R.33 had made its last flight on 23rd November 1926. The flying of British rigid airships was, in fact, concentrated into ten years, between 1916 and 1921; in 1925 and 1926 and in 1929 and 1930.

In all, between November 1916 (No 9) and November 1926 (R.33) a total of fourteen British rigid airships had flown 3,382 hours and 20 minutes* – rather less than the annual number of hours flown by one modern transport aeroplane. This small amount of British rigid airship flying compared with about 50,000 hours flown by 151 German rigid airships (131 Zeppelins and twenty wooden Schütte-Lanzs) up to June of 1929. By then the United States had flown only two rigid air-

* Appendix 3. British Rigid Airships Flying Hours.

ships; ZR-1 *Shenandoah*, built at the Philadelphia Naval Aircraft Factory in 1922 and assembled at Lakehurst, New Jersey in 1923; and ZR-3 *Los Angeles*, built at Friedrichshaven as the Zeppelin LZ.126 and delivered by air to Lakehurst in October 1924 under the command of Dr Hugo Eckener.

Because *Shenandoah* had broken up in the air over Ohio on 3rd September 1925 as a result of flying into severe air turbulence – a disaster in which fourteen lives were lost – and because R.33 had been dismantled at Pulham in May of 1928, there were in June 1929 only two completed rigid airships in existence: ZR-3 *Los Angeles* in the United States and LZ.127 *Graf Zeppelin* in Germany. There were to be only seven more* of which four met with disaster.

Relatively slim as was British experience of flying rigid airships it was, however, much more adequate in its length of involvement with lighter-than-air craft and in the total number of hours flown with all types of airship. Up to November 1926 – since when no British airship had flown – 234 British airships had completed 95,019 flying hours from sixteen bases in the United Kingdom. Out of this total, 91,637 flying hours had been by 225 non-rigids, most of them on wartime service between 1915 and 1918.

Now, in 1929, most of the surviving British airship experience was concentrated at Cardington except for the small Vickers group of the Airship Guarantee Company Limited at Howden, under 42-year-old Barnes N. Wallis, designer of R.80.†

In total lighter-than-air experience, therefore, the leading British exponents were (with the small remaining Zeppelin team under Eckener in Germany) the best qualified in the world. All of them were in the prime of life; at Cardington Colmore forty-two, Richmond thirty-six and Scott forty-one; and at Howden in Yorkshire, Wallis forty-two.

*

Those then, were the three at Cardington who stepped forward to welcome the Secretary of State for Air as an old friend and chief, happily returned to power and now on his first visit back to the

* R.100 and R.101; ZRS-4 *Akron* and ZRS-5 *Macon*, ZMC-2 and LZ-129 *Hindenburg* and LZ-130 *Graf Zeppelin II*.

† See Appendix 3 for a list of British rigid airships and their flying hours.

domain into which he had breathed new life almost five years before.

The car from Bedford drew up and Lord Thomson leaped out into the bright sunshine of the now perfect June day to shake hands with Colmore and then with his two companions. All of them were well known to him; high in his regard. Colmore and Richmond were former colleagues on Geoffrey Salmond's Airship Advisory Panel of the Cabinet Committee of 1923. Subsequently Scott had been a speaker at a number of Royal Aero Club dinner-discussions under Thomson's chairmanship. He knew how thoroughly they all shared with him a steadfast belief in the future of airships.

They trooped out of the sunshine into the shade of the fine entrance hall. A Royal Air Force ensign hung from the ceiling. The Secretary of State looked up at it. Scott followed his glance.

'We propose to stream that, sir, from the stern of R.101 on our first flight overseas,' he said.*

'Splendid,' replied the Secretary of State. 'A true-blue flag to fly from what will be the most modern and advanced of His Majesty's ships.'

He looked round at the interior of the imposing Administration Block. It had been built in 1917 by the pioneer balloon, aeroplane and airship constructors, the Short Brothers, as the intended headquarters of their future business.

Alas for Short's far-sighted plans; in April 1919, the Cardington works and aerodrome had been taken over by the Air Ministry under a compulsory purchase order to create the Royal Airship Works. The price of £40,000† had included the well-built housing-estate of Shortstown on the south side of the Shefford/Bedford road, in 1929 occupied by families from all ranks of the Royal Airship Works.

From the hall, Wing Commander Colmore shepherded his visitors towards his large, ground-floor office at the south-west corner of the block on the right of the entrance. Cups of tea and biscuits were handed round while Colmore introduced to the Minister three additional senior members of the Cardington team – all of them serving officers of Royal Air Force 'attached for special duty'.

First was Wing Commander Thomas R. Cave-Browne-Cave, CBE, (always known as 'T.R.') forty-four years old and an experienced engineer officer, now responsible for R.101's power plant installations. Next was 41-year-old Squadron Leader F. Michael Rope, Colonel

* Reported by Sir Christopher Bullock. R.101's scorched and torn RAF ensign, brought back from Allonne, now hangs in the Church of St Mary, Cardington.

† Equivalent to about £430,000 in 1980 values.

Richmond's chief assistant and widely acknowledged to be a design-engineer of genius, earlier responsible for the excellent SS Zero non-rigid airships.

Third came Flight Lieutenant Sydney Nixon, OBE, forty-two years old, chief administrative officer of the Royal Airship Works.

Introductions completed, the party of eight sat down with their tea-cups around Colmore's conference table. DDAD briefly welcomed the Secretary of State, 'the onlie-begetter of the 1924 Airship Programme.' As a start, he said, and before they went down to inspect R.101 in its construction shed, he was sure that Lord Thomson would like to hear from Colonel Richmond a brief review of progress so far, the design problems encountered on the way and how things were looking for the future.

At a nod from Colmore, Richmond moved over to speak from in front of a large general arrangement drawing of R.101 pinned on the wall. Stocky, confident and very much in command of his subject, Richmond was an excellent speaker on his own special subject. He began by saying that he knew that the Secretary of State would recall that – way back in 1924 – R.101 had been projected as a Service airship, to be designed to carry 200 troops with their equipment, weighing 30 tons in all, over long distances, or, alternatively, four or five fighter aircraft for Home Defence.* In the latter role the idea had been that the airship would be able to release the single-seat fighter at 15,000 feet to intercept hostile bombers off the British coast, reported to the cruising airship by wireless.

Attractive as this idea undoubtedly was for the mounting of a defensive patrol at height, experiments with R.33 in 1925 had shown that such a plan would be possible only with a much larger airship than the five-million-cubic-feet of R.101. Moreover, until helium was available in quantity – which it still was not – hydrogen-filled airships would remain very vulnerable in war.

The use of R.100 and R.101 as troop-carriers did, however, remain a future possibility. In the meanwhile, as the Secretary of State knew, both airships had now been designed solely as long-range passenger, mail and cargo craft. The programme had begun, as originally decided, with a comprehensive series of aerodynamic and structural tests to eliminate any chance of a repetition of the weaknesses which

* Outlined in AMSR Note to DAD, 20th November 1924 (AIR2/262/520117/29) and in AMSR Note to S of S of 24th November 1924 (AIR 8/21/06990).

had led to the loss of R.38 in 1921.* These tests, conducted at Cardington and at the National Physical Laboratory at Teddington, were satisfactorily completed by the end of 1926. One of the results had been the evolution of a new and fatter streamlined shape for R.101 which had no parallel sections in its hull and a fineness ratio (of length to maximum diameter) of 5.5 to 1.

There were, said Richmond, two important advantages in the plumper, streamlined, form of R.101 compared with the 10 to 1 fineness ratio of earlier ships. First, the relatively low drag. Secondly, that an airship of more than double the volumetric capacity of any previously built at Cardington could be assembled in the existing shed. A modest increase in the shed's dimensions had been completed during 1926 and the internal dimensions of the enlarged shed were now 812 feet long by 180 feet wide and 156 feet high, which made it the biggest building in Britain. R.101 had a comfortable clearance of 81 feet in length, 48 feet in width (24 feet on each side) and 15 feet in height.

Richmond went on to say † that initial design work on R.101 had started, as the Secretary of State would know, in July 1924. By the middle of 1925 it had progressed to the stage at which the ship's general layout had been worked out in detail and the general form of the structure had been decided.

The next step had been to build up a typical bay section of the framework of the hull (complete with a 500,000 cubic feet gas bag) to serve as a structural test-specimen. It was ready for trial loadings in July 1926. By then all the major decisions about the design of R.101 had been taken in conjunction with a strong team of consultants. Prominent among them, on structures, was Professor R. V. Southwell, late RNAS, then Head of the Aerodynamics Department at the NPL and now Professor of Engineering at Oxford University. Richmond had a high regard for Southwell and, as he said, was always ready to imbibe from him 'the milk of structural wisdom'. In addition, on aerodynamics, there was Professor Leonard Bairstow of London University and the Imperial College of Science and Technology and – on detail design and construction, J. D. North of Boulton and Paul.

For a time, valuable advice had been gained also from Herr Paul

* The Admiralty-designed, Cardington-built, British airship R.38 broke up in the air over the Humber on 24th August 1921 on its fourth test flight. Out of forty-nine on board there were only five survivors. In all twenty-eight British and sixteen Americans lost their lives including the Chief Designer C. I. T. Campbell.

† Recorded in Colonel Richmond's diary and in AIR 11/23/9848.

Jaray, formerly of the Zeppelin Company in Germany where he had been responsible for the design of the LZ-120 *Bodensee* and for the LZ-126, now the ZR-3 *Los Angeles* with the US Navy. Herr Jaray was living in retirement near Lucerne in Switzerland where Richmond had visited him on a number of occasions.

Richmond went on to run over the design philosophy and the progress of construction of R.101 including the detailed estimates of weights and performance – a *tour de force* followed with rapt attention by Christopher Thomson.*

The airship, said Richmond, was the first large enough to make possible the economical use of stainless steel in its main structure. That had resulted in a saving of about three tons in weight while achieving a marked increase in strength compared with the conventional light alloys of Zeppelin practice. About 65 per cent of the weight of R.101's structure was now of high tensile steel-strip. Tests of typical structural specimens had demonstrated that R.101 was, without doubt, the strongest airship ever built.

In addition to its advanced structure R.101 contained a number of important innovations. The most important of them were the shape of the hull, which was as near a perfect streamline form as possible; the use of diesel engines to eliminate the formation of petrol vapour, with its dangerously low flash point in tropical conditions; a new method of assembly by means of bolted joints which would substantially reduce the manufacture time of future airships; the extensive use of stainless steel; a new patented form of automatic gas-bag valves and a new and elegant system of gas-bag wiring to transfer the lift from bags to structure.

In all, R.101 was a radical departure from previous practice in directions which would advance the art and science of lighter-than-air craft, leading the way to still bigger ships. Safety first had been the watchword throughout and every departure from earlier techniques had been the subject of much thought and of extensive research and development; hence some of the delays.

There had been one major disappointment. The Beardmore Tornado heavy-oil engines had come out in practice much heavier and of lower power than had been estimated at the start and these engines had also run into some torsional vibration problems at higher

* An account of Richmond's presentation is set out at Appendix 4, compiled from Richmond's diary and notes in the Royal Air Force Museum, from Lord Thomson's journal and from notes and records in AIR 5 and AIR 11 in the Public Record Office.

engine speeds. Instead of giving 700 horsepower for a weight of 2,200 lb each, as now delivered the engines were cleared for a maximum continuous 585 bhp each and weighed more than double the original estimates.

When all this was taken into account, together with increases in the weights of the passenger accommodation and in the, now perhaps unnecessarily strong, structure the total fixed weight of the airship was now estimated at 105 tons instead of the 90 tons of the original specification. Hence the estimated disposable lift available for fuel, oil, ballast, crew, stores and passengers was reduced from 63 tons to a little under 47 tons.

That meant that, with adequate fuel reserves, R.101 could now be expected to be capable of carrying a useful load of twenty-four passengers and their baggage and supplies, together with one ton of mail or cargo, over the route to India with one refuelling stop in Egypt. Though this was disappointing compared with the original specification it was still a good deal better than could be achieved with any other existing type of aircraft and there was every hope that further development work would restore a good deal of R.101's original design load.

R.101 was, of course – said Richmond – essentially an experimental ship aimed at a substantial advance in the state of knowledge and the achievement of the high standards of safety which the Secretary of State had himself laid down as a first requirement. One of the chief lessons from the work of the past five years was that airships of 5,000,000 cubic feet capacity, though twice the size of previous ships, were still too small to meet the specified performance over long-range Empire routes.

Richmond sat down.

After this full and frank report, characteristic of Richmond's direct approach, Colmore ran through the progress of work from the start of the project in July 1924, to, now, mid-June of 1929 when the airship was almost complete with three of its five engines delivered, inflation of the gas bags due to begin early in July and the first flight expected in September.

Leaning back in his chair, the Secretary of State thanked Colmore and Richmond for their clear statements of the results of so much hard work during the past five years. Before they went down to the shed, there were just two questions he would like to ask; what powers were now expected from the Tornado engines and what was R.101's cruise performance likely to be on the latest data available?

Colmore turned to Wing Commander Cave-Browne-Cave.

'That's your province, Cave,' he said. 'Would you like to give the S of S the latest position?'

Tom Cave-Brown-Cave opened a folder.

There had been three main problems with the engines, he said: weight, crankshaft resonance and the development of suitable propellers. They were all understandable in adapting engines designed for rail-car work to airship use. He was confident that the Tornadoes could be made into good, reliable, engines, though – in case of trouble – the possibility of developing a compression-ignition version of the Rolls-Royce Condor engine, to run on heavy-oil, was being pursued. It would give about 400 bhp.

On weight, the three Tornadoes so far delivered to Cardington had been found to have a bare weight which averaged 4,733 lb each – just over two tons – which was, unfortunately, more than twice the weight forecast in 1924. As Colonel Richmond had said, this increase had been offset to some extent by coming down from the original seven engines to the present five. But the power plant weight was still excessive.

The Tornado had been redesigned to give 700 horse-power at 1,000 rpm. Unfortunately, early running on the test bed had shown up problems of torsional vibration in the crankshaft – sufficient to break it – at 950 rp upwards. The engine was, therefore, now restricted to a maximum of 935 rpm which resulted in a maximum usable output of 650 horse-power. For normal operations the maximum continuous speed had been set at 890 rpm which gave a maximum continuous output of 585 bhp. Normal cruise would be 450 horse-power at 810 rpm.

There was, however, another difficulty which, he hoped, would eventually be overcome. Originally they had intended to use steel, variable-pitch, propellers the blades of which could be turned to give reverse thrust for mooring. So far the steel blades had not stood up to the high level of vibration of the 'straight-eight' cylinder arrangement of the engine. A 16-feet diameter wooden propeller was, therefore, having to be used, and this could not so far be run in reverse.

As some reverse thrust was essential the forward starboard (No 1) engine would now have to be fitted with a special reverse-thrust propeller and therefore could be used only for going astern, at least initially. That meant that the flying trials would have to start with only four engines available for forward propulsion. He hoped that this would be only a temporary shortcoming soon to be rectified.

To answer Lord Thomson's questions, therefore: the normal cruise power initially available would be 450 horse-power from each of four engines, that was a total of 1,800 horse-power. Colonel Richmond would confirm that, with the airship in equilibrium, the latest estimates were that 1,800 horse-power would give a cruising speed of between 55 and 58 miles an hour at a fuel consumption of between 720 and 760 lb an hour. With 25 tons of fuel on board that would result in an absolute endurance of about 74 hours – say 70 hours with four hours reserve. At 55 miles an hour in still air that would mean a range of 3,860 miles, or say, 2,800 miles against a 15 miles an hour headwind.

From Cardington to Ismailia, Egypt, was about 2,570 statute miles. So it looked as if there would be adequate range in hand. It could be extended by flying a little slower in 'long range cruise' regime.

At this point Richmond came into the discussion again.

In tropical conditions, he said, the lift would inevitably be reduced by ten to twelve tons because of the thinner air. This would have to come off the lift available for fuel and payload. So a lot depended on the actual weights which would be established in the shed trials shortly. He felt reasonably optimistic about the results.

The time was now just after noon and Colmore suggested that they move down to inspect R.101 in the shed and continue the discussion there.

The Works' Humber car still stood outside the main door. Lord Thomson, Bullock, Colmore and Richmond piled in while Scott, Cave, Rope and Nixon followed in Scott's Morris Oxford. They drove the half mile down the slope, past the hydrogen plant and the fabric shop, to the side door of the great construction shed, now smart in grey paint which had replaced the original green and black camouflage of 1917.

Inside the shed, under the towering, silver, flank of R.101 the Secretary of State was welcomed by four more of Cardington's leading lights; the three officers of R.101's designated flying crew together with the veteran Works Manager, R. S. Hubbard, in charge of the assembly of the airship and of its future maintenance.

Most senior of R.101's officers – all former Naval officers in the tradition of the Airship Service – was 38-year-old Squadron Leader Ernest L. Johnston, OBE, AFC, a robust character selected to be Chief Navigation Officer of both R.100 and R.101.

Johnston, a qualified Master Mariner, had transferred to the Airship Section of the RNAS in 1916. He had commanded Coastal-type non-rigids and then the airship station at Luce Bay, south-east of

Stranraer in Scotland. Subsequently, at the Air Ministry Johnston had developed aircraft navigating techniques as the holder of licences as a First Class Airship Pilot and as a First Class Navigator. In December 1926 he had been seconded to Imperial Airways as the navigator for a survey flight over the landplane route from Croydon to India and back, flown in twenty stages by a DH66 Hercules, *City of Delhi,* carrying Sir Samuel and Lady Maud Hoare, Sir Sefton Brancker, Christopher Bullock and Sir Geoffrey Salmond who was going out to take up his post as Air Officer Commander-in-Chief, India.

Alongside Johnston was R.101's future captain, Flight Lieutenant H. Carmichael Irwin, AFC, a tall sensitive Irishman, thirty-five years old, who had joined the Airship Section of the RNAS in 1915 where he had commanded Coastal, North Sea and SS Zero non-rigid airships in Home Waters and in the Eastern Mediterranean. In 1920 he had commanded the rigid airships R.33 and R.36 and then had been posted to the Department of the Chief of the Air Staff at the Air Ministry. In 1925 he had been appointed to the Royal Airship Works where he was again in command of R.33 during trials to assemble technical data for the design of the new airships. With the laying up of R.33, Irwin had been posted to command the School of Balloon Training at Rollestone Camp, Larkhill, on Salisbury Plain during 1927 and 1928 before returning to Cardington to take command of R.101. A fine middle-distance runner, Irwin had been a member of the British Olympic Team at Antwerp in 1920.

With Irwin was his second-in-command, R.101's First Officer, Lieutenant-Commander Noël Grabowsky Atherstone, AFC, late of the Royal Navy and of the non-rigid airships SS Zero 1 and Zero 13, and formerly First Officer of R.29 before returning to service at sea with the Grand Fleet.

Atherstone had retired from the Royal Navy in 1920 to farm in Australia. From Australia, recruited by Peregrine Fellowes, he had returned to England to join the flying staff being assembled by the Air Ministry for the new British airships.

The Secretary of State warmly greeted these four and, then, conducted by Scott and Irwin, climbed up a ladder through a hatchway in the bottom of the hull of R.101 into the crew's quarters on the lower deck of the airship. R.101, still uninflated, was both supported on trestles and secured by slings from the roof.

The whole of the accommodation of R.101, except for the control cabin and power cars, was inside the hull, located on two decks between Frames No 6 and 8 approximately under the centre of buoyancy. This internal arrangement made possible a more compact layout of the accommodation than would have been possible with a long, narrow, external car. It also avoided undue disturbance of the airship's trim through movement of the passengers. Against this had to be set some reduction in window area and some sacrifice in the gas volume of the airship – about 100,000 cubic feet, equivalent to three tons of lift. On the other hand the additional drag of an external car of the same volume would have required an estimated extra four tons of combined fuel and engine weight to achieve the same speed and range for a non-stop journey between England and Egypt.

So the internal location of the accommodation could be seen as a net gain in lift and load on the long-range flights for which R.101 was designed.

The lower deck contained the crew's living and sleeping quarters, the kitchen, the smoking room, lavatories, the Captain's control room, with access to the control-car immediately below, and the wireless cabin. In all, the lower deck had a floor area of 1,750 square feet.

The upper deck was much larger, with a floor area of 6,030 square feet. It included the main lounge, flanked by promenades, lit by large, sloping, windows, a dining room, sleeping accommodation for fifty-two passengers in two-berth cabins and separate wash rooms for men and women.

No one who was shown over R.101's passenger accommodation could fail to be impressed. It set altogether new standards of spaciousness in aircraft interiors, never again to be equalled.

Scott and Irwin walked, with the Secretary of State, first through the crew's quarters, on the lower deck, with its ceiling height of 9 feet 2 inches, through the chart room and down into the control-car, with its two prominent control-wheels; that, at the front, to control the directional rudders, and that, on the starboard side, for control of pitch through the elevators. Engine telegraphs and the speaking tubes to the power-cars lined the port side.

Coming up from the control-car, they looked into the small smoking room, measuring 16 feet by 12 feet 9 inches, and furnished with settees and armchairs to seat twenty-four persons. Scott pointed out the fire-proof floor and ceiling, made of aluminium sheeting over balsa wood, flanked by asbestos side walls.

A staircase led to the upper deck and through a companionway to the magnificent lounge at the after end; a room which measured 60 feet across the beam of the ship by 32 feet, fore and aft, to give a floor area of 1,914 square feet. The ceiling height of 8 feet 4 inches was adequate, though, perhaps, a little low. The lounge was flanked at both sides by the raised 'promenade decks', each 7 feet 6 inches wide by 32 feet long with a ceiling height of 6 feet 6 inches. Full length windows, 8 feet deep, and let into the outer cover, were designed to give a fine view at an angle of 45 degrees, protected by a 'promenade rail' on which passengers could lean to survey the passing scene.

Just forward of the lounge was the dining room, 32 feet by 20 feet, and arranged to seat fifty at a time. It was served by a food lift from the kitchen below, equipped with electric ovens in a space 14 feet 6 inches by 5 feet 6 inches. Opposite the dining room were 26 small two-berth passenger cabins – each 6 feet 6 inches by 5 feet 3 inches by 8 feet 4 inches high, and, alongside them, the wash rooms. All the walls were of fabric stretched across light duralumin girders with columns, mouldings and shelves made of painted balsa wood.

The whole of the accommodation was supported between the airship's transverse frames by a series of suspension wires, cross-braced to prevent deflections.

The general impression was of elegance with a modest degree of luxury – an amalgam of the atmosphere of a pre-war steam yacht and of the Admiral's cabins aboard a warship, even though the walls were of painted fabric. The total floor area of 7,780 square feet* was equal to that of a large country house. Nothing like it had been seen in any aircraft before.

After walking round the whole of the accommodation and looking up at the great bulk of the airship in the shed, the Secretary of State and the RAW party adjourned to the Officers' Mess for lunch, joined there by other leading members of the team, predominantly Royal Navy in background, but wholly Lighter-than-Air in outlook.

Over lunch, in the middle of a long table, flanked by Colmore and Richmond, the Secretary of State found himself sitting opposite two attractive members of this remarkable team. They were Michael Rope, Richmond's Chief Design Assistant, and Noël Grabowsky-Atherstone – 'Grabby' Atherstone – R.101's First Officer.

Michael Rope, a key man in the whole R.101 project, had been educated at Shrewsbury School and Birmingham University where he graduated as an engineer. His first job had been with the British Electric Plant Company at Alloa, then with Rio Tinto, and, from thereon, locomotive engineering with the London, Brighton and South-Coast Railway. When the war came, Rope joined the RNAS

* This compares with the 5,277 square feet of a Boeing B.747–200 Jumbo Jet.

in 1915 as a technical staff officer, first at Capel and then at the Kingsnorth Airship Station, where under Tom Cave-Browne-Cave he led the design work on the SS Zero, non-rigid airship, of which 77 were built.* He went to the Air Ministry in 1919, and, from 1921 to 1924, was stationed at Hinaidi, Baghdad, as a technical staff officer, Royal Air Force Middle East Command. From there, he was seconded to assist Richmond, on design of R.101, at the Royal Airship Works at Cardington.

In the informed opinion of Sir Alfred Pugsley, then a junior design assistant at the RAW, Michael Rope was 'an outstanding and practical design genius, but so modest and retiring that he tended to efface himself and to discount the credit which was his due. Few realised how gifted and valuable he really was.'

His servo-control system for R.101, his 'parachute' wiring system for the gas bags, his automatic gas valves and his variable-pitch, wind-driven, propellers for the electric generators, were all notable advances – and 'they all worked well'.

Rope was popular with everyone and devoted to R.101. As they talked around the table over lunch, Richmond nodded across to the tall, lean, good-looking Rope. 'He got just two loves, Minister,' he said, with a smile, 'R.101 and Doreen Jolly. He and Doreen are getting married in a couple of weeks' time, but I say it's almost bigamy; Michael's married to R.101 already!'

On 2nd July Christopher Thomson – characteristically – sent Michael and Doreen Rope a telegram of congratulations. He recognised a fellow spirit. No doubt he thought, wistfully, that in one respect, Rope was more fortunate than he.

Beside Michael Rope was 'Grabby' Atherstone, 'a calm, cool, analytical man; highly intelligent, very steady, a fine wartime airship pilot, much respected by his fellows, but a bit tough.'†

Noël Atherstone‡ had been born in St Petersburg, Russia, in 1894; the son of a Scottish merchant of Polish ancestry, Carl Grabowsky, who – like his father – had been born in Glasgow and had married there a Scottish girl, Mary Manley McCarthy. Their son, Noël, had been educated at both Winchester and Charterhouse, and had joined the Royal Navy as Sub-Lieutenant Noël Grabowsky in 1912. He

* Fifty-nine SSZ non-rigids flew 34,795 hours 17 minutes in 1918, more than all other British airships put together.

† L. A. Speed of the Cardington design staff and, later, Chief Draughtsman.

‡ His full name was Noël Edward Alexander Carl Eugene Grabowsky Atherstone, the eldest of six children.

transferred to the RN Airship Service in 1917 and qualified as a pilot of SS Zero and North-Sea non-rigids. On 7th April 1918, flying in command of *Sea Scout Zero 1*, from Capel Airship Station, near Folkestone, he had stalked, bombed and sunk an unidentified German submarine.

In 1919, while still in the Royal Navy, Lieutenant-Commander Grabowsky married Miss Evelyn Susanna Atherstone Hales and changed his name by Deed Poll to Atherstone (the maiden name of his mother-in-law) – but continued to be known by all his contemporaries as 'Grabby'.

Subsequently, Noël Atherstone was First Officer under Carmichael Irwin in the rigid airship R.29, and then retired from the Service to join one of his younger brothers, Ian (likewise ex-RNAS), farming in Hamilton, Victoria, Australia.

There in 1927 he was sought out by Group Captain Fellowes, on the recommendation of Irwin, and agreed to return to England, with his wife and small son and daughter, to join the Royal Airship Works and prepare for the flying of R.101 as the designated First Officer of the airship. He proved himself to be immensely competent both technically and as a leader of men.

So now, drawn back to Cardington from so many corners of the world, the design staff, the Works force in the shed and associated workshops, and the flying staff were on the final run-up to the long-awaited first flight of R.101.

After lunch and while coffee was served, Christopher Thomson stood up in the Mess, and in a happy, informal speech, told them all how delighted he was to be back at the Air Ministry, just when all their plans were coming to fruition.

'Before so very long' he said, 'I look forward to the day when we shall fly together in R.101 down the old route of Marco Polo to India. Thereafter, we must press forward with a fleet of British commercial airships to establish regular airship services across the world.'

Vincent Richmond recorded in his diary: 'The S of S for Air is anxious to fly with the ship to India at Christmas.'

It was a prospect which they all eagerly awaited.

Lunch over, the visitors' book was produced and signed. Cardington was becoming a Mecca for leading personalities of the time. Frederick Montague, the Under-Secretary of State for Air, had been there the previous day. Above his signature was that of HRH the Prince of Wales on 11th June; that of Sir Geoffrey Salmond (home on leave from India) on 8th June; of a delegation from the Royal

Naval College, Greenwich, on the 3rd; of Members of the Aeronautical Research Committee on 31st May; and of a large party from The Royal Aeronautical Society, headed by the President, Colonel the Master of Sempill, on 25th May. It was all very encouraging, said Colmore, though somewhat of an interruption of work.

A quick look round the hydrogen plant and the Drawing office and the Secretary of State, with Christopher Bullock, went off to catch the 16.05 train from Bedford back to London, after a farewell cup of tea with the dozen chief members of the Cardington team. That evening Thomson wrote to the Princess Bibesco:

> They strike me as a most impressive community – wholly dedicated to the task in hand; living, eating and dreaming airships with a fervour which is almost a religion.
>
> I must say that I have the greatest confidence in this Cardington team. I am especially taken by their quiet enthusiasm, overcoming all problems, and by their determination to achieve with R.101 a resounding success for their country and for Empire ties.

As for Thomson, L. A. Speed* recalled of him, forty-five years later, 'We looked on him as God.'

* Formerly of Cardington Drawing Office.

CHAPTER THREE

High Summer

'O tell her brief is life, but love is long
'And brief the sun of Summer in the North.'
ALFRED LORD TENNYSON 'The Princess'

June 1929

FROM GWYDYR HOUSE in Whitehall, Christopher Thomson plunged into a continuous round of aeronautical events, as day followed day of sunshine in a summer whose brilliance matched well the enthusiasm of the, as yet, small world of aviation.

Skies indeed seemed bright as the new chief took command of a Department in which all the leading personalities were, by now, old friends and, for the most part, the same as in his first ten months of office, five years before.

Sir Hugh Trenchard was still Chief of the Air Staff; an almost legendary figure at the head of Royal Air Force at last secure from the take-over ambitions of the other Services. He and Thomson had always got on well, sharing a common conviction about the importance of air power in home defence and throughout the British Empire.

Sir Sefton Brancker, Thomson's right-hand man at The Royal Aero Club since 1926, was still Director of Civil Aviation, as he had been in 1924, and as energetic as ever. Sir Walter Nicholson was still the unruffled Permanent Under-Secretary of State as the Air Ministry's top civil servant and Chief Accounting Officer, responsible for the administrative machine. The experienced and ambitious Christopher Bullock, devoted to the Royal Air Force, was Principal Private Secretary to the Secretary of State, as he had been before. The one new face was that of the Parliamentary Under-Secretary of State for Air, 53-year-old Frederick Montague,* the stocky, good-natured, hard-working MP for West Islington, who described himself as 'self-educated newsboy, shop-assistant and freelance journalist, commissioned in the British Army in 1917, and successful Parliamentary Labour candidate in 1923'. They all served Thomson well.

At once the Secretary of State was the central figure in the surge

* Frederick Montague was created Baron Amwell of Islington in 1947. He died, at the age of ninety, on 15th October 1966.

of aviation events, which had begun on 14th June with a Tenth Anniversary luncheon at the Savoy to celebrate the first non-stop air crossing of the North Atlantic when John Alcock and Arthur Whitten-Brown had flown from Newfoundland to Ireland between 14th and 15th June 1919.

The lunch was given jointly by Vickers (Aviation) Limited and Rolls-Royce Limited, who, as builders of the airframe and engines, respectively, had shared in the successful trans-Atlantic flight by the Vickers Vimy biplane powered by two Rolls-Royce Eagle engines. The luncheon might just as well have been one of the by now familiar series held jointly by The Royal Aero Club, The Royal Aeronautical Society, The Air League and The Society of British Aircraft Constructors. The 350 people present were all the same – and, as usual, Christopher Thomson was in the chair.

The occasion gave him a chance to survey the scene from his new official stand, and to welcome his successor to the chairmanship of The Royal Aero Club: the former Under-Secretary of State for Air (his friend and Marthe's) Sir Philip Sassoon.*

'I am,' said C.B., 'embarking upon my new position in a spirit of hope and modesty. You will expect much of me. I will do my best to deliver.'

Through The Royal Aero Club, he had been brought into a close association with the leaders of British aviation.

'Now I look forward to doing all I can to harness still further all the energy, the skill and the knowledge of everyone in British aviation, not only for the Royal Air Force – the finest in the world – but, equally for the development of civil and commercial aviation as well.'

In the sunny days of that summer, the energy, the skill and the knowledge were all to be taxed to the full while there was pressed forward, in parallel with the outside events, the serious business of planning the re-equipment of the Royal Air Force for the 1930s; the negotiating of new overseas routes for Imperial Airways and a determined attempt to gain for Britain the three world's absolute records – for speed, for distance and for height, as the hallmarks of leadership in the air.

* Sir Philip Sassoon (1888–1939), the son of Sir Edward Sassoon and Aline, daughter of Baron Gustav de Rothschild, was Private Secretary to Field Marshal Sir Douglas Haig during the 1914–18 War. He served at the Peace Conference at Versailles in 1919, alongside Thomson, and was appointed US of S for Air in November 1924. Wealthy, a keen exponent of the arts, a qualified pilot and a magnificent host, Sassoon entertained widely at 45 Park Lane, at Trent Park in Middlesex, and at Port Lympne, near Hythe in Kent.

Christopher Thomson looked through the formidable list of outside engagements to which he was now committed. They started with a levée at St James's Palace on 2nd July, held by HRH The Prince of Wales on behalf of the King and attended by senior officers of the Royal Air Force. They went on with the opening of the new Heston Air Park and the presentation of the King's Cup to the winner of The Royal Aero Club's round-Britain Air Race for 1929. That was to be followed by the annual RAF Display at Hendon and then by the opening of the great International Aero Exhibition at Olympia, from which much was hoped in the way of exports. Next, would come the celebration – with Louis Blériot himself – of the twentieth anniversary of the first cross-Channel flight. All this would lead on to the climax of the season's air events – the contest for the International Schneider Trophy at Cowes in September.

Some time during that memorable summer, the two new airships, R.100 and R.101, were due to make their maiden flights and then, all being well, India by Christmas.

Among these preoccupations and expectations there would be that promised visit which C.B. so eagerly awaited.

A telegram from the Princess in Paris confirmed the dates. Then on Thursday, 4th July she was at the Ritz Hotel in Piccadilly to be welcomed by the bouquet of red roses with which C.B. invariably greeted her – and, that evening, to be carried off by him for a tête-à-tête dinner at his favourite restaurant, the Café Royal in Regent Street.

At forty-three in 1929, the Princess Marthe Bibesco was acknowledged to be still one of the most beautiful, as well as one of the most elegant and talented, women of her day. A brilliant conversationalist in French, in English, in Spanish or in German; her English spoken with an enchanting hint of accent, Marthe Bibesco revelled in an international circle of friends and relations the extent and breadth of which could hardly be equalled. First among them was Christopher Thomson.

A little over medium height, slim, with a generous mouth and large, expressive, green eyes, a mountain of auburn hair and long, slender fingers, Marthe Bibesco's smile was known to capture the sternest hearts though she was also reputed to be able to freeze an interloper with a single glance. It was said of her – and undoubtedly with truth – that 'she made any man she liked feel that he was the only man in the world for the moment'* and that one of her fasci-

* *The Times*, 30th November 1973.

nations was 'her intense gift to please men'.

For the past fourteen years the man of the moment – the man to please – had been C.B.T.

So it was that, on Thursday 5th July, after a cheerful lunch with Sir Arthur and Sybil Colefax at Argyll House in Chelsea, in company with their old friend Lord d'Abernon, C.B. and Marthe drove over Putney Bridge to take their places in the Royal Box on the Centre Court at Wimbledon in time to see Helen Wills of the United States retain the Ladies Singles Championship for the third time when she defeated her fellow Californian, Helen Jacobs, by 6–1, 6–0, on a brilliantly sunny afternoon.

Marthe recorded in French in her diary for that day:

> The two 'Helens'. Mrs Wills admirable with the air of a severe priestess – not a smile until the end of the match. The other, plump, passionate – but outclassed. The Centre Court crowd applauded frenziedly every point won by Miss Jacobs. I asked, 'Do they really prefer her?' 'No,' said Kit, 'but she is behind.'

That evening Marthe was collected from her hotel by Sir Philip Sassoon to drive her down to his sumptuous country house at Trent Park, Enfield Chase near Cockfosters in Middlesex for a week-end house-party while C.B. – on duty – performed Ministerial Saturday tasks at the formal opening of the fashionable new private-flying centre at Heston Air Park.

That Saturday morning, 6th July, the Secretary of State for Air drove from Gwydyr House to the de Havilland Aircraft Company's aerodrome at Stag Lane off the Edgware Road and, after lunch with Geoffrey de Havilland, Alan Butler and Charles Walker, flew the few miles to Heston in a DH Moth (G–AABL) of the London Aeroplane Club, to arrive at 3.45 p.m., piloted by Major H. G. Travers (ex-RNAS) the Club's Chief Flying Instructor.

> Thereafter [in the words of C. G. Grey*] Lord Thomson hied himself to the microphone on the balcony of the control-tower and in a few of his usual well-chosen words declared the new Heston Air Park officially opened and wished it long life and prosperity† Its instigator, Mr Nigel Norman, said how honoured he and his partner, Mr Alan Muntz, were to have their aerodrome opened by the Secretary of

* In *The Aeroplane*, 10th July 1929.

† The 1929 buildings were sold for demolition by the Civil Aviation Authority fifty years later in 1979.

ardington – R.101C at the nooring mast at Cardington n the afternoon of 4th)ctober 1930, with the No 2 irship Shed in the ackground. At the moment f the photograph the airship vas taking aboard 25 tons of uel for its flight to Egypt, en oute to Karachi.

dieu – a last photograph at ne foot of the moorning mast t Cardington on the evening f Saturday, 4th October, efore the final group entered ne lift to go on board R.101 n route for India. (Left to ight) Squadron Leader Ernest ohnson OBE, AFC, R.101's hief Navigating Officer; lajor Louis G.S. Reynolds Principal Private Secretary to ne Secretary of State for Air) who did not travel on the ship; rigadier-General the Lord homson of Cardington, PC, BE, DSO, His Majesty's ecretary of State for Air; ieutenant-Colonel Vincent C. ichmond, OBE, (Assistant)irector of Airship)evelopment – Technical) hief Designer of R.101 and ir Vice-Marshal Sir W. efton Brancker, KCB, AFC,)irector of Civil Aviation at ne Air Ministry.

Moment of Take-off – R.101 about to cast off from the Cardington Tower at 1936 rs BST on the evening of 4th October 1930, en route for ndia with a crew of 42, 12 passengers and a load of 27 tons of fuel on board – the last photograph to be taken of the airship, at the time the largest vessel in the world.

[PLATE V]

C.B.T. and C.Ll.B. – Lord Thomson and Sir Christopher Bullock at Gwydyr House, Whitehall, shortly after Thomson was appointed to his second term of office as Secretary of State for Air on 10th June 1929.

(LEFT) *A Key Figure* – Wing Commander Reginald B.B. Colmore (1887-1930) was appointed to the Royal Airship Works at Cardington in July 1924 as Assistant Director of Airship Development and promoted to Director (DAD) in January 1930. He was responsible for the specification of both R.100 and R.101 and administrative head of the airship development programme throughout its existence between 1924 and 1930. (RIGHT) *In Charge of Flying and Training* – Major George Herbert Scott (1888-1930) who commanded the British airship R.34 during its double-crossing of the North Atlantic from East Fortune in Scotland to Hazelhurst Field, Mineola, Long Island, U.S.A. and back to Pulham in Norfolk 2nd and 13th July 1919. Scott, who was awarded the CBE and AFC, was appointed to the Royal Airship Works in 1924. (Both photographs taken on the eve of departure for India, 4th October 1924).

State for Air and promised a hearty welcome to all good aviators and fealty and allegiance to the Air Ministry and all its regulations.

As he finished speaking, and precisely at the predicted time of 4.25 p.m., the winner of the 1929 King's Cup Air Race was to be seen coming in very low past the Southall gasometer on the north-west boundary of the aerodrome. The small, compact biplane with its large radial engine dived across the finishing line – a two-seat trainer variant of the RAF's Gloster Grebe fighter (400 hp Armstrong Siddeley Jaguar engine) flying in this civil air race oddly still in its military markings – J.7520. Aboard were two members of the RAF's High Speed Flight, now training at Felixstowe for the Schneider Trophy Contest in September. In the front seat was Flying Officer R. L. R. ('Batchy') Atcherley* navigated from behind by Flight Lieutenant George H. Stainforth – their aircraft entered by Sir Walter Preston, MP for Cheltenham, home of Gloster Aircraft Ltd.

The Grebe, a descendant of the famous SE5A fighter of the 1914–18 War, had been the thirty-ninth starter out of forty-one competitors to leave Heston for Blackpool and Glasgow on the previous morning, giving the 'limit' man some three-and-a-half hours' start. Now Atcherley and Stainforth had completed the 1,169 mile course in 7 hours 55 minutes at the respectable average speed of 147.73 mph, 31 seconds ahead of Lieutenant L. G. Richardson, RN, flying a D.H. Cirrus-Moth (G-EBPO) at an average of 100.23 mph.

So the afternoon went by; a steady stream of unsuccessful competitors landing one-by-one and in droves until by half-past-six, twenty out of twenty-two eventual finishers were safely on the ground, welcomed amid threatening – and much needed – rain-clouds by a social gathering almost a facsimile of that at Ascot three weeks before. Aviation events had suddenly become fashionable in 1929.

At the end of the day, amid much applause, the Secretary of State presented the King's Cup and a cheque for £250 to the winners, Atcherley and Stainforth, together with a special prize of £100 from Sir Charles Wakefield for the fastest speed in the race. In the front row of the group around the table with the prizes, glamorous between Philip Sassoon and the King of Spain, smiled the Princess Bibesco.

Next day C.B. journeyed to Trent Park to join Philip Sassoon's

* Later Air Marshal Sir Richard Atcherley who, with his twin brother, Air Vice-Marshal David F. W. Atcherley, made outstanding contributions to the Royal Air Force during and after World War II. David Atcherley was lost while flying a Gloster Meteor jet fighter from Egypt to Nicosia, Cyprus, on 7th June 1952. Sir Richard died on 17th April 1970 after retiring from the Royal Air Force.

social and political lunch party, a gathering typical of the circle in which Sassoon moved, somewhat odd for a Socialist peer but not unfamiliar to Christopher Thomson. It included HM the King of Spain, the Marquess and Marchioness of Salisbury, the Earl and Countess of Pembroke, Lord Charles Cavendish-Bentinck and Mr Austin Chamberlain; a group in which art, politics, aviation, country life and international affairs were primary topics of conversation.

Philip Sassoon's houses at Port Lympne, at 45 Park Lane and Trent Park were all renowned for their comfort, the splendour of their art collections and their sumptuous tables. Winston Churchill once remarked of his inclusion in a Conservative administration: 'When you set out on a long and sometimes tedious journey it is as well to attach a well furnished dining-car to the train.'

Sassoon and Thomson were on opposite sides in politics but close friends in their common interest in aviation, in the Royal Aero Club and, in the summer of 1929, in the forthcoming contest for the Schneider Trophy to the preparations for which they had together devoted much time and effort.

On the Sunday afternoon C.B. drove Marthe to London to dine together that evening with HRH The Duke of Connaught at Clarence House, St James, in the company with the Duke's eldest son, Prince Arthur of Connaught and his wife, the Duchess of Fife. They were old friends since the days when Thomson had served on a Military Mission to the King of the Belgians in company with Prince Arthur in September 1914, and Princess Bibesco had entertained the Connaughts during their visit to Roumania after the war. There was much to talk about and while Marthe enjoyed the atmosphere of Clarence House and its magnificent paintings she did not wholly approve of its architecture. That evening she noted in her diary:

> *La mediocrité* de l'escalier. La Chapelle-atelier. Les Van Dycks; les *Poussins; les Rembrandts.*

The following afternoon C.B. collected her from the Ritz for tea on the terrace of the House of Lords with the Prime Minister, Ramsay MacDonald; the PM was at his most courtly, enjoying the company and the conversation of the beautiful and sophisticated princess together with his closest friend in the political arena.

Marthe recorded that MacDonald was concerned that King George V had contracted a chill at a recent Thanksgiving Service for his recovery from pleurisy. The King was now ill again. Consequently he could not go to Sandringham whither he had invited MacDonald for three days during the summer recess 'not as Prime Minister but as a friend.' A sad disappointment.

After tea beside the Thames, Thomson and Marthe walked out together through the Cholmondley Room to the Central Lobby and down through Westminster Hall into Parliament Street and Whitehall, to walk the short distance to Gwydyr House where, in the large room running from front to back on the first floor, the Air Minister had his office. That of Sefton Brancker was at hand, immediately below.

Once in C.B.'s room – *'le bureau bleu du Ministre de l'Air'* – Marthe threw herself down in one of his comfortable armchairs and smiled up at him as he glanced through new papers in the red box on his desk. On her right large windows looked on to the Admiralty across Whitehall; at the other end of the room, beyond the conference table, matching windows offered a glimpse of the Embankment and the Thames.

'Who would have thought that we would be together thus in England?' she said wistfully. 'You the Minister of Air – a Shakespearian title – and me still adrift between Paris and Bucharest. Who would have foreseen this all those years ago when first we were together at Mogosoea and Posada? It suits you, Kit, this office – right in the heart of Empire; Lord of the Air,' and she smiled towards him.

He looked up from his papers.

'Marthe, there is something I must tell you and in the strictest of confidence,' he said, 'something which will affect us both if I accept.'*

He explained to her, walking up and down the long room, that the subject of the appointment of a successor to the present Viceroy of India had been brought to the Prime Minister during the past few days.

The present Viceroy, Lord Irwin† had been appointed by the Baldwin Government in 1926, and was due to complete his five year tour of office early in 1931, in just over eighteen months' time. Now the Prime Minister had asked him, C.B., whether he would 'go out into the Empire' – subject to the King's approval – and take up the reins from Irwin when his term was complete.

'There is no one else of our Party who could take that high office,'

* Princess Bibesco's diary for 8th July 1929.

† Edward Frederick Lindley Wood created 1st Baron Irwin in 1925, succeeded his father as the third Viscount Halifax in 1934 and was created first Earl of Halifax in 1944.

the Prime Minister had said. 'You don't have to accept yet. Think it over. You would be a heavy loss to me here but there is no more vital post than India – especially at this time.'

This would mean a return to the land of his birth in the highest office of state. But would it mean also another five years apart from Marthe? At this time of life was not that a sacrifice which was really too great even for such an office and such a duty to his country?

The one saving grace of such a distinguished exile would be a regular airship service – as now seemed possible – to bring India within five days of London.

He stopped and gazed at Marthe. What would the future hold for them?

They went out, across the Horse Guards Parade and into St James Park, absorbed in each other and unrecognised among the summer's evening strollers in the park. As so often before, their steps led them down to the lake and to the little peninsula opposite the Foreign Office where the pelicans stood, deep in thought, surrounded by busy ducks.

Those pelicans, old friends, always took them back in memory to the idyllic few days together on the yacht of the European Danube Commission in the spring of 1915. There had been pelicans on the shore – 'as solemn as pontiffs, pecking at their white fronts with avid beaks, pretending they are beating their breasts'.

'Ah, the pelicans of the Danube those bright mornings,' murmured Marthe. 'How wide is the ocean. And Kit, how far would I travel to be where you are? Kit, I suppose you must accept,' she said. 'But one day perhaps I can come too to join you there.' Unspoken was the thought that one day, somehow, she might be free.

They walked on across the park and up St James Street to the Ritz where they were to meet again later that evening to go to dine with Philip Sassoon at 45 Park Lane together with Lord d'Abernon and a cheerful theatre party.

One more day in London and then, on the morning of Wednesday, 10th July C.B. called for her at the Ritz to take her to Victoria to see her off to Paris with a promise to escape from his official duties during the summer recess to come over for a few days of holiday together in France.

Then, as Marthe went on from Paris to Bagnoles-de-l'Orne in Normandy and to Combourg in Brittany the surge of British aeronautical events of the high summer re-engulfed the Secretary of State for Air.

The Tenth Annual Royal Air Force Display at Hendon, held on Saturday, 13th July, brought a record crowd headed by Their Royal Highnesses The Prince of Wales and the Duke and Duchess of York with King Alfonso XIII of Spain, who was just back from staying with Sir Philip Sassoon at Port Lympne. The Prime Minister, the Chancellor of the Exchequer and most of the rest of the Cabinet, together with four Maharajahs and a large number of ambassadors and other diplomatic representatives, saw a magnificent afternoon of flying by 171 aircraft of the Royal Air Force in what was described as 'a real "pre-war" English July summer's day with a cloudless blue sky and a refreshing breeze.'

Next day, J. L. Garvin wrote in *The Observer*:

> The weather has given Lord Thomson a magnificent send-off at Hendon. Flying has now become an integral part of our civilisation and we are just beginning to realise the vast cultural importance of air services which are bringing continents more and more closely into touch.
>
> If we are to retain our high place in the world we must be pioneers in the use of aircraft. Nothing short of equality in the air with any nation of the World will serve our national and Imperial needs Lord Thomson left behind him a good record during his previous term of office. Now we look to him to facilitate progress and to maintain strength and efficiency.

Each year the RAF Display was seen as the culmination of the year's flying training; an occasion on which to demonstrate the high standards of skill attained by the Royal Air Force in close formation manoeuvres, in aerobatics and in mock air combat on the part of squadron pilots. In 1929 they showed their paces on seventeen types of Service biplane some of which were still of wartime vintage – Bristol Fighters and Avro 504 trainers.

The latest advances in aviation were, however, provided for all to see when HRH The Prince of Wales opened the Seventh International Aero Exhibition in London, at Kensington's Olympia, on 16th July, described at the time as 'the greatest and most up-to-date display of aircraft ever assembled under one roof'.

That boast seemed justified when the Secretary of State for Air and the Chairman of the Society of British Aircraft Constructors, the gallant and handsome Captain Peter Acland, both immaculately turned out in full morning dress with spats (still de rigueur), escorted the Prince of Wales to inspect the sixty-four aircraft from five nations gathered at Olympia; civil and military types in equal numbers.

The star of the show was, by common consent, the elegant, sturdy-looking, Hawker Hornet single-seat 'interceptor-fighter' biplane; just off the Secret List as the first military aircraft capable of exceeding 200 miles an hour in level flight.

The prototype Hornet (J.9682) was beautifully turned out, its cowlings polished to a mirror-like surface, housing the 480 hp Rolls-Royce F.XI S Kestrel engine within its 'eversharp' nose. Already there were plans to order Sydney Camm's latest design as the standard single-seat fighter for the Royal Air Force – eventually to go down in history as the Hawker Fury.

Alongside the Hornet, and looking rather like a small cousin, was a bright little blue and gold two-seat biplane; a show version of Geoffrey de Havilland's best-seller, the DH Gipsy Moth which you could buy for £595; a type of light aeroplane already owned by the Prince of Wales.

Among exhibits which had earlier demonstrated their quality was the Webster's little blue-and-white Supermarine S.5 racing seaplane (N.220) which had brought the Schneider Trophy back to Britain from Venice in 1927 and was to be made ready again as a reserve aircraft for the 1929 race in September. The S.5 was hoisted high up in a diving attitude. Beneath it sat the large, white, Short Singapore twin-engine flying-boat which Sir Alan Cobham had flown round Africa in 1928 accompanied by Lady Cobham and a two-man crew. A further contrast could be seen, slung from the roof, the slim and handsome Fairey Long Range Monoplane (J.9479) which in April 1929 had been the first aircraft to fly non-stop from England to India (from Cranwell to Karachi) in 50 hours 37 minutes.

These record-breaking monoplanes were in contrast to the fact that biplanes still predominated in 1929, and at Olympia, where only fourteen monoplanes were on show, compared with forty-eight biplanes, one autogyro and a 12-foot model of R.101. The international nature of the show was demonstrated by five German aircraft, three French, three Italian, a single Ford Trimotor from Dearborne, USA, together with fifty-one aircraft from the British Aircraft Industry and a multitude of aero-engines which ranged from 40 to 1,000 horse-power.

Aerolympia, the last of its kind, combined in a two-week event all the spit and polish of a motor-show with the glamour of flying. The ranks of glittering aircraft with their brightly-painted colours, their shiny silver cowlings and their sweet smell of acetate dope on their fabric-covered wings and fuselages, held the magic of a strange new world for the public of the day, few of whom had ever flown.

Aeronautical hospitality was lavish. It began with a lunch at Olympia for 450 exhibitors, foreign guests and VIPs on the opening day at which the Prince of Wales welcomed the diplomatic and other representatives from overseas.

It was followed on 18th July by a dinner at the Savoy to all the exhibitors, British and foreign. Lord Thomson proposed the toast to 'The Aircraft Industries of All Nations'. It was an occasion to proclaim his hopes and to register his thoughts. He hoped, he said, one day to see a British air transport system as great as the British Mercantile Marine, and that it would be supported by a British Aircraft Industry as great as the nation's Shipbuilding Industry, never to suffer from such troubles that had afflicted shipping in recent years.

In thinking over this subject he had, he said, reflected on the life story of the aircraft industry, born before the war as a sickly, puling, infant. In those early days as an officer of the General Staff, he had been sent down to Aldershot to report on Samuel F. Cody's man-lifting kites and at that dim and distant time, like most military staff officers, he had looked on aviation as mildly ridiculous.

War came and the infant industry was subjected to forcible feeding, crammed with good things. Then, with peace, the infant was left in the cold like a hot-house plant on a bleak hillside. Many who had dreamed great dreams suffered grievous disappointment.

When he looked back to the early days before the war, twenty years ago, he was delighted to see that four of the pioneer aircraft constructors of 1909 were happily present: Sir Alliot Verdon Roe (Avro); Robert Blackburn, Frederick Handley Page and Oswald Short.

He welcomed them and, with them, distinguished visitors from overseas especially His Royal Highness Prince Purachatra of Siam, brother of the reigning monarch of Siam and that country's Minister of Postal Services and Communications. He had had the pleasure of being associated with Prince Purachatra way back in 1902 when the Prince was at the Royal Military Academy and he, Lord Thomson, was instructing in military engineering there. Their positions had been reversed when the Prince, coxing the Academy's boat and roundly cursing him at bow, had shown a remarkable command of the English language.

As to the Aero Show itself, it showed the tremendous advances in aviation since the war. Moreover it was truly international in character. As an unrepentant and confirmed believer in Lighter-than-Air, as much as in heavier-than-air craft, he hoped that in future more attention would be given to airships.

Most important of all, those gathered at the banquet that evening were, he was sure, united by a bond of interest and concern for aviation and its future. They were fighting side-by-side as a progressive minority against the prejudice which always seemed to be drummed up against all innovations.

He hoped that the Aero Show would be remembered for the progress it so well illustrated; for the expanding markets it might serve and for the great advances it promised. Aviation would bring all nations closer together. That must be its continuing aim.

The 'progressive minority' was well in evidence next week-end at an appropriate entertainment for the small but widening circle of aviation folk. Examples of many of the aeroplanes on show at Olympia together with most of the people who had designed and built them were gathered at a summer's afternoon garden party at the new Heston Air Park in the warm sunshine of Saturday, 20th July. All the youth and beauty of fashionable society was there alongside official aviation headed by Christopher Thomson and sporting aviation headed now by Philip Sassoon. No such party would be complete without Sefton Brancker, monocle in eye, cheerfully supporting both.

Five days later, on 27th July, the achievements of the past were further linked with the progress of the present when small, dapper, 57-year-old Louis Blériot flew from Paris in a Blériot 127 twin-engine monoplane bomber, his chief pilot, Bossoutrot, at the controls, to land at Swingate aerodrome, near Dover, where the French party was welcomed by the Mayor of Dover and Sir Sefton Brancker.

Accompanied by Madame Blériot, who had flown to Swingate in another aeroplane, the pioneer aviator drove to Dover's North Fall Meadow for a reception there on the site of his historic landing early on the morning of 25th July 1909 – the first man to fly the English Channel.

From Swingate the Blériots flew on to Croydon to be met at the airport by Lord Thomson, Sir Sefton Brancker and George Woods Humphrey, Managing Director of Imperial Airways. The Blériot monoplane bomber was parked incongruously among the fleet of civil biplanes; a ponderous-looking military juggernaut about the inelegant lines of which everyone was too polite to comment. It seemed that Blériot of 1929 had lost that feeling for line which his little Type XI monoplane had shown twenty years before.

That evening, Monsieur and Madame Blériot were guests of honour at the closing banquet of the Aero Show held at the Savoy Hotel by the Society of British Aircraft Constructors with Lord Thomson in the chair.

The Blériots were received with cheers. Before the banquet began, after a few words of welcome in his fluent French, Christopher Thomson presented Madame Blériot with a diamond brooch on behalf of the *Daily Mail* whose £1,000 prize Louis Blériot had won in 1909.

On a specially convivial evening there were speeches 'brief and bright' from Louis Blériot in French, from Sir Arthur Whitten-Brown, from T. O. M. Sopwith, Sefton Brancker, Signor Caproni from Italy, Mervyn O'Gorman and Peter Acland (Chairman of SBAC for 1929) who proposed the health of 'our popular, hard-working and ever-enthusiastic Chairman, Lord Thomson who presides so benignly over a happy culmination of one of the happiest periods in the history of British aviation.

'Such pleasures,' he continued, 'are difficult to express adequately in English. Other languages can do it still more gracefully.'

Christopher Thomson wound up. It was an extraordinary satisfaction to the Government as well as to himself to be able to do honour to that great airman, Louis Blériot. The Government had paid him the tribute of sending a squadron of the Royal Air Force to escort him to Dover, 'twenty years after', and to mark the friendly relations between our two great nations.

He ended, 'What a difference will be made to the world if, in future, we use, and do not abuse, this great gift of aviation of such potential benefit to mankind.'

The newly cemented link between British and French aviation was further secured ten days later, when the French Air Minister, Monsieur Laurent Eynac, arrived in London, accompanied by Monsieur Cameraman, Head of the French Civil Aviation Department, for an Anglo-French Air Transport Conference. On the evening of 6th August Lord Thomson entertained the French visitors to dinner at the Ritz, a dinner attended also by the Prime Minister, Ramsay MacDonald; by the French Ambassador, Monsieur Fleurian; the French Naval Air Attaché, Lieutenant Sala; and by Sir Sefton Brancker; Air Commodore Cyril Newall and Sir Walter Nicholson from the Air Ministry.

Next day, the air transport talks were held at Gwydyr House between the French delegation and a British team led by Lord Thomson and including Frederick Montague, Sefton Brancker, Air Commodore Newall and Mr W. W. Burkett, Brancker's deputy.

In the afternoon, a joint statement was issued from the Air Ministry:

> The French Air Minister and the Secretary of State for Air have reached agreement both in principle and on certain details for the development of commercial air transport in Africa, the Middle East, the Far East and in South America based on mutual co-operation between the two countries. The conversations have laid the foundation for wide collaboration in aviation between France and Great Britain.

That evening, 7th August, the Secretary of State for Air flew with Monsieur Laurent Eynac from Croydon to Le Bourget 'on a purely private visit'.

CHAPTER FOUR

Time Out

'O Angel of the East, one, one gold look,
'Across the water . . .'

ROBERT BROWNING,
'Rudel to the Lady of Tripoli'

10th–25th July 1929

THE PRINCESS had written to C.B. immediately on her return to Paris on 10th July – and again on the 23rd from Bagnoles-de-l'Orne, the Normandy spa where she spent two weeks each summer. She told him that she would be back in Paris between 8th and 16th August, after which she would have to leave for her estates at Mogosoea and Posada, in Roumania, before returning to Paris, by way of Venice, at the end of November. Please could he come to spend a few days with her before another, inevitable, parting through the autumn?

C.B. wrote back on 25th July:

My dearest,

So many thanks for your letter of the 23rd. It was sweet of you to write in spite of my not having answered your last letter. I have been swamped with work.

Now that I know that you will be in Paris between the 8th and 16th of August, I will arrange to be there also for two days. I had intended to fly to see you at Bagnoles, on the 8th, and then on to Aix the following day, but, with a flying office, one can easily change plans. If I can get hold of Lyautey on the 3rd, it shall be done.

The French Ambassador is dining with me on Monday next and I will talk to him about it. My French vis-à-vis, Laurent Eynac, is paying me a visit and I have insisted on an interpreter for our conversations!

Thank goodness the House rises to-morrow and we will be able to get on with our work. Ministers are like horses on a hot summer's day. They exhaust their energies on the flies which sting them.

Unless we commit suicide, this Government can last longer than I care to think about. Our enemies are divided; as the old saying goes, 'they are willing to wound, but afraid to strike'.

If they killed us, they would have to form a Government themselves and they have no common policy. Both the Liberals and the Conservatives want office, but two into one won't go.

All the same, the Labour Party is not representative of all the best elements in Britain. Unless we amputate on our Left and attract on our Right not only radicals, but even more so, progressive Conservatives, we are – and will be justly – doomed.

MacDonald is sixty-four and enjoying his 'crowded hour of glorious life' and displaying prodigious powers of work. But, someone else is fifty-four and will soon go out into the Empire, rather than plough sand in sterile talk at home – and he is yours, ever and as ever, since 1915.

With the very best of love,

Kit.

The plans worked out and, as we have seen, Laurent Eynac arranged to come to London on 6th August. Thus, on Saturday, 3rd August, C.B. could write to Marthe from his flat at 5/6 Princes Street, Cavendish Square, off Oxford Circus:

I am now arranging to fly to Paris, *avec mon cher ennemi, Eynac*, on 7th August, Wednesday afternoon. I shall stay at the Hotel du Rhin, and get through such light official duties as I have to perform, on the 8th and 9th, and then find you at St Germain. I can always get back to Paris in the evenings.

I will try to see Marshal Lyautey here, but I fear greatly that I may miss him. Perhaps it will be possible, however, to see him while I am in France.

Am working hard now while Parliament is up and the gadflies cease to sting in the name of democracy and self-advertisement.

No more now – I am so looking forward to being with you – God bless you – and with all my love.

Ever your

C.B.

There was just time to fit in another visit to Cardington to check the advance of the work on R.101 before the French Air Minister arrived for the air transport conference – a Sunday on which to combine business with pleasure.

So, on Sunday, 4th August 1929, Noël Atherstone recorded in his meticulously-kept diary of R.101's progress:

Last Sunday, Lord Thomson of Cardington, S of S, paid us a surprise visit and had another quick look around the ship. He appears to have been very pleased with the progress which has been made since inflation began.

Indeed, he couldn't keep away.

Two of C.B.'s closest friends at this time were Richard Casey, the handsome, go-ahead liaison officer in London between the Australian and United Kingdom Governments, and his attractive, intelligent, wife, Maie. C.B. consulted them, and on the Sunday Dick Casey drove his open Renault tourer up the old Great North Road, past Hendon, Hatfield and Stevenage, to Bedford, with C.B. in the front and Maie and Christopher Bullock in the back. After lunch at the Bridge Hotel, they drove to the Royal Airship Works and, as Noël Atherstone noted, spent some time in R.101. By 4th August R.101's sixteen gas-bags, in which inflation had begun on 11th July, were about half filled with hydrogen.

The Lady Maie Casey wrote of this visit: *

> C.B. believed in the future of R.101 which had for him a special romantic appeal. It was to take him on its inaugural flight to India where he was born. He watched its development at Cardington like an anxious parent, walking over the structure, envisaging an ample, floating, apartment that would safely and comfortably traverse long distances between continent and continent.
>
> Dick and I accompanied him, stepping carefully through empty compartments, outlined but not yet fitted up, so wide in space after the tight little aeroplanes we had both flown in. Dick noted with misgivings that every ten yards or so a hand fire extinguisher was installed.

Meanwhile, the much smaller *Graf Zeppelin* had been showing what Lighter-than-Air could do. On 1st August the LZ.127 flew from Friedrichshafen to Lakehurst, New Jersey on its third overseas flight and its second crossing of the North Atlantic. The flight, with eighteen passengers, took 95 hours 22 minutes for the 5,185 miles, covered at 54 mph. On 7th August the German airship flew back to Friedrichshafen in 55 hours 22 minutes on the first leg of a four-sector round-the-world flight. The ship left Friedrichshafen again on 15th August and flew non-stop – but not without incident – to Tokyo, a distance of 6,984 miles in 101 hours 49 minutes. The Pacific crossing to Los Angeles took 79 hours and 3 minutes and from Los Angeles to Lakehurst, 51 hours 57 minutes, after a hazardous take-off. Thus, the round-the-world flight and back to Lake-

* In *Tides and Eddies* published by Michael Joseph, London. Maie Casey was the sister of Rupert Ryan, British High Commissioner to the Inter Allied High Commission in Cologne and there a friend of Dr Konrad Adenauer.

hurst had taken 20 days in 288 hours 11 minutes flying time for 20,363 miles. The *Graf Zeppelin* then carried on to Friedrichshafen, which was reached on 4th September. The airship had flown from Germany and back in four stages around the world in 21 days, and had taken 300 hours 20 minutes to cover 21,238 miles at an average ground speed of 70.7 mph with a following wind.

So against the background of the *Graf Zeppelin*'s third and fourth Atlantic crossings, the Anglo-French Air Transport Conference came and went on 6th and 7th August, and, on the Wednesday afternoon, Christopher Thomson accompanied Laurent Eynac back to Le Bourget in a Loire et Olivier 21, twin-engine, *Rayon d'Or* biplane of Air Union, the French national airline.

'Quel bel accident,' Monsieur Briand jokingly remarked to Marthe when he heard that the two Air Ministers were travelling together.

That evening of Wednesday, 7th August, was spent happy and relaxed, at the house of Monsieur and Madame Pierre Goujon, in the rue de Medicis in Saint-Germain-en-Laye on the Seine, fourteen miles west of the centre of Paris, where Marthe was staying for a few days. There, they could be together out of the limelight.

Of those days, Marthe wrote later :

> It was there, on a golden, late summer day, as we stood on the terrace of the house facing towards Paris, that he told me of his projected voyage in R.101.
>
> He was going to India. By some curious chance, he who had been about the world so many times, had never returned to the place of his birth. Now he would be going to see the Ganges from the prow of his almost mythological ship.
>
> His exultation seemed natural. Could one imagine anything more appropriate?

Four sunny days together and, on the evening of Monday, 12th August, back in London, C.B. wrote from Princes Street :

> Just a line to say that I got back at 2.30 this afternoon after a glorious flight of 360 kilometres at a height of 4,000 feet. The Channel was deep blue in colour and, in places, we could see right to the bed of the ocean. I sat beside the pilot and enjoyed myself hugely – except for leaving you.
>
> There is no peace for the wicked. I have returned to an accumulation of letters, official and otherwise, and I have an audience with my Sovereign at 5.30 p.m. tomorrow. Things look a little better at the

Hague. Here, Philip Snowden is the hero of the hour and he won't lose his head in spite of that.

A long letter has just arrived from Lady Barclay that was; now Beatrice de Candolle.* She is at Geneva.

You should get some roses this evening. There were only a few red roses of the right colour. They will come with all my love.

Don't forget the suggestion about 'M. of M.' Baedeker for soul-starved Americans. There is money as well as a good service for suffering souls in the idea.

With the very best love, as ever, and so many thanks for golden days.

In England, official events still came thick and fast. Following a long meeting at Gwydyr House, Air Ministry orders were agreed for production of the Hawker 9/29 as the Hart 1 Day Bomber, for the Handley Page 13/29 Hinaidi Night Bomber and for the Fairey 17/29 as the Seal Torpedo/Spotter/Reconnaissance (TSR) aircraft for the Fleet Air Arm.

Towards the end of August, the Secretary of State made a flying visit to units of the Royal Air Force in the various Home Commands of what was then designated the Air Defence of Great Britain under Air Marshal Sir Edward Ellington, KCB, CMG, CBE, psa, Air Officer Commanding in Chief. The tour included units of the Fighting Area at Northolt and Tangmere; the Coastal Area at Gosport, Lee-on-Solent and Calshot; the Wessex Bombing Area at Worthy Down; the Inland Area (No 23 Group) at Grantham and visits to No 1 School of Technical Training (Apprentices) at Halton and to the Headquarters of the Home Defence Force at Uxbridge.

The Air Minister's tour was designed to see as much as possible of the day-to-day work of the Royal Air Force in the space of six full days. It began at Northolt on the morning of Monday, 26th August, where No 41 (Fighter) Squadron, under Squadron Leader Robert S. Aitken, MC, AFC, put up an impressive flying display with its twelve Armstrong-Whitworth Siskin 3a single-seat fighters; sturdy 150 mph radial-engined biplanes, of which the prototype had first flown way back in 1919. Only a few days before, Thomson had authorised an

* Lady Barclay, widow of Sir George Barclay (d. 1921), British Minister to the Court of King Ferdinand of Roumania, was an old friend of Christopher Thomson and the Princess Bibesco since the days of 1915 and 1916, when Thomson was Military Attaché in Bucharest. Lady Barclay was the youngest daughter of Henry Chapman of New York. She married Major-General Raymond de Candolle, CB, in 1922. He died in 1935.

order for a much more advanced replacement for the Siskin; the Hawker Fury biplane; the first RAF fighter to exceed 200 mph.

From Northolt, the Secretary of State flew in a Fairey IIIF of No 24 (Communications) Squadron, piloted by Squadron Leader David S. Don, to RAF Station Headquarters, Tangmere, near Chichester, commanded by his old friend, Wing Commander John B. Graham, MC, AFC, who had accompanied him on his tour of Royal Air Force Middle East, Transjordon and Palestine, and Iraq Commands in 1924.

Thomson lunched at Tangmere with Graham and officers of the two fighter squadrons stationed there: No 1 and No 43 Squadrons, both equipped with Siskins.*

After lunch, escorted by aircraft from both Tangmere squadrons, the Secretary of State flew down the coast on a brilliantly sunny afternoon to land at RAF Gosport, the Headquarters of No 10 Group, Coastal Area. Lined up there were two units of the Fleet Air Arm from the aircraft carrier HMS *Furious*; No 447 (Fleet Spotter Reconnaissance) Flight** with Fairey IIIFs and No 481 (Fleet Torpedo Bomber) Flight,† flying Blackburn Darts, together with the Torpedo Training Flight.

From Gosport, the Secretary of State was driven the short distance to the Headquarters of No 10 Group, Coastal Command at Lee-on-Solent (Air Commodore T. C. R. Higgins, CB, CMG) to watch a demonstration of torpedo dropping by No 481 Flight's Darts and to inspect the RAF School of Naval Co-operation†† – a direct outcome of the Air Ministry/Admiralty altercation in 1924.

There was a convivial dinner that evening in the Mess of Coastal Area Headquarters with the Air Officer Commanding RAF Coastal Area, Air Vice-Marshal Charles L. Lambe, CB, CMG, DSO, in the chair, enthusiastic about the loan to Coast Headquarters of the elegant steam-yacht *Karen* by Major Colin Cooper (ex-RFC and

* No 1 (Fighter) Squadron was at that time commanded by Squadron Leader Eustace O. Grenfell, MC, OBE, AFC, and No 43 (Fighter) Squadron by Squadron Leader Cyril N. Lowe, MC, DFC.

** No 10 Group with HQ at Gosport was under Group Captain Charles D. Bresse, AFC. At Gosport, No 447 Flight was commanded by Flight Lieutenant Lewin B. Duggan.

† No 481 Flight was commanded by Flight Lieutenant F. W. R. Clarke (Lieutenant-Commander, RN).

†† The RAF School of Naval Co-operation was under Wing Commander George F. Pretyman, DSO, OBE. This was still the time when the Commanding Officers of almost every first-line squadron in the Royal Air Force held decorations of RFC or of the RNAS gained during the 1914–18 War.

now a city magnate), to act as flagship of the Schneider Trophy Contest General Committee, under Charles Lambe, for the forthcoming events.

Next morning, 27th August, another bright, sunny day, the Secretary of State spent some further time at the School of Naval Co-operation before embarking at the Lee-on-Solent slipway to board a Supermarine Southampton, twin-engine flying-boat of No 201 (FB) Squadron, moored in the somewhat ruffled waters of the Solent under a stiff south-westerly breeze. He was welcomed on board by Wing Commander David Donald, DFC, AFC, CO of the squadron, seated in the front of the two open cockpits forward of the wings and now signalling the hand-starting of the two Napier Lion engines.

Here was one of the more delightful aspects of the flying of those times; reserved for the fortunate few. A brilliant sunny morning, blue waters, the distant prospect of the Isle of Wight, the rumble of the Napier Lions warming up above and behind; the whole scene savoured not from the closed confines of a narrow cabin, but from the fresh spaciousness of an open cockpit in which even a helmet and goggles was set on one side in the warmth of the sunshine behind the substantial windscreen.

Slipped from its buoy, the Southampton was taxied out in a wide sweep to warm up its engines and then headed into wind across the Solent towards Cowes Road, bespeckled with sail.

The roar of the uncowled Lions between the wings behind, the rush of white waters on each side as the flying-boat climbed up on to its 'step', the long dash across the waters to gain flying speed, held an exhilaration not experienced in any other form of aviation. Suddenly, the rough motion from wave to wave became satin smooth as the boat lifted off the water at some 50 miles an hour. Instead, the salt air swirled past the open cockpit, draught-free unless one raised one's head a fraction clear of the shelter of the screen.

Conversation was not possible against the combined noise of the engines and the airflow. With a tap on the knee, Thomson was motioned down to a catwalk, alongside the cockpits, into the front gun-cockpit right in the bows of the ship. There, a little behind the forward edge, the Secretary of State could stand out of the 80 miles-an-hour breeze to survey the passing scene.

Once in the air, the Southampton was flown towards Cowes, at the mouth of the Medina River, and then, at 200 feet above the water, along the coast of the Isle of Wight, past Ryde Pier, the starting line of the 31-mile diamond-shaped Schneider course.

The first leg, of 3¼ miles, lay along the shoreline to a yellow-and-black chequered pylon mounted on a destroyer off Seaview. A 30-degree turn to port, on to the second leg of five miles, towards Chichester, brought the Southampton to a sharp 110-degree turn, over open water, at a pylon on another destroyer anchored in Spithead, opposite Hayling Island. Another five-mile run towards the Hampshire shore came to a gentle turn at a pylon on the end of Southsea pier, and a 10¼-mile leg over Gilkicker Point to the final and sharpest turn over a destroyer moored in the middle of the Solent, half-way between Cowes and Calshot. So back 7¼ miles to Ryde Pier.

The Southampton completed a first lap of the course in 23 minutes at 81 miles an hour, and set off, still at 200 feet, on a second lap, the crew waving back to holiday crowds thronging the beaches.

On the second time round, instead of completing the West Cowes turn, Wing Commander Donald put the boat down off Calshot and taxied to the slipway extending from the Spit at Calshot Castle, now the temporary home of the RAF high-speed flight training for the Schneider Race.

A somewhat tousled Lord Thomson was received on the slipway by Group Captain Eric R. C. Nanson, CBE, DSC, AFC, OC, Calshot. The Secretary of State lunched with the Schneider team headed by Squadron Leader A. H. Orlebar, AFC, psa (with whom the Italian team was delighted for his resemblance in profile to Dante) and including among its members, George Stainforth and Dick Atcherley, to whom the Lord Thomson had presented the King's Cup at Heston six weeks before. It was a happy augury.

The Schneider Contest was now only eleven days off, but the Italian team had still not arrived though 'incessantly expected'.* On the British side, the two new Supermarine S.6 racing seaplanes had been delivered to Calshot together with one of two new Gloster Napier VIs. The first of the S.6s (N.247) had made seven brief flights, totalling 2 hours 34 minutes, and the other (N.278) one flight of 23 minutes. The Gloster VI (N.249) had managed only one flight of 30 seconds, cut short by engine trouble.

* The Italian team arrived at Calshot on 29th August, with their five pilots and six aircraft; two Macchi 67s, two Macchi 52s, one Fiat C.29 and one Savoia Marchetti S.65.

It was all being cut very fine. The three British racing monoplanes, together with the second Gloster VI, in course of assembly after arrival by road from Cheltenham, were on their beaching trolleys in the long hangar at Calshot with two Gloster IV biplanes* from the 1927 race at Venice (and now regarded as comfortable hack aircraft for practice flying) and two 1927 Supermarine-Napier S.5s as reserve aircraft for the race. They were all shown to the Secretary of State with much pride and some disappointment that the increasing breeze had made the water too rough for any practice flights.

An RAF motor launch took the Secretary of State back to Gosport in the afternoon, whence he flew to Worthy Down Aerodrome north of Winchester – a grass field on which the longest run for take off was 600 yards, to visit No 7 and No 58 (Bomber) Squadrons, both equipped with Vickers Virginia twin-engine night bombers (the largest and slowest of their day) under Wing Commanders Edye R. Manning, DSO, MC and Ernest W. Norton, DSC, respectively. That night, the Air Minister was treated to a demonstration of night take-offs and landings along flare paths by aircraft of No 58 Squadron. Next morning he flew with No 7 Squadron on a bombing run, before lunching with Air Vice-Marshal Sir John Steel, KBE, CB, CMG, commanding the Wessex Bombing Area, and returning, by air, to Northolt.

Next day, 29th August, the Secretary of State continued his tour by flying from Northolt to Grantham, where he inspected No 3 Flying Training School, with its Avro 504s, Altas and Siskin aircraft, and the Headquarters of No 23 Group, Inland Area. On 30th August he flew to Halton to inspect the RAF Apprentices at the No 1 School of Technical Training.† To round off an exhausting week, on Saturday, 31st August, Lord Thomson drove to Uxbridge to visit the Headquarters of the Air Defence of Great Britain and to lunch there with the Air Officer Commanding-in-Chief, Air Marshal Sir Edward L. Ellington, KCB, CMG, CBE, psa, and his staff.

This six-day tour of RAF Home Commands – 'Britain's First

* The fuselages of the Gloster IVs were designed on the basis of a low-drag airship form developed from R.101 and tested in the wind-tunnel at the National Physical Laboratory. So airships and the Schneider Contest were closely linked.

† No 3 FTS at Grantham was then commanded by Wing Commander Alan M. Jackson, psa; No 23 Group by Air Commodore Bertie Drew, CMG, CBE, psa; the No 1 School of Technical Training at Halton by Air Commodore Ian M. Bonham Carter, CB, OBE – all names renowned in the history of the Royal Air Force.

Line of Defence' – had shown the Secretary of State both the strengths and weaknesses of the Service; as he underlined it in a letter to Ramsay MacDonald.* The strength of the Service was the quality of its officers and men; many with the fighting experience from 1914–18 and subsequent Middle East and North-West Frontier skirmishes, the new generation reared on the traditions of the old.

The weakness was in the performance of its aircraft, few of which displayed any significant advance on those of eleven years before, at the end of the war. Certainly engines were more reliable and gave rather more power and the aircraft were stronger and somewhat better equipped; perhaps most important, with parachutes. The one bright spark was the High Speed Flight, with its glistening, streamlined monoplanes and prototype engines, which, in service form, might be expected to achieve an important increase in power for the next generation of fighters and bombers.

The old Ten Year Rule still held sway – that there would be no major war in Europe for another ten years. How long could that hold? The year was 1929. Unless strenuous efforts were made, would the Royal Air Force have much more advanced aircraft ten years from now?

The world's Speed Record of today ought to be the everyday performance of ten years hence. Yet, the new Service aircraft just ordered, good as they would be, were 100 miles an hour slower than the latest Schneider racers. We still had a long way to go to provide aircraft of ten years from now to match the men to tomorrow and, perhaps, an unknown threat.

A week later – as if to rub in these facts and his forebodings for the future – Thomson persuaded the Prime Minister, with his daughter, Ishbel MacDonald, to accompany him to the Hampshire coast, there first to visit the British and Italian teams at Calshot and then to board HM Aircraft Carrier *Argus* (14,000 tons), lying off Portsmouth, and due to be moved to moorings in Cowes Roads, near the Western turning point, with an assembly of the great and the good, to watch the Schneider Trophy contest.

The event began on the fine afternoon of Friday, 6th September, with the navigability trials of the three British and three Italian racing seaplanes; two new Supermarine S.6s and one Supermarine S.5 of 1927; two new Macchi 67s and one Macchi 52 of 1927.

All the aircraft successfully completed the two mandatory take-

* MacDonald Papers, PRO, 2nd September 1929.

offs and landings on the waters off Cowes followed by a taxying test on the water 'of not less than half a mile at not less than twelve knots' and then a third take-off and alighting before being moored out for a six-hour flotation test. The proceedings were enlivened by HMS *Iron Duke*, veteran of Jutland and flagship of the C-in-C Portsmouth, steaming slowly straight across the approach path of one of the Italian Macchi 67s flown by Lieutenant Monti coming in for his second alighting (he touched down just past the stern of the warship) and by the fact that Atcherley's S.6 (N.248) sprang a leak in its port float and barely survived its six hours in open water.

That evening, the party on board the *Argus* was joined by HRH the Prince of Wales with Marshal of the Royal Air Force Sir Hugh Trenchard and Lady Trenchard, together with Sir Sefton Brancker, General Italo Balbo from Italy and representatives from the Aero Clubs of France, Germany and a host of other countries.

The Saturday of the race was once more brilliantly sunny and warm with a gentle breeze from the south-east. There were only two teams in the contest: British and Italian. Challengers from France and the United States had fallen by the wayside; the French after an accident to one of their pilots and disappointing results with the Bernard-Ferbois mid-way monoplane, while the American Mercury-Packard seaplane, built at the US Naval Aircraft Factory at Philadelphia failed to get off the water.

All was now set for the last of the great international seaplane races; the end of an epoch. In a positive way it laid the foundations for the winning of the Battle of Britain, eleven years later.

On this summer's morning of 7th September pontoons for the six competing seaplanes were moored at the mouth of Southampton Water near the Hampshire shore.

At two o'clock precisely the first machine, Waghorn's S.6 (N.247) was launched from its pontoon with 108 gallons of fuel, calculated to be sufficient for 50 minutes flying; 42 minutes on the seven laps of the 31-mile course and eight minutes for take-off, turn on to course and 'to get you home again'.

Enormous crowds on every beach on both the Hampshire and Isle of Wight shores looking on to the course saw in the distance, or heard on the BBC broadcast, Waghorn's long take-off run of more than a minute. The Supermarine S.6, with its 1900 hp Rolls-Royce 'R' engine, was loaded to the greatest weight yet lifted by the type – some 5,652 lb.

As the aircraft left the water the starting gun was fired. Waghorn turned left and swept down at 300 feet over the starting line at Ryde Pier and disappeared eastwards towards Seaview and the Spithead turn.

From the flight deck of the *Argus* the blur of smoke from the exhaust of the Rolls-Royce engine could next be seen approaching Portsmouth from the east, then the machine swung round the Cowes turn and thundered back over Ryde and away again. Everyone waited to hear the times and speed announced. When it came: '5 minutes 45 seconds: 324 mph' a great cheer went up. It was a new world speed record; five miles an hour faster round the four-sided course than the previous best over three kilometres straight.

The second lap was better still: 329 mph – another new record; and the third still faster, 331 mph.

By this time the first of the Italians had started; Warrant Officer Tomaso Dal Molin in the 1927 Race Macchi 52 bis, noticeably quiet compared with the roar of the S.6. Dal Molin swept on to the course on his first lap well in front of Waghorn coming round in the S.6 on his fourth. As they disappeared towards Southsea the S.6 was closing up fast and, by the time they came round to the Cowes turn the Supermarine was hard on the Macchi's heels.

Dal Molin took the acute turn close and very low. Waghorn was a little higher, close behind. On the turn he swung up a little and straightening up dived ahead of Dal Molin and crossed the Ryde Pier well in front.

Of course the new S.6 was expected to be faster than the older Macchi 52 but it was a magnificent sight for the packed crowds of spectators all along both shores – and the only straight fight in the race.

Fifteen minutes after Dal Molin had started, Flight Lieutenant d'Arcy Greig took off in the second British machine, the 1927 'reserve' aircraft, the Supermarine-Napier S.5 (N.219). There were now three aircraft on the course; Waghorn having just completed his fifth lap at 330 mph and Dal Molin his second at 287 mph. The interest, for the moment, concentrated on whether the British S.5 or the Italian M.52 aircraft of 1927 was the faster.

Greig's first lap time gave the answer: 284 mph against Dal Molin's 285 mph for his third lap. The Italian was just two seconds faster over the 31-mile circuit. It showed how fortunate Britain had been to win the 1927 race at Venice with the S.5 at 281 mph.

Now Waghorn' seventh and final lap time came up: 331 miles

an hour; to complete the 217-mile course in 39 minutes 43 seconds; an average of 528.75 kmh for 350 km or 328.62 mph.

But Waghorn had miscalculated the number of laps he had flown and went on for an eighth. His engine spluttered as he rounded the Cowes turn and then cut as he went ahead and alighted short of the Ryde finishing line.

A disconsolate Waghorn climbed out of the cockpit on to the top of the fuselage, convinced that he had failed to complete the course, although the whole ship's company of the *Iron Duke*, close by, was cheering him. The RAF boat's crew quickly ended Waghorn's misery as they took the S.6 in tow and told him he was home and dry at 328 miles an hour with a world's record in hand.

By now the second Italian was due to start on the first of the new 1929 Macchi 67s which, just possibly, had the heels of the S.6s.

Tenente Remo Cadringer got away with a long run and almost at once the engine began to throw out clouds of black smoke. He came round very wide over Cowes, flying higher than the earlier competitors and apparently in some trouble. When the speed for his first lap came up it was a disappointing 284 miles an hour at exactly the same time (6 minutes 34 seconds) as d'Arcy Greig's first lap in the old S.5.

When Lieutenant Cadringer came round again, trailing smoke with the characteristic high-pitched snarl of the 18-cylinder Isotta-Fraschini engine/propeller combination, he was seen to be right off course and obviously giving up. He alighted close to Dal Molin who had just completed the course at 284.2 miles an hour in the Macchi 52 bis; two miles an hour – or 21 seconds – faster than d'Arcy Greig for the 350 km course.

Later, it turned out that a lubricating-oil pipe in Cadringer's Macchi 67 had burst and covered both machine and pilot with very hot oil. He had done well to get down safely.

Now it was Atcherley's turn to start in the second Supermarine S.6 (N.248) His take-off raised momentary worries among the knowledgeable because of the violent 'porpoising' – or bouncing progress – at the early part of the long run. Eventually the S.6 bounded into the air, tail down, and almost stalled; then pulled away and lined up for Ryde pier, flying a somewhat erratic course. The fact was that, in pushing up his goggles to clear them of take-off spray, Atcherley had put his head into the slipstream and they had blown off. Thereafter he had had to keep his head well down, a position which further restricted his already limited forward view.

He was too pre-occupied to bring up the second pair of goggles which each pilot wore around his neck and, in consequence went over, instead of round, the Seaview pylon and, on his first lap, flew some three miles extra and returned the disappointing time of six minutes nine seconds or 302 mph. Shortly after came the announcement that he was disqualified.

Knowing nothing about this Atcherley went on to complete the course at an average speed of 325.54 miles an hour and to set up the fastest lap speed of the race at 332 mph. In the process he collected new world's records for 50 km and 100 km at 331 mph.

There was still the third Italian competitor to fly: Lieutenant Giovanni Monti in the second of the new Macchi 67s.

While Atcherley was on his third lap Monti took off, climbed steeply and then dived down across the starting line. He flew lower than any of the others – no more than 150 feet – on his first lap. It was completed in 6 minutes 11 seconds at the speed of only 301 miles an hour. He had swung wide on his turns and coming back over Ryde he was too far to the south, right over the pier pavilion. Near the end of his second lap Monti put down on the water off Hayling Island, blinded by steam from a broken joint in a coolant-water pipe. The boiling water scalded his legs and forced him to abandon the race.

The final result of the 1929 Schneider Contest was, therefore:

First: H. R. D. Waghorn – S.6 (N. 247) at 328.63 mph.

Second: Dal Molin – M52 bis at 284.20 mph.

Third: D. d'Arcy Greig – S.5 (N.219) at 282.11 mph.

So ended the last of the ten contested Schneider Trophy races which had spanned sixteen years since 1913. The final event – in 1931 – was a fly-over by Flight Lieutenant John Boothman in a Supermarine S.6B at 340.8 mph – in the absence of any foreign competitors.

The full drama of the 1929 race was not, however, generally known until afterwards. After the 'navigability trials' and before the speed contest on the following morning while the overnight inspection was in progress on Waghorn's S.6 – N.247 – some white metal was found on a sparking plug and, when the cylinder head of the Rolls-Royce 'R' engine was taken off, one of the piston heads was found about to break up. A few minutes more flying and it would certainly have failed.

The rules forbad an engine change and the only solution was to fit a new piston and cylinder block – never previously done outside the Derby works. Fortunately the Rolls-Royce team who had built the engine were in Southampton to watch the race. They were rounded up after midnight, worked till dawn and a satisfactory engine run was completed in time for the aircraft to go out on its pontoon.

But for this the 1927 Macchi 52 bis flown by Dal Molin would have won the contest at 284 miles an hour. Thus narrow was victory.

There was a celebration dinner that evening on board the Royal Aero Club's headquarters ship for the contest, the *SS Orford*, anchored off Ryde.

After dinner to the competing teams and assorted distinguished visitors Sir Philip Sassoon congratulated both the winners and the losers and spoke especially of the splendid sportsmanship of the Italian team who 'had made it a damned close run thing' (how close only a few knew at that time). Until that day Italy had held the world's speed record. Finally he turned to the Secretary of State for Air, Lord Thomson – 'their mentor and chief; my revered and respected predecessor'.

'He was the perfect Chairman in his time at the Royal Aero Club; the driving force behind Britain's entry and today's victory. We all thank you for having done so much to make today's contest possible, enjoyable and successful.'

General Italo Balbo, for Italy, added his congratulations and hoped that, in due course, they would meet again with other nations to renew the competition.

Lord Thomson read a telegram from His Majesty the King conveying his warm congratulations to Flying Officer Waghorn and the Royal Air Force – and he added his own as well to the Italian pilots for their high courage and skill.

To wind up, the Prime Minister 'pledged the British nation to defend the trophy. We assure our Italian friends that we are going to do our best to win again.'

'A wonderful day,' wrote C.B. to Marthe. 'It lacked only one thing: your presence.'

The Schneider victory on a glorious summer's day over the blue waters of the Solent was, indeed, a fitting climax to a vintage summer for aviation in all its aspects. It was rounded off on 10th September by the raising of the world's Speed Record to 336.3 mph by George Stainforth flying the Gloster-Napier VI (N.249) followed, two days later, by Harry Orlebar at 357.7 mph in the Supermarine-Rolls

Royce S.6 (N.247) the aircraft which had won the race.

And that – for the time being – was that. The seeds of both the Spitfire and the Merlin engine had been sown.

CHAPTER FIVE

The Big Ships

'The life so short, the craft so long to learn
'The essay so hard, so sharp the conquering'.
Geoffrey Chaucer:
'The Parlement of Foules'

July/August 1929

NOW, at last, the airships were coming.

With fading summer and after five years of endeavour, R.101 was almost ready at Cardington. At Howden, R.100 was not far behind.

Inflation of R.101 had begun on 11th of July – a long and intricate process primarily undertaken by the airship's future flying crew. It was prolonged by the inadequacies of the output from the hydrogen plant, by the need to make sure that each of the sixteen bags fitted snugly within its wiring system without folds or tears, and by a number of frustrating incidents. They included not only the patching of gas bags after sundry tools had inadvertently been dropped through them from the top of the shed but also one instance of a rigger falling through No 13 bag – fortunately without serious injury – after missing his footing when working in the top fin.

As inflation went forward, bag by bag, there was a long process of testing and clearing each valve and then, because of the ingress of air, of purging the bags to bring them up to an acceptable standard of gas purity.

Even at this early stage, the pre-doped portions of the outer-cover began to show signs of trouble. Between 8th and 12th August a number of splits appeared on the upper panels on the port side, caused by changes in humidity. At the same time the riggers were busily fitting and doping the cover at the nose and stern where the amount of double curvature made pre-doping impossible. Doubts were beginning to arise about the wisdom of pre-doping, intended to reduce the amount of work by men suspended in slings around the ship.

With this slow progress, inflation to 85 per cent capacity was completed by 21st August, forty-one days from the start. By then all five power-cars had been attached to the structure; the last, the aft car to Frame 11, on 15th August.

A month later, by Saturday, 21st September, the ship was partially freed and made airborne in the shed for the first time at 95 per cent filled and at a gas purity of 97 per cent. There was an unpleasant surprise when it was found to be seriously tail-heavy.

The outer cover had by now been fitted and doped on the fins, rudders and elevators. That done, the weight of these tail surfaces – at 6.78 tons (15,187 lb) – turned out to be almost double the estimates. This accounted for the tail heaviness, compounded by the fact that the fine, streamlined, shape of R.101 meant that the gas-bags beneath the tail surfaces were of relatively low capacity.

When water ballast had been discharged aft, the ship came into equilibrium with only 34 tons of fuel and half a ton of water ballast on board. There were glum faces all round, though there was general agreement in the Design Office that the true facts about disposable lift could not be gained until the bags were filled to 100 per cent capacity and the 'Lift and Trim' taken properly. Even so the outlook was clearly serious.

Two days later there was some cheer, however. Noël Atherstone recorded in his diary for Monday, 23rd September, 1929:

> Preliminary run on No 1 engine car took place this morning and everyone is agreeably surprised at the smoothness and almost complete lack of vibration with the engine running at all speeds. In fact, at full speed vibration was at a minimum. This afternoon the first official airborne engine test was successfully carried out; the Tornado engine being run for one hour at cruising speed, quarter of an hour at full speed and the same at slowest speed. The engineers say the car is the smoothest they have ever been in. Everybody is all smiles today which is rather a contrast to facial expressions last Saturday.

On 29th September he recorded: 'No 5 engine successfully passed its official trial this morning so that clears all engines.'

Now came the culmination of all the design work and of all the estimates; the 'Lift and Trim trials' with the gas bags filled to 100 per cent capacity.

All was ready on Monday, 30th September. After a full day of gassing and checking, at 1800 hours the ship was floated clear of the trestles and the top restraints were let go. Amid a good deal of anxiety the ship was ballasted-up to equilibrium and the amount of fuel and of water on board carefully checked. Then, at 1900 hours the securing tackles were replaced, top and bottom, and the ship made fast once more to the trestles and to the roof of the shed.

The results when worked out in a preliminary way next day, 1st October, were acutely disappointing. At 100 per cent gas capacity and

96.15 per cent purity, the gross lift, on a day of low air density, was 148,564 tons or 332,783 lb. Adjusted to standard temperature, density and gas purity at sea level the theoretical gross lift would be 151.6 tons – exactly three tons more – at a total gas capacity of 4,893,740 cubic feet contained in the 16 bags. (See Appendix 6)

The critical issue was to determine how much of this gross lift could be used in the form of disposable load: ballast, fuel, oil, crew, passengers, baggage, stores, food, drink and goods.

On 17 June 1929 the fixed weight of R.101 had been estimated at 105 tons, made up of structure, bags, cover, engines and installations. Now the detailed tally of weighted items added up to 110.1 tons. Fuel and water ballast piped on board to bring the ship precisely into equilibrium came to 35.2 tons. This disposable load, added to the fixed weight of 110.1 tons, resulted in a gross weight of 145.3 tons and showed a discrepancy of 3.3 tons compared with the estimated gross lift, on the day, of 148.6 tons. The only explanation was that dust, dirt and unweighed details must account for the difference and that the true weight of the ship on 30th September 1929, must be taken as 113.4 tons. The melancholy fact was that this was an increase of 26 per cent – or 23.4 tons – over the original target weight of 90 tons – optimistic as that had been.

In standard atmosphere the theoretical gross lift was 151.6 tons so that the nominal disposable lift could now be set at 38.2 tons. This compared with the forecast of 46.8 tons made on 17th June and the original specification of 63 tons.

Clearly substantial weight reductions must be sought, together with any possible increases in lift. For the moment, however, there was another preoccupation; an official Air Ministry invitation to the press to inspect R.101 in No. 1 Shed at Cardington before its first flight.

Two hundred Press men and women from the national, the provincial and the overseas newspapers and from the technical press, descended upon Cardington at 11 o'clock on 2nd October.

Ahead of that a special effort had been made to spruce up, to paint and to furnish R.101's passenger accommodation for the visit. Chairs, tables and galley equipment which had been used in R.36 were taken out of the Cardington store and Flying Officer Maurice Steff, R.101's Second Officer, who had reported to Cardington on 26th September,*

* Flying Officer Maurice H. Steff had been a sub-lieutenant in the Royal Navy where he had served on board H.M.S. *Inflexible* at the Battle of Jutland in 1916 at the age of twenty. He had transferred to the RNAS in 1918 and, after serving under Irwin at the RAF School of Balloon Training at Larkhill had joined him at Cardington.

was sent off into Bedford to buy 'moderate quality linen table-cloths of suitable size for the dining room' * as well as blotters for the writing shelves at the forward end of the lounge. He was instructed also to hire a dozen deck chairs for the promenade decks 'for Wednesday's Press visit only'.

To add a homely touch, Colonel Richmond's Secretary, Miss Foster, was 'to be asked to undertake the arrangement of flowers on the dining room table and ferns for the lounge'.

Noël Atherstone's diary notes for 2nd October 1929:

> Up to 1100 hours the frantic furnishing went on and by that time the inside of the ship looked very fine indeed. Come the Press and they swarmed like earwigs over everything. The shed doors had to be partly opened so that pictures could be taken – but I had them closed again very soon as there was a stiff breeze blowing and the ship was beginning to get restive. The Press were finally disposed of at about 1330 hours and then all and sundry in the Works came to have a look. Fifty parachutes were put on board this evening and after a few experiments I managed to find out a way of slinging them in their stowage positions with a quick release consisting of a short piece of dural tube and some string.

At 11.30 that morning Reginald Colmore had welcomed the assembled journalists at the Administration Block and had told them that they would be divided into groups of twenty for the visit to the ship. He then invited Richmond to speak briefly about R.101.

Richmond began by saying that R.101, which they would be seeing shortly, was now virtually ready for flight. It was the biggest airship ever built – 732 feet long, 140 feet high, with a diameter of 132 feet. The outer cover had an area of 214,000 square feet; or almost five acres. An idea of the airship's size might be gained by comparing it with the largest ocean liner in the world, the SS *Leviathan* of the White Star Line. If the airship's hull were to be submerged in the sea – which God forbid – to a depth equivalent to the draught of the *Leviathan*'s hull it would have a displacement of about 100,000 tons compared with the *Leviathan*'s 59,957 tons. The beam of the *Leviathan* was 100 feet; that of R.101 was 132 feet. If the airship's hull were to be wholly submerged likke a submarine, its displacement would be about 160,000 tons.

Thus, R.101 was at present, by quite a wide margin, the largest vessel in the world.

* PRO AIR 11-23 XM 08109 of 29th September 1929.

Another comparison was that the *Leviathan* had 34 horse-power per passenger. R.101 had 56 horse-power per passenger to achieve about double the cruising speed.

A crew of fifty had been recruited for R.101. It included four officers, a Chief Coxswain, a First Engineer, twenty flight engineers and twenty-four riggers. In service there would be two watches each of seventeen men with a spare watch of sixteen drawn from men engaged in construction to take charge when the aircraft was moored at the mast.

Richmond recalled that Lord Thomson's airship programme of 1924 had been based on a policy that aeroplanes would be developed for operation over distances of up to about 700 miles non-stop and that rigid airships would be planned for the longer-trans-ocean and trans-continental, flights of up to about 3,500 miles.

Under the 1924 Programme, R.101 had been designed and built at the Royal Airship Works with the remit to incorporate, on the basis of detailed research, such new (and, in some respects, experimental) features as would build on previous practice and advance progress towards safe and regular long-distance flights in all climates. By contrast, R.100 built at Howden by the Airship Guarantee Company to a fixed price contract, was to be a relatively conventional project, though refined and improved in many ways compared with previous practice.

Among R.101's new features were advances in streamlined form, heavy-oil engines, a new system of controls, new forms of gas valves and gasbag restraining wires and a complete steam-heating and cooling system for the living accommodation within the airship's hull.

There was, said Richmond, no question of direct competition between the two ships, R.100 and R.101. The purpose behind both of them was to gain sufficient experience to report to the Imperial Conference in London in the autumn of 1930 when Phase Two of Lord Thomson's Imperial Communications Policy would be one of the chief items on the agenda; looking towards regular commercial air services to and from the Dominions overseas with a fleet of British airships.

R.101 was, in fact, a prototype – an experimental ship – from experience with which later commercial airships could be developed.

Colonel Richmond emphasised a point which he believed was not generally understood; that all commercial airships had to be flown over long ranges at low levels.

Airships must, if possible, be held down to a cruising height of no more than 1,500 feet to achieve the best economic results. 'For every additional 1,000 feet that an airship rises it has to lose one-thirteenth (3.33 per cent) of the total volume of its lifting gas by valving because of the expansion of the gas with height. For R.101 this is the equivalent of the loss of about five tons of lifting capacity for every extra 1,000 feet of height; that is one ton per 200 feet. So we must fly low.

'That is just one of the problems of airship operation and it is all part of the day's work. One can add that flying low in an airship offers a splendid view of the ground or the sea.'

There was a further point of great importance. In the average English climate R.101 had a gross lift of about 150 tons. In Egypt, because of the lower density of the air, the lift would be some ten tons less; about 140 tons. At the equator there would be a loss of a further two tons or so of lift. So, in flying from England to India, R.101 would have to reckon on losing about 12 tons of lift even while keeping to low levels of flight.

Richmond went on to say that the main features of the airship would be explained to each group as it went on board. Before that he would touch upon just one or two other main points.

First, R.101 was powered by five Beardmore Tornado diesel engines each giving 585 horse-power for maximum continuous cruise. There were five separate power cars; two forward of the control-car, two aft of it, midships, but further out from the ship's centre-line and one right aft, ahead of the ventral fin and designed to maintain good directional control by blowing its slipstream over the rudder.

Of the two forward engines, that on the port side (No 2 engine) had, for the present, a reverse-pitch propeller for going astern. Later, when a variable-pitch, reversible, propeller was cleared for installation, that engine would also be used for forward flight as originally intended. Meanwhile only four engines would be available for normal cruise. Even so the airship's cruising performance ought to be met. The important question of the precise speed and range would, of course, have to await trials and they should not now be long delayed.

Because the engines were steam-cooled, water in the cylinder jackets was allowed to come to boiling point and then the steam was condensed in three small triangular radiators attached to the hull above the forward and the aft power cars. Steam from the cooling systems of the two midship's engines would be condensed in a retractable radiator and used to heat the passenger accommodation. This system made possible the running of the engines at relatively high thermal

The Royal Airship Works – Cardington's Administration Block built by Short Brothers (Rochester and Bedford) Limited in 1917 as the headquarters of their aeroplane and airship business. The Cardington Works was taken over by the Government on behalf of the Air Ministry in 1919 and thereafter became the Royal Airship Works where R.30, R.31, R. 38 and R.101 were built.

Airshipmen – (Left to right). Flight Lieutenant H. Carmichael Irwin (1894-1930), Captain of R.101; Lieutenant-Colonel Vincent Crane Richmond (1893-1930), Chief Designer at the Royal Airship Works and Major G. Herbert Scott (1888-1930) Assistant Director of Airship Development (Flying and Training) at the RAW.

[PLATE VII]

The Innovator – Squadron Leader F. Michael Rope, Assistant Chief Designer at the Royal Airship Works; responsible for some of the most original and successful advances in the design and construction of R.101. Michael Rope had led the design team which produced the outstanding non-rigid SS, Zero airships at Kingsnorth in 1917.

(CENTRE) *Pillar Hall* – Awaiting furnishing, the 60 feet by 32 feet lounge of R.101 looking across the beam of the airship to the 32 feet wide by 8 feet deep windows let into the sides of the outer cover at an angle of 45 degrees. The ceiling height of the lounge was 8 feet 4 inches. The pillars were duralumin encased in balsa wood; the walls painted fabric. (BOTTOM) *Smoking Room* – With an aluminium floor and ceiling and asbestos walls, the smoking room on the lower floor of R.101 measured 16 feet by 12 feet 9 inches and had seats for 24 persons at a time.

efficiencies and helped towards what was hoped would be very low specific fuel consumptions.

R.101 was equipped to carry 9,408 gallons of fuel weighing 35.5 tons in forty-nine tanks. There were twenty-three fuel storage tanks, each of 224 gallons when filled to 95 per cent of their rated capacity. These tanks, with a total of 5,152 gallons, were located within the main frames of the airship. They fed by gravity to a single transfer tank of 224 gallons from which fuel was distributed by compressed air to ten feed tanks (two of 224 gallons and eight of 112 gallons); two above each engine car. This total of 6,720 gallons in thirty-four tanks and weighing 25.38 tons, should give R.101 an ample margin of fuel on the flight to Egypt even against a continuous headwind. Each of the 224 gallon feed tanks was designed to hold sufficient fuel to supply its engine for eight hours at normal continuous cruising power.

There were, in addition, fourteen jettison tanks (eight of 224 gallons and six of 112 gallons) to provide a further 2,464 gallons which, with one auxiliary tank of 224 gallons brought up the fuel capacity to 9,408 gallons which could be used when loads permitted. The whole of the 224 gallons from each jettison tank could be discharged through a twelve-inch diameter pipe in eleven seconds.

For very long ranges, six ballast tanks, of 224 gallons each, could be switched from water to fuel to add a further 1,344 gallons weighing 5.1 tons. The absolute maximum fuel tankage was, therefore, 10,752 gallons or 40.6 tons – calculated to be sufficient for a still air range of more than 6,000 miles if no passengers or cargo were carried. But, of course, sufficient lift had to be available to carry such a load.

The airworthiness requirements for airships * laid down that ten per cent of the lift must be available as water ballast. In R.101 there was thus provision for 3,584 gallons (16 tons of water) in 28 containers. They comprised eight tanks of 224 gallons (one ton) each, twelve bags of 112 gallons (half a ton) each and eight bags of 56 gallons (quarter of a ton) each. As he had already said, six of the large ballast tanks (1,344 gallons) could be turned over to fuel. All the bags could be discharged quickly to release a total of eight tons in a few seconds in case of need.

Pipe lines ran throughout the ship to connect up the various containers and to link all of them to the 'pressure' – (or 'transfer') – tank from which fuel could be blown by compressed air to any part of the ship so as to feed any of the engines from any of the fuel tanks as well as to adjust trim as might be required.

* AIR 11/155.

Safety, said Richmond, was *the* cardinal feature in the design of R.101. This was the primary reason for the selection of the Beardmore compression-ignition engines. The high-speed diesel engine, burning heavy-oil fuel, was the engine *par excellence* for airship work even though the weight of the engines was a good deal greater – and their power a good deal less – than had been hoped for when the design of the airship began.

From the safety viewpoint there was no doubt that diesel oil was at the top of the league and petrol at the bottom. R.101's engines ought to set new standards of safety from fire in tropical climates as well as make possible low fuel consumptions and offer prospects of exceptional reliability in service.

The safety argument went, said Richmond, that normal aviation gasoline used in petrol engines, gives off inflammable vapour even at normal air temperatures. The rate at which this – easily ignited – vapour is given off increases rapidly as the ambient temperature rises.

An airship flying between England and India must expect to encounter atmosphere temperatures of up to 130 degrees Fahrenheit, that is 55 degrees Centigrade. The whole of the fuel and all metal and fabric surfaces of the ship would at times reach such temperatures.

In those conditions the generation of petrol vapour would be very rapid indeed and any flame or spark would ignite it instantly. But the heavy oil developed for R.101's engines had a flashpoint * of about 210 degrees F (99 degrees C). It would thus give off virtually no inflammable vapour in tropical conditions and to that extent it was a relatively safe fuel which at atmosphere temperatures, even in the tropics, could not be ignited by the application of a flame.

This was the decisive factor in the choice of engines for R.101. There was, however, a penalty to pay in weight – one of the most precious commodities of an airship.

Though the weight of the engines had been reduced dramatically compared with their original form for Canadian rail cars they were still very heavy though on a fifty-hour flight the total weight of R.101's engines and the fuel they required would be less than had petrol engines been installed.

On fuel costs, the heavy oil-engines showed a very substantial advantage. Not only did the diesel engines burn less fuel than did petrol engines of the same power output, but fuel-oil cost only £5 a ton

* Flashpoint is defined as the minimum temperature at which an inflammable mixture of fuel vapour and air gathers above a pool of fuel. For petrol (gasoline) it is about −20 degrees C.

($4\frac{1}{2}$d a gallon), compared with £23 a ton (18s 5d a gallon) for petrol.* The fuel cost of a flight to Egypt by R.101 would, thus, come to about £110, compared with about £690 had the airship been powered with petrol engines.

Richmond concluded by saying that when they had inspected R.101, he hoped that the Press would agree that it gave promise of being a thoroughly comfortable, safe and economic ship – which, together with a good performance in the air, were the objectives for which everyone at Cardington had worked for so long.

With that, after one or two questions, the various groups set off to walk the few hundred yards to the shed; there to be taken, in batches of twenty, over the airship by way of a hatch and ladder, leading into the crew's quarters on the lower deck.

Clearly, the Press visit was a great success. Nothing like R.101's accommodation had been seen in any aircraft before, Miss Foster's flowers and all. Next day, Cardington had a good press.

Now, there set in ten days of poor weather, with winds blowing at right-angles to the shed doors. Such conditions ruled out any attempt to perform the delicate task of moving R.101 out of its shed for the first time.

The four different groups of airshipmen – drawing office, works, flying staff and administrators – were all busy in their separate ways, while, with them, the newspapers anxiously awaited the launch of R.101 and raised eager expectations in the public at large.

At the Air Ministry, the first flight was no less eagerly awaited. Meanwhile the Secretary of State and his Under Secretary, Frederick Montague, MP, were both active in rebutting hostile comments from the vocal anti-airship brigade which sought to make capital out of the delays and out of rumours of excessive weights.

Typical was an attack on R.101 by Colonel L'Estrange Malone, Labour MP for Northampton. Malone had written to the Secretary of State on 27th September ridiculing the whole airship policy. He suggested that the Air Ministry should cut its losses and scrap both R.100 and R.101 as useless.

He had said that to go on would throw good money after bad and 'increase a cost of £2 million to more than £2,500,000 without results, while £20 million would be needed before regular services could be mounted with a fleet of airships'. Aeroplanes would be a much better bet. He quoted the naval architect, Mr E. F. Spanner's views

* In 1980 values, fuel oil would cost 26p a gallon, compared with 108p a gallon for aviation petrol on the same basis.

about the danger of airships. Give up now rather than go on to inevitable disaster.

Christopher Thomson replied to Malone on 3rd October from Gwydyr House, in a letter which underlined his own philosophy and that of the Air Ministry as a whole.

> It seems to me quite clear that when we have just reached a point at which the two new airships are to be put to practical test that there can be no question of 'cutting our losses' and scrapping these vessels. The time for deciding on our future airship policy will surely be when R.100 and R.101 have been adequately tried out.
>
> In these circumstances I will not reply to your letter in great detail but will merely give you my personal observations on certain points you raise.
>
> 1. The experimental programme on which we have been engaged has not yet cost us £2 million and even by the end of the current financial year we shall not have spent anything like £2,500,000.
>
> 2. I can see no reason for thinking that anything approaching £20 million would be required to a regular commercial airship service.
>
> Considerable capital expenditure would, of course, be necessary as it has been for every other form of modern transport. The provision of docks and harbours for the mercantile marine and of the permanent way for rail transport has entailed a capital outlay running into many thousands of millions of pounds without counting the cost of railway rolling stock and a fleet of modern buses. Relatively speaking an airship base is cheap and the air, fortunately, gives us a permanent way free of charge.
>
> 3. I do not agree that heavier-than-air craft are today able – or likely for very many years to come to be able – to do what will, undoubtedly, be well within the capacity of airships provided that the present experimental programme is brought to a successful issue.
>
> 4. Obviously the airship is still in its infancy and no one would claim that the present airships are the last word in airship design or other than experimental. I see no reason at present for the smallest misgivings on technical grounds and when you suggest the appointment of a technical committee you are presumably unaware that the designs of both airships and the actual results are subject to the scrutiny of a panel consisting of two independent scientists of the highest standing with great aeronautical experience.*
>
> Mr Spanner is doubtless a very able naval architect but I am not aware that he can claim in any way to be an aeronautical expert. The disaster to R.38 which – in the inevitable absence of sufficient aeronautical data – was designed and constructed in accordance with naval rather than aeronautical techniques, well illustrates the

* Professors Leonard Bairstow and Sutton Pippard.

fundamental importance of the latter when dealing with the air as opposed to marine craft.

As regards the danger to personnel, I take it that no one would suggest that the operation of experimental aircraft is free from risk. It is easy, however, to exaggerate that risk; witness the very successful flights of the *Graf Zeppelin* across the Atlantic and round the world, and – such as it is – should in my opinion be faced in the interests of aeronautical progress.

I may add that I personally hope, if it can be arranged, to make an early flight to India in R.101.

Yours sincerely,
Thomson.

In fact, up to the beginning of October 1929, the airship programme as a whole had cost a total of £1,986,000 since the start on 1st July 1924. Out of this R.101 had cost £527,000 or 26 per cent of the total.*

Design costs at the Royal Airship Works, had, so far, amounted to £43,000 and at Boulton and Paul to £7,000. The total expenditure upon materials for R.101 had been £108,000. The five Tornado engines had cost a total of £41,000 – or £8,200 each.

At Howden, R.100 had, so far, cost £337,000.

Looking ahead, the policy for future airships was beginning to occupy the attention of the Air Ministry. An outline specification had been drawn up for two new, and much larger, airships – R.102 and R.103. Preliminary design studies were in hand at both Cardington and Howden.

At the Royal Airship Works, R.101 had been turned over from the Works Staff to the Flying Staff at 1500 hours on 4th October. Provisioning the ship had begun with the stowage of emergency and tinned rations. The fresh-water tanks were filled and all bags were gassed up to 97 per cent of their capacity. Meanwhile the Met Office at Cardington reported the depressing likelihood that the bad weather would continue.

As if to emphasise the frustrations of waiting, the official 'Permit to Fly' was received from the Air Ministry on 8th October.

R.101, being airborne, was now behaving in the shed in precisely the opposite manner to the behaviour of airships in the open air. During the day the 'glasshouse' effect of the roof and sides of the shed caused the warmed air to push the ship down. After dark the temperature inversion disappeared and the ship rapidly became lighter. Thus,

* Appendix 6. Expenditure on the 1924–30 British Airship Programme.

water ballast had to be discharged in daytime and then taken on board again with evening as the ship became successively heavy by day and light at night.

By 9th October the flying staff were hard at it, practising drills, working in simulated watches, starting and running engines in the shed and familiarising themselves with stations for ground and tower landings and generally 'shaking down' into the ship's routine.

Meanwhile the Works Staff busied themselves with provisioning spares, assembling a spare engine and completing the mooring gear against the time when the ship would be riding at the mast.*

In the Drawing Office the design staff under Vincent Richmond and Michael Rope concentrated upon the working out of the results of the Lift and Trim trials, on the completion of the weight analysis and on the calculation of the consequential flight performance. On 7th October a Design Office meeting went through a long list of items for saving weight and increasing lift. At the end of the day improvements of 5½ tons were thought to be possible.

How necessary this was had become clear from the results of the Lift and Trim trials on 30th September. They had established the fact that the volumetric capacity of the airship with all gas bags filled was 4,893,740 cubic feet, constrained by the gas-bag wiring. The capacity was, therefore, now some 2.13 per cent less than the five million cubic feet originally intended. The loss of capacity added up to a deficiency of about 7,337 lb (3.275 tons) compared with the original estimates.†

The immediate and natural reaction was that the gas bag wiring ought to be let out to make possible the squeezing of an additional 107,000 cubic feet or so of gas into the bags. Squadron Leader Rope was, however, of the opinion that in some places exactly the opposite course should be adopted, that the wiring should be pulled in where the bags were in danger of chafing against projections from the structure of the airship hull. But there was no gainsaying the need for extra lift.

The 4,893,740 cubic feet of gas capacity, under standard conditions of air temperature and pressure at sea level would give a nominal gross lift of 337,913 lb (150.854 tons) at a gas purity of 97.6 per cent. When reduced to 97.0 per cent of gas capacity to give a pressure height of 900 feet before gas would need to be valved as the airship ascended, the practical lift would come down to 327,777 lb (146.3 tons) at 55 degrees F and an atmosphere pressure of 989 millibars. The airship

* Appendix 7. High mast mooring of airships.

† AIR5/1407/1929.

would then have volumetric space available to allow it to rise to 900 feet before gas would have to be valved.

The basic fixed weight of the ship had now been confirmed at 254,464 lb (113.6 tons) so that the lift available for crew, ballast, fuel, oil, stores and payload – the 'disposable lift' – was 73,313 lb (32.73 tons). Out of this about 20 tons would be needed for crew, ballast, spares, lubricating oil and essential stores. The weight of the 'Airship Prepared for Service' – the 'APS Weight' – would be, therefore, about 133½ tons. At a practical gross lift of 146.3 tons, the margin of lift available for fuel and payload would then be no more than 13 tons.

Clearly there was no prospect of operating the ship on the route to India even for a demonstration flight with only this amount of disposable lift available.

Something drastic must be done.

Richmond commented regretfully to Rope that he wished that they had taken fuller advantage of the size of the enlarged shed at Cardington to make R.101 bigger, rather than holding its gas capacity to the round figure of five million cubic feet. Here it was, however. More lift and less weight were essential.

Perforce, to increase the lift, Richmond had to admit that the gas bag wiring must be let out to allow the gas bags to expand within the restraints. An additional 100,000 cubic feet of the capacity would be worth another three tons of lift. Michael Rope demurred but acknowledged that the lift was badly needed and that somehow the bags must be protected from rubbing against projections in the radial frames and the longerons.

As a result of a further long discussion a list of some twenty items on which weight might be saved was set down for detailed examination by the Drawing Office. In particular, the heavy triplex windows could be replaced by plexiglass; twelve out of the twenty-eight sleeping cabins might be dismantled (along with two lavatories) to reduce the passenger sleeping accommodation to thirty; two water ballast tanks might be deleted and some savings might be made in the weights of accessories in the power cars. In all a saving of perhaps two and a half tons might be looked for from these items. Add to this three tons increase of lift from letting out the restraining wires and there seemed a prospect of bringing the disposable lift to a little over 38 tons. That would make possible the uplift of about 18 tons of fuel. It might be just enough.

By now the weather was on the mend. The moment was at hand for R.101 to be taken from its shed. The Permit to Fly had arrived from the Air Ministry on Tuesday, 8th October, and a preliminary warning was sent out for the handling party to be ready for 0600 hours next morning. By evening the weather forecast was for increased wind – so at 21.30 the handling party was 'stood down' until the same time on 10th October, only to be cancelled again at 0230 hours.

Uncomfortably high as the wind was for precious adolescent airships it was not such as to prevent the Secretary of State for Air from landing at Cardington on the morning of 10th October on his way to the opening by HRH Prince George of Hull's new Civic Aerodrome at Heedon – the first Municipal Aerodrome in the British Isles, the result of civic initiative by Councillor Benno Pearlman, the Lord Mayor of Kingston-upon-Hull.

Lord Thomson arrived at Cardington in the forenoon, accompanied by Sir Sefton Brancker in a Handley Page W.8 of Imperial Airways; he inspected R.101 in its shed and then flew on to Heedon.

Strong westerly winds continued for another twenty-four hours until, on the afternoon of Friday, 11th October, the wind began to drop and a calm, warm day was forecast for the morrow. Accordingly, at 1530 the handling party was warned to be ready for action at 05.00 hours on Saturday, 12th October.

The ship was being alternately ballasted up and lightened again by piping on board and then discharging a ton of water at a time, as night – with its increase of lift in the shed – followed day when the ship lost buoyancy. With the promising forecast for the morning and so as to be ready for an early start, Atherstone, Johnston and Steff slept on board R.101 that night.

At 0400 hours they turned out to start the normal airship routine of 'juggling with water ballast' so that the ship would be about four tons light to lift a crew of 36 with one ton of lift to spare.

Was it an omen that 1930 days from the start of the project, R.101 was ready for launch?

CHAPTER SIX

Launch

Until at last it came to be
For length and breadth the bigness which you see

JOHN BUNYAN, 1628–1668, born at Eltham, near Cardington.

12th October, 1929

A FAINT GLOW of dawn gave just sufficient light to show up the outline of the two great airship sheds against the eastern sky as an expectant crowd lined the Bedford–Shefford road, waiting for R.101 to appear. The westerly wind was blowing at a gentle four miles an hour at ground level and at 12 miles an hour at 250 feet. Ideally, the breeze was on to the shed doors, facing the road and nearly along the north-east south-west axis of the sheds.

By 0515 hours the great four-ton doors of Number One shed slowly opened. A bath of light seemed to immerse the silver curves of the nose of the airship lurking inside.

Three parallel lines had been painted on the grass. White bags hung as pointers from nose and stern of the ship and were to be kept along the middle line; those on each side were to be followed by the handling party so that the axis of the ship could be kept parallel with the sides of the shed.

The 400-strong handling party had been regaled with cups of tea since 5 a.m. The party was made up of 200 men from the construction team at the Royal Airship Works, 150 RAF men from Halton and fifty unemployed from Bedford Labour Exchange.

Noël Atherstone wrote in his diary for 12th October 1929:

> The crew were ordered on board at 0515 and went to their landing stations. Irwin and I were the only officers on board; Johnston and Steff being in charge of the stern and bow respectively, while Scott took charge of the whole outfit. The ship began to move out of the shed at 0545, on the order 'walk the ship ahead' – dead on time – and took 3½ minutes to get clear of the shed. There was a wind of, now, about nine mph blowing more or less into the mouth of the shed and I felt distinctly nervous about the stern getting damaged while clearing the doors but my fears were groundless.
>
> As we cleared the shed the assembled onlookers raised a very fine

cheer which was most comforting. The ship was then walked more or less sideways to port until we got to leeward of the mooring tower. During this amble, crabwise, over the aerodrome the ship got very light aft and much valving was necessary. Having got to the mooring point I went forward to work cables and Scott got into the control car and took over from Irwin. I lowered and paid out to the full length the ship's main and yaw guy wires which were then shackled to their respective wires from the mooring tower. The trail rope and standing guys were secured to the ship and she was then let up on these until the strain was taken by the main wires. At about the height of the mooring tower the ship hit the inversion* and it took the discharge of two tons of water to persuade her through it.

By this time the ship was on all three wires and she was about 600 feet up and rode absolutely steadily. She was then hauled down and made contact with the mooring arm very gently. In a few minutes she was secured and all service connections made with the tower. The balance weights were hooked on to the wires which were dropped from the ship but almost at once started to give trouble. Wires kept on carrying away and as each one broke it was replaced by a heavier one. It was also seen that the weights would have to be modified to be of any use. Altogether things were pretty hectic for several hours but the panic gradually subsided. Luckily the weather remained calm or there would have been no weights left in a very few minutes.

Irwin and I have decided to keep watch together until we have worked the bugs out of things and I have gained experience. Johnston and Steff are working together as our opposite numbers. I was on watch in spasms from 0800 to 1700 with relief for meals. During the afternoon engines were run and functioned all right. Water ballasting arrangements work well but the discharge from the tanks is too slow. Went home for dinner and got back to the ship about 2100 hours and turned in, very tired but happy about the ship as she promises well.

C. G. Colebrook, Aeronautical Correspondent of *The Times*, saw it a little differently though no less favourably. He wrote from Cardington on 13th October 1929:

The predominant impression made by the State airship R.101 as she now swings at the masthead at the Royal Airship Works, silver in colour and imposing in her size and sweeping lines, is that she looks right and in a way in which the elongated-cigar ships of the war period never did.

The long parallel sections of those airships made them appear weak but the fat curves of the R.101 are insistent with the strength associated with the beam girder of the engineer and she should be a

* A layer of warmer air lying at about 200 feet on top of the cold ground layer. The warm air reduced the buoyancy of the ship.

> stout and serviceable ship. She carried with her when, at dawn, she came smoothly out of her shed, the hopes of four years of patient work and the obvious enthusiasm of several hundreds of men and women who have contributed to her construction. The cheer which came from the scattered groups as the tail swung clear of the shed, and again as the nose was clamped home on the arm of the tower, expressed a pent-up emotion entirely unsuspected and was rather a fine tribute to the enthusiasm inspired in all ranks by the little band of technical officers at the Royal Airships Works originally headed by Group Captain P. F. M. Fellowes and now led by Squadron Leader R. A. B. Colmore.
>
> With a punctuality not always associated with events in aviation, at 5.49 a.m. before the sun had risen the silver nose of the airship moved silently out of the huge grey bulk of the hangar and by its size made the handling parties underneath appear like a swarm of busy ants.
>
> Megaphones and whistles controlled the many groups of workmen holding down the central car, the aft car and the handling bars and guy ropes as the R.101 was led, floating as lightly as thistledown, over the mile of aerodrome to the mooring tower. There seemed no effort in the process of walking the ship and it was only the head of a winch attendant [Atherstone] peering out of a nose hatch 80 feet up and a diminutive figure standing up in an engine car which restored a sense of proportion and gave an air of solid reality to the moving mass – the largest airship afloat in the world today.

Next day – Sunday, 13 October – Atherstone recorded that

> Things are shaping very well and the hands are beginning to shake down very quickly considering that only a nucleus have been in the air before. The ship was let up on the wires to about 900 feet this afternoon to test out the mooring gear. When we were up we could see that the main road to London was literally jammed with cars as far as the eye could reach and this state of affairs persisted until long after dark. The mooring gear works very well and wholly according to plan, which is most gratifying.

That morning, Sir Hugh Trenchard accompanied by Lady Trenchard and their son, visited Cardington and spent forty-five minutes on board the airship and had coffee served to them in the lounge. During the afternoon Sir Sefton Brancker inspected the ship and was on board when – 'gleaming silver in the setting sun' – it was allowed to rise above the masthead like a salmon on a line.

On Monday, 14th October, R.101 left the mast on the first of its dozen flights in what seemed as a happy omen, a light breeze and a warm and sunny atmosphere. In the words of *The Times*:

> Her designers and crew could have wished for no better send-off than the widespread interest and admiration aroused by her serene and majestic appearance at every stage of her course over the counties of Bedford, Herts and Middlesex, in conditions as pleasant for the spectators on land as they were favourable for the trying out of the latest product of British skill and enterprise.

R.101 slipped from the Cardington mast at 1117 hours, circled Bedford, flew back over Cardington and then set course for London. The ship passed over Buckingham Palace, Westminster, St Paul's and the City, flying at 1,500 feet, and then headed back to Cardington and was locked back at the mast at 1650 hours; a flight of 5 hours 40 minutes for 108 miles covered.

The forty-eight persons on board under Herbert Scott's command included the four officers and thirty-four crew together with fourteen official passengers headed by Sir John Higgins (AMSR), and Air Commodore Fellowes (the former DAD), together with Colmore, Richmond, Rope, Cave-Browne-Cave, and Giblett (Chief Met Officer at Cardington), Messrs Collins, Fitzgerald, Leech and Wells of the Cardington staff, Lieutenant-Commander Wann, former Captain of R.38, Squadron Leader Booth, Captain-designate of R.100 and Flight Lieutenant Durrant, a wireless expert at the Air Ministry.

The Times wrote cheerfully:

> Altogether R.101 seems to have shown triumphantly that she possesses all the good qualities claimed for her before she was taken out of the shed. In comfort and in the extent of her accommodation as well as in safety from the employment of furnace-oil instead of petrol as the source of her motive power (at one fifth of the cost) she has a marked superiority over any of her predecessors. Her passengers in yesterday's voyage of five hours testify that there is no vibration and, in the sleeping berths, less noise than in a ship at sea. According to Major Scott 'everything went off extraordinarily well. The ship handled very well and answered her controls with the greatest ease.'
>
> In what is likely to be a new era in airship construction R.101 has made a good beginning. Her designers and constructors have worked well and wisely in determining, even at some possible sacrifice of speed and carrying power, to put before all other considerations the indispensable and much abused attribute of Safety First.

Noël Atherstone was equally pleased.

> I couldn't feel the engines start or hear them from the winch platform. The first thing which impressed me about the ship was the almost

LOG OF INITIAL TEST FLIGHT OF HMA R.101
Monday, 14th October 1929

Slipped from Mast: 11.17 hrs. Secured: 16.57 hrs. Duration 5 hrs. 40 mins.
Distance flown: 108 statute miles (Cardington–London and return).
Gassed to 97.5 per cent full. Mean gas purity 97 per cent.
Volumetric capacity used: 4,773,176 cubic feet.
Barometer: 1,028.6 mbs. Temperature 56 degrees F. Gross lift 144.9 tons.
Weather: Cloudy. Slight Mist. Wind 15/20 m.p.h. from 260 degrees.
Water ballast at start: 13.5 tons. Water ballast at end of flight 8.5 tons.

DISPOSABLE LIFT

At departure	Tons	Lb.	Imp. Gallons
Fuel oil	11.39	25,513	3055 (at 8.35 lb/gall.)
Lub. oil	0.996	2,232	248 (at 9.0 lb/gall.)
Petrol (starting)	0.25	563	75 (at 7.5 lb/gall.)
Water ballast	10.00	22,400	224 (at 10.0 lb/gall.)
Crew (39)	2.78	6,240	(at 160 lb each)
Passengers (14)	1.00	2,240	(at 160 lb each)
Food	0.36	803	(at 15 lb per person)
Drinking water	0.36	800	(at 10.0 lb/gall.)
Spares	0.13	300	
Miscellaneous	0.54	1,200	
Useful load	27.806	62,291	3,682

Journey Log:

		Speed knots	Height feet
Slipped, Cardington Tower 11.17 hrs.			
11.30 hrs:	Midwing engines 700 rpm ½ speed	30	900
11.50 hrs:	Midwing engines 810 rpm	38	1,300
12.50 hrs:	Leighton Buzzard	45	1,500
1330–1335:	Over London	49	1,500
14.00:	Set course for Cardington	34	1,500
16.25:	Cardington. Stopped all engines		
16.57	Secured to Mast		

Approximate consumptions: Fuel used 250 gallons (2.087 lb) Fuel remaining 2,805 gallons (23,206 lb) Petrol used 15 gallons (512 lb) Lub. Oil used: 12 gallons (108 lb) Water ballast used: 5.0 tons.

complete absence of vibration and no creaking. Speed was down to 40 knots at first with No 3 and 4 (midships) engines only running. No 5 (aft) engine was started up later and speed increased to about 49 knots (56 mph) with three engines at cruising speed. No 5 engine had to be stopped owing to the breaking of a water pipe, so we carried on with No 3 and 4 engines only.

For the first hour or so of the flight I was busy going all round the ship looking for trouble but I couldn't find any in the hull, cover or gas bags. In fact the cover is wonderfully good and doesn't move anywhere. The whole ship feels immensely strong and gives one a wonderful feeling of security and confidence.

We passed over London during the flight and created immense interest. I took the steering and elevator wheels in turn and found the ship answered very well indeed to both controls and that very little help and practically no effort were required to keep her steady for direction and height. The air was quite bumpy but we really felt only two bumps which made the ship pitch slowly about three degrees up and down. A good landing was made at 1650 hours and the ship was secured to the tower without incident.

The Times remarked: 'A more perfect operation could not be desired – a happy augury for the future.'

CHAPTER SEVEN

The First Commission (Seven Flights, 1929)

'To travel hopefully is a better thing than to arrive and the true success is to labour.'
ROBERT LOUIS STEVENSON (1850–1894) – *Virginibus Puerisque*

14th October, 1929

AMONG the enthusiastic crowds who watched R.101 fly over London on 14th October 1929, was the author of it all, the Secretary of State for Air.

Lord Thomson, with Sir Sefton Brancker and Christopher Bullock, went up to the roof of Gwydyr House at one-thirty before going to lunch in the House. At last a dream had come true. The airship, elegant of form and droning serenely not more than twice its own length above Westminster and Whitehall, was a magnificent sight.

Back from lunch, Thomson sent off a telegram of congratulation to await Colmore on his return to Cardington. It was – he said – the start of a new era for British airships and Empire ties.

Impatient to take the air in R.101 himself, he followed up his telegram with a note to Sir John Higgins who was on board for the first flight – a note clearly dictated 'thinking aloud'.

AMSR

I imagine you have no intention of taking any 'outsiders' on the R.101 at present. During these first flights if you should contemplate taking anyone I should be obliged if you would first refer the proposal to me.

I myself should like to make a flight as early as possible. I will take Mr Bullock with me. Probably US of S would like to come as well.

If it is possible for me to make this flight this week or early next before the House meets, when I shall be very fully engaged, it would be preferable.

I cannot manage Wednesdays as Cabinets are normally held on this day and I also have an important committee next Monday which I must attend.

Then, with characteristic generosity to his predecessor:

When I have made a flight I would like the previous Secretary of State, Sir Samuel Hoare, to be given the opportunity of making one before any other outsiders are taken up. I know that Sir Samuel Hoare would like to be given as long notice as possible in advance.

I am glad to see that already Press comment is becoming much more favourable.

I think it would have a very good effect all round if it is possible for me to announce in the very early future and before hostile critics start asking questions in the House about weights, that it is proposed at an early date to give up to 100 Members of Parliament an opportunity for making a flight.

I think it more than probable that nothing like 100 will come forward though I have received a good many applications, but the effect of offering to take up 100 simultaneously – an offer which will undoubtedly receive wide publicity – ought to do a great deal to disarm potential critics.

I shall further, in all probability, suggest to some of my colleagues in the Cabinet – particularly the Prime Minister and the Chancellor of the Exchequer – that they should make a flight. This, however, I will discuss with you later on.

T.

Here was Thomson, in the first flush of enthusiasm at seeing R.101 in the air, proposing a series of passenger flights for propaganda purposes before the trials had really begun. Yet it was very much in the tradition of the times in which Handley Page's first W.8 air liner had carried passengers from Croydon to Paris on its third flight· in a tradition in which the *Graf Zeppelin* had crossed the Atlantic with fare-paying passengers on its seventh flight and in which R.101 had been flown over London with most of the senior personnel from the Royal Airship Works on board, together with top officials from the Air Ministry, on the very first occasion on which the airship had left the Cardington mast. It was asking for trouble – but wholly in the character of the day.

The Secretary of State had, of course, on his mind the potential criticism, (earlier a feature of the, now generally friendly, Press) that R.101 was substantially overweight – as Richmond had underlined to him at Cardington in June. Meanwhile he wanted to see for himself.

On the following Friday he had his wish. On 18th October he was at Cardington by 07.30 in the morning accompanied by Christopher Bullock and Louis Reynolds from his office. They went up in the lift of the mooring tower escorted by Colmore, Richmond, Cave-Browne-Cave, Rope and by Wing Commander Pressey from the Royal Canadian Air Force who was to be in charge of the mooring mast at

St Hubert, Montreal, for next year's visits planned for the two new British airships.

In all twelve passengers were on board by 0805 in the morning. The ship was slipped from the tower at 0812. Course was set up wind which took R.101 over Olney (0850), Northampton (0911), Coventry (1013) and Birmingham (1040) where the ship circled for some time on turning trials at 1,500 feet. R.101 was on course again at 1111 to reach Lichfield at 1125 and Nottingham at 1200.

At this point the condenser pump failed in the steam cooling system so that a great deal of engine cooling water began to be lost. Course was accordingly set back to Cardington by way of Newark (1227) and Leicester (1322), reaching Bedford at 1400. Meanwhile an excellent hot lunch was served in the dining room and afterwards the smoking room below was used for the first time for cigars over coffee.

Also, for the first time all four engines available for forward flight were run together at cruising speed, at 810 rpm to give 450 bhp each. A level speed of 55 knots (63 mph) was recorded – precisely the designed speed for this power – 1,800 bhp.

R.101 arrived back over Cardington at 1430 hours. The airship was not, however, locked on the mast for another 2 hours and 20 minutes, at 17.50 hours, after 9 hours 38 minutes off the tower. The reason for this delay was that, as became increasingly obvious as time went on, Herbert Scott not only insisted on taking over for the landing from the ship's Captain, Carmichael Irwin, but also made a mess of the job. First, he upset the trim of the ship by the unnecessary discharge of water ballast and then allowed R.101 to over-ride the tower so that the side guy-wires fouled the main wire. Mooring was completed in the last few minutes of daylight and the passengers, unaware of the problems, 'went ashore'.

Irwin and Atherstone were thoroughly upset by this display of ham-handedness on the part of Scott, in keeping with his record of damage to R.33 and R.36 during his earlier landing operations. It was the first evidence of discord among the team.

Atherstone was, nonetheless, delighted with R.101 itself.

> I am more than ever impressed with the feel of the ship [he wrote] and she is behaving like a ship and doing things I would expect a ship on the water to do. She feels confident, strong and alive.

R.101 had covered 210 miles, excluding the turning trials, in 7 hours and 18 minutes on course. Some 460 gallons of fuel had been used out of 3,145 gallons on board at the start. It all augured well.

For 18th October, HMA R.101's Flight Log recorded:

SECOND TRIAL FLIGHT

Friday, 18th October 1929

Slipped 0812. Secured 1750. Duration: 9h. 38m.
Distance covered: 210 statute miles.
Gassed to 96 per cent full. Mean gas purity 97 per cent.
Volumetric capacity used: 4,707,426 cubic feet.
Barometer 1011.0. Temperature 45 degrees F.
Weather: Fair, then cloudy with showers. Cloud base 1500 ft. Wind 10/20 mph. Pressure 298.
Water ballast: 8 tons at start.

On departure:				
Gross lift:	142.9 tons		320,105 lb	
Fuel oil:	11.7 tons	3,145 galls.	26,296 lb	(at 8.36 lb/galls.)
Lub. oil:		248 galls.	2,232 lb	(at 9.0 lb/galls.)
Petrol (starting):		75 galls.	563 lb	(at 7.5 lb/galls.)
Ballast:	8 tons	1,792 galls.	17,920 lb	
Crew:		38	6,080 lb	(at 160 lb each)
Passengers:		12	1,920 lb	(at 160 lb each)
Food:			803 lb	
Drinking water:		80 galls.	800 lb	(at 10.0 lb/galls.)
Spares:			400 lb	
Miscellaneous:			1,200 lb	
Useful load lifted:				
	26.00 tons		58,214 lb	
Fixed weight of ship:				
	113.6 tons		254,464 lb	

Fuel expended during flight	460 gallons	3,846 lb
Petrol	17½ gallons	131 lb
Lubricating oil	22 gallons	198 lb
Total	499½ gallons	4,176 lb

Ballast used: 4.5 tons; Ballast remaining 3½ tons, 7,840 lb
Fuel remaining: 10.0 tons; 2,685 galls, 22, 447 lb

The Air Minister was surrounded by the Press when he came down in the lift to the foot of the mooring mast.

'Yes,' he said, 'I have enjoyed the flight immensely.' *

* Report in *The Times* of 19th November 1929.

'I have rarely had a more pleasant experience. There is a feeling of complete detachment and security. Certainly I have never been so comfortable on any ocean vessel and the motion is comparable to that of a river steamer, only the airship is less noisy. There is room to walk about and to admire the view with always about fifty acres of countryside to be seen. We watched a pack of foxhounds running across country. I signed a good many documents and I found that an airship is a particularly congenial place in which to do work requiring study.'

He went on to say that he had done virtually a normal day's work on board, had breakfast, read the papers, chatted as easily with the ship's officers in the lounge as in his own room, had luncheon followed by coffee and cigarettes in the smoking room and tea later. The ship's heating had maintained a pleasant temperature everywhere in the passenger quarters.

Much work and many further trials remained to be completed of course. 'I want to make it perfectly clear,' he said, 'as I did when I introduced the programme in 1924, that it is all tentative and experimental. So far the experiment has amply justified itself.'

'So long as I am in charge no pressure will be brought to bear on the technical staff or on anyone else, to undertake any long distance flights until they are ready and all is completely in order. Subject to this I hope that perhaps it may be possible for me to travel to India in R.101 during the Parliamentary Recess at Christmas. But whether this is possible or not, the whole policy of the airship programme is "Safety First" – and "Safety Second" as well.'

There was certainly an air of confidence about but no doubts about the need to gain more lift and a great deal of thought was being devoted to this problem at Cardington. More lift, such weight saving as might be possible and additional masts at Malta and at Basra or Bahrein would be needed for a regular service to and from India. It was all very embarrassing but, for the moment, the basic soundness of the airship must be proved in flight.

So, the first two flights, enthusiastically received, were the start of a period of relatively intensive flying in R.101's brief career. Between 14th October and 18th November 1929 the airship completed seven test flights and 73 hours 27 minutes in the air. In addition by 30th November R.101 had spent 865 hours 14 minutes in the open at the mast. But India still seemed far away.

In public and in the Press airships were now very much a centre of interest and of debate. While R.101 was at the mast the roads around Cardington were continuously blocked with cars – there were estimates that from mid-October to mid-November more than a million people made the journey up and down the Bedford road.

In *The Aeroplane* of 23rd October 1929 Charles Grey wrote in his more caustic vein:

> Those who are responsible for R.100 and R.101 deserve to be regarded as honest, straightforward experimenters to whom all of us wish success. But the Air Ministry ought to make quite clear to the Press the fact that no sort of performance is expected from R.101 and that she will probably be wrecked in the course of the next few months, either because her engines break down and she alights out of control or because, being 25 tons over weight she cannot carry enough ballast to save herself in a forced descent.
>
> To let people think that she is going to take a hundred passengers to India in a month's time, or in three months or in six months is simply humbugging the public.
>
> The two ships both have so many original features that they must both be looked on as entirely experimental. If they perform journeys without accident the designers and the constructors and the crew will deserve full credit for their success. If they are wrecked nobody will have any cause to grumble because pure experiments must be expected to wreck themselves.
>
> But pople who argue that airships are perfectly useless and are a disgraceful waste of taxpayers' money are just as wrong as are the people who say that airships are the only possible solution to the problems of long distance air transport.

At that date the solutions were hard to find.

R.101 had been taken from the mast and returned to the shed on 21st October because of a storm forecast, the need for modification of the starting engines (to make them easier to start themselves) and a replacement required for an elevator control cable. Again Scott took charge from Irwin and, again, so confused the issues that a process which should have been simple in an almost flat calm took 2¾ hours from 1730 hours to 2015 hours. It finished in the pitch dark – fortunately without damage.

The work on R.101 was completed by 31st October and the ship taken out again at 0540 on Friday, 1st November. It was locked on the tower head at 0745 hours and slipped for its third test flight at 0940 hours. Sir Samuel Hoare, Lord Thomson's predecessor at the Air Ministry, was on board by express invitation of the Secretary of State for Air, accompanied by fourteen other passengers and a crew of thirty-nine to make fifty-four in all.

R.101 left the tower with 3,890 gallons (32,520 lb) of fuel on board and a useful load of 34.3 tons out of a gross lift of 148.2 tons.

The airship flew north-east and circled low over Sandringham at 1133 where Their Majesties King George V and Queen Mary came out on the gravel path to watch and wave; they could be seen clearly from the air.

Thereafter R.101 circled back over King's Lynn and then up to Cromer and on over Overstrand and Sidestrand before turning to fly over Sir Samuel Hoare's home at Templewood, Northrepps, just south-east of the town. Thence, south to Norwich where R.101 flew over the Boulton and Paul Aircraft Works so that the staff there could admire their handiwork. From Norwich to Thetford, Newmarket and Cambridge and back to Bedford.

R.101 was overhead Cardington at 1500 hours and locked home at the mast after 7 hours 15 minutes in the air; 5 hours 20 minutes on course and a flying distance covered of about 260 statute miles.*

On the way the ship had made a short maximum continuous speed run on four engines at 890 rpm and 585 bhp each. The indicated air speed had been confirmed at 59.5 knots, or 68.5 mph.

Noël Atherstone commented in his diary:

> Sir Samuel Hoare seemed very pleased with his experience of airship flying. The weather was beautifully fine with the remains of fog, floating in woolly masses at about 1,000 feet. The ship was flown at 1,500 feet so we were in the nice calm air above the inversion.
>
> The landing was done by Irwin but it was very nearly ruined by 'F.T.' [Scott] who let go ballast from 12 and 13 frames unknown to Irwin. This let the stern go and the ship would have dived on to the ground if she had not been so light. The tower took fifty minutes to haul us down because, I found later, the winch would not heave round fast enough as the steam engine had not been properly drained and had water in the cylinders.

Apart from the run up to the tower the flight had been a great success. The engines had run well and there was no leakage from the gas-bags; the ship landing light.

A longer time in the air was scheduled for the fourth test flight on the following day. By then an 'Air Log' – designed by Michael Rope – had been sent from the Royal Aircraft Establishment at Farnborough and installed in R.101 to give a more accurate measurement of true air speed and a check upon the ship's airspeed indicators. The Air

* Appendix 11: Flights of R.101 and AIR 3/63.

Log was to be lowered on a wire beneath the airship to measure the air speed free of turbulence around the airship's hull.

R.101 slipped from the mast at 2012 hours on the evening of Saturday, 2nd November, carrying a crew of thirty-nine, together with twelve passengers and 4,220 gallons (35,295 lb) of fuel. The useful load was 33.5 tons and under the prevailing conditions the gross lift 147 tons.

There were minor problems from the start. Noël Atherstone noted in his diary:

> The gas bags were a little too full and we got to pressure height at 1,000 feet and the ship became tail heavy – as usual. A little later we went to 1,500 feet and she got heavy over all as well. At 45 knots air speed she was carrying 15 degrees down helm and was pitched five degrees down by the tail so things were not so good for a few hours.
>
> London looked wonderful but we missed the West End as we made rather a bad shot and hit the place too far east.

R.101 entered cloud at 1,500 feet just south of Croydon Airport and those in the control car did not glimpse the ground again until a blaze of brilliant lights marked Portsmouth Harbour and the edge of the cloud cover.

The chief purpose of the flight was to make a series of runs over a triangular course of forty-three miles. The speed course ran from the Needles to Durlston Head, then south of Swanage, on to Poole and back to the Needles. These places were all easy check-points, even in the dark.

Flying south from overhead Portsmouth, R.101 crossed the Solent to Newport on the Medina River in the Isle of Wight. The airship flew on to turn west at St Catherine's Point on the south coast of the island. By 2140 hours the ship was on course from the Needles to Durlston Head, seventeen miles to the west, flying on four engines at 825 rpm delivering 475 bhp each. The speed was checked on the Air Log at 55 knots (63 mph).

As R.101 turned north towards Poole, to start along the second leg of eight miles, the aft (No 5) engine had to be shut down because of a broken water-pipe in its cooling system. The run to the Needles was continued on three engines. A speed of 51 knots (58.7 mph) was recorded.*

Over Poole No 5 engine was restarted after a temporary repair with rubber hose and the ship flew back to Durlston Head to make a run

* AIR 3/63.

from there to the Needles in the opposite direction to the first run. Halfway along the leg, however, No 3 engine had to be stopped because of yet another cracked aluminium water-pipe. Water from the steam-cooling system began to leak seriously.

Irwin turned R.101 out over the Channel to cruise slowly in the smooth air over the sea while repairs were attempted. By 0530 hours the ship was back off the Needles heading for Southampton having abandoned further speed runs. As dawn came R.101 was overhead Reading in turbulent air at 1,000 feet, just on top of the temperature inversion. An increase in height to 1,500 feet brought the ship into calm air and clear skies with early sunlight picking up the autumn tints in the trees.

Cardington was in sight at 0900 hours on 3rd November and Irwin quickly and expertly linked up with the mooring cable. A rapid coupling seemed in prospect until Scott intervened – and another mix-up occurred. It ended with the ship running on to the mast, with minor damage to one of the reefing booms at the bows. By the time R.101 was secured at 1014 hours the airship had been in the air for 14 hours and two minutes – the longest flight yet. The airship had been 13 hours on course and it had covered about 380 statute miles. Fuel consumption was 814 gallons.

Repairs were set in hand at the tower. A spare reefing girder was installed while the defective aluminium water-pipes were replaced by copper pipes. That satisfactorily solved that problem. The problems of Scott's impulsive actions and poor judgement were, however, a good deal more intractable. They were to cause growing concern to Irwin and Atherstone.

R.101 had now completed four flights for a total of 36 hours and 35 minutes in the air. Not enough data had, as yet, been gained to make any reliable estimates of fuel consumption although the cruising speed on four engines seemed to be up to expectations. A reversing gear was badly needed so that the ship would no longer be limited to 80 per cent of its available horse-power, nor have to carry around 3½ tons of useless weight and drag solely to provide reverse thrust near the mast.

The fair-weather flying so far, had shown that the airship was stable, though somewhat tail heavy because of the greater than expected weights of the tail surfaces (the fins, rudders and elevators) and their servo operating gear. R.101 answered the controls well – lighter on the rudder than on the elevators. Though the engines had run satisfactorily, the water cooling system had been unreliable – a problem that had not been difficult to cure through the substitution of copper for aluminium pipes. The chief difficulty remained inadequate lift.

Another fact was also clear, at least to the flying staff. Four officers were not enough for a flight programme interspersed by days and nights moored at the mast. During all that time one of the four officers had to be in charge during successive four-hour watches. The result was that by early November they were all suffering from fatigue. It was beginning to tell in their attitudes to the work and to each other.

Officials at the Air Ministry remote from the flying base – chiefly Sir John Higgins and Louis Reynolds in his office – were, however, primarily occupied at this time in attempting to create the right atmosphere for the airship programme as a whole, both in Parliament and among the Press. Substantial public funds had been expended on the new airship programme over the course of the past five years. There was an urgent need to demonstrate that the sums had been wisely spent and now there was a prospect of achieving value for money. Hence pressure on Cardington to superimpose upon the test flying programme the 'joy-riding' of a steady stream of important passengers.

An example of this ill-judged policy came on Friday, 8th November, when, after a night of heavy rain, a flight was arranged for the Mayor and Deputy Mayor of Bedford together with members of the Aeronautical Research Council and officials of the Air Ministry, the Admiralty, the National Physical Laboratory, the Post Office, the Royal Aircraft Establishment and the Treasury; forty in all. These, with forty-four crew, brought the live-load up to eighty-four persons; the greatest number flown by any British aircraft up to that time – and not exceeded for more than twenty-five years.*

The need for more experience and more experiment was rubbed in on this, R.101's fifth test flight, as was the fact that both airship and mooring tower were prototypes still under test.

An ingenious system had been devised in the nose of the airship and at the mast-head whereby both the load on the coupling, and the amount by which the ship was 'heavy' or 'light', was indicated on instruments in the airship's control-car and at the mast head.

On the morning of 8th November this Lift and Trim device showed that the ship was heavy because of the weight of rain on the cover and because the barometer was low. Consequently two tons of fuel had

* The first British aircraft to carry more than 84 persons after this flight of R.101 was a Bristol Britannia turbo-prop airliner which lifted 98 people on 9th September 1955.

to be offloaded and R.101 slipped from the tower at 1347 hours with the 84 persons on board together with 11 tons of fuel, eight tons of water and 'a pressure height of 500 feet' – which meant that the bags had been filled to 98 per cent and that gas would have to be valved at that height, and then ballast dropped, to maintain equilibrium.

In this state, R.101, in Atherstone's words:

> staggered round the vicinity of Bedford and Market Harborough for a couple of hours on three engines and then came in to land at 1615. The landing was done by Irwin with his usual skill. I gathered that the passengers were very pleased with what they saw and enjoyed the trip.

but it was clearly a flight which ought not to have been made and invited disaster. The ship was locked home at the mast at 1651 hours after three hours and forty minutes away on its fifth flight.

On Sunday, 10th November, with the ship at the mast, the wind began to rise rapidly and a whole gale was forecast. 'Storm Routine' was put into operation at the mast head. By the following morning, 11th November, the eleventh Armistice Day, the wind was gusting at up to 60 miles an hour accompanied by heavy rain and squalls. Noël Atherstone noted:

> As there was no place for the ship to go to get out of the wind it was decided to remain at the tower till torn off. By noon the wind was averaging 45 to 50 mph with gusts over 60 mph, it was also raining and becoming squally. The ship was remarkably steady.

The peak of the storm came during the Monday afternoon. At 1507 the wind reached 75 mph; at 1539 it touched 81 mph and at 1600 it registered 83 mph. Then it began to swing and in a cold front line squall, with torrential rain and hail, the wind changed direction through 135 degrees in the space of little more than a minute while gusting up to 89 mph accompanied by a drop in temperature of ten degrees from 56 degrees F to 46 degrees F. The airship swung with the wind and, throughout all these extreme conditions 'the ship rode comfortably at the tower without violent movement,' wrote Colmore.

'At the bow everything was snug and secure,' noted Atherstone in his diary. 'The ship rolled up to five degrees occasionally but the motion felt natural and not in the least laboured.'

During all this time the greatest horizontal force or 'pull' on the bow had measured 15.6 tons; well within the design limit of 30 tons. By 0740 hours next morning the wind had died down to 20 miles an hour and continued to decrease. 'Storm Routine' was officially washed out.

R.101 had ridden out a storm which had scattered shipping and had caused wide-spread damage on land. No airship had encountered such high winds or such torrents of rain in the open before. There was much to be pleased about – but two disturbing points.

First, the rain had come in badly at Frame 7 and at a number of other places where the cover was imperfectly sealed and especially where wires passed through. A good deal of water found its way into the passenger accommodation.

That was bad enough, though relatively easy to remedy. Much worse was that the rolling of the ship, and the consequential surging of the gas-bags, had caused the bags to chafe on the sharp corners of the radial frames. This chafing had caused a great many holes to be rubbed in the bags which, up till now, had held gas well.

R.101 had to be gassed around midnight on 11th November and again during the following afternoon 'to keep it afloat' while the holes were repaired at the mast. It was a task which occupied 48 hours. The serious thing was that, in spite of the amount of clearance provided between the bags and the structure by Rope's ingenious 'parachute wiring', in really turbulent air the surging of the bags brought them in rubbing contact with the radial girders.

On 12th November R.101's Chief Coxswain, George W. Hunt – 'Sky Hunt', a reliable veteran of R.33 and R.36 – reported in commendable detail:

> During Storm Routine it was observed that a very considerable movement of the gas bags was taking place, from side to side as much as four inches and, at times, the surge of the bags in a fore and aft direction was considerably more.
>
> Where the bags are touching, or nearly touching, the radial struts around the main frames and longitudinals the plates on top of the radial struts rubbed and chafed the bags and in places (such as at No 8 starboard fore-end) tore the bag nine inches in a jagged tear. No 8 bag thus became deflated to 60 per cent.
>
> The holes on top of No 14 bag were caused by the bag bearing hard on the topmost girder where several nuts project. This, combined with the movement of the bag, caused punctures.

It was disturbing news. As Colmore recorded: 'The ship was at all times well under control but there was, however, a slow rolling motion at higher wind speeds, reaching a maximum of six degrees and there was a certain amount of chafing of gas bags due to roll.'

While the implications of these facts were being digested under calmer conditions, the bags were repaired and preparations were made for the sixth test flight.

The flight was intended to be another short cruise in the vicinity of Bedford to make sure that the modifications to the cooling-system and the repairs to the gas-bags had been satisfactorily completed before the flight arranged for the large number of members of both Houses of Parliament. It had been scheduled for 16th November, two days later.

Even so, ten Members of Parliament were invited for this test flight. They were all MPs who had special aviation connections. In addition there were twenty-two other passengers, all from the Air Ministry, headed by the Director of Civil Aviation, Sir Sefton Brancker who was flying in R.101 for the first time. The forty-two crew members brought up the total number of board to seventy-four; the second largest number ever to be carried in free flight by a British airship.

Unfortunately, the lift and trim indicators at the top of the mooring tower had, unknown to anyone, become badly out of adjustment as a result of the strains induced by the recent storm. The bow indications were that the ship was riding comfortably in trim some four hundredweight (448 lb) light at 1330 hours. By this time everyone was on board, together with 2,944 gallons (11 tons) of fuel and 15 tons of water ballast.

The tower indicators were, however, badly out. As soon as the nose coupling was unlocked, at 1334 hours, the bows of the ship

> shot up into the air with a sickening lurch which left R.101 tail down at an angle of 25 degrees.* There was a vast crashing of crockery in the dining saloon where the ten Members of Parliament were having lunch, and many of the passengers were liberally spattered with coffee and drinks.
>
> Only Irwin's swift call for full power from the idling engines enabled the ship to pull upwards clear of the mast while it was brought on to an even keel by the discharge of a 225 gallon emergency water ballast bag between Frames 12 and 13;

the loss of a ton out of the 15 tons of ballast on board.

* Noël Atherstone's diary for 14th November 1929.

The subsequent flight of just on two hours away from Cardington took R.101 over Hitchin, Hatfield, St Albans, Hemel Hempstead, Luton and Ampthill and so back to the mast, a distance of sixty-two miles flown sedately in fine weather, a hazy mist and practically no wind.

Sir Sefton Brancker spent most of the flight in the control car with Scott and Irwin and even took a turn at the wheel. 'How steadily she answers her helm', he remarked, 'but what a pity to have to carry that forward starboard engine just for going astern.' Its stationary propeller was glaringly obvious from the control car. Scott assured him that work was in hand on a reversing mechanism.

The flight, short as it was, had its moments. On departure from Cardington, the 'superheat' effect of the mid-November sunshine had brought the gas bags to become 100 per cent filled at the mast. Consequently gas had to be valved immediately after the ship had levelled out and R.101 was 'heavy' throughout the flight even though more water ballast was dropped.

The result was that dynamic lift had to be gained by flying pitched nose-up. Even so, Irwin made a perfect approach to the tower under very difficult conditions and at the expenditure of eight of the nine tons of water ballast by then remaining on board.

The mooring wire was joined up at 1542 but the ship was not secured until 1643 hours, after three hours nine minutes in the air. The delay came because the, still inexperienced, ground party tangled the yaw guy wires so that yet another hole was torn in the bow cover against the top of the tower.

Noël Atherstone commented: 'I was in the control car while the ship was being landed and I had a horrible feeling of nervous tension owing to the shortage of ballast and the inadequacy of the storage ballast system.'

The purpose of test flights is, of course, to reveal inadequacies and, following R.101's sixth flight, modifications were set in hand for the quicker discharge of ballast and more satisfactory means of adjusting trim.

Next day preparations were begun for the projected flight by a hundred Members of Parliament (twenty from the House of Lords and eighty from the Commons) arranged for Saturday, 16th Novembed. The lift and trim indicators at the mast head were recalibrated and the passenger accommodation in R.101 swept and polished ready for the MPs to come on board.

Throughout Friday, however, the weather went from bad to worse;

the barometer fell, the wind rose and steady rain set in. Accordingly there was a general sigh of relief at Cardington when the Parliamentary flight was cancelled early on the Saturday morning as the MPs assembled at Westminster ready to leave by coach.

Instead, the weather charts showed much more promise of clearer skies and lighter winds for at least two days ahead. Consequently the next test flight on the programme was brought forward; an endurance flight scheduled to be of not less than thirty hours and, if all went well, to be extended up to forty-eight.

During Saturday R.101 was gassed to 98 per cent and refuelled to 4,328 gallons, or 16.23 tons, together with 2,296 gallons (10.25 tons) of water ballast, 239 gallons (2,151 lb) of lubricating oil, 90 gallons (578 lb) of petrol for the starting engines and 140 gallons (0.625 tons) of drinking water. All this added up to 7,093 gallons (28.325 tons) of fuel, oil and water out of a disposable lift of some 32.66 tons. This long-awaited flight, to measure consumption and to show the capabilities of the airship for continuous day and night flying, was the more cheerfully welcomed by the crew in the knowledge that they would be living on the delectable fare and drinks brought on board for the Parliamentary flight.

On Sunday, 17th November, R.101 slipped from the mast at 1033 hours for its seventh flight and by far the longest of its career. In Noël Atherstone's words it was:

> the best tower leaving that has ever been done. Irwin was, of course, in control and when he slipped, the ship was only just light and came off the tower smoothly and gently. A half ton of emergency water ballast, released from mid-ships, gave us a bit more buoyancy and then the engines were started. We had only RAW people on board, acting as observers, and no other passengers.

There were thirty-seven crew for the endurance flight together with Cave-Browne-Cave, Rope, Collins, Giblett, Booth, Leech, Wells and McWade but not Colmore nor Richmond who had been called back at the last minute to work on a report required urgently by AMSR, Sir John Higgins, on R.101's lift and range capabilities and the modifications thought necessary to achieve adequate performance for a flight to India at the earliest possible date.

Thus, with forty-five on board, R.101 set course to the north at 1040 on 17th November, flying at 1,000 feet at 51 knots (59 mph) on four engines at 810 rpm and a total of 1,800 bhp.

The weather was fine with a light northerly wind and the prospect of fog coming down on the following morning. Atherstone summarised the flight in his diary:

> Our course was up the East Coast and we passed over Lincoln, York, Darlington and Durham which was reached at 1603 hours. I had the afternoon watch from noon, with Steff as a very able assistant and as our track lay close to Howden I deviated a little and paid the Airship Guarantee Company a visit.*
>
> York and Durham both looked beautiful and as the weather was fine it made things very enjoyable.
>
> Johnnie took the Dog Watch at 1600 hours and flew the ship over Newcastle and then out to sea and up the coast to Dunbar and North Berwick reaching Edinburgh at 1906 hours. I came on again for the First Watch at 2000 hours when the ship was over Glasgow and flew her down the Clyde. There was a bright moon and the snow on the mountains showed up well. We passed over Greenock and Gournock and then headed for Holy Island off Arran. The air was a bit bumpy as far at the 'Tail of the Bank' but after that it got much smoother.
>
> After Holy Island we hit a patch of fog but picked up Ailsa Craig all right at 2102. Then we flew to Corswall Point and from there to Belfast over where we arrived at 2203. Course was then altered southwards to Dundrom Bay and then eastwards to the Calf of Man where we arrived at 2332 and commenced turning trials.
>
> I turned over the ship to Johnnie at midnight for the Middle Watch and promptly turned in.
>
> Turning trials were carried out during the Middle Watch and when Steff and I came on again for the Morning Watch at 0400 we were just approaching Blackpool. Having turned the ship over the bright lights of this North Country pleasure resort I set course for the Skerries Light off Holyhead and got there at 0532. Course was then set for the Wish Light Vessel but an unexpected south-easterly wind drifted us a bit to the north and we actually picked up Rockabill Light at 0645. We then steered south and arrived off the entrance to Dublin Bay at 0800.
>
> Here Irwin took over for the Forenoon Watch and flew the ship in majestic sweeps over his native land and city! Up de Valera!! We wandered down the coast to Bray and gave the Kingstown–Holyhead mail boat a good start and then hooshed past her headed for the same place which we reached at 1006.
>
> We then did a tour of Anglesey and made for Chester, reaching there at 1155. When I took over at noon the weather became gradually thicker until we were flying over the thick fog by the time we had reached Stafford. From there we did not see the ground but only the

* This flight over R.100, still in its shed at Howden, was criticised as 'bad form' but it was intended as a friendly gesture.

wireless masts of Rugby sticking about 100 feet above the fog until we got somewhere south of Cardington.

We flew about in circles trying to locate our position as we got odd glimpses of the ground through holes in the fog – but it wasn't until the Chief Cox'n recognised a petrol filling station on the top of Hammer Hill that we definitely knew where we were. Shortly after this the fog thinned a little and we saw lights on the sheds and mooring tower. So we went to landing stations right away – at 1540 hours.

Irwin made a most perfect landing in very bad visibility and the mooring was carried out a little quicker than before. The ship was secured at 1714 the flight having lasted 30 hours 41 minutes.

Throughout the flight the engines ran well but the Romac fuel transfer pump packed up several times and gave a lot of trouble.

As it was Johnnie's and Steff's turns for duty Irwin and I cleared off as soon as things were made secure for the night and I took over again from Johnnie at 0800 next morning, 19th November, and kept an uneventful Forenoon Watch until relieved by Irwin.

So, in two days and one night of continuous flying in good weather, R.101 had covered 997 nautical miles (1,148 miles) not including the distance during the turning trials.* The airship had spent 29 hours on course out of 30 hours 41 minutes away from the mast. The ground speed, excluding the three hours spent on the turning trials, worked out at 38.3 knots or 44 mph against a headwind which averaged about 12 mph. Some 2,336 gallons (19,622 lb or 8.76 tons) of fuel had been consumed so far as could be measured· that is a mean consumption of about 80.55 gallons (676.6 lb) of fuel an hour. The ship had landed with 1,992 gallons of fuel remaining in the tanks, sufficient for a further 24 hours of flying at the same rate of consumption.

At the foot of the tower, Major Scott explained to the assembled newspaper reporters that the original plan of staying away from base for 48 hours had been cut short so as not to land in the dense fog forecast for the following morning.

'R.101 has behaved splendidly,' he said with conviction. 'She has finished her acceptance trials after 73 hours in the air. Any further trials will be to assemble additional data. On this flight we set out to go through certain turning trials; that is measuring rates of turn at different angles of the rudders. The airship passed these trials admirably'; – maybe with a thought back to R.38 which had broken up in the air during turning trials.

* The log of R.101's longest flight is set out in Appendix 8. The distance covered was recorded in the ship at the time at 939.5 nautical miles which is some 57 short of the actual distance run; the difference being the result of an error in calculation between Greenock and Ailsa Craig.

'We had plenty of sleep during the flight and we enjoyed a very good meal at midday today. It consisted of soup, roast chicken and bacon, castle pudding, cheese and coffee.'

Remarking upon the chill November fog on the ground, he added: 'It is far colder here below than it was in the air. We were always very comfortable and we enjoyed wonderful views – including snow on the hills in Scotland. We flew at heights which varied between 1,000 and 1,700 feet and we certainly seemed to create quite a stir when we passed fairly low over towns where we could easily see people in the streets.

'Throughout the flight our position and compass headings were reported by wireless back to Cardington. On the whole the weather was fine except for a few rainy patches and the thickening fog this afternoon which gave us no trouble at all.'

Scott was always bland and optimistic. One of the purposes of test flights is, however, to uncover troubles so that they may be corrected and it would have been odd if the first long flight of a new airship had not revealed some deficiencies; and it had. Richmond recorded with satisfaction that the substitution of copper pipes for the original aluminium system had cured the troubles encountered during the speed trials on its fourth flight and

> no appreciable engine trouble was experienced although the Romec pump broke down several times and must now come out of the ship for thorough examination.
>
> Serious vibration was reported on the lower tongue strut on the starboard car at Frame 9 (Car 3). Something must be done to steady this. One of the main lift bridles at the bottom of Frame 12 parted just before landing. The failure was apparently caused by excessive chafing where the bridle passes over the pulley. I am having all other bridles examined.
>
> Fuel consumption appears to be surprisingly low. I am having this examined also. The ship handled well in spite of bumpy weather over Scotland.
>
> Turning trials were carried out but the yaw could not be measured as the markings in the eye-piece of the drift sight could not be seen at night.

Then, next day, 19th November; 'It was discovered that one of the main bearings of the starboard engine at Frame 9 had completely failed, all the white metal having been hammered out.'

In fact, a few more minutes of running and the No 3 engine would

[PLATE IX]
(RIGHT) *The Princess Marthe Bibesco* – Photographed during her visit to London in July 1929, Princess Marthe Bibesco (1886-1973) had met Christopher Thomson in Bucharest in March 1915, when the then Colonel C.B. Thomson was attached to the British Legation there as Military Attaché. The Princess moved in a wide literary and social circle in Roumania, France, Spain and England with a legion of devoted admirers, foremost among them, C.B.T.
(BELOW) *Opening Day* – The welcoming party for the Secretary of State of Air when Lord Thomson arrived in a D.H. Moth of the London Aeroplane Club to open the new Air Park at Heston on the afternoon of Saturday, 6th July 1929. (Left to right) Lieut-Colonel Maurice O. Darby of Rolls Royce Ltd.; Mr Nigel St. V. Norman, founder Chairman of Airwork Ltd., owner of Heston Air Park; Lieut-Colonel Mervyn O'Gorman, Vice-President of the Royal Aero Club; Mr F. Alan Muntz, Managing Director of Airwork Ltd.; Lord Thomson, Secretary of State for Air; Sir Sefton Brancker, Director of Civil Aviation; and Major Robert H. Mayo, Technical Adviser to Imperial Airways Ltd.

(TOP LEFT)
The Schneider Team – The Secretary of State for Air, Lord Thomson, with the High Speed Flight of the Royal Air Force at Calshot on 27th August 1929, eleven days before the Schneider Trophy Contest of 1929. (Left to right): Flight Lieutenant George H. Stainforth (who set up a new World Speed Record of 336 mph flying a Gloster-Napier VI on 10th September); Flying Officer H.R.D. Waghorn (who won the 1929 Schneider Contest on 7th September at 328 mph, flying a Supermarine Rolls-Royce S.6); Flight Lieutenant D. D'Arcy Greig (third in the Schneider Contest at 282 mph in a Supermarine-Napier S.5); Lord Thomson of Cardington; Squadron Leader A.H. Orlebar (who set up a World's Speed Record of 357 mph on 12th September flying a Supermarine Rolls-Royce S.6); Flying Officer R.L.R. Atcherley and Flying Officer T.H. Moon.

At Calshot – On the eve of the Schneider Trophy Contest, on 6th September 1929, the British and Italian teams at Calshot were visited by the Prime Minister and the Secretary of State for Air on the brilliantly sunny Friday afternoon. Here, they walk past the hangars housing the racing seaplanes with Calshot Castle in the background. (Left to right), Lord Thomson, Mr Ramsay MacDonald and his daughter Miss Isabel MacDonald.

Schneider Winner – The Supermarine S.6 (N.247) with a 1,900 bhp Rolls Royce R12 cylinder Vee engine, which won the Schneider Trophy Contest on 7th September 1929, flown by Flying Officer H.R.D. Waghorn, RAF.

[PLATE X]

have seized up completely with a big end bearing failure. Inspection of other bearings showed signs of trouble and repairs and modifications were started at once.

Meanwhile the log of the flight was examined in detail. Out of precisely twenty-nine hours under power, eighteen hours had been spent cruising on four engines at 810 rpm each for a total of 1,800 bhp, seven hours had been spent on three engines at 810 rpm for a total of 1,350 bhp; one hour had been on four engines at 700 rpm for 1,180 bhp and the remaining three hours had been at varying powers.

Speeds had been recorded as follows:

1. Ship trimmed neutral: four engines: 1,800 bhp. 55 knots (63 mph)
2. Ship heavy: nose up: four engines: 1,800 bhp. 52 knots (60 mph)
3. Ship heavy: nose up: three engines: 1,350 bhp. 48 knots (55 mph)
4. Ship neutral trim: four engines: 1,180 bhp. 48 knots (55 mph)

Tom Cave-Browne-Cave, in charge of R.101's machinery, worked out the fuel consumption in detail from the readings taken on each engine 'every hour on the hour' and recorded in the Engine Logs. They showed that 106 engine hours had been run during the 29 hours of flight. In all, 45,950 horse-power-hours had been generated to which a further 1,595 had to be added for low-power manoeuvering near the mast. This gave a total of 47,545 horse-power-hours in the air.*

So far as could be measured a total of 2,336 gallons of fuel had been burned, weighing 19,622 lb or 8.76 tons, at an average, therefore, of 0.413 lb per bhp hour; a very good figure indeed.

Cave-Brown-Cave calculated that during 29 hours of cruising flight R.101 had consumed some 18,963 lb of fuel at a mean True Air Speed of 51 knots (58.7 mph) on an average of 1,583 bhp. At that speed and power the mean fuel consumption was 654 lb an hour or one ton of fuel used in every 3.42 hours of flight. In other words, the 16.23 tons of fuel uplifted would have been sufficient for 55½ hours of flight to dry tanks and for a still air range of 3,220 statute miles at 58 miles an hour true air speed.

* Wing Commander T. R. Cave-Browne-Cave's analysis of the 29 hours of non-stop flying showed that the 106 engine hours could be subdivided as follows:

Engine hours		*BHP per engine*		*RPM*		*Total BHP hours*
2	at	500	at	840	=	1,000
74	at	450	at	810	=	33,300
20	at	435	at	800	=	8,700
10	at	295	at	700	=	2,950
106	at	433	at	798	=	45,950

Though these were somewhat rough and ready calculations, and depended very much upon accurate measurements of both the amount of fuel burned and the average powers and speeds maintained, they were encouraging figures which, in the end proved to be just about right.

While the results of the duration flight were being assembled and debated, the subject of the Parliamentary flight was brought up again by Major Louis Reynolds from AMSR's office at the Air Ministry. One hundred and fifty-nine Members of Parliament had put their names down as anxious to fly – more than could be accommodated – and a ballot had been held from which a hundred had been selected including twenty Peers and forty Labour MPs, thirty-one Conservatives and nine Liberals. The selected party included three women: Lady Astor, Dr Ethel Bentham and Dr Marcia Phillips. The postponed flight was rescheduled for Saturday, 23rd November.

There was dismay among the flying staff at this intention. The ship had not completed its trials, despite Scott's statement; it had not received a Certificate of Airworthiness and it had not sufficient lift to carry more than a total of about eighty passengers and crew with safety. To attempt more was asking for trouble – though at this stage the true situation was clearly not understood at the Air Ministry.

Scott, in charge of the flying programme, ought to have put his foot down. On this, as on other major issues, he showed that his judgement left much to be desired.

When Atherstone checked the available lift of R.101 on Friday, 22nd November, he found that it was about 3½ tons short of the requirement to carry 100 MPs and a reasonable amount of ballast and fuel. With the barometer going down and temperature going up things would be worse next day, including the expectation of heavy rain.

Irwin and Atherstone accordingly agreed that all fuel should be drained out of the ship except for 5½ tons in the feed tanks – sufficient for about 18 hours flying.

'If we fly tomorrow,' wrote Atherstone, 'it will be takiing an absolutely unjustifiable risk with practically nothing to gain and everything to lose.'

Next day, 23rd November, R.101 was gassed to 97 per cent, to provide a bit more lift, and all the emergency stores and parachutes were taken off the ship. The wind was rising, rain was falling and it was increasingly clear that a flight would not be possible. By 1100 hours, however, quite a crowd of Lords and Commons had congregated at the foot of the mast and at 1130 they began to come up the lift. About half way through, the lift shed a balance-weight and stuck halfway up the tower and was out of action for about three hours. So the rest of the party had to walk up 170 feet – which, gallantly, most of them did.

When all were on board there were 148 people in the ship and, but for a 45 mph wind giving dynamic lift, there would not have been sufficient buoyancy to carry them.

An excellent lunch was served in two sittings of fifty-seven each. A number of the visitors 'got drunk and were convinced that they had been taken for a longish flight' – an impression enhanced because engines were warmed up in case of need and to provide heating in the passenger accommodation.

During the second lunch-sitting a line squall passed over the aerodrome with very heavy rain and hail, while the wind veered 30 degrees and the temperature dropped four degrees. Just before the front passed the wind reached 56 miles an hour – and then began to moderate. At sea, shipping was scattered and the Saturday afternoon's sport was widely disrupted. R.101 rode comfortably at the mast.

By 1600 hours the last of the visitors had gone, well pleased with the whole affair though the ship had, fortunately, remained locked to the mast-head throughout the proceedings.

For the next few days the November weather lived up to its reputation for high winds and rain. Storm Routine was maintained on board the airship. Gas purity in the bags began to drop seriously through the alternate drenching and drying of the fabric. At the same time the long spells of watch-keeping at the tower were proving to be exhausting for the four officers who had to share the continuous manning between them; four hours on watch, followed by twelve hours off, and another four hours on, in every twenty-four hours, every day of the week – with no regular pattern of sleep.

Eventually, at 1700 hours on Saturday, 30th November, the wind dropped sufficiently for the handling party to be assembled and for the ship to be taken off the tower and housed in the shed by 1830 hours that evening.

Next day R.101 was stripped of all portable gear, rations and bedding, all of which was returned to stores ready for the start of a refit. The ship was turned over to the Works Staff under Mr. A. Gerrish. Deflation was scheduled to begin on 5th December.

So ended R.101's first commission which had lasted for fifty-four days from 7th October to 30th November 1929. During that time

R.101 had spent 938 hours and 41 minutes (39 days) at the mooring mast, 73 hours and 27 minutes (three days) away from Cardington during the course of seven flights, and twelve days in the shed. The distance flown had been 2,220 statute miles in 63 hours 20 minutes flying time at a mean ground speed of 35 mph against fairly light average headwinds. About 4,810 gallons (40,212 lb or 18 tons) of fuel had been used at a mean consumption of about 635 lb an hour or a little better than two miles to the gallon.*

The average gross lift had been 145.8 tons and the average disposable lift 31.08 tons. The greatest amount of useful load carried by the ship had been 34.3 tons. The average number of passengers on each flight had been nineteen with a maximum of forty. But a matter for some concern was that, in the fifty-four days since 7th October, a total of 4,667,380 cubic feet of top-up gas had been needed; some of it on account of the necessary valving on climbing to more than 'pressure height' but a good deal of it through leakage, especially in rough air during Storm Routine at the tower.

The rate of gas consumption had been 86,433 cubic feet a day, equivalent to the loss of 2.6 tons of lift during each twenty-four hours. On the other hand, the purity of the gas had been maintained at a satisfactory level; 97.3 per cent at inflation and 95.6 per cent on deflation on 5th and 6th December.

All-in-all there seemed good prospects that R.101 could be developed into a sound and useful ship. Most of all an additional fifteen tons of lift was needed so that the projected flight to India could be accomplished successfully with one refuelling stop at Ismailia.

The flight to India remained the primary objective, to be accomplished as soon as R. 101 could be declared fit for this first pioneering long-range flight to India and back attempted by any airship.

* Appendix 13. R.101's flights.

CHAPTER EIGHT

More Lift

Take her up tenderly;
Lift her with care.

THOMAS HOOD: 1799–1845.

18th November 1929

MODIFICATIONS to R.101 to increase its lift and to reduce its weights wherever possible, so that the India flight could be launched with real prospects of success, remained the dominant question hanging over the Airship Programme.

On 18th November 1929 Colonel Richmond recorded in his diary: 'Report to AMSR of Lift of R.101 was sent off today.' *

He and Colmore had spent much of the past weeks reviewing the situation in the light of the Lift and Trim trials and of R.101's initial performance in the air.

It was now clear that, with fixed weights amounting to 113.6 tons the disposable lift under standard conditions and at a gas purity of 97 per cent had to be accepted as 38 tons or 85,120 lb.

The minimum weight required for crew, ballast, spare parts, fittings and stores for the India flight amounted to 20 tons which would leave only 18 tons for fuel, passengers, food, etc.

With what reluctance Colmore and Richmond had therefore to write:

> It would be impossible to attempt to operate the ship on the Indian route even for demonstration flights with only this disposable lift available. At Karachi during the mid-summer months the air density is such that there would be a loss of about seven per cent of the gross lift of the airship, or say 11 tons, which would reduce the disposable lift to 27 tons. Even if crew, ballast, etc. were reduced to a minimum the ship could not leave the ground in Karachi with more than, say, seven tons of fuel and even then it would be necessary to leave at the best time of the day.

Seven tons of fuel would, on latest evidence, be sufficient for about

* Note to AMSR (Sir John Higgins) through S.9 (Major Louis Reynolds) dated 18th November 1929, signed jointly by Colmore and Richmond. (AIR2. 349/522039/29 – 06130.)

twenty-four hours of flying to dry tanks, equivalent to about 1,390 still air miles without reserves. It was less than half the safe range required.

Here was a grave situation indeed, starkly faced for the first time. From the earliest discussions way back in 1924 the purpose of a new British Airship Programme had been to mount regular commercial services between United Kingdom and the chief centres of the Empire, starting with the route to India.

Now Colmore and Richmond, who had together drawn up the original specification, had to admit that their earlier calculations had been wrong. With the power-plants available, with the cautious approach to hull-strength to ensure no repetition of the R.38 and *Shenandoah* disasters, and with the commodious passenger accommodation, airships of 5,000,000 cubic feet capacity were now seen to be too small to be able to make the long non-stop journeys required – and to cope with tropical conditions.

What could be done?

Weight must be saved and lift increased if R.101 was not to be written off as an expensive failure after five years of effort. It would be small consolation to know that enough had been learned for a new and larger ship – R.102 – to be laid down which would meet all requirements. Not for the first time in aeronautics – nor for the last – prototypes which had been designed and built 'on the frontiers of knowledge' had proved at once inadequate but pointing the way ahead. Design staffs and works engineers would understand but the problems were much more difficult to explain to lay politicians, Treasury officials, the Press and to the public at large.

Back to the drawing board.

A meeting had been held in the Cardington Drawing Office as long ago as 7th October, before the ship first flew, at which possible weight savings amounting to $5\frac{1}{2}$ tons had been discussed without taking account of any drastic changes to the passenger accommodation or any savings in the weight of the tail surfaces.

Discussions between Reginald Colmore (acting Director of Airship Development) and Sir John Higgins (AMSR) at the Air Ministry in Kingsway on 23rd October had ranged over the whole subject of lift and weights, including the possibility of replacing the cantilever monoplane tail-surfaces, weighing 4.7 tons, with much lighter braced biplane fins, rudders and elevators suggested by Mr E. F. Relf, Superintendent of the Aerodynamics Department of the National Physical Laboratory at Teddington.

Since then a schedule of possible weight-saving and lift-increasing modifications had been drawn up at Cardington. Now, an increase in disposable lift of about six tons was estimated to be possible at a cost of about £6,000 and these improvements might be completed by the middle of February 1930, if additional Drawing Office staff were authorised. Colmore and Richmond gave as their opinion that, with these alterations to R.101, a demonstration flight to India and back could be scheduled for the second half of March 1930.

A list of twenty specified items for savings in weight was drawn up, most of them of a minor nature. The more important were as follows:

		Approximate Weight-Saving
1	*Removal* of reefing girders from the bow to Frame 1 and between Frames 13 and 15	1,280 lb
2	*Deletion* of 12 out of 28 double-berth passenger cabins	1,200 lb
3	*Reduction* in weights of miscellaneous engine-accessories	1,000 lb
4	*Removal* of two water ballast tanks	900 lb
5	*Savings* in weights of gangways and ladders	700 lb
6	*Removal* of servo steering gear	680 lb
7	*Reduction* in weights of electrical system	620 lb
8	*Substitution* of plexiglass for glass in lounge windows	380 lb
9	*Reduction* of kitchen equipment	200 lb
10	*Deletion* of two lavatories	120 lb

These ten items added up to a total of 7,080 lb (3.16 tons) estimated reduction in weight.

A further valuable contribution towards increased disposable lift could be gained, according to Richmond, by letting out the gas-bag wiring so as to increase the capacity of the bags by allowing them to occupy more completely the space within the structure ring frames of the airship's hull.

At the Lift and Trim trials in the Cardington shed on 30th September 1929, R.101's sixteen gas bags had been estimated to have a volumetric capacity of 4,893,740 cubic feet, restrained to this amount by Michael Rope's elegant 'parachute' wiring. This capacity gave a lift of 148.564 tons (332,784 lb) under standard atmospheric conditions and at a gas purity of 96.15 per cent to give a lift of 68 lb per 1,000 cubic feet.

By letting out the wiring an additional 104,761 cubic feet of gas capacity could be gained worth 7,124 lb (3.18 tons) of extra lift. The increased capacity of the airship would then amount to a total of 4,998,501 cubic feet and a lift of 151.74 tons (339,898 lb) under standard conditions.*

The projected increase in lift together with the estimated savings in weight would achieve an addition to the disposable lift of 6.34 tons (14,204 lb); a valuable increase but still not enough. The original specification of 1924 had prescribed a disposable lift of 62 tons as necessary for the route to India. Even with the proposed modifications the disposable lift would still amount to only some 44 tons.

Colmore and Richmond came up with further significant suggestions to improve performance. They wrote:

> A suggestion has been made that the Beardmore Meteor heavy-oil engine should be used in place of a Tornado for astern power of the ship which would effect a saving of approximately half a ton. It is not clear yet whether the Meteor can be modified to prevent the water pump fouling the suspension struts of the power car. If this can be done the suggestion appears practical.
>
> In considering this suggestion it should be mentioned that a saving of approximately 1½ tons could be obtained if a Rolls-Royce Condor was fitted in place of a Tornado. The petrol for this engine could be limited to about three times the amount which is already carried in each power car for the Ricardo starting engines and it could be arranged for the whole of this petrol to be carried in the car. I do not know whether this would be considered as contrary to the decision in regard to the use of petrol engines in hot climates but there is no technical objection and we do not think that there would be any more risk of fire than from the existing petrol tanks in the power cars.
>
> If it is decided that petrol carried outside the hull of the ship is permissible as a temporary measure, then there is the still more attractive proposition of making use of two German Maybach airship engines which we should like to examine.
>
> Any of the above alterations in addition to saving weight has the merit of providing additional spare Tornado engines at an earlier date than those at present on order.
>
> In relation to the future operation of the ship as distinct from the

* See Appendix 6. When the lift and trim of the airship was taken on 26th June 1930, after the gas bag wiring had been let out, because the purity of the gas had been brought up, exceptionally, to 98.35 per cent the actual lift was found to be 155.206 tons; an increase of 6.6 tons compared with 30th September 1929, and 3.5 tons more than under standard conditions.

first demonstration flight to India, an important improvement could be obtained by inserting an additional bay in the structure. We do not see any design or operational objection to this.

It is considered that a net gain in lift of nine tons could be obtained by this means, the whole of which would be available for fuel and freight. This would enable the ship to operate on the England/Egypt section throughout the year carrying a useful load which would also be possible during a number of months of the year on the Egypt/India section.

If it is decided to adapt this proposal, although it could not, of course, be carried out for the first flight as now proposed, it would appear desirable to incorporate the new section in the ship as soon as possible afterwards. The earliest date at which the material could be provided and the components got ready for the assembly of the ship is considered to be June 1930, but to do this it will be necessary to provide additional staff almost immediately. If all components are ready beforehand it would not be necessary to lay up the ship for more than eight weeks.

RAW 18-11-1929. R. B. B. COLMORE.

Three days later, on 21st November 1929, Sir John Higgins addressed a long minute to the Secretary of State,* weighing up

the present position with regard to R.101 and possibilities of a flight to India in the near future.

Briefly, the position is that the disposable lift of the airship at the present moment under standard conditions is 38 tons, and that by certain alterations and improvements, including a reduction in the number of cabins to sixteen, the disposable lift can be increased to 44 tons. We have already said that useful load has been sacrificed to the inclusion of experimental features, with a view to future development, and this is a perfectly just statement of the position. The airship includes a number of features which have never appeared before in an airship and on practically all of them the weight can be saved in later airships. These features include the heavy oil engines, the large passenger lounge and smoke room, the reefing girders, the new type of gas valve to deal with an ascent at the rate of 4,000 feet a minute, the parachute type of gasbag wiring. The unbraced ring in R.101 is also entirely novel, but the Airworthiness Authority has expressly said that, as regards weight, there is little to choose between this and the braced ring of the traditional type as used in R.100.

The improvements to give this increased lift of 6 tons are set out

*Air2. 349/522039/29 – 06130.)

in detail in Enclosure 1A; the most important is the enlargement of the gasbag wiring so as to enable the gasbags to fit closer up to the outer framework of the airship than they do at present; this accounts for something over 3 tons. These alterations can hardly be carried out before the middle of February, even if RAW put in a certain amount of overtime; that means that the airship will hardly be able to get out to India before the middle of March. This is disappointing, but we must not undertake a flight of this kind without a reasonable reserve of fuel, and, on present information with regard to fuel consumption, it looks as if we should have little margin even with a disposable lift of 44 tons. You will see that of these 44 tons, 20 tons must be allowed for the crew, furniture, ballast, spares and allowance for getting to flying height, leaving 24 tons for fuel and freight. Allowing 1 ton of passengers that leaves us 23 tons of fuel, under standard atmospheric conditions. There should be no difficulty in obtaining these conditions at Cardington, but we have to reckon on a loss of lift both in Egypt and Karachi owing to the lower density of the air in those places. The figure we should reckon on for Egypt from November to April is 3 tons, and 5 to 7 tons at Karachi. Consequently, during these months the airship should be able to leave Egypt with about 20 tons of fuel, and Karachi with, say, 17 tons of fuel.

During the summer months the loss of lift at both places will generally be higher, possibly as much as 7½ tons in Egypt and 10 tons in Karachi.

I should perhaps explain why this loss of lift occurs in Egypt and Karachi. The useful lift of an airship is determined by the weight of the air she displaces less the weight of her structure, etc. The air in Egypt and India is, generally speaking, lighter than it is in England (owing mainly to differences of temperature) and consequently weighs less for a given volume. Consequently as the *volume* of air displaced by the airship must be taken as constant, the airship has less lift in hot climates. As is always the case with aircraft, this loss in lift has to be borne by the useful load, since the structure weight of the aircraft is a fixed quantity.

We have not yet got accurate figures of the fuel consumption of the Tornado engine in the air, nor do we know definitely what is the most economical speed for R.101 to cruise at. Certain data should be available as the result of the flight of the 17th/18th but the figures have not yet been worked out and may not be available for a day or two. I think, however, with the weights of fuel given at the end of paragraph 3 there should be no doubt about our being able to fly R.101 to India in selected weather conditions; and I think we may reasonably count on being able to fly back provided we avoid the most unfavourable time of the year at Karachi and Ismailia. I am therefore proposing that we have the alterations set out in 1A put in hand as soon as R.101 can be put back into the shed after the MP's flight. The only one which seems to be at all doubtful is the alteration in the heating system and this is necessitated quite as much in the interests of engine reliability as of saving weight. I shall therefore ask DDAD if they cannot arrange for some method of heating the passenger quarters if the present system is done away with.

At the same time it looks as if even with these alterations R.101 can hardly be expected to operate with any degree of regularity on the Indian route; we must therefore find other means of increasing her lift. A good deal can I hope be saved on the weight of the engines, by the use of aluminium instead of steel for the lower half of the crankcases, and in other ways which are now being investigated. If successful this should save about 2 tons. If the variable pitch propeller with solid alloy blades proves successful we shall also be able to give the airship 5 engines capable of running ahead instead of only 4. This will simplify the operation of landing and give her a greater margin of power for dealing with adverse weather conditions. It will, however, not improve matters with regard to disposable lift – indeed variable pitch propellers with their hubs are rather heavier than the normal wooden type.

We must therefore consider the question of putting an extra bay into R.101, as suggested in the penultimate paragraph of minute (1). This is estimated to increase the lift by 9 tons, and this together with the lightening of the Beardmore engines and the auxiliary engines should give a disposable lift of approximately 55 tons, or 5 tons less than was originally estimated would be required for the Indian route. The question of this extra bay is one that will have to be very carefully gone into but I do not see that it presents any insuperable difficulty, nor would it necessarily reduce the aerodynamical efficiency of the airship to any serious extent. The small German airship *Bodensee* was lengthened in this way and flew just as well afterwards.

To sum up therefore the position is that on present information certain minor alterations have got to be made to R.101, which will take some three months or so to carry out, after which she will have a disposable lift under standard atmospheric conditions of 44 tons, with which she ought to be able to fly to India and back in selected weather provided the flight is carried out before April or after September 1930. Every possible effort will be made to ensure the former date. Even then R.101 will probably not be suitable for regular operation on the Indian route, for which purpose it may be necessary to insert an additional bay into the airship, increasing her capacity to 5½ million cubic feet. This is a point which I should prefer to deal with when I put before you proposals for the next stage of airship development which I hope to be able to do in two or three weeks. In the meantime I propose to proceed with the necessary alterations to R.101 as soon after the MPs' flight as she can be replaced in the shed.

JOHN HIGGINS, AMSR

21.11.29

Lord Thomson replied on 28th November.

AMSR.

I am of the opinion that no good and quite possibly some harm might be done by a flight to India in the early months of 1930. The best course would I think be :

(a) To make the various alterations you suggest in paragraph 3 of your minute.

(b) To insert the extra bay.

(c) To make every effort for a flight with 55 tons disposable load to India and back at the end of September 1930.

28.11.29. T.

Thus 'Lord Thomson was determined to adopt the course of prudence even at the expense of further delay'. He had been bitterly disappointed – though not wholly surprised – by the facts which were now revealed to him. The possibility of a voyage to India at Christmas was clearly out of the question. He consoled himself with the knowledge that R.101, much more than R.100 which had yet to fly, had been deliberately constructed to try out new ideas 'on the frontiers of knowledge'; and that a research project must by its nature be expected to show up problems. There still seemed reasonable hope that the modifications to R.101 would give it the performance to fly to India and back successfully – and in time for the Imperial Conference of 1930.

The decision taken to set in hand the necessary improvements, Louis Reynolds in Sir John Higgins' office sought and gained from Colmore an estimate of the likely total costs of the proposed work, including the new bay and additional Drawing Office staff. The sum came to £20,000 in round figures. Reynolds set about gaining the necessary Treasury authority and drafted letters to be sent to the India Office and to Sir Geoffrey Salmond, now Air Officer Commanding, Royal Air Force, India.

Sir John Higgins followed up with a note to the Secretary of State to set the situation in perspective. It was headed :

Disposable Lift of R.101

1. You may like to have a few notes on this point in amplification of our conversation the other day.

2. As you know, the difficulty is that the fixed weights of R.101 (i.e. the framework, fixed furniture, power units, fuel and ballast systems) have come out a good deal heavier than was expected, whilst the gas capacity is rather less. The gas capacity is being increased, by the alteration to the gasbag wiring, and should then be up to the estimated figure, i.e. something over 150 tons (this does not take into account the lift of the extra bay). Certain minor alterations are also being made to reduce the fixed weight by two or three tons; but there will still be certain features in the airship which will weigh a good deal more than similar features in other airships, and as the extra weight is in all cases due to their being experimental, I think you should have some figures about them. We have always stressed the fact that R.101 is highly experimental, and justly so.

3. If we compare R.101 (5 million cubic feet) with the *Graf Zeppelin* (3¾ million cubic feet) we find that the *Graf Zeppelin* is only designed to carry 20 passengers, and provides for them only one room for eating and living in, and no facilities for smoking; whereas R.101 was designed for 100, has cabin installation at present for over 50, provides dining room for 50, a lounge big enough for 100, and a smoking room. The actual weight of the passenger accommodation now in R.101 is 11.6 tons, and it is clear that this must be much higher in proportion than that of the passenger accommodation in the *Graf Zeppelin*. It may be that we have aimed too high in R.101; if so, that is one of the things which the present programme will have taught us.

4. Another great difference in weight is to be found in the power plant. The *Graf Zeppelin* has five 560 hp Maybach engines using Blau gas and petrol, and, according to information from the Zeppelin Company, the whole weight of the power installation is rather less than 7 tons or about 6 lb per horse power. R.101 has five 585 hp Tornado engines, four of which can at present be used for thrust ahead. The total weight of the power installation, which includes auxiliary engines for starting, and generating power, is about 17.3 tons, the weight/horse power ratio for each power unit complete coming out at nearly 13 lb per horse power. This represents the price we are paying for the greater safety of the compression ignition engine with its low flashpoint fuel, and absence of magnetos, high tension leads and sparking plugs. It must be remembered that the present Maybach engine is the result of many years of intensive development, including wartime development, whereas the Tornado is the first of its kind. In the next edition of the Tornado we hope to save something between 800 and 1,000 lb per engine, or, say, something over 2 tons on the whole installation, and I am confident

that it is only a matter of time before we shall get a compression ignition engine of a lower weight still.

5. Another item on which a good deal of weight has gone in R.101 is the system for tautening the outer cover, by pushing it out with 'reefing girders'. These are small girders between the main longitudinals. In R.100 the outer cover is pulled in by means of a system of wires. It remains to be seen which method is the better and whether the reefing girder system is worth the sacrifice in weight which it entails, which is 5.7 tons as R.101 is at present and something over 5 tons after the proposed alterations are made. I am, however, sure that the experiment was worth making in any case.

6. In the above paragraphs I have only mentioned certain items which are in themselves experiments, but it must be remembered that the whole design and construction is experimental, and that weight can certainly be saved on the structure generally in another airship. This view has been expressed by the Airworthiness Authority. By taking the above items only I think it is not unreasonable to say that some 5 to 8 tons may be regarded as being due to these particular experiments.

7. Incidentally, they are features which do not figure in R.100, except in a modified degree in case of the passenger accommodation. The passenger accommodation in R.100 has been very ingeniously designed to meet contract requirements, but the designer has been compelled to resort to four-berth cabins and has not provided a large lounge and separate smoking room as in R.101. I mention this for your information, but I deprecate any public comparison in this respect if it can be avoided.

2.12.29 AMSR

India, and the projected flight there by way of Egypt, still filled the minds of the Cardington team as the objective of all their work. Colmore wrote to Louis Reynolds at the Air Ministry on 20th December:

It should be possible to add the new section to R.101 in time to carry out the first flight to India during the middle of September, but I am not quite clear as to the importance attached to the flight taking place during this month. The break up of the SW monsoon occurs about this time, generally about the middle of September, and squally conditions must be expected from then on to about the middle of November.

It would be preferable to carry out the flight either before the break of the monsoon, or say, late in November, as the conditions on the route are likely to be considerably more favourable.

From this point of view I think there is no doubt that late November would be a better time for the flight than September.

I am afraid it is impossible to promise to complete the section in time for a flight in August, that is before the monsoon breaks, even if considerable overtime is worked.

20.12.29 COLMORE

So, R.101 could be ready to fly to India by mid-September 1930 but the meteorological advice was that the flight ought to be postponed until the end of November because of the monsoon in India. That would not, however, affect the route through Europe and the Middle East; and further discussion showed that the earlier that R.101 could get away the less the weather problems – especially on the return flight across Europe. All in all, the end of September would be the most favourable date at which to aim.

Meanwhile the Secretary of State for Air pursued his six main objectives across the wide front of the Air Ministry's work: new aircraft for the Royal Air Force as Britain's First Line of Defence; new aircraft for Imperial Airways; project contracts to the chief companies in the British Aircraft Industry; new help for the Flying Clubs; new aerodromes for both military and civil aviation – and the steady advancement of the airship programme.

He had covered some of this ground on the evening of 4th November when he opened an after-dinner discussion at The Authors Club on 'Air Facts and Problems'.*

Christopher Thomson began by saying that he was convinced that air power and the development of aviation would be one of the dominant factors in a peaceful world of the future. Linked with air power was air transport – to bring peoples closer together. Governments could help by providing research and development and ground organisation and leave operations to commercial undertakings.

That was – he said – his plan for airships. As a Sapper (he did not claim to be an aeronautical engineer) he had been since 1923 a firm and unrepentant believer in Lighter-than-Air and especially as a solution to the problem of Empire communications.

The requirement was – he said – long-range, ample accommodation, a cheap fuel and a well-found craft; safe, comfortable, reliable and economic. Five years ago, after four months of sweating work on a Cabinet Committee against much opposition, he had succeeded

* *Air Facts and Problems* was the title of a book written by Lord Thomson in 1926 and published by John Murray in 1927. It included a survey of the whole aviation scene and chapters incorporating articles written by Lord Thomson for *The Chicago Daily News*, *The Observer* and *Airways* during 1925 and 1926 when he was out of office.

in getting Parliament and the Treasury to provide funds for 'The Great Experiment' – the new airship programme.

Fortunately the men for the job had been 'happened on' just when they were wanted: Colmore and Scott, Richmond and Rope, Wallis and Teed. He believed in their competence to conduct the experiment and, eventually, to achieve results even if the way might be hard.

The tests had started, he said; not yet searching but a start. The potential was there. Here at last was a new advance towards Empire communications. He had confidence in the two new airships: Richmond's R.101 under its Captain 'Bird' Irwin; soon to be joined by Wallis' R.100 under its Captain Ralph Booth.

'During my recent flight in R.101 I was able to work, free from distraction, in exhilarating surroundings and do better work than ever before. There was room to walk about; the flight was smooth and there was very little noise and no vibration.'

R.101 was an advance. But it was a prototype and still experimental. There would be bigger and better airships to come. On the whole his people at the Air Ministry were enthusiasts for heavier-than-air machines rather more than for airships. He was enthusiastic for the aeroplane too. But there was room for Lighter-than-Air as well, especially for the longer ranges and across the oceans.

'We have open minds about the future – minds open to every form of progressive development. At the same time we must have boundless energy. Those are recipes for authors too – authors who are, likewise, always experimenting.'

So – he concluded – authors and aeronauts had much in common; as he himself had found out long ago.

'The two new airships are giving me a lot of interest,' he wrote to Marthe on 22nd November from his flat at 5/6 Princes Street, Cavendish Square.

> They are my children, born of the airship programme I introduced in 1924. And now they are leaving the nursery. One ship – R.101 – has done seven flights. I transferred my 'bureau' to her on the second flight and did my day's work in the upper air.
>
> Tomorrow – if weather permits – R.101 is taking 150 people up on her eighth flight; probably the last before Christmas. It will be the largest number ever taken up in any aircraft.

But that eighth flight had to be postponed until after the airship's refit. In the meanwhile, R.101 had to be content with a record for airships of eighty-four people lifted on its fifth flight on 8th November.

With December the turn had come for R.100 from Howden to enter the scene.

CHAPTER NINE

R.100

> Here is an opportunity to combine private enterprise and State support to produce results which will lead the world.
>
> LORD THOMSON OF CARDINGTON, June 1924

1924

RIGHT FROM THE START, one of the main planks of Lord Thomson's airship programme of 1924 had been that the two new airships would be built in parallel by two different teams. The best features from each of these prototypes would be selected, after trials, to be incorporated in a series of production airships designed to operate regular scheduled services in the 1930s – plying from England to India and on to Australia, to Canada, to South Africa and, eventually, between England and the United States.

So the arrangements were completed for R.101 to be designed and built by the Royal Airship Works at Cardington and for R.100 to form the subject of a quite separate, fixed price, contract between the Air Ministry and Sir Dennistoun Burney's Airship Guarantee Company Ltd., a subsidiary of Vickers Ltd.

Both the new airships were to conform to the same broad specification – to be of 5,000,000 cubic feet capacity and to be designed to carry 100 passengers in comfort for a still air range of 3,590 statute miles non-stop in 57 hours to dry tanks flying at 63 miles an hour.

Lord Thomson had stressed in the House of Lords* that there would thus be a deliberate policy of setting up two separate design and manufacturing teams so that the nation would have the benefit of two different approaches to the difficult technical problems posed by airship development.

'If one of the first two ships crashed we would still have the experts in Government employ, we would still have the air highway and we would still have the power of employing people and giving out contracts. We would still have the determination to go on.'*

Though the requirement to be met was the same for both airships the approach was to be quite different. Whereas R.101 was intended to advance the state of the art in every worthwhile direction with bold

* Lord Thomson on 21st May 1924 (See Page 32 and Appendix 2).

innovations in structure, power-plant, gas valves and gas-bag wiring; R.100 was to follow established practice, refined in detail, as an essentially commercial craft. The airships had, however, one thing in common. Both were of a plumper streamlined form than had ever been tried before. Virtue had to be made of necessity to build airships of more than twice the capacity of any before them in sheds designed for an earlier generation of craft and therefore limited in length.

So it had come about that a contract was signed on 22nd October 1924, by Sir Trevor Dawson for Vickers Ltd and C. R. Brigstocke, Director of Contracts, for the Air Ministry, for design, construction and delivery of a commercial airship (R.100) at a price of £300,000 plus £50,000 for capital expenditure, at the former RNAS airship station at Howden. The sum of £150,000 was paid as 'launching aid' with the signature of the contract and this would be followed by progress payments as work proceeded. The intention was that £150,000 should be paid in the financial year 1925–26 in five instalments of £30,000 each against certificates of progress with a further £50,000 in 1926–27 of which £25,000 would be paid on the completion of acceptance tests and £25,000 after a successful flight to India and back.

Against all this, £50,000 would be retained by the Air Ministry until the airship had demonstrated its specified performance in all respects with a penalty clause of £1,000 for each half-mile-an-hour by which the airship failed to meet its required maximum speed of 70 mph at 5,000 feet.

An additional £10,000 would be offered towards the cost of the necessary gas installations and the supply of gas and consumable stores for acceptance trials in England; to be paid when the airship had passed its inflation tests.

Finally, the Airship Guarantee Company Ltd would have the option of buying back R.100 from the Air Ministry at half the price paid for it by the government provided that the Company established an approved British airship transport service.*

The centre of all this, the Airship Guarantee Company had been registered as a private company on 20th November 1923, with a nominal capital of £5,000 of which 2,993 shares of £1 each were issued; all held by Vickers Ltd from which Sir Trevor Dawson, Sir Herbert Lawrence and Mr Douglas Vickers assumed financial responsibility.

* AIR 8-74-06990; Cmd 2147 and AIR 5-975.

The authorised capital of the AGC Ltd was increased to £100,000 on 1st August 1924, but none of it called up until 24th August 1926, when 21,720 shares of £1 each were allocated to Vickers Ltd.

Meanwhile, on 1st July 1924, Barnes Neville Wallis was confirmed as Chief Designer to the AGC and began to collect a team at the Vickers Works at Crayford. Wallis had taken part in the design of the first Vickers Rigid, No 9, in 1914 and 1915 and No 23 and No 26 in 1915 and 1916 – and had initiated the design of R.80 at Vickers in 1917 where he had shown technical competence combined with an imaginative flair and a mental arrogance – and gained a reputation for both erudition and for being as difficult to get on with as he was unwilling to delegate.

Thomson's hope of collaboration – or, at least, co-operation – between the Cardington and AGC teams was largely stultified by Wallis who rebuffed Richmond's overtures; Richmond, the kindly autocrat who did not believe that 'he knew it all' and was ready to delegate to design collaborators of competence such as J. D. North and F. M. Rope.

The former Royal Naval Airship Station at Howden,* on the north bank of the River Ouse in Yorkshire, was bought from the Air Ministry for £40,553, as the base for R.100's construction. Howden's chief asset was the large wartime airship shed, which measured 750 feet long by 150 feet 9 inches wide and 140 feet 4 inches high. To an extent, this set the dimensions of R.100. The airship was built with an overall length of 719 feet 10 inches, a diameter of 133 feet 4 inches and a height of 136 feet 8 inches, dimensions which gave just sufficient clearances within the shed for both construction and launch.

The staff of the Airship Guarantee Company moved from Crayford to Howden on 6th April 1926 – the bleakness of the site softened by the appearance of spring. Mr James Watson was appointed Works Manager, with W. A. Dove as his assistant. In 1924 Nevil Shute Norway, then aged twenty-five, had joined from the de Havilland Aircraft Company as Chief Calculator while Mr W. Horrocks came as Chief Draughtsman.

By November 1927, after delays caused by strikes for better pay and conditions in the previous year, the structure of R.100 was beginning to be assembled at Howden with the hope that the airship

* The location of the Howden Airship Station, now cleared, is on the north side of B1228, ESE of Newsholm Farm (National Grid reference 747/327). The nearest habitation is the little village of Spaldington, about one mile to the north-east of the former RNAS airship station.

would be completed by April 1928. Already, the original intention of running on hydrogen-kerosene engines had been abandoned as impractical in the time available. The proposed power-plant was changed – 'temporarily' – to six Rolls-Royce Condor, 12-cylinder, Vee petrol engines of 650 hp each, and the contract amended accordingly. This change ruled out R.100 from any flight to India, or to tropical climes. A trial flight to Canada was substituted in the contract instead. Thereafter regular flights would be made between Cardington and Montreal in summer months and flights to Ismailia, in Egypt, during the winter when cooler conditions would exist.

In November 1927, the data given by Burney for R.100 were:

Length: 719.8 ft; *maximum diameter*: 136.7 feet.
Displacement: 5,000,000 cubic feet. *Gross lift*: 156 tons.
Fixed weights: 83 tons. *Disposable lift*: 73 tons.
Fuel capacity: 30 tons.
Payload: 24.5 tons (100 passengers plus 10 tons of mail and cargo).
Range: 3,750 statute miles in 50 hours at 75 mph.
Maximum speed: 82 mph. (71 knots).

These figures were even more optimistic than the performance being estimated for R.101 at this time. Clearly rivalry influenced slide rules.

At the same time details were given about R.100's structure which was to achieve a form with a fineness ratio of 5.27 to 1 – even tubbier than R.101.

R.100 had been designed to follow established Zeppelin practice in its general principles – 'with many novelties' and inspiration from Wallis' original approach. In particular R.100 was assembled from only eleven standard components of different size. There was an ingenious method of jointing for quick replacement of parts.

The structure was wholly of duralumin. Sixteen curved main longitudinal girders of triangular section were built from dural strip and kept in place by fifteen transverse ring-frames built up of shallow girders. The frames were braced by a system of sixteen wires to an axial girder, which ran from end to end of the ship in a tunnel through the centre of each of the fifteen gas-bags. The largest of these bags, No 7, weighed 1,580 lb and had a capacity of 551,890 cubic feet to give a gross lift of 16.7 tons. In cross-section the ship was thus a 16-sided polygon without the intermediate longitudinals used in R.101 to give a more rounded form.

R.100's fuel capacity of 10,500 Imperial gallons of petrol, weighing

36 tons, was contained in forty-two tanks of 250 gallons each. There was provision for 18 tons of water ballast in eighteen bags of one ton (224 gallons) each. Fresh water for drinking and washing was contained in four tanks holding one ton each.

Living quarters in R.100 were, like those of R.101, entirely inside the hull but arranged on three decks and rather less commodious than the accommodation of the Cardington ship.

R.100's lower deck was devoted to officers and crew with roomy and comfortable space and a larger control car than R.101. Above it, the main passenger deck consisted of a large saloon which served both as a lounge and dining room, seating fifty-six at a time, and flanked on each side by nine four-berth sleeping cabins. Passageways led from the saloon past the blocks of cabins to two promenade balconies 40 feet long, some way in from the large windows let in to the outer-cover.

Two staircases led to the upper floor made up of a gallery running round the sides of the saloon below, open in the middle and flanked with seven two-berth cabins on each side. Passages led to two additional balconies opening on to the same windows as those below. The advantages of having two passenger levels was offset by the fact that little daylight could get through into the saloon which had to be lit continuously by electricity. There was no smoking-room as in R.101. The thirty-two passenger cabins (eighteen four-berth and fourteen two-berth) provided sleeping accommodation for one hundred passengers.

R.100's six 650 bhp Rolls Royce Condor III engines – (superannuated and reconditioned after service between 1924 and 1926 in Avro 549 Aldershot III single-engine bombers of No 99 Squadron, RAF, stationed at Bircham Newton) – were installed in pairs in three power-cars; two in wing cars 130 feet behind the passenger quarters and the third power-car on the centre-line about 40 feet forward of the ventral fin.

In each power-car, the forward engine drove a tractor airscrew, 17 feet in diameter, and the aft engine a pusher propeller of 15 feet diameter. The three pusher propellers could run in reverse through massive gear-boxes, each of which weighed 952 lb, or, for the three of them, 1.257 tons.

R.100's six engines, with their cars, three Bristol starting engines and accessories, weighed, in all, 10.2 tons (22,324 lb), compared with the 17.96 tons (40,225 lb) of R.101's five-engine installation. This weight penalty of 7.76 tons (17,401 lb) incurred by R.101 was a heavy price to pay for the use of the safer fuel and, in the end, more reliable engines. The fact that, at comparable cruising speeds (63 mph) R.101 burned some 220 lb less weight of fuel an hour, only partly offset the discrepancy in engine weights. A flight of more than 75 hours would be required to even-up the additional installed weight of the power-plant. In 60 hours flying, R.100 would burn about 61,200 lb (27.3) tons of fuel against the 48,000 lb (21.14 tons) of R.101. For 60 hours flying, R.100's power-plant and fuel weighed 37.5 tons; those of R.101 a total of 39.2 tons.

The two airships were, in fact, very different solutions to the same broad requirements; R.100 largely conventional in construction. R.101 a radical departure from standard Zeppelin practice, especially in its deep, unbraced, transverse frames and its diesel engines.

By the end of 1927 the framework of R.100 was almost complete in the Howden shed, whereas at Cardington only the first of the transverse ring-frames had been delivered from Boulton and Paul at Norwich and assembled and lifted into position in the shed.

The methods of assembly and the shed organisation under R. S. Hubbard and Arthur Gerrish at Cardington were, however, a good deal better than those under John Watson and W. A. Dove at Howden where strikes also held up progress so that in the end R.101 was completed and flown three months earlier than R.100.

Inflation of R.100 started on 30th July 1929,* and was completed, except for No 15 bag at the tail by 9th August. R.100's engine trials were begun in the Howden shed on 25th September with the nose of the airship secured to the aft door while, at the other end, the shed's front doors were opened to let out the slipstream. The engine noise reverberating around the corrugated metal panels of the shed was reported as 'indescribable', while the dust thrown up by the propellers – their tips only twelve inches from the floor – created almost a fog. But all was well.

R.100's Lift and Trim trials were completed in the Howden shed on 11th November 1929. When finally worked out, like those for R.101, they showed disappointing results.

The volumetric capacity was established at 5,156,000 cubic feet, to give a standard gross lift of 156.5214 tons,† which was almost eight tons more than R.101, and, to that extent, as Michael Rope remarked, 'quite uplifting'. The 'fixed empty weight' (the structure, engines and

* Inflation of R.101's 16 gas-bags was started with No 2 bag on 11th July 1929, reached the halfway mark on 4th August and was completed (to 85 per cent capacity) on 21st August 1929 — an inflation period of 41 days.

† Calculated at 68 lb of lift per 1,000 cubic feet at 97 per cent gas purity.

THE NEW BRITISH AIRSHIPS – 1924–30

Leading statistics of R.100, R.101A and R.101B

			R.100	R.101A	R.101B
1.	Date of Lift and Trim Trials		11.11.1929	30.9.1929	26.6.1930
2.	Length, overall	ft. ins.	719′ 9½″	731′ 3″	731′ 3″
3.	Diameter (max.)	ft. ins.	133′ 4″	131′ 3″	131′ 8″
4.	Height (with bumper bag)	ft. ins.	138′ 8″	140′ 11″	140′ 11″
5.	Volumetric Capacity	cu. ft.	5,156,000	4,893,740	4,998,501
6.	Gross Lift (at 68 lb. per cu. ft.)	tons	156.521	148.564	151.740
7.	Fixed weight	tons	105.520	113.600	111.300
8.	Disposable load/lift	tons	51.001	34.600	40.440
9.	Normal maximum fuel capacity	Imp. galls.	10.500	9,184	9,184
10.	Normal maximum fuel weight	tons	35.906	34.686	34.686
11.	Water ballast tankage	Imp. galls.	4,032	3,584	3,584
12.	Water ballast capacity weight	tons	18.0	16.0	16.0
13.	Power Plant		6 x Condors	5 x Tornados	5 x Tornados
14.	Maximum power for forward flight	b.h.p.	3,900	2,600	2,600
15.	Power Plant weight installed with starting motors		10.1893	17.9575	17.2187

		R.100	R.101A	R.101B
16. Power Plant installed weight per max. horse-power lb./b.h.p.		5.8523	15.4711	14.8346
17. Fixed weight of ship less Power Plant		95.3307	95.6425	94.0813
18. Airship prepared for Service Weight (incl. Crew)		123,520 Tons	131.600	129.300
19. Weight available for fuel and payload		33,001 Tons	16.964	22.440
20. Normal cruising power	b.h.p.	1,800	1,800	1,800
21. Cruising speed at normal power	knots	55.6	53.4	53.4
22. Consumption per hour at cruise speed	lb.	1,050	745	745
23. Fuel required for 1,000 still air n. miles (no reserves)		8,430 Tons	6.228 T	6.228 T
24. Cost of fuel for 1,000 n. miles		£177.12s.0d.	£31.2s.6d.	£31.2s.6d.
25. Weight available for fuel with three tons payload		30,001 Tons	13.964 T	19.440 T
26. Absolute duration with three tons payload at 55.6 knots		64.0 hrs.	41.98 hrs.	58.45 hrs.
27. Still air range to dry tanks with three tons of payload		3,558 n.m.	2,242 n.m.	3,121 n.m.

equipment) was found to add up to 105.52 tons (236,365 lb) which – though eight tons less than R.101 with its heavy diesel engines, was 15½ tons (or 17 per cent) more than the specified figure and seven tons more than the recently estimated 98½ tons.

When the respective weights of the power-plant installations were subtracted from the basic empty weights of the two airships, there was virtually no difference in the structures and installations of the two airships.* Though R.100 had come out with three per cent more volumetric capacity, hence three per cent more lift, than R.101, the really significant difference between the two ships lay in the fact that, by meeting the requirement to install diesel engines, R.101 had accepted a penalty 7¾ tons of additional power-plant weight; an increase of 76 per cent in engine weight at a reduction of 20 per cent in available power.

To have achieved an equivalent weight of structure and installations in the highly original R.101, went a long way to vindicate Richmond, Rope and J. D. North's structural innovations. Once established, they could be substantially lightened in a repeat airship.†

By the end of November 1929, with R.101 back in the No 1 Shed at Cardington and the weather showing signs of improvement, the Cardington team of Colmore, Scott, Rope, Johnston, Steff and Giblett prepared to travel to Howden for R.100's first flight. They arrived at Howden on 15th December to join R.100's designated Captain Ralph Booth and First Officer George Meager.

On the following day, a detachment of the York and Lancaster Regiment arrived in the early hours to help to walk R.100 out of its shed. It emerged, tail first, at 07.20 hrs on 16th December. At 07.53, R.100 was 'let go' and rose into a clear morning to leave Howden for ever, under the command of Major Scott with Squadron Leader Booth, R.100's official Captain, acting on this occasion as First Officer. There were 57 on board; five officers, 38 crew and 14 passengers, with 5,500 gallons (18.85 tons) of petrol included in the disposable load of 40.62 tons.

The airship circled the village of Spaldington and the town of Howden and was back over its shed at 8.15 hrs, setting course to the north-east, over Bubwith, to circle York at 09.00 hrs, with all six engines running 'slow ahead'. A few minutes later, the No 3 (port

* See Table on page 171.

† When both airships were first flown in 1929, R.101's fixed weights less power plant came to 95.642 tons; R.100's fixed weights less power plant to 95.331 tons. Weight saving efforts reduced R.101's weight less power plant to 94.081 tons in June 1930.

TABLE OF COMPARATIVE WEIGHTS

	Air Ministry Specifications 22/24 & 24/24 1924	R.100 at 16th December, 1929	R.101A at 12th October 1929
1. Volumetric Capacity Cu. Ft.	5,000,000	5,156,00	4,893,740
2. Standard Gross Lift Tons (at 68 lb. lift per 1,000 Cu. Ft.)	151,7857	156.5214	148.5600
3. Fixed Weights Tons	90.0000	105.5200	113.6000
4. Disposable Lift Tons	61.7857	51.0014	34.9600
5. Airship Prepared for Service Weight (with Crew, Stores & Equipment tons)	110.0000	125.5200	133.6000
6. Lift available for Fuel Oil and Payload tons	41.7857	31.0014	14.9600
7. Lift Available for Fuel with 10 Tons of Payload tons	31.7857	21.0014	4.9600

Weight Analysis tons			
1. Hull Structure	NOT DETAILED	49.8434	50.3752
2. Power Plant Installation		10.1893	17.9575
3. Gas Installation — Bags Valves & Wiring		13.7070	13.7601
4. Envelope — Outer Cover & Empennage		7.8060	7.5774
5. Passenger Accommodation & Control Car		14.1659	13.9376
6. Systems & Miscellaneous Items		9.8084	9.9922
7. Total of Fixed Weights	90.0000	105.5200	113.6000

forward) engine had to be shut down, with a cracked water jacket. It was the first of frustrating instances of unreliability on the part of the five-year-old, secondhand, Condors.

R.100 cruised around York for fifteen minutes, at 1,700 feet, on, now, a fine, sunny morning. Scott then set course for Cardington, helped by an 11 mph tail wind from the north-west.

Flying by way of Selby (at 0925) Newark-on-Trent and Melton Mowbray, R.100 was overhead Kettering at 1050, over Bedford at 1112 and circled the Cardington area for an hour-and-a-half before dropping the mooring wire at 1245. The airship was secured to the mast at 1340 with the two aft engines of the forward cars running in reverse and the front engine of the aft car running ahead to blow slipstream on the rudder. There had been some minor problems as was to be expected. Apart from the trouble with No 3 engine the lower parts of the gas-bags were being blown about by the rush of air through the ventilators in the bottom of the outer cover. They had to be blanked off. At the same time the cover developed pulsations and ripples in some of the large unsupported panels. They started to split in several places.

The first flight had lasted 5 hours and 47 minutes for a distance of 150 statute miles. A total of 380 gallons (2,812 lb) of petrol had been burned, out of the 5,500 on board at departure. Three-and-a-half tons of water ballast had been used leaving 4¼ tons remaining in the tanks.

The fourteen passengers on this initial flight included the whole of the senior staff at Howden – a hostage of fortune regarded as normal in those days. On board were Sir Dennistoun Burney, Barnes Wallis, Nevil Shute Norway, Major Philip Teed, Hessell Tiltman, James Watson, William Horrocks, F. Lansdon, P. Clusloe and J. B. Moseley from Rolls Royce, together with Reginald Colmore and Michael Rope from the Royal Airship Works, and Wells and Cox-Walker from the RAE.

R.100's second flight took place next day, 17th December, and completed British airship flying for 1929.

With thirty-seven crew and seven passengers on board, R.100 slipped from the tower at 0947 under a clear blue sky and a light wind from the north-west. The ship carried a disposable load of 36.48 tons which included 5,650 gallons (19.37 tons) of petrol, 8.3 tons of ballast, two tons of food and drinking water and 2¾ tons of spares.

The intention had been to fly to London and back. Shortly after departure, however, a large strip of fabric came loose from the ventral fin. As a precaution, therefore, the flight was confined to a gentle cruise in the vicinity of Bedford before mooring at 1616 after 6 hours 29 minutes in the air for a fuel consumption of 450 gallons.

The flight had been used to test the controls, which were found to be light and positive, and to measure speeds with Michael Rope's patented Air Log; the ingenious apparatus made up of a streamlined form lowered about 50 feet below the control car. The Air Log was driven by a small propeller at the aft end, turned by the airstream. Electrical impulses measured the revolutions per minute to give an accurate calibration of the true air speed. Precisely 60 mph (52.1 knots) was recorded with all engines running at 1400 rpm giving a total of 1,800 bhp.* Later, with the ship in equilibrium (pitched neither up nor down) somewhat better figures were recorded.

The chief problem with the ship continued to be the outer cover on which some of the restraining tapes broke while the mooring cable developed a kink and had to be scrapped.

Next day, R.100 was taken from the tower and housed in No 2 shed at Cardington – the shed transferred from Pulham – so that work could begin on modifications to the outer cover to make it more secure. A maze of 'zigzag' wires which held the cover in position were, in addition, chafing on the transverse frames. They all had to be altered; all 3,600 of them. It was a tedious job which took R.100's crew till 11th January to complete.

On that same day, 18th December, Frederick Montague, Under Secretary of State for Air, announced in the House of Commons Lord Thomson's decision that a new bay of 500,000 cubic feet gas capacity would be installed in R.101 – 'to add 15.5 tons to the gross lift for an estimated increase of five tons in structural weight'. In addition, said Montague, the capacity of R.101's existing gas bags would be increased by letting out the wiring, to give an additional 130,000 cubic feet of gas and to increase the lift by four tons. Various steps would be taken to lighten the airship's weight, such as the removal of the servo motors on rudders and elevators because they had been shown to be unnecessary.

All this work, except for the new bay, would be completed in three months. Both R.100 and R.101 must be seen as prototypes pointing the way towards future commercial airships.

So ended 1929 in which R.101 had made seven flights for a total

* Cruising on 1,800 bhp in equilibrium, R.100's measured true air-speed was recorded as 55.6 knots (64.02 mph) whereas on the same power R.101A recorded 53.4 knots (61.49 mph) somewhat retarded by the drag of the inoperative No 1 engine which, at that time, could be run only at reverse thrust. In these circumstances the hull drag of R.101 was slightly less than that of R.100.

of 73 hours 27 minutes in the air and R.100 two flights totalling 12 hours and 16 minutes.

With the end of the year the Air Ministry could congratulate itself on the fact that the long awaited 'Approach Towards a System of Imperial Communications' was off the ground. The two new airships, the largest in the world had both flown. There were problems of course. They were attempting things which had never been tried before. The high promise could be fulfilled if 'boundless energy' was applied.

In the course of six years a gleam in the eye had been translated into hardware. Now there were just 290 days before the 1930 Imperial Conference was due to open in London. Among other things it would review the next stage of airship development. By then R.101 should have flown triumphantly to India and back and R.100 should have made a return crossing to and from Canada.

Those voyages, eastwards and westwards, would prove that faith could move mountains even if they were Lighter than Air. A new era in Empire communications would have begun.

So a New Year and a new decade – the Thirties – were about to dawn. On every side there was the hope that it would see a vindication of all that had begun to be put to the test with R.100 and R.101.

CHAPTER TEN

New Year in Paris

'Resistless Paris calls'
G. M. Carrel, 1873–1904

100 Ashley Gardens
Westminster,
London SW1

4th January, XXX

My dearest,
1929 is well over and 1930 has come in with wind and rain [wrote Christopher Thomson from London to Marthe Bibesco in Paris].

I have not written earlier to thank you for your two long letters – so gratefully received – because my plans were most uncertain. Now I can say something *défini*; the fact that I shall arrive in Paris on the 7th, Tuesday next, and stay at the Embassy. The excuse for my visit is the dinner of the British Chamber of Commerce; the real reason is to get a glimpse of you. The dinner is on the 9th and I shall have to return on the 10th as the Cabinet is meeting several times during the following week.

I spent Christmas with my mother in Devon.* She is now eighty-five and wonderfully well and insists on my sending her the official reports of my speeches in The House of Lords.

Your new home at the Rue de l'Université sounds attractive. I shall be ringing up 'Littre 020c' on Tuesday evening in the hope of getting you even if then only to hear your voice.

This is my new address – 100 Ashley Gardens near Victoria. It consists of a large drawing room, a dining room, a hall, a commodious kitchen, three bedrooms, a bathroom and servants quarters. Rather too much for me but better than a tight fit. I hope you will see it soon.

My cook is called Rose, my house parlour-maid is named after the *bienheureux* Martin, my man † (valet and chauffeur) after St James of Campostella and I have a black cat called Sambo. He was needed in my last flat at Princes Street where there were mice. He came as a

* Mrs Emily Thomson had been born Emily Lydia Birdwood on 29th January 1844, the daughter of Major General Christopher Birdwood of the Indian Army and of Lydia Juliana, daughter of the Reverend Joseph Taylor of the London Missionary Society of Belgaum, India. On 26th January 1867 she married Major General (then Captain) David Thomson of the Royal Engineers, Indian Army. In 1930 Emily Thomson was living at Higher Brimley, near Teignmouth in Devon. She died, aged almost ninety, on 26th December 1933.

† James Buck who accompanied the Secretary of State on the flight to India.

kitten and was altogether charming. I lost my heart to him then and now he is the most important person in the household.

He is sitting on this writing table now, watching me with yellow eyes and I don't think he likes me much – no doubt he has heard from feline sources of the hard things I said about poor Pat.*

On the sterner side, it may be that the Government will be defeated on the Coal Bill shortly in which case I shall be 'in a position of greater ease and less responsibility' than I am at present. I doubt however whether the Tories or the Liberals want a dissolution. The former have no policy and no leader. The latter have several policies and all the Liberal MPs are leaders – or think they are.

Our Coal Bill, on which I must speak, is far from perfect but it is the best possible in the circumstances. It will be presented in the Commons by one of our ablest men – William Graham † – who may achieve any position in this country in due course; none is too high for him. He is what we call 'emergent' and on Tuesday his testing time will come. My money is on this good Scottish horse. He is not a flier but he gets over the ground!

Well, no more till we meet. It will be such a pleasure to see and hear you again. How long are the hours apart; how fast they go when we're together.

All my love as ever,
And for ever your
Kit.

The debate on the Coal Bill delayed him a day. He telegraphed Marthe that he would arrive about three o'clock on the afternoon of 8th January. She wired back that she would invite a few friends for drinks at the Rue de l'Université and then they would dine *à deux*.

Lord Thomson left Croydon Airport on the 1110 Imperial Airways scheduled service to Paris (Return Fare £10 18s 6d) with thirteen other passengers aboard a twenty-passenger Armstrong-Whitworth Argosy three-motor biplane (G-EBLF) due to land at Le Bourget at 1340 hrs. There was fog over France and the Argosy had to turn back over Beauvais and land, in rapidly falling visibility, in a field near Abbeville; fortunately without damage or injury.

Later, the Princess Bibesco recorded: ‡

* Princess Bibesco's Persian cat of 1915 brought back to Roumania from Isfahan in 1904.

† The Right Hon William Graham PC, MA, LLB (1887–1932) Labour Member of Parliament for the Central Division of Edinburgh from 1918 to 1931 and President of the Board of Trade between 1929 and 1931. His death at the age of forty-five robbed him of the great future which Lord Thomson predicted for him.

‡ In *Lord Thomson of Cardington – A Memoir* (Jonathan Cape, 1932).

·ace and Lightness – The
ınge of R.101 measuring
feet by 32 feet, furnished
the Press visit in the
rdington shed on 2nd
tober 1929. The tables and
airs had been used
eviously in R.36; Flying
ficer Steff had been sent to
y table cloths in Bedford
d Miss Foster, Colonel
chmond's Secretary, had
en asked to 'undertake the
rangements of the flowers
the dining room table and
ns for the lounge'.

rst Appearance – At 0545
ı the morning of Saturday,
2th October 1929, R.101
nerges from the shed at
ardington for the first time;
valked out' by a handling
arty of 400 men from the
oyal Airship Works, from
AF Henlow and from the
edford Labour Exchange.

Graceful Silver Shape –
.101 being hauled down to
e moored for the first time to
e arm at the head of the
ardington tower at 0723 hrs
ı the morning of 12th
ctober 1929. The time taken
move the ship from its
ed, traverse it across the
erodrome and to secure it to
e mast head on this launch
as one hour and 38
inutes.

PLATE XI]

Out and About – R.101A moored at the Cardington tower for the first time on the morning of 12th October 1929. [PLATE XII]

Secure – R.101 riding at the Cardington mast on the morning of 14th October 1929, before slipping at 11.17 hrs for its first flight – of 5 hours 40 minutes to London and return. The No 2 Shed at Cardington, hiding the shed in which R.101 was assembled, is in the background.

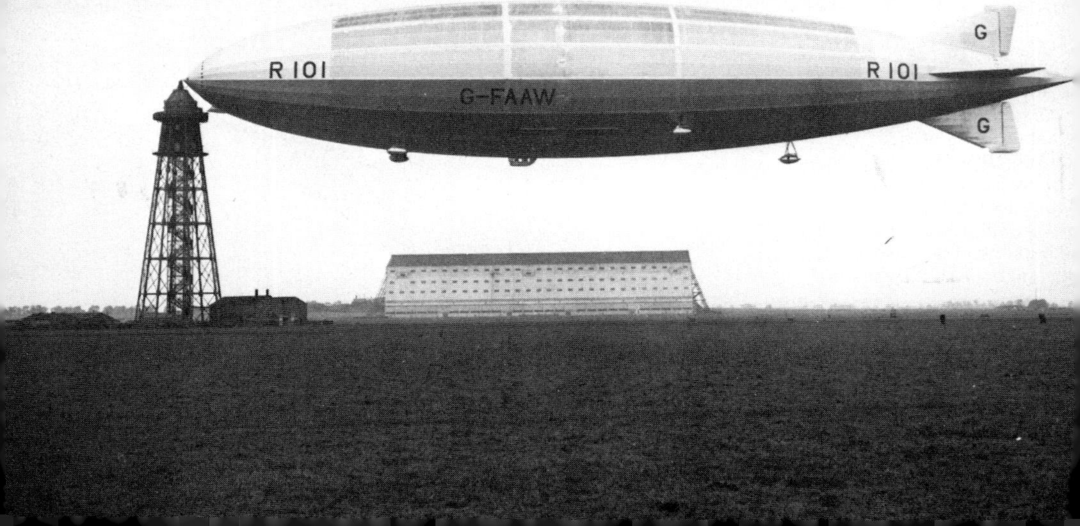

First Flight – R.101 flies over St Paul's in the City of London on the early afternoon of Monday, 14th October 1929, on its first flight of 5 hours 40 minutes duration from Cardington to London and back carrying 39 crew and 14 passengers on board. [PLATE XIII]

Wonders Above – Lord Thoms S. of S. for Air (right) and Lieutenant-Colonel Vincent C. Richmond, Chief Designer of R.101, look up at the airship fr the foot of the mooring tower Cardington on the early mornir of 18th October 1929, before going aboard for the R.101's second flight – a circuit of the Midlands of England.

Ministerial Flight – Lord Thoms entering the lift at the base of t mooring tower at Cardington o the early morning of 18th October 1929, to fly aboard R.101 on its second flight in which it covered 182 n. miles ir 6½ hours flying time over the Midlands of England. On the lef is Herbert Mann, the Royal Airship Works lift attendant.

[PLATE XIV]

I had gathered a few of his dearest friends to meet him. There was the Comtesse de Castellane, Mme Pierre Goujon, Princess Lucien Murat my cousin – the Abbé Mugnier and, also, his companion through good and evil days when the English fleet was to pass the Dardanelles, Edouard Tavernier. We expected C.B. at five o'clock. But the aeroplane must have been unexpectedly delayed. What could it be? Vainly we telephoned Le Bourget. At seven o'clock there was still no sign of him.

And then, suddenly, there was his tall figure framed in the doorway. He came in rubbing his hands which were frozen by the journey, bracing himself, shaking the night from his shoulders as he used to do long ago when, in bad weather, he arrived at Mogosoea near Bucharest.

He had been forced to come down in a fog in the middle of a field near Abbeville. He had spent an hour in a poor woman's cottage until a car had been found to take him – and other passengers – to the train. His hostess, an aged peasant woman, had offered him a place by the fireside and a glass of warm wine, and had refused to accept anything in exchange for her hospitality. She had had the pleasure of his talk she said.

'How happy you are to be able to live here, Madame,' he had said when he was leaving. 'You have peace.'

'Say solitude rather,' she answered.

'That's your people. Ah! My dear Tavernier,' he said, turning to our friend, 'you belong to a good and a beautiful country. Some day I shall come to die in France.'

The moment he said it, the phrase affected me like a warning bell. But the warning was drowned by a confidence in a life which was so vitally present. Nonsense! He would come to die in France when he was an old man – as his father, seeking a little warmth in his last summer, had retired to die on the hill-sides near Florence.

I knew that his heart was charmed by the kindliness and the sweetness of things French, by the magic of a tongue which he loved and understood down to its finest shades. The country between the Seine and the Loire was really for him 'the most beautiful garden in the world' and he felt no stranger in it. We had explored it together in years gone by.

I often used to say to him that, without knowing it, and from that Gallic divinity he bore within him from a marriage long ago,* it was still France he loved in Roumania. With this completely civilised man it was the love of French literature and French speech which drew him to our house in Bucharest in my father's time.

* On his mother's side Christopher Thomson could trace his ancestry back through an old Huguenot line to Ambroise Paré (1510–1590), physician to King Henry II. Paré was born in Laval, Marne, and invented ligatures for arterial wounds.

On that January evening when he left his aeroplane in the Plain of Amiens, he brought his sympathy to France for the last time.

He talked to us of the Naval Conference which was opening in London a few days later. Ramsay MacDonald was to preside – and CB was one of the members. 'Talk about blue waters,' he said. 'What about grey air, blue skies?' With his mind on the far-away future he was already imagining the great air forces which might keep the peace.

The hour grew late. My guests departed – all except Tavernier. He and I went on listening to the beautiful, familiar, voice as we used to do when only we were left beside the fireside at Mogosoea in the Old Days. Now that voice talked to us of the future – the sky was vast and the world would be small; inhabited by men who could fly like gods even above the clouds.

Before he left Paris on 10th January C.B. wrote to Marthe from the British Embassy.

Just a line before starting, so say *au revoir, à bientôt* and, as ever, to express to you my grateful thanks for all your kindness. It is always an inspiration to be with you. I leave you refortified for life's battle.

The Banquet last night went off all right. The atmosphere was most cordial but the room was very hot and I spoke last. Fortunately my written remarks – about Anglo-French air transport and the need to put safety in the air at the top of the list – had been given to the Press so that after that it didn't matter so much what I said or how I said it. After-dinner oratory is a special, and most uncertain, form of self expression.

Please come to London soon – before Easter in mid April. And then I can pledge myself to be in Paris again at the end of April. What delight that will be again with you.

Well; no more now. I leave Le Bourget in a few minutes.

I shall look forward to meeting Kellogg * in London again and for your sake principally but also for all our sakes, *Je le mangerai.*

God bless you dear one. Ever, your Kit.

With the New Year of 1930 the Secretary of State for Air was becoming overstretched. The Prime Minister had come to rely increasingly on his advice and piled task upon task on a willing horse. Thomson was speaking up and down the country at each week-end, running his Department, attending Cabinets, active in the House of Lords, taking part in the Naval Conference, which ran from 21st January to 15th April, and helping on foreign affairs.

* Frank Billings Kellogg (1856–1937) United States Secretary of State from 1925 to 1929 and formerly US Ambassador to United Kingdom, 1924–25. Negotiated the Kellogg Pact, a multilateral treaty to outlaw war signed in Paris 27th August 1928.

Nor did he neglect the sidelines. In addition to his official briefings every week he read the two aviation weekly magazines, *The Aeroplane* and *Flight*, from cover to cover. As the year began he sent a warm, handwritten, message of congratulation to Stanley Spooner, Editor of *Flight*, on the completion of twenty-one years of publication, appreciating specially its value as the official organ of the Royal Aero Club and the help it had given to one of his pet schemes, the Light Aeroplane Clubs – 'a movement calculated,' he wrote, 'to be one of the greatest importance to the future development of aviation in this country.'

Since he and Sefton Brancker had launched the Light Aeroplane Club £10 subsidy for flying licences in 1924, a total of ninety clubs had been formed of which fifty-seven now qualified under the subsidy scheme.

Events crowded in. On 13th January the Secretary of State spoke on Air Power at the 209th Annual Banquet of the City of London Tradesmen's Club. On Sunday, 19th January, he escorted the Japanese Naval Delegation, led by Admiral Takarabe and Baron Abo and including Commander Yamamoto, to lunch at Hanworth Air Park in company with Sir Sefton Brancker and Sir Henry Segrave, followed by a flying display.

Next Saturday, 25th January, Lord Thomson drove to Hendon Aerodrome to inspect the offices and air survey laboratories of the Aircraft Operating Company Ltd and went on to lunch at Heston Air Park with Alan Butler, Chairman of both the AOC and the de Havilland Aircraft Company Ltd. The occasion was the launching of a new special photographic aeroplane designed and built as a joint venture by de Havillands and Gloster Aircraft Ltd. The new, twin-engine biplane – the D.H.67/Gloster AS31, registered G-AADO – was to be flown to Bulawayo in Northern Rhodesia by Alan Butler and his wife, Lois, to undertake there the air survey of 63,000 square miles at £4 a square mile.

Before lunch, Lord Thomson, Fred Montague and Sefton Brancker flew in the new machine piloted by Howard Saint, Gloster's Chief Test Pilot accompanied by Alan Butler. On landing, Christopher Thomson gave an enthusiastic interview to British Movietone News, appearing before the new combination of microphones and cameras for the first time.

At a convivial lunch at Heston afterwards, Alan Butler said that this day marked the birth of a new industry – air survey – and that the Secretary of State had paid them the honour of being the first passenger in the first aeroplane built expressly for air survey work. He congratulated Lord Thomson on the good example he set by flying everywhere.

In his reply, C.B. said that he flew as every progressive person ought to do – whether by landplane, seaplane or airship. When, flying to Paris recently, their aircraft had had to land at Abbeville because of fog, with five lady passengers, two middle-aged men and seven youngsters; one red-headed boy had told him that this was his fifth, and most interesting, flight and added, 'I always make Daddy send me to school this way.' The air-minded boy had been especially interested in the official people who came to fuss over him and asked him who he was? He had replied in the words of a friend of his, a pavement artist – 'The 'ead of all the air', a position he was proud to hold. Nothing abashed, the boy had replied with great courtesy: 'Well, sir, if you'll excuse me saying so, like my father you are getting a bit thin on the top."

The way to make an enterprise pay was certainly to be a bit thin at the top – and what better example could there be than the Chairman of the air survey company, Alan Butler, holding down overheads by making the delivery flight of the Gloster Survey plane himself. A commercial flying enterprise which was financially self-supporting was like a spring in the wilderness. He wished everyone connected with the survey the best of luck. The Air Ministry would always be ready to help those who helped themselves.

Two weeks later he picked up air survey again as a new and valuable service which aviation could offer to Empire countries, when he spoke at a lunch of the British Empire League at 12 St James Street on 7th February. He continued the theme at Cardiff on 14th February. Air Power and civil aviation in all its aspects must march forward wing-tip to wing-tip; the Royal Air Force as the mainstay of Home Defence, civil aviation as the airways of the Empire.

On the military side there were now seventy regular squadrons, including the equivalent of twelve squadrons in the Fleet Air Arm, together with twelve non-regular squadrons of the Auxiliary Air Force. The policy must be to provide a sufficient Air Force to act as an adequate deterrent to any hostile government.

In the civil field there were now 191 private owners of light aircraft in Britain with, between them, 206 aircraft· the beginnings of a new mobility in the air.

Over long ranges the airship seemed likely to be the most suitable and comfortable means of transport but there had been wild exaggerations about the money spent on airships. The total sum spent on the whole programme since 1924 was just over £2 million and most of this on research. The two ships built so far had added much to the sum of knowledge and a country such as ours must pay for its research.

So far 427 people had flown the Atlantic and 402 of them had done so by airship. Not a bad record.

The new, but still experimental, R.101 had shown itself able to stand up to any gale. This ship, which was contributing so much to the research and development programme for better airships to come, had cost a total of some £610,000. R.100, which had fewer new features had, naturally, cost rather less. Four mooring masts had been built – at Cardington, near Montreal, at Ismailia in Egypt and at Karachi – at a combined cost of about £150,000. It was all good value for money.

Knowledge was being gained which would be invaluable for the future. 'Given a good skipper and a good crew I would cheerfully go out in R.101 in any weather.

'Indeed I believe that, when fully proven and tested, with its immensely strong structure, its heavy-oil engines and by then an experienced crew, R.101 will be safe as a house except for the millionth chance.'

Plans would soon be developed for new commercial airships to follow on. They would carry forty passengers to Karachi or to Montreal in about three-and-a-half days. It was now clear that they could be built for about £600,000 each.

Quite soon the public would be able to decide whether airships were going to be a sound commercial proposition.

'At any rate Great Britain has now given a lead to the rest of the world. The prospects are golden. We must keep the lead in the air in linking the distant parts of the Empire as well as providing with the RAF our first line of defence.'

A week later Thomson went down with a sharp bout of 'flu. On 24th February, still recovering, he wrote to Marthe from 100 Ashley Gardens:

> I have not been well for the past few days which may account for a feeling of intense depression, deepened by being away from you. It is hard to be optimistic when human affairs are so muddled.
>
> We are living in hectic times and the weeks pass in a turmoil without much being achieved. This is especially true of the Naval Conference which drags on and I imagine that the change in the US

government may mean that Mr Kellogg* will not come at all.

What a mess we are in. There is no such a thing as a plenipotentiary in these days. No one can act and publicity only adds to the confusion. Tardie † and Miasset both dined with me. But we did not get very far at this meeting as, indeed, I did not expect we could. And then Tardieu left to give way to 'the unknown football player'.

Please forgive a dull and despondent letter. I will write again when a happier mood returns.

But dull and despondent or bright and of good cheer you know that I am as ever

Your
C.B.

Marthe wrote back three days later – a letter full of solace and encouragement, of love and concern about his health. He was overworking. He must take time off. Come to Paris. The spring was near. Surely the Naval Conference must finish soon? This spring would be fifteen years since that spring when first they'd met in Bucharest. Perhaps he might arrange to accompany her to Roumania for Easter? They would cruise down the Danube again and re-live old memories and create new memories even more fragrant than the old. Come anyway to be with her. They had been too much apart.

Christopher Thomson replied:

My own dear love.

Please forgive the delay in my reply to your sweet and charming letter of 27th February. Of course I long to be with you. But I have a tremendous rush of work on the Air Estimates, the Conference continues, and I have several Committees in addition to my work in the House.

I will make a big effort to get over to Paris before 1st April. But honestly I cannot promise anything – unless we get turned out, in which case I shall be a free man at long last.

As for Roumania, I'm afraid that is out of the question – at Easter anyhow. Perhaps you can come here instead? Please manage it if you can.

No more now. Please forgive a hurried line.

With all my love my dear, dear, Marthe

Ever your,
Kit.

* Mr F. B. Kellogg ceased to be US Secretary of State on the election of President Herbert Hoover in the autumn of 1929 and was appointed in 1930 a judge in the Permanent Court of International Justice.

† André Pierre Gabriel Amidée Tardieu (1876–1945) French Premier 1929–30 and 1932, a member of the Paris Peace Conference 1918–19 where Christopher Thomson came to know him well.

At the Air Ministry, the Secretary of State was now putting in many hours with his officials on the annual Air Estimates – the aviation budget for the year ahead, 1930–31. The Estimates were issued, with a memorandum from the Secretary of State for Air, on 8th March.

Thanks to long and persuasive discussions by Christopher Thomson with the Chancellor of the Exchequer, Philip Snowden, the Air Ministry was once again able to count an increase in the Estimates; up by £910,700 compared with the previous year to a total of £20,923,000 after a national 'super-cut' of £150,000 as an obeisance to the hardness of the financial times.

'These Estimates,' wrote the Secretary of State, 'allow for a broad continuation of air policy, although the approved increases of the Royal Air Force are of small dimensions and designed to demonstrate once more the earnest desire of His Majesty's Government to avoid disastrous competition in air armaments.' He pointed out, however, that the United States expenditure on air defence had now risen to the equivalent of £33 million even though North America was out of range of any potentially hostile air force.

In all the circumstances the Air Ministry proposed the formation of one new regular fighter squadron for Home Defence, one squadron of reconnaissance flying boats and four auxiliary squadrons, while a number of squadrons would be equipped with new aircraft. The total manpower of the Royal Air Force, exclusive of India but including Aden, would be 3,338 officers, 5,760 non-commissioned officers and 19,375 men who with cadets and apprentices would make 32,000 in all.

Some £6.64 million would be required for aircraft and engines and £500,000 for civil aviation of which £348,000 was for subsidies to Imperial Airways for European services and the route to Karachi, £20,000 was for light aeroplane clubs and £7,000 for improvements to Croydon Airport.

On the airship side, the shed at Karachi had just been completed and the mast there was now ready to receive R.101 later in the year. The policy to be pursued was a policy of 'safety first'. To that end a new bay was about to be inserted in R.101 'before the flight to India and back was undertaken even though such a flight might well have been carried through successfully without this addition'.

The intention was that R.100 would embark upon a series of flights to the tower at Montreal between May and September 1930 and to Ismailia and back between October 1930 and March 1931. R.101 would fly to Karachi and back in September or October 1930 and again to Karachi in December 1930 or January 1931 and then be used on experimental operations on the India route to obtain the technical, operational and financial data necessary before a commercial service could be launched.

Meanwhile, to make easier and more reliable the handling of airships at the main base, Cardington, a mechanical transportable tripod stub mast, 80 feet high, was to be built which could be used to move airships from the mast to the sheds. It would cost £33,000.

In all, airship work would cost £345,000 in the year ahead including £12,000 for work on new designs and £82,000 for Research and Development.

On the personal side Lord Thomson paid a special tribute to the first Marshal of the Royal Air Force, the new Lord Trenchard of Wolfeton, who had been elevated to the peerage in the New Year's Honours on his retirement after eleven years as Chief of the Air Staff; the 'Father of the Royal Air Force'. 'The Royal Air Force will always remember the inspiration and guidance it received from him during the difficulties of its early years.'

From the same date, 1st January, Wing Commander Reginald Colmore had been confirmed as Director of Airship Development to carry forward the airship programme from the experimental to the commercial stage.

The Air Estimates were approved – but the Naval Conference went on.

CHAPTER ELEVEN

Dinner at Eight

'If only you knew how much I want to see you,
to hear your voice and to tell you things
one does not write.'

C.B. to Marthe Bibesco, 29th August 1921

ON 31ST MARCH 1930, instead of C.B. travelling to Paris, Marthe arrived in London. If he could not come to her at least she could come to him. The more he became tied down with official affairs the more she felt the need to be with him. It was as if she knew that time was running out.

She had telegraphed her plans but he was at a Cabinet Committee when she came in to Victoria. Instead of meeting her, a note from 100 Ashley Gardens, with a large bouquet of red roses, awaited her at the Ritz Hotel.

Welcome to London.

I am so much looking forward to seeing you tomorrow night, Tuesday 1st at 8 p.m.

Shall I send the car to meet you and bring you here?

He was on the telephone – and next evening James Buck arrived at the side door of the Ritz to bring her to Ashley Gardens.

He was waiting outside Block 6 of the large row of red brick apartments in Thirlby Road, beside Westminster Cathedral.

As she stepped out, a radiant figure in her flowered evening dress of rose-coloured satin, he raised her hand to his lips as he always did in greeting. Around her neck she wore her magnificent emeralds. 'The drops of Forest, Hope and Sea', for both of them full of memories of their first meeting fifteen years before.

Together they went up in the lift to the second floor.

'Welcome to Ashley Gardens,' he said as he opened the door. 'Prime Minister : the Princess Bibesco.'*

Ramsay MacDonald came forward, a mane of grey hair above his

* Recorded by Princess Bibesco in her diary and in a letter to the Abbé Mugnier in Paris.

handsome, bespectacled face.

'My dear Princess,' he said, rolling his 'r's'. 'Now C.B. will be happy. He's been like a cat on hot bricks till you were here.'

Thomson turned to the fourth member of the party – a tall, elegant, woman of about fifty. 'Princess, you know Lady Wilson, don't you?'

Barbara Wilson was indeed an old friend of Marthe's* and they greeted each other with delight. When they went in to dinner it was to find that Rose the cook had excelled herself while C.B. had – as always chosen the most delectable of wines. Both he and the Prime Minister were in excellent form. The dinner party was an undoubted success.

After dinner and over coffee in the drawing room, the story of the magnificent Persian carpet on the floor was related with enjoyment. It had been presented to C.B. at Suliamaniya in Khurdistan in September 1924, when he had flown there in a D.H.9A of the Royal Air Force during an official visit to RAF Iraq Command. Extraordinarily, it closely resembled a carpet which Marthe Bibesco had brought back from Achrat Abbas near Isphahan in 1905. That carpet had tender memories for Marthe and C.B. together at her 'Castle in the Mountains' at Posada in Roumania in 1915. At the end of that year the castle and the carpet had been burned in a disastrous fire.

The Suliamaniya carpet was like a Phoenix arisen from the ashes and, like a Phoenix it had been flown back to Baghdad, strapped to the side of a D.H.9A.

'I look on this carpet from Suliamaniya as a talisman,' said C.B. 'So long as I have it I can be sure that Marthe will always come to me.'

At the end of the evening Ramsay MacDonald renewed his invitation to the Princess to come to spend a week-end with his daughter, Ishbel, C.B. and himself at Chequers.

'The gardens there will soon be looking their best,' he had said. 'Once this Naval Conference is over, can you come?'

'Perhaps in July, Prime Minister. Will you ask me then?' Marthe had replied.

Writing to the Abbé Mugnier next day, she related:

> He looked at me through his big spectacles with that penetrating gaze of a man who is obliged by his vocation to form rapid judgements about those with whom he speaks; a look as that of one of the great medical consultants.

* Lady Barbara Wilson, wife of Colonel Sir Matthew Wilson, was the daughter of Thomas Lister, Lord Ribbesdale and the niece of Reginald Lister, a former First Secretary at the British Embassy in Paris. Her sister was Lady Lovat with whom the Princess Bibesco and the Abbé Mugnier stayed in Inverness in 1927.

'You think then that I shall still be Prime Minister in July? Four months is a long time in politics at present.'

With my usual outspokenness I replied:

'My Conservative friends tell me so.'

MacDonald laughed. 'They have ready tongues,' he said. 'But you really can't depend on anything they say. Not even that.'

Two days later Marthe left for Paris. C.B. was at Victoria to see her off.

'Do try to come to Paris for Easter in two weeks time,' she had said as she leaned out of the carriage window. 'Kit, you need a rest so badly and you can have it there. We can be alone together and just walk and talk.'

'I wish I could,' he had replied. 'But I doubt if the Conference will be over and there are the Egyptian Treaty talks coming up as well. I'll do my best but if I can't get away, please do come back in July. The PM really meant that about Chequers. Please come, Marthe.'*

The Naval Conference did, in fact, continue until after Easter.† Instead of coming to Paris, Christopher Thomson wrote to Marthe on Good Friday, 18th April, from 100 Ashley Gardens.

> I am spending part of the Friday known as 'Good' – no doubt because it marks the supposed anniversary of a crime often repeated since by narrow reactionaries – in writing to you to whom I owe many letters and so much inspiration.
>
> Yes. That evening here at Ashley Gardens is still close in my memory – as is so much else of you. At least I can now feel your presence here; and Sambo has more than an air of poor, departed Pat.
>
> You made a profound impression on the P.M. – as you always do on everyone.
>
> I am jealous of your pious feelings for my political chief. 'Pity is akin to love' – and for me, it seems, you have no pity! Alas. Poor Yorick. Why should I not be pitied? Is it because the blasts of cruel fortune – the most cruel to be apart from you – have not even made me appear to button up my coat? I felt them – I feel them – all the same. Perhaps I am too contemptuous of opinion. But at least I do not confound publicity with fame and I am not in the least bit disappointed that I get neither.

* Letter from Princess Bibesco to the Abbé Mugnier.

† The Naval Treaty was signed at St James Palace on Tuesday, 22nd April 1930. One of its provisions was to limit new aircraft carriers to a maximum of 10,000 tons displacement and to guns of not more than 6.1 inch calibre. It was to remain in force until 31st December 1936.

My disappointment is of a different kind.

As one of our delegation in the Egyptian Treaty I have been spending days and nights with Copts and Moslems. I watch their faces across the table and think of Edfoo and Komento.

They want some things which they can never have in our lifetime and, if they got them they would find they had got nothing. The logical outcome of Democracy is Anarchy – tempered by the instinct of self-preservation. The bastard democracy which Egypt might enjoy would be merely a dissolvent.

No more now, I fear. I long to see you but I cannot leave England before August – and in September I go to India.

All my love, my dearest Marthe. You are very precious to me.

Ever your Kit.

Now, with plans hastening fast, airships were beginning to take the centre of the stage. They were claiming an increasing part of the Secretary of State's attention – and all was not going well.

CHAPTER TWELVE

Airship Summer

'High endeavours are an inward light
'That makes the path before him always bright:
'Who, with natural instinct to discern
'What knowledge can perform, is diligent to learn.'

WORDSWORTH, 'The Happy Warrior'

January 1930

IN THE Lighter-than-Air realms of buoyant flight the final decade of the rigid airship was launched by R.100 with three flights in January 1930. The last of them – R.100's fifth flight – was an endurance trial of 54 hours and 52 minutes on 27th to 29th January, chiefly over the West Country and the English Channel in and out of cloud at between 1,000 and 3,800 feet.

In the course of the flight R.100 covered a distance of 1,780 nautical miles at an average ground speed of 33 knots or 37.3 miles an hour, carrying eleven passengers and a crew of forty-five.* Among them was Noël 'Grabby' Atherstone flying as a second navigator in company with Ernest Johnston. His comments were interesting.

> During the flight [he wrote in his diary] I tried the ship on both rudders and elevators and I found her to be very sensitive to small angles of helm and quick to respond.
>
> She handles better than R.101 and seems much lighter on the controls. This is probably due, to a large extent, to the fact that the ship is neither tail heavy nor bow heavy but is nicely balanced about her centre of buoyancy.
>
> The outlook from R.100's passenger coach is very bad but the control car is roomy, has plenty of windows and there are no draughts even when the windows are open.

R.100's five flights and 87 hours 29 minutes in the air had yielded valuable data. A maximum speed of 80.5 miles an hour (69.9 knots)

* During its flight R.100 consumed 4,790 gallons (35,446 lb) of petrol at an average of 87.3 gallons or 646 lb an hour which was almost the same rate of fuel consumption as that of R.101 on its own endurance flight two months before. But R.101 had maintained a higher air speed on a higher power.

had been recorded; exceeded up to that time among rigid airships only by the German LZ.120 *Bodensee* which had attained 82.4 mph (71.6 knots) in 1922.*

There had been disturbing problems however. The fabric of R.100's outer cover had flapped and torn. Rain water had leaked through the cover on to the gas-bags, which had split, and into the electrical system which had fused. The secondhand Condor engines had developed a number of faults. In particular the gearboxes had given trouble and several supporting studs on the exhaust pipe system had failed and had damaged the propellers of the pusher engines by going through them. Cooling water had syphoned out of the vent pipes in the radiator system. On one occasion the elevator controls had jammed – possibly through ice – and there had been a very nasty minute or so until they were freed.

An added problem had been found when ice formed on the outer cover and lumps of ice, broken off, had damaged two of the propellers and had been thrown through the cover.

Modifications were clearly needed. So on 30th January R.100 was put back into its No 2 shed at Cardington for work to be done under shelter. In the shed the cover was carefully examined and found to be in such a poor state that Major Percy Bishop of the Aeronautical Inspection Directorate ruled that it was no longer airworthy. Though the strength of the fabric had not deteriorated seriously, the cover was no longer waterproof and R.100's gas-bags were soaked in every shower of rain and were now in a poor state.

A series of modifications and repairs were set in hand, including sealing tapes on the outer cover and patches on the gas-bags.

A day later – on 30th January – Vincent Richmond wrote a long letter to Sir Geoffrey Salmond, AOC in C Royal Air Force, India, to bring him up to date on the status of R.101 and the happenings of the past few months. The letter well summarises the then position and Richmond's hopes and expectations together with his objective view of the problems of the time. He wrote from Cardington:

30*th January* 1930.

Dear Sir Geoffrey,

I am afraid you will think I have been very remiss in not replying

* The fastest speed attained by any rigid airship was 87.2 mph (75.7 knots) by the US ZRS-5 *Macon* in 1933.

before to your very kind letter of congratulations. I had hoped to get an opportunity of really digesting the results of the trials before writing to you. Unfortunately I went sick with a bad attack of influenza in December, and I have been away endeavouring to recuperate ever since from the somewhat unpleasant after effects.

Please accept my best thanks for your kind thoughts and good wishes to my wife and myself.

Before the ship left the shed, we knew, of course, that in several ways she had not fully come up to our expectations. I refer principally to the engines both as regards their weight, performance and the unsatisfactory arrangement for reversing, but also our estimated weights had been exceeded in other directions so that the fixed weights of the ship, instead of being 90 tons as we had hoped, actually worked out at about 110 tons.

With regard to the actual trials, however, I think I can fairly say that the results if anything exceeded our expectations. They were carried out under extremely trying conditions, the weather during the latter part of October and the beginning of November being some of the worst on record. We had many gales and torrents of rain at the mooring tower. In one gale, the gusts of wind rose as high as 83 mph and the storm finished up with a line squall in which the wind changed direction through 135 degrees in about a minute. The instruments we had fixed to record the force on the bow, worked very well, and showed that at no time did this force approach at all dangerously near to the calculated strength. No damage occurred, except that the ship got a very severe wetting, and we discovered that more care would be necessary in sealing off holes in the cover as a considerable amount of water got into the passenger accommodation. Fortunately it did no damage, and the crew seemed to have no difficulty in keeping the ship on an even keel during the storm.

In flight, the ship behaved exactly as we expected from the aerodynamic tests which the NPL had made for us. There was absolutely no trace of the instability which certain people (Mr Spanner included) had predicted because of the short fat shape. The ship was easily controllable, and so much so that we have now taken out the auxiliary Servo steering gear, which we had put in as an insurance in case the efforts required to steer the ship had proved unduly great. With the maximum power available from four engines, we reached a speed of approximately 70 mph which confirmed our predictions with regard to the resistance of the ship. The outer cover behaved very well, and the large number of visitors we took up seemed to find the passenger accommodation very comfortable and not unduly noisy.

As you may remember, one of my worst fears was that with regard to the suspension of the power units, knowing that the torque fluctuation in the engines was sufficiently severe to break up the metal airscrews even when not running on a resonance period. We were also afraid of the very severe kick which the engines gave if a bad start was made. However, none of this proved serious, although we had taken pretty severe precautions in the design of the suspensions in order to avoid it with quite considerable expenditure of weight in consequence.

The main engines themselves did not actually give us much trouble during the trials except that towards the end, one of them ran a big end bearing and the subsequent examination of others showed that the bearings were still unsatisfactory. This proved a trouble earlier, but we hoped that it had been cured. Further modifications are now in hand. We have fixed up one of the cars in the Testing Gantry, and propose to put it through a 200-hours-run in order to see if we have now got the final solution to troubles of this kind.

Other troubles occurred in the power cars, many of them due to the failure of the aluminium pipes, which I had used for the fuel and oil installation. I do not think we must blame the engines for this. We have now changed over to copper pipes. The petrol starting engines were really a source of greater anxiety than the main engines, and had to undergo several eleventh hour modifications. I do not think that even now they can be regarded as likely to be reliable for sufficiently long periods. Other small petrol engines are being considered as a substitute, and also a small compression ignition engine. The first of these has been made, and has done a little preliminary running. The principal difficulty at the moment seems to be finding a satisfactory method of starting it.

The position with regard to the provision of astern thrust seems to be as gloomy as ever. Farnborough made us a pair of solid duralumin blades to put in the variable pitch hub, which looked as though they ought to be able to stand up to anything. However, after a comparatively short run in a car on the gantry, they failed completely. The present ideas seem to incline towards interposing a spring drive between the engine and the propeller by which it is hoped to prevent the severe torque fluctuation from getting to the propeller. Such an arrangement would add very considerably to the weight, especially if it has to be built on to the existing engines. In any new engines I understand it could be incorporated inside for less weight. Fortunately for us the wooden airscrews we are using seem to be able to take the punishment from the engines without failure. They have had to be built, however, with very large bosses, and consequently very heavy steel hubs to carry them.

This question of the fixed weights is one of our chief anxieties with regard to the present ships, though I do not think we need take a gloomy view with regard to what can be done in the future.

R.100 is now carrying out her trials, and we are taking the opportunity with R.101 to carry out certain modifications which we hope will give the airship another five or six tons lift. The principal one is the enlargement of the gasbag wiring which will give us more volume. This is made possible by the fact that we now find that the clearances we had allowed between the wires and the girders in order to be on the safe side, are unnecessarily large. Later on, we intend to cut the ship in the middle and add an additional bay which should give an additional 8 or 9 tons *disposable* lift.

There are, of course, many other details of interest, but I find that this letter has already grown unduly long, and doubtless Colmore will deal with some of the points I have left out.

Yours sincerely,

V. C. RICHMOND.

Already on 27th December 1929 Sir John Higgins, AMSR at the Air Ministry, had written to his predecessor, Sir Geoffrey Salmond, in India, to let him know the situation 'about R.101's flight to Egypt and India'.

As a result of the trials we have found that although the airship is very strong and able to stand up to heavy wind, her disposable lift is considerably less than we had hoped for. This can be increased, however, by a number of minor alterations such as the enlargement of the gas bag wiring (there is at present an unnecessary large margin between the gas bags and the outer cover) and the removal of a quantity of fittings and equipment which the trials have shown are not necessary. Unfortunately these alterations can hardly be completed in time for the airship to fly to India before March or April next and at that time of year she will lose a large amount of lift at Ismailia and Karachi, probably as much as seven tons at Karachi. There will be no difficulty in her flying out to India in March or April 1930, but the loss of lift at Karachi is a serious matter when it comes to the return flight.

Accordingly, in order to increase the lift of the airship still further, we are proposing to put in another bay during the summer of 1930 and then to carry out the first flight along the India route after the airship has done further flying trials with the new bay fitted.

Thus, by early in 1930 the Air Ministry in London and the Royal Airship Works at Cardington had reaffirmed as a primary objective the flight to India by R.101 in the autumn of the year. The one essential requirement was the achievement of adequate lift in the airship – to be gained by a reduction in the basic weight, by letting out the gas-bags and by inserting an extra bay.

Richmond was, however, by now clear in his mind that an airship of five million cubic feet of gas capacity was not big enough to meet the requirement for flights over long-range Empire air routes while carrying sufficient payloads. With Michael Rope he was working on a new and substantially larger design project – R.102.

On 22nd January 1930 he explained his views in a note to Sir John Higgins.*

> The airworthiness regulations set a new standard in hull strength but nobody could foresee what this would mean in weight. Now we know pretty definitely and, in playing for safety, I do not think that we have exceeded appreciably the weight implied in these regulations.
>
> The hull weight of the two ships – R.100 and R.101 – are almost identical and Bairstow says that they are of equal strength.
>
> In future, we may save a few tons by refinement of design such as in the fins and the passenger quarters. The deliberations of the new Stressing Committee may also ease the strength requirements, but I do not look for too much in that direction.
>
> The important question of the effect of choice of machinery on weight and consumption is well known. R.101 with five engines and good propellers would attain about the same speed as R.100 and any limitation of range caused by the weight of the engines is compensated for by decreased consumption. The range of neither ship is sufficient, certainly with any payload, during the bad months.
>
> What all this comes to is that a five million cubic feet airship, according to current British ideas, is not inherently suitable to carry 100 passengers over journeys to the East of 2,500 miles non-stop, at all times of the year.
>
> This may be a serious conclusion in view of what the British Public have come to expect but I do not think that the particular airships which have been produced can be blamed nor do I think the position could have been wholly foreseen, in view of the highly experimental nature of the job.
>
> Either we must have our stopping places closer together or our ships larger – or both. I am afraid there is no doubt about the second – larger ships – but I naturally hope very much that the size will not have to be so large that a lot of fundamental research work has to be undertaken. One would like the chance to perfect the present type.

Richmond saw, therefore, a double task ahead. First, to modify R.101 to achieve with it sufficient lift for a flight to India and back by the autumn of 1930; and, second, to define the size of new airships now seen to be required to meet the original objective of long-range commercial air services along Imperial routes.

He and Michael Rope had, in fact, been engaged on design studies for a new, larger and more advanced airship, R.102, since

* AIR 5/14 – XM 08109.

8th March 1929, when a preliminary discussion had been held with Colmore and Major G. P. Bulman, Director of Engine Development at the Air Ministry to discuss future engine requirements. On 18th March 1929 an initial specification was drawn up for an airship of seven million cubic feet capacity, designed to have a still air range of 5,000 nautical miles carrying 15 tons of payload, including 50 passengers and 5 tons of goods at a cruising speed of 65 knots (75 mph).* The preliminary layout of R.102 was for a length of 800 feet and a maximum diameter of 147 feet. Some problems were foreseen, however, of how to construct such a large vessel within the existing No 1 shed at Cardington; 812 feet long, 157 feet high and 180 feet wide.

The required disposable lift of the new airship was set at a minimum of 70 tons including 20 tons for crew, stores and ballast; 15 tons of payload and 35 tons of fuel and oil.

Meanwhile work was going forward on R.101 in the No 1 shed at Cardington to provide the essential extra lift for the India flight, including the design of the structure, the wiring and the gas bag for the new bay. In No 2 shed efforts were being made to solve the problems which had shown up in R.100's outer cover. During three flights totalling 74 hours 46 minutes, between 16th and 29th January 1930, R.100's cover had flapped even more vigorously. Because of all this work both airships were grounded during the late winter and spring of 1930 until R.100 emerged from its shed on 24th April.

The two outer covers were indeed giving much cause for concern. As things stood neither could be considered airworthy.

On 20th January Michael Rope made a detailed survey of R.101's cover with Mr J. W. W. Dyer, the head of the RAW's fabric section. They found substantial damage and serious deterioration along the top of the ship where water had lain and where construction crews had walked while the airship was in its shed.

New strips of fabric, a metre wide, were made to reinforce the top of the cover throughout the length of the ship.

On the same day, R.100's fourth flight, a cinema film of that airship's undulating cover was taken from a Bristol Fighter of the Royal Aircraft Establishment flying over Farnborough. It showed a most disturbing series of ripples in the cover which increased with speed. This fact, together with the obvious lack of watertightness of the envelope so that the gas-bags were constantly soaked led to the hold-

* AIR 2/349 – 522,017/29.

ing of a meeting at the Air Ministry on 7th February at which a decision was taken to ask the Airship Guarantee Company to submit a design for a completely new and more satisfactory cover for R.100. Meanwhile the airship was not to be flown until strengthening, re-doping and modifications to the attachments of the outer cover had been completed.

By this time – from 1st January 1930 – Barnes Wallis had left airship work on his appointment to be Assistant Chief Aircraft Designer (Structure) under Rex Pierson at the Vickers Aircraft Works at Weybridge.* Modifications to R.100 were placed in the hands of Alfred Hessel Tiltman and Nevil Shute Norway of the AGC.

In addition to the worries about the outer cover of both ships at Cardington, much thought was being given to the problems of reverse thrust for R.101 so as to overcome the serious disadvantage of having to carry around one heavy engine solely for use during mooring operations.

Early in February, Richmond was considering the possibility of installing two 240 hp Armstrong-Siddeley Lynx seven-cylinder air-cooled radial petrol engines in the nose of the two forward cars to provide reverse power and to be used also as starter engines for the main Tornado engines in those cars. That solution was discarded as adding too much weight. Then, on 7th March, Major A. Graham Forsyth of the Air Ministry put forward a proposition which not only added virtually no weight but also gave a rapid answer to the problem of reverse thrust in an entirely satisfactory way. By relatively minor modifications to the camshaft of each Tornado engine, the inlet side of the induction system could be changed over to become the exhaust side, and vice versa, so that the crankshaft would run in the opposite direction. Although the fixed-pitch wooden propeller, running backwards, would be relatively inefficient it would be wholly adequate for reverse thrust and two engines could be modified quickly. Thus full forward-power from all five engines could be restored to R.101 simply, cheaply and without much delay. Everyone agreed that this obvious answer ought to have been thought of before.

* Barnes Wallis had become increasingly at loggerheads with the Managing Director of the Airship Guarantee Company, Dennistoun Burney. Both Wallis and Burney were forceful characters and by the middle of 1929 were hardly on speaking terms. At the same time Wallis (who – like Richmond – had become convinced that airships of five million cubic feet gas capacity were too small for long-range services with adequate payload) had decided that military aircraft design and development provided more opportunities for his talents. Wallis left Howden with relish.

This was a major breakthrough. Now, with the design of the new bay well in hand some design effort could be diverted to the new project. Accordingly, a folder was drawn up on 14th March setting out plans for an advanced new airship to incorporate all the lessons learned from R.101; a meeting was held on it at the Air Ministry on 17th March and a preliminary specification for R.102 was prepared by Richmond and Rope on 18th March 1930 – a year after an initial design study.

Much more arduous requirements had been set during the course of the Air Ministry meeting on the previous day in an effort to define the specification for production airships capable of tackling the Indian and the Atlantic air routes in all weathers carrying commercial payloads.

For the new ships accommodation was now sought for fifty passengers and eight tons of cargo and mail; a capacity payload of 15 tons including food and drinks.

A cruising speed of 65 knots (75 mph) was specified so that the Cardington–Montreal route of 3,000 n. miles could be flown in 63 hours against a 20 mph headwind, or from Cardington to Ismailia (via Brest, Barcelona and Malta) a distance of 2,440 nautical miles in 51 hours against a 20 mph headwind. In each instance a 30 per cent fuel reserve was to be provided. On the Atlantic route fuel would therefore be required for a duration of 82 hours while fuel for 66½ hours would be needed on the India route, where at Karachi the available lift would be reduced by eight per cent.

The outcome of this stiff requirement was a design study for a new airship of 9,500,000 cubic feet capacity; (Project 'G') nearly double the size of R.101 and some 36 per cent bigger than Richmond's initial ideas for R.102. Such an airship would have a gross lift of 287 tons, a useful lift of 104 tons and would require a fuel capacity of 62 tons, or 16,420 Imperial gallons, of diesel oil.*

The whole project began to look somewhat impractical when the dimensions were worked out at a length of 906 feet and the maximum diameter at 163 feet. A new shed would be required in which to build, and subsequently to house, an airship of this size – or, at the very least, the existing No 1 shed at Cardington would have to be extended in length by 100 feet and in height by 15 feet.

Had Lighter-than-Air services eventually proved practical, how-

* Though massive for its day this was, of course, modest compared with the 42,500 Imperial gallons (138 tons) fuel capacity of a Boeing 747 Jumbo Jet.

ever, even bigger airships would have eventually appeared. As things were, serious discussions followed at Cardington and at the Air Ministry to explore whether the requirements could be eased. After much deliberation the most practical answer was agreed to be the addition to Imperial routes of three additional airship mooring towers; one, at Monkton in New Brunswick to serve the Atlantic route and one at Malta and one at Basra on the Indian route. Those masts, with two at Cardington to make eight in all, would make possible extra refuelling stops when flights encountered particularly adverse winds. The longest distance would then be 2,100 nautical miles from Cardington to Nova Scotia. In any case such extra masts would provide valuable additional safety as ports of call in a storm or in emergencies as well as offering opportunities for increased payloads from sector traffic.

When these possibilities had been fully assessed and agreed, yet a further study was set in hand at Cardington (Project 'H') for R.102 and R.103 reduced in size to a capacity of 7,500,000 cubic feet; a 50 per cent increase on the size of R.100 and R.101. All this was to take, however, a further four months to resolve.

Meanwhile, during March 1930 a Tornado engine installed in one of R.101's power-cars taken from the ship, successfully completed a 200 hour full-power test-run on the engine gantry at Cardington to the great satisfaction of everyone concerned. At the same time the gas-bag wiring was being let out by three inches all round to give another 130,000 cubic feet of gas capacity, worth almost four tons of lift – a bonus which was too good to lose.

In Cardington's No 2 shed, the doping of the top half of R.100's outer cover was put in hand during March and work began on the changing of the six original, second-hand, Rolls-Royce Condor IIIA engines for six new Condor IIIBs; with larger and more robust crankshafts, stiffer crankcases and a good deal more reliability. The Airship Guarantee Company's parsimony in buying the first set of reconditioned Condors at knock-down prices was now made good by the Air Ministry in a proper insistence upon the best possible equipment for the projected Atlantic crossings.

As a further contribution to those flights, R.100's passenger accommodation was being lightened by eliminating some of the cabins and by reducing the number of bunks in others. While this was going on, a bad tear 20 feet long developed in No 7 gas-bag through the bag being caught up in radial wires during a night-time temperature-change in the shed. The bag was deflated and repaired by a patch

26 feet long by five feet wide. No sooner was the bag back in position and re-inflated than Coxswain Thomas Hobbs, working on the top of the ship, put his arm through the top surface of the bag. The foot-long split was quickly mended – but the accident served to show that the single-ply, skin-lined bags used for both R.100 and R.101 were less robust than the earlier two-ply fabric bags of R.33 and R.36.

A provisional date of 11th May had now been set for R.100's Atlantic flight to Montreal and, as R.101 could not be in the air before June, Lieutenant-Commander Noël Atherstone was detailed to take charge of R.100's third (stand-by) watch to go by sea to Canada to await the airship's arrival as a relief crew during the ship's scheduled two weeks there.

Work on R.100 was complete by 24th April 1930, and at 1330 hours that day the ship was walked out of its Cardington shed by a 300-man handling party, largely from RAF Henlow. Just as the stern was level with the doors a gust of wind from the south swung the tail out of control so that the starboard elevator was damaged against the shed door.

There was no help for it but to take the ship to the tower because the wind had by now risen too much to return it to the shed – although up to the moment of the first gust there had been almost a flat calm. There followed several days of early morning fogs dispersed by gusting winds so R.100 had to remain at the mast, damaged elevator and all.

So it was that R.100 was riding at the Cardington tower on a fine spring evening when on 26th April the German LZ.127 *Graf Zeppelin* arrived at the British airship base to pick up its commander, Dr Hugo Eckener, who had been dining with the Secretary of State for Air in London the previous night.

The great Dr Eckener, then aged sixty-two – short, thick-set, the Napoleon of the Zeppelin Company and the maestro of rigid airship flying – had three days previously arrived in England on 23rd April by sea from the United States where he had been seeking to negotiate a supply of helium gas for future German airships.

On the evening of 25th April, Christopher Thomson had arranged a private dinner in Dr Eckener's honour at the Ritz Hotel in Piccadilly. All the leading protagonists of Lighter-than-Air in Britain were there – from the Air Ministry, from the Royal Airship Works and from Vickers Ltd; Sefton Brancker and 'Josh' Higgins, Reginald Colmore and Vincent Richmond, Dennistoun Burney and Barnes Wallis, Herbert Scott and Carmichael Irwin, Tom Cave-Browne-Cave and Ralph Booth.

It was a lively evening and talk went on far into the night. Among Dr Eckener's contributions to the debate was his conviction that, for the future, helium gas would be a necessity for commercial airship services. With an experienced crew, he said, there was not the slightest danger from the use of hydrogen. But the time was rapidly coming when there would be many airships in service. Their crews would not have the necessary experience to guarantee the absolute safety which was so essential. So, in spite of the extra cost of helium and its reduced lift compared with hydrogen, it would have to be used to avoid any risk of fire.

A salutary and prophetic warning from the most experienced airship captain in the world.*

Next day they were all at Cardington to watch the *Graf Zeppelin* come in. In a light breeze the German airship arrived, impressively under precise control without the help of a mast. It had left Friedrichshafen at 0602 hrs, carrying twenty-two passengers, and it had flown by way of Basle, Dijon, Paris, Brighton and London to Cardington in 11 hours 13 minutes.

The airship was on the ground with the passenger-gangway lowered within fifteen minutes of coming into sight from the south at five o'clock in the afternoon. It was away, with eighteen people embarked, in the space of thirty-five minutes. The smooth competence of the whole affair was an eye-opener; a splendid demonstration of airship expertise, helped by the fact that there was no more than a light wind and a steady temperature.

Among the seventeen British passengers who embarked for Germany with Hugo Eckener were Sir Sefton Brancker, Colonel Richmond and Squadron Leader Rope; guests of the Zeppelin Company as leading protagonists of rigid airships.

'Grabby' Atherstone noted in his diary on 26th April:

> Dr Eckener arrived this morning from London and he was shown all over the works and both ships. He appeared to be very impressed with the steadiness of R.100 at the mooring tower and, from his remarks, gave one the impression that our method of mooring to the High Tower was very much better than he had been led to expect.

* Use of helium did not, however, prevent the loss in 1933 and in 1935 of the two new United States rigid airships, *Akron* and *Macon*, each of 6,850,000 cubic feet capacity. They were both wrecked over the sea in bad weather partly because they were being flown by inexperienced crews.

He was keenly interested in R.101 as, since it was empty, he was able to get a very good idea of the structure of the hull. He liked the position and shape of the control car and he was particularly struck with the design of the bow. After lunch he was taken round the shops, inspected the engine cars and engines and was even shown test samples of a new type of girder in which he showed great interest.

During the day we received various reports from the *Graf Zeppelin* indicating her progress towards Cardington. She was over London at 1630 BST and appeared low down to the south of the aerodrome here shortly after 1700 BST. The weather was overcast with a WNW wind of about 15 mph so conditions for landing were fairly good. She did not fly over the aerodrome before landing but worked round to leeward and approached the landing party up wind, gradually getting lower and lower until finally she was stopped with the bumping bag under the control car about two feet above the ground and right over the landing 'T'.

It was the most perfect landing I have ever seen and a fine demonstration of the marvellous control which Lehmann has over that ship. She was a little heavy as she was approaching the landing party but a little ballast from forward and aft was all that was required to correct that.

Once the ship was down she was immediately surrounded by the very large crowd which had gathered to witness this historic event and it was a bit difficult at times to keep them clear of the ship. There was no time lost in disembarking the passengers and embarking others for the return journey to Friedrichshafen. The whole thing was done in about half an hour without any fuss or bother.

Dr Eckener was the last to get on board and after the ship had been ballasted up and found to be just light he made a speech thanking us all for the reception he had been given here and extending a very warm invitation to visit Friedrichshafen in one of our ships as soon as possible. The engines were then started and the order given to let go.

Some ballast was dropped forward, the ship slanted up and with the engines opened out she climbed straight off the ground and, in a few minutes, was out of sight.

She landed at 1720 and left at 1750 BST. The whole show from beginning to end went off without a single hitch and can only be described as perfect.

This was the *Graf Zeppelin*'s 60th flight and it completed the Cardington–Friedrichshafen journey in 12 hours and 52 minutes.

The intention had been to keep the landing party together to take R.100 into the No 2 shed after the *Graf Zeppelin* had departed. Unfortunately it came on to rain heavily and the wind began to gust so the men were dismissed. By morning the wind had dropped, the party reassembled and R.100 was kennelled at 0600 hours, bow first – with three members of the *Graf Zeppelin*'s crew observing from the control car. They were design staff doubling up as members of the flying crew to gain experience – a point not wasted on those at Cardington.

Another result of the *Graf Zeppelin*'s visit, and the smart appearance of the members of the German airship's crew in their naval type uniforms, was a decision from the Air Ministry to issue dark blue uniforms to the flying staffs of R.100 and R.101 – a reversal of an earlier edict. Up till now both British crews had flown in mufti.

A tailor was sent up from London and all was bustle to get R.100's crew kitted out before the flight to Canada began.

Excellent impression as the *Graf Zeppelin* had made at Cardington, something of a furore was raised in the rest of the country when, on the way in to Cardington, the airship had flown over the Association Football Cup Final at Wembley.

The *Graf Zeppelin* crossed the Stadium at a couple of hundred feet, stopped the game, made headlines and incurred the wrath of the sporting columnists. Cup Finals in Britain were then, as now, too much akin to religious festivals to be so desecrated.

Once back in the shed the repairs to R.100's elevator took longer than had been expected and the airship did not appear again until 21st May.

For several days previously the handling party had been awaiting a flat calm to move R.100 at first-light from shed to tower. There would then be one final test flight before the airship left for Canada. Each night – following a weather forecast – the move was cancelled, by the simple device of shining a searchlight into the sky from the tower for an hour between 2130 and 2230 hours. There was no searchlight beam on 20th May and shortly after 0300 hours next morning R.100 was transferred from shed to mast without mishap.

The airship's sixth flight was planned to last 24 hours specifically to test out the modifications to the cover, the installation of the new engines, the repairs to No 7 gas-bag and the operating range of the radio equipment. For the first time Scott was not on board, leaving Ralph Booth, R.100's captain, in sole charge.

If all went well the plan was that R.100 would leave for Canada on 25th May and return to Cardington early in June. Lord Thomson would make the journey both ways in the ship and take the opportunity to visit both Ottawa and Washington DC while the airship was engaged in local flights from the Montreal tower.

With that prospect ahead, R.100 slipped from the Cardington tower at 1802 hrs GMT on 21st May with sixty-five on board (forty-one crew and twenty-four passengers) including Fred Montague, US of S for Air. The disposable lift was 48.4 tons, the highest figure yet achieved.

Over Buckingham Palace the airship dipped in salute to King George V and Queen Mary, on their way to the opera at Covent Garden, and then turned east to fly up the Thames Estuary into the North Sea and north to Flamboro' Head during the night.

The weather was poor with a good deal of low cloud and some heavy rain in cloud at the airship's flying height of 2,000 feet. The downpour once again soaked the gas-bags and their restraining nets. During the following morning, over the Midlands the airship was run for three hours on all engines at 1,600 rpm to give a true air speed of 70 miles an hour.

Over Liverpool, Sir Dennistoun Burney who was on board as a representative of the AGC, asked Booth to open the throttles up to full power for ten minutes to see how the modifications to the cover would stand up to maximum speed. At 1,900 rpm from all engines, at a total of 3,900 bhp, R.100 again reached 80 miles an hour while heading back to Cardington.

But all was not well. When R.100 came in sight of the tower, Flight Lieutenant Brian H. Cook, the Officer-in-Charge, radioed that the pointed tail of the airship, aft of the rudder and behind the last transverse frame, was hanging down in a most unbecoming attitude. The upper structural members in the tail had failed during the high speed run. This conical tail section was a lightweight streamlined fairing which contained no gas-bags. Fortunately its collapse had not fouled the control surfaces.

Conditions for mooring were difficult in squalls and thunder showers and the tail tended to drop when the engines were run in reverse. After some manoeuvring 'while Flying Officer Cook played R.100 as an angler plays a fish' the ship was successfully locked home to the mast at 1657 hrs after 22 hours 55 minutes in the air and 691 nautical miles flown.

While R.100 was still in the air the Canadian Department of National Defence had cabled to Lord Thomson to ask that the flight to Canada should be postponed until after the Canadian Parliament had risen at the end of May, because many Ministers and Members of Parliament wanted to be in Montreal to see the arrival of the airship. Lord Thomson cabled back at once, to agree to this postponement.

The news was a relief to the airship's officers when they landed. The embarrassment of the failure of the tail section – although not serious in itself – would have been compounded had it been the cause of a further delay to the Atlantic flight. There were more troubles, however. The cover was still not waterproof, the gas-bags were once again very wet and a split had opened up in the underside of the cover behind the aft power-car.

In these circumstances the flight to Canada was put back until early July when the forthcoming Canadian elections would be over and Members of Parliament there would be free to come to Montreal again. Meanwhile, R.100 rode at the tower awaiting good weather so that the airship could be man-handled into its shed.

On 24th May a different flight was successfully completed – a flight which had been followed eagerly at Cardington. On that day Amy Johnson arrived at Port Darwin in Australia, having crossed the Timor Sea in her D.H. Moth to complete the first solo flight from England to Australia by a woman, in 19½ days from Croydon.

On Monday, 26th May, R.100 was back in the No 2 shed for repairs to be undertaken by the Royal Airship Works there. New detailed drawings were prepared for the tail section in consultation with Professor Bairstow while N. S. Norway was kept in the picture as a representative of R.100's constructors, the Airship Guarantee Company. Burney, however, raised contractual problems in his usual difficult way so that all work was held up until these could be resolved.

After much to-ing-and-fro-ing Sir John Higgins authorised work to go ahead at the end of May on Colonel Richmond's proposal that R.100's original lightweight, pointed, tail-fairing should be replaced by a short, dome-shaped section to be constructed at the Royal Airship Works. This decision was directly contrary to Norway's plans for the replacement of the failed fairing by a facsimile structure of heavier construction.

Norway was overruled – and he didn't like it.*

* The disagreement between Norway and Richmond on this subject led to Norway's gradual build-up of hostility towards Richmond, R.101 and the Royal Airship Works. This was eventually expressed in his otherwise attractive autobiography *Slide Rule* (William Heinemann Ltd., 1954). The disagreement over R.100's tail, and Richmond's subsequent refusal to offer Norway employment at Cardington, led Norway to write distorted and inaccurate accounts of the airship programme and those concerned with it and to unsupported suggestions of

Work went ahead on both airships at Cardington where, now, some 800 men and women were employed.

During the first ten days of June, therefore, the Royal Airship Works had in hand five separate projects:

1. A new tail-section for R.100.
2. The installation of a Rolls-Royce Condor petrol engine and reverse gearbox in one of R.101's spare engine cars so that a full speed trial on five engines could be run at an early date. The Condor would replace the reverse-thrust Tornado in the starboard forward (No 1) engine car of R.101.
3. Discussions with Burney and Norway of the AGC on the design and construction of a new bay and gas-bag for R.100 to increase its gas capacity and lift, on the same lines as the new bay now agreed for R.101.
4. Examination of the possibility of converting R.100's Condor engines to run on fuel-gas.
5. Preliminary design work on R.102 and R.103 in two forms; 9½ million cubic feet capacity (Project 'G') and 7¾ million cubic feet capacity (Project 'H').

Significantly the outer covers of both ships were still causing anxiety. The proposed new cover for R.100 was being detailed at the Airship Guarantee Company's Design Office at Howden but it could not be ready much before the end of the year. The cover of R.101 was now suspect also, following its soakings at the mast in October and November 1929, and its subsequent six months in the humid atmosphere of No 1 shed at Cardington.

Michael Rope, at Richmond's right hand, was especially concerned. On 2nd June he addressed a serious note to Richmond and Colmore on the results of a series of tests which he had made with the fabric of the cover of R.101 to compare its existing strength against the original requirement for a breaking strength of 700 lb per foot run.

incompetence and prejudice at Cardington and the Air Ministry. These views have been widely quoted as fact in later years. Chapters Three to Seven of *Slide Rule* have, regrettably, become a source of a great many erroneous impressions and statements about the 1924–30 airship era in Britain. Norway eventually regretted this himself and said so to the author in Australia in 1953.

He was greatly bothered about what he had found. He wrote:*

> Measurements of the cover of the ship have failed to find anything over 85 lb per foot whereas straight flight at 76 miles an hour is calculated to result in tensions of 143 lb per foot.
>
> Laboratory experience cannot account for the deterioration and the only explanation which suggests itself is that, throughout the history of these covers, permanent extension has taken place locally rather than uniformly along the weft. Calculations and small scale laboratory experiments may be liable to large errors – but what scant information is available from these and other observations in the shed suggest that there is no margin of safety for flight in rough atmosphere.
>
> It is for consideration as to whether the risk involved in sending either ship on a long overseas flight is – or is not – greater than is justified by the need to fulfil public expectation.
>
> Is it not conceivable that a public statement could be made which would satisfy the people who matter – to the effect that overseas flights have been postponed for, say, six months on account of improvements which have been shown by test flights to be desirable before long flights could be undertaken with the reliability required of convincing demonstrations – etc – etc?

Here was a clear warning of potential disaster. Everyone held Rope and his technical competence in high esteem.

Later the same day, 2nd June, a further inspection of R.101's cover, prompted by Rope's note, revealed tears in the cover towards the bows caused, it was thought, by the high humidity in the shed, going up to 95 per cent, coupled with very high summer temperatures.†

The result was that on Friday, 6th June, at a meeting in Reginald Colmore's office, a decision was taken to replace the pre-doped cover of R.101 with a new cover made of cotton and linen and doped after fitting into place; a reversal of the previous procedure. In the interim, for the small amount of flying expected to be undertaken before the new cover could be ready, the existing cover would be reinforced with fabric bands.

On 11th June, at the personal request of the new Chief of the Air Staff, Air Marshal Sir John Salmond, R.101 was detailed to fly

* Note from Squadron Leader F. M. Rope to DDAD now among Wing Commander T. R. Cave-Browne-Cave's Papers at the Imperial War Museum. This note was not quoted at the Public Inquiry into the loss of R.101.

† Colonel Richmond's diary, 2nd June 1930.

over the Royal Air Force Display at Hendon in just over two weeks time, on 28th June. That would mean that there would not be sufficient time before the Hendon Display to replace the reverse thrust, Tornado diesel engine, in the forward-starboard nacelle, with the new Condor petrol engine capable of providing both forward and reverse thrust through its reversing gearbox.

Next day, 12th June, the Secretary of State for Air paid another informal visit to Cardington to discuss progress. He expressed to Colmore and Richmond once again his hope that, if possible, R.101 could be made ready to fly to India before October – 'but not,' he said, 'at any risk of the necessary work being rushed. Safety is our first consideration.'

He had, indeed, stressed these points in a speech at the House of Lords in an Aviation Debate nine days before, on 3rd June. In it he made clear his positive attitude to the airship programme, its experimental nature and that, whatever else was done, no risks would be taken.

He had begun by saying that *'the airship programme which I had the honour to introduce in this House in 1924 was then described by me not once but several times as experimental and tentative.

'It was never claimed that we were going to build two commercial airships which would at once take the air as their native element and fly to all parts of the globe. That was not so at all. If you will allow me to re-enumerate the stages of that programme, I think you will see that there is some justification for that remark.

'The programme was to begin with experiments and research and then go on with full scale flying with an old airship, reconditioned.

'There was then to be the construction of an experimental section. The fourth stage was meteorological investigation of the most searching character on the Indian route. The fifth stage was the provision of the ground organisation and the sixth stage was the building of two ships. The building of these two airships was started only three years ago.

'So, I repeat that this programme is an experiment and the final stages of that experiment will be made only this month in regard to one ship – R.100 – and in September in regard to the other – R.101. If there is any blame to be imputed on the slowness of this matter I would like to take it at once. This is one of the most scientific experi-

* House of Lords *Hansard*, 3rd June 1930, Columns 1359/1363.

ments that Man has ever attempted and there is going to be no risk while I am in charge of the thing being rushed or of any lives being sacrificed through lack of foresight. It is far too scientific and important a matter for that.

'I am graeful to the Press for the advertisement they have given to the programme but, as a matter of fact, this scientific experiment really does not need advertisement. Sometimes undue publicity is a little inconvenient as when people get the impression that we were building an airship which will take one hundred MPs all over the country. To suggest that was the greatest error of all my life. I was punished by weather which made the flight impossible.

'These airships are still scientific experiments. They will not cease to be experimental until their overseas tests have been completed.

'One will take place at the end of this month of June to Canada and the other to India in September. I regret the recent disappointment which has been caused in Canada by the delay. Hopes were raised in Canada that, early this month, there would be R.100 over Montreal where the Canadian Government, at their own expense, have built a mast. No one deplores more than I do this disappointment but it cannot be helped. We are not going to send that ship while there is anything perceptibly wrong with her. And, though what was wrong was of a very minor character, it is going to be put right before that ship goes overseas.

'Until the test flights have been finished it is impossible to give an idea of what the future airship policy is going to be. If the almost unthinkable happened and these ships failed on their final flights, there might be a strong movement towards scrapping the programme. That would be heart-breaking. I cannot tell you what magnificent work has been done by the airship men. There are officers in that work today who have sacrificed promotion and many other advantages in order to get on with this great experiment. We have a body of enthusiasts at work there and only the worst of luck could interfere with the success of the experiment.

'If we do have bad luck I should have thought it was in the British character to stick to it still and triumph over temporary adversity.

'Supposing – and I think it is very likely to be the case – that these two final test flights prove satisfactory, it may be urged that the construction should then proceed forthwith of a 7,500,000 cubic feet ship as that would be about the size of an airship to be a commercial

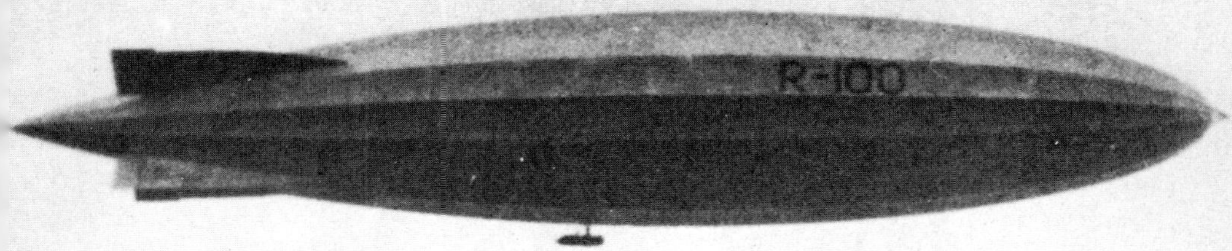

R.100 – HM Airship R.100 arriving at Cardington after its first flight from Howden on 16th December 1929 – a flight of 5 hours 37 minutes carrying 45 crew, 12 passengers and 6,500 gallons of fuel of which 380 were used.

As it was – R.100 riding at the mooring tower at Cardington on 17th December 1929, shortly after the Howden-built airship had flown south from Yorkshire to Bedford. The structure of the original, pointed, tail section behind the tail surfaces, shown in the photograph, failed during R.100's flight on 22nd May 1930. It was then replaced by a less vulnerable rounded stern. The concave fabric panels of R.100's outer cover, pulled in with tapes, are clearly seen.

[PLATE XV]

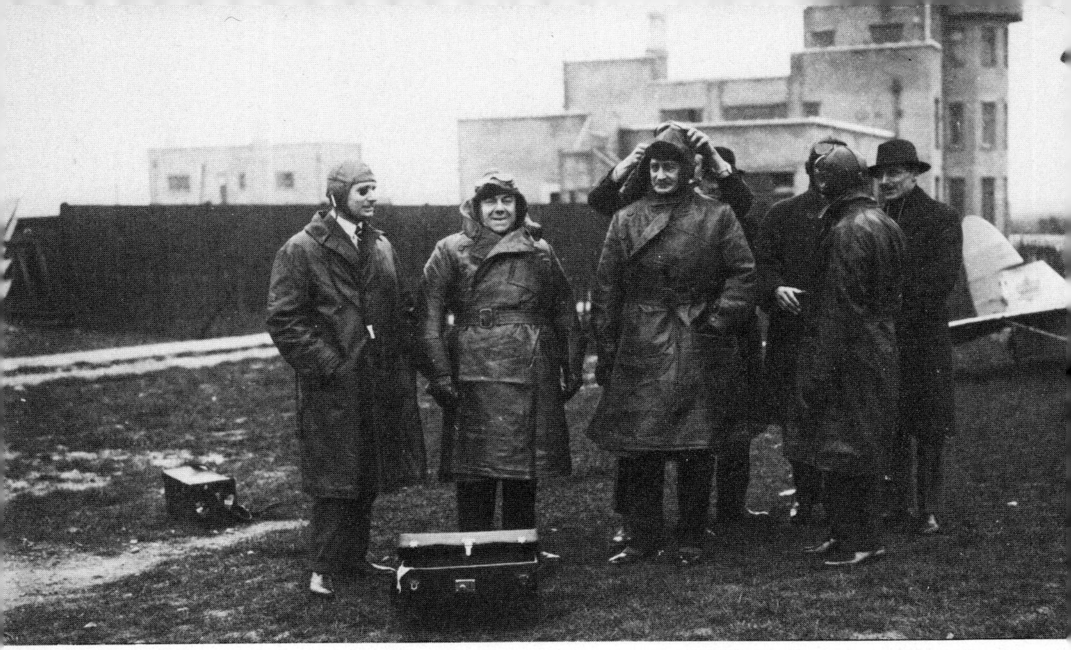

Preparing for Flight – At Heston Air Park on 25th January 1930, the Secretary of State for Air is fitted out with leather flying coat, helmet and goggles ready to take the air in the open cockpit of the D.H./Gloster A.S.31 Air Survey twin-engine biplane piloted by Howard Saint, Chief Test Pilot of Gloster Aircraft Ltd. (Left to right), Major Hemming of the Air Survey Company, Ltd., Fred Montague, MP, (Under Secretary of State for Air), Lord Thomson, Howard Saint and Alan Butler (Chairman of the de Havilland Aircraft Co., Ltd.).

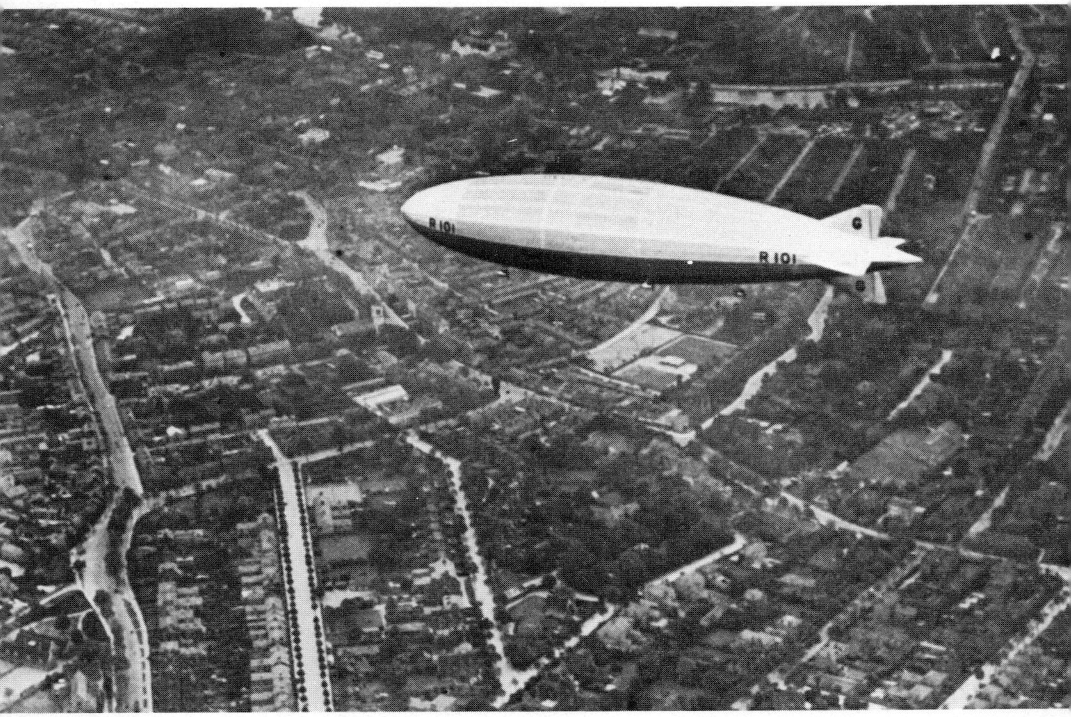

Home-Town – R.101B flying over Bedford on 26th June 1930, on its eighth flight and its first appearance after its lay-up for the letting-out of the gas-bag wiring and the reduction of fixed weights. The disposable lift had been increased by almost $5\frac{1}{2}$ tons to some 40.4 tons compared with its original condition in the autumn of 1929. The photograph shows R.101 flying over the north bank of the River Ouse with Bedford's Prebend Street bridge above and behind the tail (top right) and the Clapham Road in the foreground (bottom left). The nose of the airship approaches the fork of Dame Alice Street and St Loges Street. [PLATE XVI]

proposition. Whether that ship would be constructed at Cardington by the Government constructors is still an open question. It is still one of the policies open to the Government.

'An alternative policy would be to invite the collaboration of the big shipping companies. If, for example, during the course of this year several flights take place over great distances and careful records are kept of the expenditure, of the running costs, of the risks to be avoided and of the methods of navigation to be pursued, I have many good reasons to believe that the great shipping companies would be interested – and not only shipping companies at home but shipping companies on an international basis. I have always regarded airships as an international means of navigation.

'My object has always been to make British airships the best and the safest in the world because I feel that we ought to take the lead in this matter as, essentially, an Empire problem.

'You may well ask why it is that the *Graf Zeppelin* can go round the world and make enormous flights while our two ships spend their time mainly in their sheds. That is a very easy question to answer.

'The Germans have thirty years' experience of building airships. They ran commercial ships from Lake Constance before the war. I do not wish to decry the virtues of our own people but, in view of that experience, I think it is only natural that we should accept the fact that there are very few Dr Eckeners in this world. We have not yet had time to produce our Dr Eckener – one of the most remarkable men I have ever met. They are not found in every generation.

'Then again, though I wish to say nothing in depreciation of the *Graf Zeppelin* as a ship I do not think that you should compare it as a structure with R.101 or even R.100. We have started out on the design of these two ships on the basis of first principles.

'We have not copied any well-known model but we have produced two ships which are the strongest in the world. One of Dr Eckener's principal experts – Dr Ludwig Dürr – after seeing R.101 said to me, "That is the safest conveyance on land or sea or in the air." He was enthusiastic about its qualities.

'During the construction of these two ships we have never hesitated to sacrifice a gain in lightness to solidity. We have produced two almost unbreakable ships capable of flying in any foreseeable weather provided they are properly navigated.

'There are improvements to be introduced into both of those ships which will considerably increase their ability to lift loads but these improvements will be on the basis of "Safety First" – exactly on the same basis as that on which the ships were originally constructed.

' "Safety First" may not be a very paying proposition in regard to politics* but I am sure that it is the right thing in regard to airships.

'What have we achieved? We have those two ships and we have a magnificent ground organisation and design staff. We have spent £2,300,000 odd on the whole programme. I read in the newspapers about the "million pound ship". There is no £1,000,000 ship. The true figures are certainly interesting.

'The cost of the two ships together has been £894,000. Research cost £305,000. Then we have a ground organisation which includes two of the largest sheds in the world, three mooring masts – one of the great developments due to British enterprise – and, last but not least, an immense amount of knowledge. Personally, I believe the knowledge we have gained – which we could not have compiled unless we had built the ships and absorbed practical experience – is alone worth £2,300,000.

'There are various improvements in sight and if we can only keep our design staff going I have not the least doubt that our expenditure will be amply justified by future developments.

'There are three conditions of safety (as my advisers tell me) in airship navigation. *The first* is sound construction and first-rate materials. Those, I think, we have. *The second* is good navigation and well-trained crews. Good navigation is necessary to avoid thunderstorms and electric discharges,† and well-trained crews are as necessary on airships, or even more necessary, than on any sea-going ship. It is amazing how our men are training in airship practice. They are taking to airships like ducks to water; perhaps I should say, "like eaglets to the air". They seem to be at home in it at once.

'The *third* condition of safety for an airship is good meteorological information, communicated promptly and systematically. That also is being attended to and I think is satisfactory.

'I have always been an enthusiast in all these matters. I take a special pride in the fact that I introduced this programme and I want to see it through to ultimate success. I can only add this: that in

* Mr Stanley Baldwin, leader of the Conservative Party, had fought the General Election of 1929 on the slogan 'Safety First' – and lost.

† Lord Thomson clearly had in mind the loss of the French airship *Dixmude* (formerly the German Navy Zeppelin L.72) which was struck by lightning off Sicily in December 1923, and lost with all hands.

spite of many setbacks and some disappointments, my faith is absolutely undimmed. I am sure that the people of this country will be right in continuing these experiments. Even if they were to continue them on the basis of research alone they would find that their money had been well expended.

'But it is unnecessary for me to point out what the airship may mean to our race and to our Empire. It is essentially a vehicle for going over wide open spaces – the oceans. It is not so much for going over land. But it will link up our Empire in a way which, so far as I can see, no other means of transport can approach. It combines speed, safety and indeed amenities because, to travel in an airship is by far the most delightful form of travel that I, personally, have ever experienced.'

This speech was by far the most comprehensive statement on the airship situation in Great Britain which had so far been made in public. It was also a personal reaffirmation of faith and, at the same time, a realistic summing up of the 'state of play'.

Lord Thomson had stressed the experimental nature of the whole programme and how much depended upon a successful flight to Canada by R.100 at mid-summer, followed by a flight to India by R.101 in the autumn. But nothing was going to be rushed. Neither ship was going to embark upon an overseas voyage until those in charge were satisfied that everything was in order.

If, in spite of all precautions, things went wrong then, nothing daunted, the ultimate goal should still be pursued. There was almost a hint of a premonition in this – 'we have produced two almost unbreakable ships capable of flying in any foreseeable weather *provided they are properly navigated'* – and 'well trained crews are as necessary on airships, and even more necessary, than on any sea-going ship' – but 'we have not yet had time to produce our Dr Eckener.'

' "Safety First" may not be a very paying proposition in regard to politics but I am sure it is the right thing in regard to airships . . .

'We have a body of enthusiasts at work and only the worst of luck could interfere with the success of the experiment. If we do have bad luck I should have thought it was in the British character to stick to it still and triumph over temporary adversity.'

The policy was clear. It was firmly stated and as firmly pursued.

Pioneering has always had its hazards – not least in the air.

CHAPTER THIRTEEN

Hendon Occasions

'I do but wait a time and fortune's chance: oft many things do happen in one hour.'

Sir Thomas Wyatt

23rd June 1930

As June moved on towards July time was clearly getting short. Nothing was going to be rushed, but the Secretary of State and everyone else concerned with the airships were acutely aware of the need to be seen to perform and – however deliberate the progress – for the airships to demonstrate their capabilities.

Early on 23rd June 1930, while R.100 was being prepared for Canada in No 2 Shed, R.101 was at last taken from No 1 Shed after nearly seven months of incarceration and modifications. In spite of a stiffish breeze, fortunately blowing exactly along the fore and aft axis of the shed, at 0500 hrs R.101 was walked out of the shed in four minutes. Moored to the mast in a freshening wind the upper side of the outer cover – not yet replaced but strengthened with fabric bands – began to ripple from bow to stern with a converging undulation coming forward from the tail.

Shortly after 0600 hrs the cover suddenly split along the starboard side from Frame 3 to Frame 6, a rent 90 feet long. There was a hasty conference. A decision was made to repair the cover at the mast and that additional strengthening-bands should be stuck along the length of the ship between Frames 3 and 12.

A. E. Gerrish, the shed manager, with Flight Sergeant G. W. 'Sky' Hunt, R.101's Chief Coxswain, mustered a repair party of riggers at once but, while the work was going on another split, about 40 feet long, appeared on the port side between Frames 6 and 7 – depressingly confirming the rottenness of the cover. By dint of hard work both these tears were repaired by the afternoon of Wednesday, 25th June, ready for a 'refit' trial flight on the following day, with the RAF Display only two days ahead.

On the airship's eighth flight, and its first of the year 1930, R.101 left the Cardington mast at 1551 hrs to cruise in the vicinity of Cardington, carrying 4,760 gallons of fuel oil weighing 18 tons, with a crew of thirty-seven and nine passengers. The ship was locked back at the mast at 2026 hrs after 4 hours 35 minutes in the air. In good weather with some heat-turbulence from the ground R.101 had behaved well and the controls were described as 'powerful and fully adequate'.

By the time that R.101 returned to the mast the results of the latest Lift and Trim checks, made before the ship left the shed, were available for consideration. They provided a measure of the value of the various modifications completed during the past seven months.

In the first place, the letting out of the gas-bag wiring was found to have increased the volumetric capacity of the bags by 104,760 cubic feet – a little less than had been hoped. The result was that R.101 (now 'R.101B') had a volumetric capacity of 4,998,500 cubic feet.* The nominal gross lift with full bags was thus increased by 7,115 lb, or by 3.176 tons, to a total of 151.74 tons at 68 lb of lift per 1,000 cubic feet.

At the same time the weight-saving exercise had reduced the fixed weight of the airship by 5,152 lb, or by 2.3 tons – a reduction from the original 113.6 tons in October 1929, to 111.3 tons in June 1930.

The net result was, therefore, an increase in useful lift of 5.476 tons from the original 'disposable lift' of 34.964 tons of R.101A on 11th October 1929, to the rather less unsatisfactory figure of 40.44 tons on 22nd June 1930.

This improvement in lifting capacity was worth in terms of payload an additional 35 passengers with their baggage, food and drink or, alternatively, an extra 650 n. miles of range, cruising at 55 knots in still air.

In detail the achieved weight reductions had included: †

Main structure: lightened by 1,168 lb to 49.854 tons;
Passenger Accommodation: lightened by 1,890 lb to 12. 456 tons;
Crew Accommodation: lightened by 665 lb to 3.128 tons;
Systems: lightened by 2,235 lb to 5.476 tons;
(Controls, hydraulics, electric and fuel systems)
Power Plant: lightened by 1,657 lb to 14.515 tons.

These weight savings, which added up to 7,615 lb or 3.4 tons, had been partly offset by some increases in the weights of the extended gas-bag wiring and by the reinforcements to the outer cover which, together, came to 2,463 lb or 1.1 tons. The net weight reduction was, therefore, 2.3 tons. When the increase in lift of 3.176 tons was taken

* See Appendix 6: 'Volumetric Capacities of R.101; 1929–30'.
† See Appendix 10: 'R.101's Comparative Weights: 1929–30'.

into account the additional disposable lift added up to 5.476 tons – or a worthwhile 15.6 per cent.

As soon as the airship returned to the mast following its 'refit' flight these results from the Lift and Trim trials taken on 23rd June were announced from the Design Office – and received with a good deal of satisfaction. Even so, everyone understood that the increase of disposable lift, to the new figure of 40.44 tons under ideal conditions, was still not enough to make possible the flight to India and back with any tolerable safety margins on lift or range, understanding that in the hot temperatures at Karachi, the disposable lift would be reduced to no more than 30 tons.

The new bay to add a further net lift of about nine tons would be essential before the India flight could be seriously contemplated. There was just about three months in which to get it all completed.

Meanwhile, the immediate objective was to make sure that the appearance of R.101 at the Royal Air Force Display at Hendon could be mounted successfully two days ahead, on 28th June.

There was, in addition, one further matter to be sorted out. At the end of the short test flight on 26th June R.101 had been found to be flying somewhat heavy and, in order to lighten the ship before mooring, two tons of fuel oil had been jettisoned from the control car as the ship approached the mast. The reason for this heaviness at the end of a five hour flight was put down to the fact that when the ship left the mast at four o'clock, local time, the heat of the sunny afternoon had 'super-heated' the gas in the bags to give a substantial amount of extra lift. By nine o'clock in the evening the gas had cooled with a consequent reduction in lift. Even so, the heaviness seemed somewhat excessive for such a relatively short time. It needed further investigation.

To make things more difficult, 'Bird' Irwin was having to fly the airship without the skilled and perceptive assistance of 'Grabby' Atherstone, his First Officer, on whom he had come to rely. Atherstone's absence in Canada was sorely missed during the three flights of R.101B (the airship's second commission) during the last week of June 1930.

In fact, Atherstone had been summoned back to Cardington from Montreal on 12th June in order to take over as second-in-command of R.101 on its Hendon flights. Problems of gaining a berth on a Canadian Pacific Liner out of Quebec held up his arrival at Cardington until R.101 was returning to the mast from the Hendon Display on 28th June.

Atherstone's absence meant that Irwin was backed up by officers from R.100 who were unfamiliar with R.101 and so not able to report in the informed manner on the airship's behaviour which Atherstone would have done.

Meanwhile, preparations were made for the RAF Display.

Ahead of that, Friday, 27th June was the occasion of the 'dress rehearsal' for all aircraft taking part in the Display. On that day R.101 was out again, this time for 12 hours 33 minutes; slipping from the mast at 0722 GMT and returning to lock home the 'dewdrop' at 1955 hrs. On this rehearsal flight the airship carried a crew of 36 and five passengers, including Squadron Leader Ralph Booth standing in for Noël Atherstone. R.101 left the mast with 4,592 gallons (17.3 tons) of fuel and 12½ tons of water ballast on board, at a gross lift of 148.05 tons with 36.75 tons of disposable lift.

The flight took place in generally hot and turbulent conditions below cumulus cloud in intermittent sunshine. Colonel Richmond recorded in his diary:

> During rehearsals for the Pageant the ship had to be pulled rather suddenly out of a steep dive and, shortly afterwards, one of the lifting bridles at the starboard side of Frame 8 snapped. This was apparently due to the wear caused by passing over the small diameter pulleys in the chestnut fittings.
>
> It was noticeable that the ship – which was flying at a comparatively low speed – had to be pitched up considerably by the nose and was rather unstable in pitch generally. A considerable amount of water ballast had to be dropped on landing.

Ralph Booth added:* 'I noticed that the ship was getting heavier than was consistent with the temperature and the height at which we were flying. During the day we had to let go some ballast as the height coxswain reported that she was difficult to keep at the flying height of 1,000 to 1,200 feet. For the whole day we expended nine tons of ballast and, assuming that we used about two tons of fuel, that makes up 11 tons of ballast that we expended. I remarked on this to Flight Lieutenant Irwin and he, I think, was under the impression that the gas valves were giving trouble owing to the slackness of the outer cover.'

Booth reported also that the dive mentioned by Colonel Richmond had been a deliberate dip in salute over Hendon Aerodrome when R.101 had been brought down steeply to 500 feet from the ground – lower than intended – and that the height coxswain had then pulled

* In evidence at the R.101 Inquiry.

up the ship rapidly on the other side of the aerodrome. At that moment the bridle, just aft of the passenger accommodation, at Frame 8 had snapped with a jerk which was distinctly felt in the control car a hundred feet further forward; not a serious matter but a subject to be put right.

Clearly the ship was not handling as well as it had done during the seven flights of its first commission in October and November 1929. Atherstone's absence meant that Irwin had to form his own judgement of the possible causes.

Meanwhile, on that same Friday, Lord Thomson addressed the Fourth Imperial Press Conference in London, gathered to attend the RAF Display next day. Again it gave him an opportunity to show his complete absorption with aviation and to express his belief in its widening future.

The Secretary of State began by welcoming the Empire delegates and by saying that because of its responsibilities for the defence of the greater part of the British Empire as well as that of the United Kingdom the Royal Air Force bore a heavier load than did the military aviation of any other country. As tomorrow's Display would show, the Royal Air Force made up in efficiency and in mobility what it lacked in numbers.

'We have 778 machines in England and throughout the Empire, including 80 on the North-West Frontier of India; the French have 1,350 aircraft mainly concentrated in France· the Italians have more than 900 aircraft in the Mediterranean area and the Russians have 600 aircraft in their own territory and parts of Asia.' *

High quality of personnel and mobility of squadrons were two of the RAF's great attributes. The annual RAF Display was the culmination of the year's training but though the standard of flying would convince everyone who saw it of the efficiency of the Service, those taking part were not specially trained but drawn from squadrons doing their normal tasks.

'I remember driving a great French ace from Hendon down to Chequers after the Air Display in 1924 and he had said: "Your average is equal to our top-notch". Well, you will see tomorrow.'

For still further mobility air routes must be extended throughout the Empire. As a start, Imperial Airways now ran across Europe, through Egypt and Arabia to Delhi in India.

'Now we want to extend that route on through Calcutta, Rangoon, and Singapore to Australia. A survey is now complete for a line run-

* AIR 5/904/07090.

ning from Cairo to the Cape. I hope to see that line through next year. It will be a tremendous achievement.

'At present air transport has to be subsidised. But in due course we shall put civil aviation on a commercial footing. It will pay its way. In that belief I, for one, am willing to go on with present subsidies. But in ten years time there will be a transformation in the whole development of air transport both in terms of performance and in its economics. Of that I am sure.

'Even the Atlantic will be spanned safely and regularly by aircraft – just as Kingsford Smith has done this week as a carefully planned but still exceptional performance. Eventually, I look for air services from Croydon to Chicago through Canada.'

But the most promising form of long distance air transport was, undoubtedly, the airship.

'R.101 has been over London during this morning. R.100 is at Cardington. It was to have gone to Canada some time back. On that occasion we disappointed the Canadian government. We have been slow but we shall be very sure. I assure you of this – that the men responsible for this very great experiment with airships are not going to be rattled into doing anything which is not strictly in accordance with the principle of Safety First; not while I am where I sit. We are not going to allow that ship to start while there is the least little defect at all perceptible.

'Now, to meet the wishes of the Canadian Government we have postponed the flight until after the date of the Canadian General Election on the 28th of July. Then, in three and a half days or so, R.100 will fly to the mast at Montreal which the Canadian Government has erected.

'Then, in the month of September R.101 will go to India by way of Egypt. I am looking forward to going in her myself. It is, of course, no particular adventure. One of Dr Eckener's experts having seen R.101 not long ago said, "This is the safest conveyance on land, sea or in the air that human ingenuity has yet devised." R.101 is certainly a superb example of applied science and engineering skill.'

The building of these two airships had provided much valuable practical experience which could have been gained no other way. We still needed to build up our flying experience and that was coming along well. The Germans now had about thirty years of experience of airship operations. 'Before the war they were plying for hire from Lake Constance. They carried, before the war started, some 200,000 passengers who paid for their trips. They flew over two million miles.

In Dr Eckener they have one of the most remarkable men of the century; a man of great vision, immense enterprise and very wide experience. I have not the least doubt that we shall train up several men as good but you cannot reach the pitch of experience of Dr Eckener over-night. We now have some magnificent skippers training in this country who can handle their ships in the most appalling conditions and we have, of course, Major Scott who was the first man to fly the Atlantic in both directions in command of an airship – R.34.

'Certainly we have in these airships the embryo of a most wonderful form of transport. I defy anyone who has ever travelled in an airship to travel in any other conveyance with equal pleasure.

'I transferred my office to R.101 on her second flight. There was a lot of wind blowing but you do not feel the movement. You are sitting there as part of a bubble up in the blue with a feeling of detachment and serenity with which I have had no experience to compare. I tried to persuade my colleagues in the Cabinet to have a Cabinet Meeting there. I believe our deliberations would have been more detached and quiet.

'For the airship the harbour is the mooring tower; the dry dock is the shed. At the tower the most considerable repairs can be effected as they can be done even in flight. Dr Eckener mended a fin of the *Graf Zeppelin* in the air somewhere over the Arctic Circle. For the past three days we have been mending a big rent in the R.101's cover at the mast.

'What the Empire can do is to give us those harbours – those mooring masts. I can see the time when airships will start from Cardington, fly to the Cape Verde Islands, tie up to a mast and refuel with hydrogen or helium, then on to the Cape and from the Cape through "the Roaring Forties" and, I hope, find a mast at Perth two days out from Cape Town. They would not come back that way against the winds but will come north of the Line by Colombo and Europe.

'So, as an Empire, we have to have a means of crossing great ocean spaces – 3,000, 5,000 6,000 miles of ocean – because that is how we lie as an Empire, far flung, scattered all over the world and divided by the Seven Seas. That problem the airship will have to solve.

'The final tests of our two airships take place this year and I hope to be able to submit to the Imperial Conference in October some tangible results. I do appeal to you to lend a sympathetic ear and, more important still, ply a sympathetic pen on behalf of airships because, the more you study this great experiment the more you will realise that it represents the most scientific of experiments and that their success is a matter of incalculable significance to us all. If we can bring Australia within ten or eleven days of England and Canada within three days by means of regular airship services offering every comfort and amenity to the traveller, we shall indeed have accomplished much.

'What I am sure we should aim at is this. We know that British aviation is, and must be, an Empire proposition. This little island with its railways and its race tracks for motor cars, cannot develop aviation by itself alone. It must use the wide spaces of the Empire and we have it in our power, acting together, to create an Empire-wide air transport system around the world and to feed and sustain that system with an Empire-wide aircraft industry. I believe that system would possess the finest and most reliable machines in the world whether aeroplanes or airships. I believe that it would provide what one might call the arteries and veins of the Empire and that it would be of incalculable value in linking us up together; in providing export trade; in reviving that Empire spirit which some pessimists think is drooping; I see no sign of it myself.

'It would enlist that incalculable force, speedy communication, frequent direct intercourse. If we get the support of our Dominions and Colonies for all these objectives we shall soon be able to make civil aviation "fly by itself". We shall soon establish that new industry on a sound commercial basis and we shall do what has been the dream of poets in the past; we shall put a veritable girdle of air communications around the earth.'

That fluent, lucid and forward looking speech, made from a few notes, was received with enthusiastic applause. Taken with his House of Lords' speech of three weeks before, it reaffirmed Christopher Thomson's passionate belief in the future of British aviation with which he had now been so closely concerned for seven years.

Progress towards the achievement of some of his ideals was dramatically illustrated at Hendon on the following day in the presence of HRH Prince George, later the Duke of Kent, and more than 100,000 other spectators of the Eleventh Royal Air Force Display.

Among the pilots of 180 aircraft from more than forty units of the RAF was Pilot Officer Frank Whittle, then an instructor at No 2 Flying Training School at Digby, Lincoln who, with Flying Officer G. E. Campbell, DFM, performed Crazy Flying on Lynx Avro 504Ns, descended from the classic trainer of pre-war days.

Even then Frank Whittle was beginning to set down the germs of his ideas about jet propulsion which were to change the whole aspect of aviation from the 1940s onwards. At Hendon on this Saturday afternoon, however, he was engaged with George Campbell in demonstrating remarkable skill in low flying with the vintage Avros.

Then, promptly at 1550 hrs BST, precisely on schedule, R.101 droned over the Hendon arena at 800 feet, dipping in a salute to Prince George in the Royal Box. The Display Committee had, earlier, shown their lack of airship experience when they proposed that R.101 should make two runs over the aerodrome within the space of ten minutes and dive to 500 feet at 70 miles an hour. Colmore hotly rejected this and stressed the risks of such a dive at so low a level with so large a craft.

The airship had slipped from the mast at Cardington at 0722 hrs GMT that morning carrying a crew of thirty-six and five passengers from the RAW, a fuel load of 4,592 gallons weighing 17.3 tons and 13 tons of water ballast.

In fine, sunny weather the airship flew to London, then down the Thames to Southend and out over the Thames Estuary beyond the Tongue Lightship to fill in time before being due to arrive over Hendon in the mid afternoon.

George Meager, R.100's First Officer, who was standing-in for Noël Atherstone during his absence in Canada, wrote of that flight:*

> Conditions over the sea were delightful. The contrast with flying over the land where there were bumps below cumulus cloud, was very pleasant; the ship riding beautifully. She certainly flew straight and level in stable conditions over the sea and she was easily handled both in regard to steering and on the elevators.

From the Tongue Lightship off Southend, R.101 flew back over Southend pier and cruised slowly up the River Thames, running on three engines, to reach Hendon Aerodrome on schedule at 1450 hrs GMT. On the way the ship ran through a narrow 'cold front' marked by a long ridge of low-level cumulus cloud across the flight path. Meager wrote of it:

> As we nosed into the cloud the bow of the airship came under the influence of the strong upward current; the whole ship being carried bodily upwards. In a few seconds the gust reached the tail and up it went so that the airship was being carried upwards with her nose down at an angle of about ten degrees. The ship rode through the front in good style.

* In *My Airship Flights 1915–1930* by Captain George Meager AFC (William Kimber; London 1970). Lord Baker remembers George Meager as 'the first and only grown man I ever saw in a nightgown.'

We then proceeded to Hendon without further event and punctually at 1550 she was dipped in salute as she flew past the Royal Box.

Having done their piece satisfactorily everyone trooped off to tea leaving me on watch in the control car with Cox'n Oughton on the elevators; I forget who was steering. We passed through two or three heavy run squalls and were being bumped about a bit. This was to be expected in the type of weather through which we were passing.

After a time, however, after passing Luton I began to get worried about the behaviour of the airship. She would go into a short sharp dive and then the cox'n would get her nose up and we made a long, slow, climb back to our flying height. Directly the cox'n levelled his elevators in order to maintain his height, down would go the nose again and the procedure would be repeated.

After this had occurred several times I remarked to Oughton that the ship seemed to be a little heavy.

'It is as much as I can do to hold her up, sir,' he replied; the while his cheery round face was simply streaming with perspiration.

This rather alarmed me so I at once turned on the valve to release water ballast. After letting go what I judged to be about a ton, I asked Oughton if there was any improvement and he said she was much easier.

When Irwin came down to relieve me at 1800 hrs I told him I thought the ship was heavy and as we could see the sheds at Cardington at the time I suggested making a landing straight away as the evening was drawing in with the consequent cooling down of the gas to make us still heavier.

However, he would not hear of it, saying the ship's behaviour was entirely due to bumps. When I informed him that I had let some ballast go he added that he would have preferred to have kept the ballast for landing.

So I went up to the saloon to enjoy a well-earned cup of tea during which time the airship carried on northwards for an hour or so. We eventually made our landing to the tower at Cardington at 2050 hrs using something like ten tons of oil and ballast to get into equilibrium before landing.

Clearly something was wrong. The airship had been 12 hours 21 minutes in the air and, cruising at low speeds, about two tons of fuel had been consumed. In the normal way the ship ought to have been light at the end of such a flight. On this tenth flight, however, as on the flight on the previous day, ten tons of ballast had to be released as the ship came up to the mooring tower in order to bring it into equilibrium for locking home.

As soon as all connections were made some 33,000 cubic feet of gas had to be piped into the bags.

At 0330 hrs next morning, Sunday, 29th June, R.101 was taken off the mast by the handling party and transferred to the No 1 Shed.

Colmore called a meeting early on the Monday morning with Richmond, Rope, Irwin, Gerrish and McWade; the last was the representative of the Aeronautical Inspection Directorate at Cardington.

All the symptoms evident on the three June flights were carefully analysed and everyone agreed that the ship must be losing lift at a substantial rate. It must be either through holes in the gas-bags or because the gas valves were being opened by the excessive amount of flapping which was still taking place in the outer cover.*

On the following day, 1st July, Colonel Richmond recorded in his diary:

> Searched for holes in the gas bag, which revealed the fact that there was a very large number due to the fact that now the gas-bag wiring had been let out, the bags were pressing on sharp projections on the longitudinal girders and radial struts.
>
> Decided to institute extensive padding. The question of the valves is being looked into.

Here was a dilemma indeed. The extra lift gained by letting out the gas-bags would still be essential even after the addition of the new bay. But now there could be no doubt that the letting out of the gas-bag wiring had allowed the structure of the hull to rub a large number of holes in the gas-bags during the first flight after the refit. Padding between gas-bags and structure was common practice on German Zeppelins but it added weight – especially when wet – and was a laborious process both to install and to inspect. There might be an alternative – to run the Tornado engines on a lighter combination of fuel oil and gaseous fuel. The mixture would be ten per cent diesel oil and 90 per cent fuel gas; a solution which would save perhaps 12 tons of weight on a flight to India at some cost in volumetric space to house the inflammable gaseous fuel.

Richmond discussed these possibilities with Reginald Colmore, Michael Rope and Tom Cave-Browne-Cave on 3rd July. There were certainly opportunities for substantial savings in fuel weights which would make possible the pulling in of the gas-bag wiring again. The possibilities were, however, long-term rather than short, and extensive tests would have to be run with a Tornado engine on the ground before authority could be given for use of this new fuel mixture in the air.

* Recorded in Colonel Richmond's diary for 30th June 1930.

Meanwhile, one further modification was set in hand. On the endurance flight between 17th and 18th November 1929, and again on the flight on 27th June for the RAF Display rehearsal, one of the heavily-loaded lifting bridles, which transferred lift from the gas-bags to the framework of the hull, had snapped as a result of wear where the bridle-wire passed over a 3½ inch-diameter pulley. These bridles were important because they not only distributed the lift into the structure in a carefully stressed manner but restrained the gas-bags from surging in rough air. Richmond decided to replace the wire bridles with chains and to increase to 14 inches the diameter of the pulleys to give more bearing area.

The problem of the bridles could thus be resolved fairly easily at a small increase in weight. The holes in the gas-bags remained serious.

Colmore, Richmond and Rope reviewed the subject further on 2nd July. Colmore recalled that after the storm on 11th November 1929, when R.101 was at the mast, he had reported to the Air Ministry that though the ship had ridden comfortably and without violent movement, 'there was, however, a slow rolling motion at the higher wind speeds reaching a maximum of six degrees', and there had been some chafing of the gas-bags on the structure which he had not regarded as serious at the time.

On the same day, the Chief Coxswain, 'Sky' Hunt, had, however, drawn attention to a substantial movement of the gas-bags, the rubbing of the bags on some of the longitudinals, the tearing of a nine-inch hole in No 8 gas-bag and a number of other holes in No 14 bag.

When R.101 had been brought back to the shed at the end of November 1929, all the gas-bags had been deflated, removed from the ship, and carefully inspected. Every gas-bag except one, was found to be holed though many of the holes were too small to lead to a serious loss of gas. There were, however, exceptions and among them gas-bag No 5 with 57 holes and gas-bag No 11 with 103 holes.

In consequence, after the bags had been repaired and reinstalled and before they were re-inflated, no less than 4,000 pads had been wrapped around sharp projections on the structure wherever they might press against the bags. These 4,000 pads had been in place for all the three flights in June.

A careful check of the records showed that between 7th October and 29th November 1929 – a period of fifty-three days – a total of 4,667,380 cubic feet of hydrogen had been pumped into the bags. This worked out at an average of 88,064 cubic feet of gas used each day. While the ship had been in the shed there had been an average loss of 51,070 cubic feet a day. This was excessive. A normal loss through seepage would have been expected to be about 14,250 cubic feet a day; less than one-third of the loss in the shed.

Richmond was worried. After the meeting on 2nd July Vincent Richmond addressed a Minute to Herbert Scott who had not been present. He wrote:

> In connection with the reported loss of lift of R.101 during flight I have been investigating the effect of holes in the gas-bags.
>
> I find that, if it be assumed that the average height of holes is equal to three-quarters of the height of the ship, then the rate of loss of lift is about one ton per square inch of opening in twelve hours. This calculation assumes quite a pessimisitic co-efficient of discharge such as would be applicable to a triangular hole.
>
> In my opinion this result is somewhat startling and emphasises the great importance of guarding against holes in these present ships. Even if the holes are in the form of slits which have comparatively little effective areas of discharge, it seems to me quite conceivable that we may have had leaks in both ships amounting to four or five square inches of area.
>
> In order to enable me and my staff to determine what should be done to check holes in the present ships and also to guard against their occurrence in future ships I should be glad if you could arrange to let me have full particulars of the positions, approximate sizes, etc. of the holes found by the crew from time to time.
>
> V.C.R.

Colmore and Richmond agreed that the calculations showed a very serious state of affairs. The leakage of gas was threatening to undo all that was in hand to build up adequate lift for a long flight. Now, with R.101 in the shed, Carmichael Irwin was asked to conduct a minute inspection of all the gas-bags immediately.

Frederick M. McWade, the AID Inspector in Charge at Cardington was no less concerned. On 3rd July he addressed a confidential letter to the Director of Aeronautical Inspection, Lieutenant-Colonel H. W. S. Outram, CBE, at the Air Ministry through the Chief Inspector (Aircraft) of the Directorate, Major P. Bishop, OBE.

McWade wrote:

> *HMA R.101*
>
> *Airworthiness of the above ship*
>
> On 26th June 1930 I handed over the 'Permit to Fly' dated 20th June 1930, valid until 19th July 1930, to DAD.
>
> Owing to the modifications which have recently been carried out on the wiring system, the gas bags are now hard up against the main

longitudinals and rubbing very hard on the nuts and bolts positioning the stirrup into which the tie rods are screwed. Further, the gasbags foul very badly the heads of the taper pins at the joints of the main and intermediate struts at the inner ridge girder ends. This matter, in my opinion, has become very serious as the points of fouling occur throughout the ship and amount to thousands.

Padding has been resorted to by wrapping fabric over the parts mentioned above and this is the usual recognised method used in isolated cases. Padding to the extent now necessary is, in my opinion, very unsatisfactory because the bags move when the ship is in flight and the padding becomes loose and the projection complained of is again exposed.

Although the gasbags have recently been reconditioned and were in good order when placed in the ship a few weeks ago, there are now many holes in them.

The next point is that where the fabric is wrapped round a joint it may be difficult to know what is happening underneath the wrapping (I have in mind the corrosion question). The fabric will become damp and in many cases wet when the ship is in flight; therefore, there will be alternate processes of wetting and drying of the fabric which must be detrimental to the metal underneath.

I am fully aware that to remedy the faults complained of is in the nature of a large undertaking and it may be necessary to remove the bags from the ship. Until this matter is seriously taken in hand and remedied I cannot recommend to you the extension of the present 'Permit to Fly' or the issue of any further Permit or Certificate.

F. McWade.*
Inspector in Charge, AID at RAW,
Cardington, Bedford.

3rd July 1930
RAW

This was a damning indictment. It meant that in the view of Britain's most experienced and capable airship inspector, R.101 was not airworthy.

* Frederick M. McWade (1872–1947), Resident Inspector, Aeronautical Inspection Department at the Royal Airship Works, was born in Glasgow and trained at the School of Military Engineering at Chatham. His long career in British Lighter-than-Air activities began when he joined the School of Ballooning, Royal Engineers, in 1895 and transferred to Farnborough in 1897. He took part in the building of the airship *Nulli Secundus* in 1903 and was engaged in the construction of the Army non-rigid airships at Farnborough Beta, Gamma Delta and Eta up to 1914. From 1915 to 1918 he was posted to the Kingsnorth Airship Station, RNAS, on the building of non-rigid airships and from 1920 to 1924 he was an AID officer at the Marine and Armament Experimental Establishments at the Isle of Grain and Felixstowe. From 1924 till 1931 he was Resident AID Inspector at the Royal Airship Works, Cardington, responsible for inspection of R.101 during and after assembly and for R.100 after that airship came to Cardington. He flew to Montreal and back in R.100 in 1930. Frederick McWade died on 23rd February 1947.

Moreover, in his opinion, R.101 could not be made airworthy unless the decision to gain extra lift by letting out the gas-bag wiring was countermanded. He believed that padding was not a satisfactory solution. The three flights in June had given ample evidence of the need to keep the gas-bags clear of the structure.

Colonel Outram, Head of the AID, received McWade's letter on 4th July. He consulted Wing Commander Colmore, the Director of Airship Development, at once. In view of the seriousness of the statement he felt that he would have to submit the letter to Sir John Higgins, the Member of the Air Council responsible for airships. Before he did so, what were Colmore's views?

Colmore, the usually cautious, wrote back to Outram a rejection of any suggestion that padding was not an adequate remedy. This, despite Richmond's concern; despite the fact that Rope's complex but elegant wiring system had been designed specifically to keep the bags clear of the structure and that Rope had wanted to pull in the wiring rather than to let it out. The need for more lift seems to have over-ridden all else in Colmore's mind. He wrote to Outram on 8th July:

> I feel sure you will agree that we cannot accept, as a matter of principle, that the gasbags in an airship should be clear of all girders. Also, I expect you will agree that we can accept padding as being a satisfactory method of preventing holes forming in gasbags from this cause.
>
> As far as we can trace at present there have been remarkably few rips in the gasbags of R.101 and that the holes which have occurred are due to the bags fouling girders. We have little doubt that padding will be a permanent remedy and, if this is accepted, then it is certainly not a large undertaking to put the matter right. In fact, we hope to complete the necessary padding in R.101 by the end of the present week or, at any rate, sometime next week.
>
> We do not think that any objection can be taken to wrapping padding round obstructions of this nature.
>
> The above covers our views but I should be very glad to discuss the matter with you when you are next at Cardington.
>
> R. B. B. Colmore
>
> RAW DAD
>
> 8-7-30

Here was a direct contradiction of McWade's opinion. Colonel Outram knew little about airships. Colmore was the Director of Airship Development and, to Outram, his view must have been expressed with the knowledge and support of Colonel Richmond, the Chief

Designer, and of Major Scott in charge of flying. Should Outram ignore this advice and uphold his Inspector?

Clearly he should have done. At the very least he ought to have referred the matter to Sir John Higgins and through him to the Secretary of State. He did not do so. He telephoned Colmore and after a long discussion decided that Colmore was right; that 'the situation was not so serious as he had first supposed'. He and Colmore agreed not to bother Sir John Higgins with such detail of airship operation.

Neither the Secretary of State nor any member of the Air Council were told of the gas-bag problem or of the divergent views about them. It is difficult to believe that had Lord Thomson been fully informed, he would not have taken a serious view in accordance with his declared policy of 'Safety First'.*

In the event McWade was overruled.

Sir John Simon wrote in March 1931 in his 'Report of the R.101 Inquiry' (Cmd 3825):

> The incident was closed so far as McWade was concerned, by a Minute from Colonel Outram of 11th July communicating the joint view of himself and Wing Commander Colmore. The Minute includes the following paragraph:
>
> 'Of course I fully realise the necessity of avoiding the contact of these damaging points to the gas-bags. I have taken the matter up with DAD who is in agreement but, as you yourself realise, it is impossible to alter the hull structure of the ship at this stage. The only expedient at the moment is to pad and it is your duty to see that every point which may lead to damage is padded in a proper manner.'

McWade had done his duty. He had said that in his view padding was not an effective remedy. He had said that he was unwilling to recommend the issue of a further 'Permit to Fly' until a satisfactory solution was found, fundamental as that might have to be. His superior had taken a different view after consultation with the Director of Airship Development at Cardington.

McWade accepted that he could do no more against the weight of technical opinion. His recommendations, based on long experience,

* The Report of the Public Inquiry glosses over this subject (Page 51) by saying that: 'It has always to be remembered that the officers and experts specially acquainted with every detail of airship construction were concentrated at Cardington and that the officials at the Air Ministry in London were almost without exception men whose training and experience had been gained in the course of service with aeroplanes as distinct from airships. There was therefore a natural tendency at the Air Ministry to rely on the advice and judgement of the airship experts who were congregated there.'

had been overruled. Normally an AID inspector's opinion was taken as law. In this instance the Director of the Aeronautical Inspection Department had ruled otherwise.

There was still the question of the gas valves. On 1st July Carmichael Irwin had written a report to Major Scott on the three flights in June calling especial attention to the fact that the outer cover just forward of Frame 13 (that is near the tail) was flapping considerably more than on previous trials and recommending that an air-inflated mattress should be fixed at this point inside the cover to steady it. He added: 'The cover was also flapping all along between Panels C and D and it is considered possible that the gas valves may have been affected as, even allowing for the numerous holes which are now being found in gas-bags where they have rubbed on protruding nuts of Main Longitudinals, the loss of gas would not have accounted for the heaviness of the ship during flight on Friday and Saturday.'

A special conference was held in Wing Commander Colmore's room at Cardington to consider this report. A decision was taken that Frederick McWade should be instructed to make further tests on the gas valves before R.101 flew again.

When these tests were run, McWade found that the valves remained closed up to and beyond the angle of roll of three degrees specified as a reasonable limit by the Design Office. The valves did, however, begin to open at five degrees of tilt so that, in extreme conditions of turbulence, with the airship rolling heavily, some loss of gas through the valves would be likely.

The test suggested, however, that the loss of gas during the June flights had been caused almost entirely by the holes in the bags and not by any deficiency of the valves.

Further analyses suggested that in moderately bumpy conditions which would be sufficient to cause some gas-bag surge, R.101 had suffered a loss of gas equivalent to about a ton of lift an hour – that is a loss of some 33,000 cubic feet of gas an hour – through holes with a total area of about twelve square inches. That was about three times the size of Colonel Richmond's earlier estimate.

On a flight to India such a leakage could result in a reduction of about 48 tons of lift; a totally unacceptable figure which would be only partly offset by the consumption of about 20 tons of fuel. Calculations had shown that by flying pitched 'nose-up' on cruising power, R.101 could count on 'dynamic lift' to maintain flight while about 13½ tons 'heavy' – but no more.

Thus, unless something was done, the loss of gas, the consumption of fuel and the maximum amount of dynamic lift, when combined, would result in a forced landing after a flight of about 23 hours' duration.

So padding *had* to be effective. There was no other solution to the problem of getting the ship to India before the end of 1930.

Since R.101 had been put back in No 1 Shed at Cardington on 29th June work had been going forward on the gas-bags, the cover and on various modifications while awaiting orders to start the installation of the new bay. The ship had been bedded down on trestles at Frames 4, 6, 8, 10 and 12 with the stern about three feet from the end of the shed, so that when the airship was parted at Frame 8 only the forward half would have to be moved.

The self-adjusting bridles were being modified by substituting chains for wires over pulleys increased in diameter by four times from 3½ to 14 inches. New panels of the outer cover were being fitted in place as they were delivered, one by one, from the RAW Fabric Shop under J. W. W. Dyer. The new cover of strong aeroplane-type linen would be doped in place and follow established German practice.

Noël Atherstone, back from Canada, was – like Rope – especially worried about the cover. On 2nd July he examined some of the fabric which had been taken off the top of the ship and found it to be completely rotten.

> It can be torn without using any force at all [he wrote] just like paper. I don't understand how any of it stayed on the ship at all. The cover is very much worse than I expected it to be and to have ordered the ship to fly with a cover in such a rotten condition was, in my opinion, a totally unjustified risk. The new doped-on covers are being fitted as far down as 'F' longitudinal on each side and as far aft as Frame 7. The rest of the old cover is being left on for the present but is having three-inch transverse bands doped-on the outside, spaced three feet apart.
>
> I don't think this scheme is going to add any strength at all to the pre-doped covers as the adhesive dope which is used for sticking on the bands seems to make the fabric of the pre-doped cover absolutely rotten. So I don't see where the gain in strength comes in.

By 12th July the more extensive padding of the framework of the airship wherever the surface of the gas-bags would touch the structure was almost complete. The new pads were made from cotton-fabric bandages which had the merit of being thicker and of remaining securely in place. All this work on the cover and on the gas-bags was recorded in detail at Cardington where the defects had caused such deep concern. None of it was, however, reported to the Air Ministry.

Thus, to Sir John Higgins, to the Secretary of State and to the Air Ministry as a whole, R.101's flying in June had seemed to confirm their confidence in the ship and the belief that, with the extra bay, it would be able to fly successfully to India and back. This confidence was based on the fact that R.101 had now built up more than 102 flying hours in ten flights, the airship had appeared over Hendon precisely on schedule two days in succession and it had ridden out the November gales at the Cardington mast without serious problems. It all instilled a belief that R.101 would be capable of running through the most severe weather likely to be encountered.

For the Secretary of State the summer's round of aeronautical events was now in full swing.

On 5th July, the Saturday following the Royal Air Force Display, the Royal Aero Club's ninth King's Cup Air Race began and finished on a cloudless summer's day at Hanworth Air Park, Feltham. It attracted the biggest entry ever; 88 aircraft started and 63 finished. The winner was Miss Winifred Brown of the Lancashire Air Club flying an Avro Cirrus-Avian (G-EBVZ). She averaged 102.7 mph for the 750 mile course, by way of Hamble, Bristol, Birmingham, Liverpool, Manchester, Leeds, Newcastle, Hull and Leicester, navigated by her fiancé, V. R. Addams.

Miss Brown and Addams arrived back at Hanworth almost exactly ten hours after taking off. They passed the finishing line twelve minutes ahead of Alan Butler, Chairman of the de Havilland Aircraft Company in a D.H. Gipsy Moth. This first lady winner of the prestigious air race beat, on handicap, such distinguished pilots as H. R. D. Waghorn, A. H. Orlebar and R. L. R. Atcherley of Schneider Trophy fame, Captain Geoffrey de Havilland and his son, Geoffrey, Edgar Percival, F. G. Miles, Nigel Norman, Arthur Marshall, Harold Balfour, Caspar John, Hubert Broad and Winifred Spooner.

On 3rd July Lord Thomson with Sir Sefton Brancker had escorted Prince and Princess Takamatsu of Japan on an inspection of Croydon Airport. On 7th July he was at the opening of the new Handley Page aerodrome at Radlett, Herts, performed by HRH Prince George, later to be the Duke of Kent, who arrived by air in a Westland Wapiti of No 24 (Communications) Squadron RAF escorted by a flight of three Armstrong-Whitworth Siskin fighters of No 41 Squadron RAF.

Among the aircraft on view, celebrating the 21st anniversary of Handley Page Ltd. was the prototype of the new H.P. Heyford bomber and the fuselage of the first Handley Page H.P. 42 *Hannibal* 38-passenger air liner for Imperial Airways (G-AAGX).

Walking through the fuselage with Frederick Handley Page, Christopher Thomson remarked that here, in the new H.P. 42, was the landplane equivalent of the rigid airship for Imperial Airways' Empire Services – offering four-engine safety and a higher standard of passenger comfort than ever before. He looked forward to travelling on one of the first scheduled services when *Hannibal* began operations.

Meanwhile, ignorant of any gas-bag problems, Sir John Higgins had addressed a note to the Secretary of State about R.101's programme. He wrote:*

> According to the programme to which DAD is working, R.101 should be completed with the new bay by 22nd September. One trial flight should be necessary before the airship leaves for India so the end of September is the earliest date at which the flight can commence. This leaves no margin for eventualities and assumes that the trial flight will be completely successful.
>
> The outward flight to India should be completed in five days and for the return journey six days must be allowed.
>
> On this first trip I do not think that it is advisable for the airship to spend more than a few days in India and I suggest a week or ten days as the maximum.
>
> I contemplate that, on her return, the R.101 should be occupied on the Egyptian route with R.100 and that she should proceed to India again next spring for a series of local experimental flights.
>
> 30-6-30. *J.F.A.H.*
> AMSR

Sir John Higgins, acting on advice from Cardington, was thus confidently proposing that a single trial flight would be likely to be sufficient to prove the extended R.101 after the insertion of the extra bay, before the end of September, that thereafter India should be reached in five days and regular services to and from Egypt should begin by the end of the year.

Buoyed up by these positive proposals, Lord Thomson wrote on the bottom of the Minute on 2nd July,

> I should like to count definitely on starting for India during the week-end of 26th-28th September. I ought to be back by 16th October.
> 2-7-xxx T.

* AIR 2–346/523024/30–06130 of 30th June 1930.

Higgins noted:

> S.9 [L. G. S. Reynolds]
> Please inform DAD – *AMSR 3/7/30.*

This would allow eleven weeks to prepare R.101 for its test flight with the new bay, five more days to make it ready for India, and three weeks for the flight to India and back so that the Secretary of State could return to London before the Imperial Conference reached the aviation items on its agenda.

On 14th July Sir John Higgins, still confident that all was going well, addressed a further minute to the S of S about arrangements for the two airships.

> DAD has now applied for authority to commence work on parting R.101 tomorrow. Until, however, R.100 has completed her next trial flight and shown that the modifications to the cover, etc, are satisfactory, a possibility still remains that the R.101 may be required for the Canadian flight.* In the circumstances I do not think that R.101 should be put out of action for the present even at the risk of delaying the Indian flight.
>
> I propose, therefore, if you agree, to hold up work on R.101 for a week in the hope that R.100's flight will be completed by then. In the meantime DAD will press on with such work as is not dependent on opening up the airship.
>
> *14-7-30* *J.F.A.H.*
> AMSR

Lord Thomson replied:

> So long as R.101 is ready to go to India by the last week in September this further delay in getting her altered may pass.
>
> I must insist on the programme for the Indian flight being adhered to as I have made my plans accordingly.
>
> 14-7-xxx T.

This statement has been widely misinterpreted as a dogmatic insistence of R.101 starting for India, willy-nilly, at the end of September. Examination of Lord Thomson's notes, speeches and a knowledge of

* See Appendix 11: 'R.101 for Canada'. If R.101 had been required for the Canadian flight in place of R.100, an operational assessment shows that it would have been able to complete the flight, in the prevailing weather, in about the same duration as was taken by R.100, carrying the same – small – 'payload' for the same duration of flight.

his attitude to the airship business, shows clearly that when he wrote 'programme' he meant a programme of projected action and not that the airship should depart whether it was ready or not. This is made still more clear in later communications.

Sir John Higgins informed Colmore of Lord Thomson's hope that the proposed programme could be met. On 17th July Colmore replied:

> Every effort is continuing to be made to complete R.101 to enable the flight to India to take place at the end of September. Work on girders and parts (for the new bay) is very well forward and a reliable estimate can be made of the time required to complete these parts and to assemble the frame.
>
> It is on the parting of the ship and the insertion of the new bay, however, that delays may occur which cannot be foreseen and, in our opinion, if we are to work to this programme it is absolutely essential that work on the parting of the ship should commence immediately.
>
> In the event of further trouble with R.100 during her re-trial flight I am afraid that the only alternative would be either to abandon the Canadian flight until these troubles have been overcome or to carry out the flight with R.101 and postpone the Indian flight until November.

Reginald Colmore was, indeed, far from happy about the state of R.100. The cover was even more unsatisfactory than that of R.101, and R.100's gas-bags had shown such a propensity to get nipped and torn by the forest of radial bracing wires that while the airship had been in the shed, it had 'consequently swallowed an enormous amount of gas'.* For all R.101's troubles with holes in the gas-bags, R.100 had been worse.

Colmore was, likewise, concerned about the reliability of R.100's

* Noël Atherstone's diary for 12th July 1930, and AIR 5/906–X/M.08109. In fact, in the forty-five days between 27th May and 10th July, 1930, while in the No 2 Shed at Cardington, R.100 had to be topped up with 3,531,600 cubic feet of hydrogen; an average daily loss, through leakage, of 78,480 cubic feet. Again, in the twenty-eight days between 18th August and 18th September 1930 (when measurements were taken), R.100 received a further 2,239,500 cubic feet of gas in the shed; an average daily loss of 79,982 cubic feet. These losses of gas by R.100 compare with the 1,467,960 cubic feet of gas which R.101 required (including the new bag) in the sixty-five days between 24th July (when gas-bag repairs were completed) and 26th September 1930, while the airship was in the No 1 Shed following the Hendon flights. The average daily amount of gas used by R.101 was, therefore, 22,584 cubic feet or – (excluding the 510,300 cubic feet required to inflate the new bag) – R.101's gas leakage in the shed was running at 14,733 cubic feet a day; which was less than one-fifth of the rate of leakage of R.100 during a comparable period.

engine installations and about some of the structural members which had shown signs of corrosion. Colmore telephoned Sir John Higgins with these problems.

Sir John went to see the Secretary of State to discuss the whole situation with him – specifically the uncertainties with R.100. Neither Sir John nor Lord Thomson still had any knowledge of the problems which had been encountered with R.101's gas-bags during the flight in June. R.101 was, to all appearances, the more reliable ship. Certainly, it had encountered fewer obvious problems with cover and structure on its trial flights. And by now the gas-bag problem was being resolved.

Following their meeting on 21st July, Sir John Higgins addressed a further minute to the Secretary of State. He wrote:

> *S of S*
>
> 1) I understand from our conversation this morning that you do not approve of the proposal that R.101 should be kept as a standby for the Canadian flight and that, if the modifications which have been made to R.100 do not prove satisfactory, the flight to Canada will have to be put off until satisfactory modifications are completed on this ship.
>
> 2) I propose, therefore, to issue instructions to DAD to part R.101 immediately so as to give the best chance of it being ready according to the programme date which is being worked to.
>
> 3) Every endeavour is being, and will be, made to keep to this programme date but, as stated in my loose Minute of 30th June, this date does not leave any margin for unforeseen circumstances.
>
> 21-7-30 *AMSR*

Lord Thomson replied next day:

> *AMSR*
>
> The first paragraph of your Minute states the position correctly.
>
> I note and approve the course of action proposed in your paragraph 2.
>
> As regards paragraph 3, I am sure everything possible will be done and I am not unduly pessimistic.
>
> 22-7-30 T.

The Secretary of State's colours were nailed firmly to the airship's mast. He had an absolute reliance upon his official advisers at Cardington and in Whitehall – and they seemed full of confidence. The time was at hand for the two airships to show what they could do – 'the one to Canada, the other to India'.

Alongside them there were other plans and other hopes as well.

CHAPTER FOURTEEN

'We Said Good-bye — for Life'

'This sweet memory
'Across the years will come to me . . .
'In my heart will ever lie
'Just the echo of a sigh
'Good-bye.'

NOEL COWARD, *Bitter Sweet*, 1929

18th/20th July 1930

AT LONG LAST, after six months apart, on Friday, 18th July, Marthe Bibesco was back in England. Once again – and for the last time – C.B. met her at Victoria Station.

The Princess Bibesco described the first week-end of this summer visit in a letter to the Abbé Mugnier in Paris, written from the Prime Minister's country house at Chequers.

Chequers,
Princes Risborough.

Sunday, 20th July 1930 Telephone: Wendover 7677

Monsieur l'abbé, mon bon oncle,

I am spending my Sunday with the Prime Minister in the company of Lord Thomson and with Miss Gordon-Cumming, a niece of the Marquise de Breteuil.

But I am also as it were the Cymbeline whose heritage this is. 'England is all Shakespeare' – as our Monsieur de Combourg * would say. Happy are the people who have had among them such a poet.

The park here contains the oldest yew tree in the Kingdom. Lime trees, hundreds of years old, line the fields. They stand isolated in their splendour, like monuments of another age. Now they are in flower and our General Trestrignel† compares them with those lines of 'Sleeping Waters' which – as you will remember – is the name he calls my Palace of Mogosoea near Bucharest. Those lime trees at

* Vicomte Francois René de Chateaubriand (1768–1848) French statesman, writer and poet, was Ambassador of France to the Court of St James in 1822 and French Minister of Foreign Affairs in 1823. His autobiographical *Memoires d'Outre-tombe* (1849–50) was a favourite book of Marthe Bibesco.

† General de Trestrignel – Lord Thomson: a pet name commemorating an idyllic month which Christopher Thomson and Marthe Bibesco had spent with a house party at Trestrignel near Perros Guirrec in Brittany in the summer of 1919; visited at the same time by the Abbé Mugnier.

Mogosoea are equally beautiful but the time of their flowering is over. So, to have come to Chequers has delightfully brought to me the scent of limes twice in the same year.

In the front court of the house there is a South American tulip tree – a fragrant form of Magnolia – which, because it is also now in flower, greets the guests when they arrive. I have never seen a tree of this species before; except in dreams. I enclose one of the sweet-scented blooms which I picked for you.

About it the Prime Minister has just told me a story that, when poor Eve was chased out of the Garden of Eden, she held on to the foliage of the last tree which hung over the Garden's wall. It was a tulip tree and the leaf seized by Eve in her despair fortunately remained in her hand and so made her respectable!

Such is Mr MacDonald's account and, as you see, he is well up in botany as well as in the Old Testament! I find him the most amiable – and teasing – of Scottish hosts.

This mansion at one time belonged to one of the daughters of the Lord Protector, Oliver Cromwell, and portraits of Cromwell's daughters and sons-in-law decorate the walls of the Great Hall. They do not supplant, however, the earlier portraits of the Stuarts which, likewise, hang on Chequers' walls. For instance, a portrait of Queen Henrietta-Maria, daughter of that great and good King Henri IV of France, painted by Van Dyck, hangs over the great fireplace in the immense hall which is, simply, a roofed-over interior courtyard.

As you can see I shall have a lot to tell you about the treasures of art and the treasures of history which are housed here. For the moment I don't have time to do that because once I start to describe them I would not be able to resist going on and on at length.

Last night, after dinner, Lord Thomson recounted to the Prime Minister the story of your visit with me to the Houses of Parliament at Westminster in 1927 when – as he said – he had the honour to act as our guide. He quoted with relish your remark when we went into the House of Lords: 'Now I understand why English politics are treated as a religion – because they are practised in a cathedral'. He remembers those words and you who said them, with both pleasure and affection.

As for me, of course I think of you always, *Monsieur l'Abbé*, as the best of friends and the best of travelling companions.

Adieu, mon cher oncle. Of course, you are always in my thoughts.

Marthe.

That week-end at Chequers passed all too quickly. C.B. was happy in the company of two people for whom he had so much affection and regard.

They had a last week together in London – Christopher and Marthe – in among his other duties. They walked in the park, they went to watch cricket at Lords – a mystery to a Princess brought up in Paris – they visited the Royal Academy, they dined with friends and they went to Noël Coward's new light opera at His Majesty's Theatre, *Bitter Sweet.*

On the warm Tuesday afternoon of 28th July, the day before the Princess's departure, they sat together at tea on the Terrace of the House of Lords overlooking the River Thames. That morning the newspapers had written of a subject which they had discussed again at Chequers – that C.B. was to be the next Viceroy of India from the spring of 1931. He would return to the land of his birth in its highest office. It would be the greatest challenge of his career.

Once again he talked with Marthe about the Prime Minister's now pressing offer. It would mean going far away from her. It would mean leaving MacDonald's side.

'I fear that he would be crucified by his Left Wing of the Party,' he said. 'If I do go he will find himself very much alone in the face of mounting difficulties and it will be for a long time – for five years.'*

'I know that such an appointment in India ought to be a joyful climax to ambition. But, to leave two of those who mean most to me in the world would be more than ambition ought to have to bear. I know how solitary a Prime Minister's life can be – and he has come to take solace in our talks together. And you – dearest Marthe? Where does my duty lie?'

Marthe wrote:

> At the most earnest moment of our discussion he was called away to take part in a Division in the Upper House. He had to vote. He would not be long. He begged me to wait.
>
> I stood there alone, leaning on the balustrade above the Thames. It was high tide. The wavelets lapped softly against the stone wall of the Terrace. I heard the sad cry of sea-birds.
>
> In measured and magnificent tones Big Ben struck six o'clock from the tower telling me that it would soon be time to go. As I was gathering my things together I saw him coming back, hurrying towards me with great strides. He was not going to let me slip away like that. The rites had to be performed. We must go out through Westminster Hall together once more.

* Recorded by the Princess Bibesco in her diary and in her essay 'Le Destin de Lord Thomson of Cardington' published in *La Revue Universelle*, Paris, 15th February 1932.

Through the Central Lobby and down the famous steps we went. We stopped for a moment where Charles I had been judged; where, ever since, the coffins of kings, his successors, had lain in State to receive the absolution of the British people. The latest who had rested there under that roof had been King Edward VII.

To the darkening radiance of a summer's evening which was already gliding beneath the arches, there was added the sense of being at the heart of England.

Once again he repeated, in the tones he always used when he was dreaming aloud, the same words he had used when first he had brought me down those steps – 'Isn't this the most beautiful and the most historic Gothic room in all the world?'

He led me as far as the centre of that place which was for him the Ark of the Lord.

So it was that in Westminster Hall we said good-bye for life – but not for eternity.*

Early next morning R.100 left for Canada.

* *'C'est à Westminster Hall que nous somme dit adieu pour la vie – mais pas pour l'éternité'* – Princess Marthe Bibesco, 'Le Destin de Lord Thomson of Cardington, *La Revue Universelle*, 15th February 1932.

CHAPTER FIFTEEN

The North Atlantic

'Far as the breeze can bear, the billows foam,
'Survey our Empire and behold our home.'

BYRON, *'The Corsair'*

July 1930

WITH July running out, only two clear months remained before the Imperial Conference was due to start at the Foreign Office in London, the opening by the Prime Minister now set for Wednesday, 1st October 1930. All the Commonwealth Prime Ministers would be there. On the agenda was the future of the Imperial Airship policy on which Christopher Thomson had set his heart. At the end of the Conference the name of the next Viceroy of India would be announced.

To Christopher Thomson it was clear that an immense amount of work remained to be done. Time was running short – but still the airships must not be rushed.

Even so, a triumphant return from India in R.101 while the Conference was still in session would set the seal on his vision of future Empire communications.

So far the chief accomplishments of the new airships, in the nine months since the launch of R.101, had been two flights round Britain; a 53-hour flight of 2,050 statute miles by R.100 and a 31-hour flight of 1,148 statute miles by R.101. The airships were still on trial; they had still to fly overseas. Unless something impressive was achieved – and soon – there was always the danger that, despite Thomson's cordial personal relations with Philip Snowden, the Chancellor of of the Exchequer, the Treasury might propose that, as an economy measure, the whole programme should be ended.

It was fair to say that, so far, the Treasury had been helpful and co-operative and there were sound reasons for believing that money would be forthcoming for the follow-up programme once it had been given a fair wind by the Imperial Conference. He must come back to it from India, as from Olympus – a runner bearing a torch.

But first – the North Atlantic.

A successful flight to Canada and back by R.100 to be followed by the longer journey to India by R.101 would open the way for an endorsement by the Conference of his proposal for the laying down of two new airships as the next stage of the programme. Then, with the payloads of both R.100 and R.101 increased by the lift from extra bays, they would begin to make a reality of the dream of safe, comfortable, long-range air-links half way round the world at more than twice the speed of surface transport.

It would all be set out in his new Memorandum for the Conference: 'The Progress of Imperial Communications'.* He had already begun to draft it in his own hand.

Meanwhile, first things first. The existing airships really must be pushed along. The programme was now just six years old and doubts were being raised in influential places. Neither airship had, as yet, ever docked away from Cardington; all the sixteen flights so far completed had been out from the mooring tower at the main base and back again to it.

The Secretary of State had intended to fly in R.100 to Montreal when the flight had first been planned for May.

Now that the programme had been so long delayed there were pressing events at home – not least the Coal Bill in the House of Lords – which would prevent him from undertaking the journey to Canada in R.100. All the more reason to fly to India in R.101 as soon as that airship could be made fit and ready for the voyage.

First R.100 had to show its paces. And so, at 0915 hours on the morning of 25th July, following a sudden dropping in the gusty wind and the promise of a fine, warm day, a scratch handling-party was assembled from employees of the Royal Airship Works and R.100 taken out of No. 2 Shed and put on to the tower. Thanks to the discipline and enthusiasm of the rapidly gathered handling-party of airship workers, it was all done quickly and easily in – now – a very light wind from the north-west on the axis of the hangar.

With R.100 at the tower, being readied for a final 24-hour test flight before departing for Canada if all went well, Noël Atherstone was detailed to act as mooring tower officer in the absence of Flight Lieutenant Brian H. Cook, the regular tower officer seriously ill in hospital.

R.100 duly left the tower at 1902 hrs GMT on 25th July for its

* AIR 5/908/308.

LATE XVII]

IGHT) *Radlett Opening* – Sir ederick Handley-Page elcoming HRH Prince George ater the Duke of Kent) on his rival by RAF Westland Wapiti rcraft to open the new Handley-age aerodrome at Radlett, erts, on 7th July 1930. Lord homson of Cardington, ecretary of State for Air, is etween Sir Frederick and HRH.

ELOW) *'Amy, Wonderful Amy'* – t nine-thirty p.m. on the evening f 4th August 1930, at Croydon irport, when Amy Johnson (light oat) was welcomed home from ustralia by Lord Thomson, ecretary of State for Air, with eft to right), Lord Strabolgi, Mr red Montague, US of S. for Air, ord Wakefield of Hythe, who ad sponsored Miss Johnson's olo flight to Australia, and Miss Margaret Bondfield, Minister of abour and the first woman nember of the Cabinet.

Ready to Go – The officers of H
Airship R.101 pose with Major
Scott, in front of the control ca
the morning of Saturday, 27th
September 1930, before the ai
was taken out of the shed for t
final trial flight. (Left to right):
Squadron Leader Ernest L. Joh
(Navigator); Flight Lieutenant H
Carmichael Irwin (Captain); Ma
Herbert Scott (Assistant Direct
Airship Development, Flying an
Training); Lieutenant-Comman
Noël Grabowsky Atherstone (F
Officer); Flying Officer Maurice
Steff (2nd Officer). In front is N
Atherstone's dog, Tim.

Eve of Departure – Officers and
Crew of HM Airship R.101 at
Cardington on 30th September
1930, four days before the ship'
departure for Egypt.

Airshipmen – Twenty-four mem
of the crew of R.101 photograp
in the No 1 Shed at Cardington
28th September 1930, by Richa
Blake, the engineer allocated to
power-car in the No 1 Watch an
detailed to act also as the ship's
photographer on the voyage to I
Blake added himself to the grou
(extreme left in the back row) fro
another photograph. Seated in fr
are (left to right): Charge Hand
Engineer G.W. Short; Chief Engi
W.R. Gent; Chief Coxswain G.W.
Hunt and Charge Hand Engineer
Scott; Galley Boy J.W. Megginso
extreme right.

[PLATE XVIII]

seventh flight, carrying a crew of thirty-seven and eighteen passengers including Air Vice-Marshal Edgar Ludlow-Hewitt and E. L. Pickles, Assistant Director of Contracts at the Air Ministry, J. G. Gibson of the Finance Branch and Mr. J. Durward, Superintendent of the Met Office in Egypt. The disposable lift was 39.7 tons including 6,665 gallons (22 tons) of petrol.

R.100 flew a meandering course by way of Bedford to the Rolls-Royce Works at Derby, Birmingham, Gloucester, Bristol, Lundy Island, Ilfracombe, Lynton, Minehead, Cardiff, Bristol, Bournemouth, the Isle of Wight, Cherbourg, Guernsey (at 1504 in bright sunshine), Weymouth, the Needles, Southampton Water, Portsmouth, Southsea, Dorking, London (Westminster and Islington), Elstree, Luton and Bedford back to Cardington at 1809 – some 805 statute miles, covered in 23 hours 7 minutes at 37.7 mph – to be locked home at 1918 hrs after 24 hours and 16 minutes in the air.

Running through heavy rain, the cover again showed itself to be far from watertight though, on the whole, there was less flapping near the engines thanks to the fitting of extra supports. There was still, however, a good deal of concern about the strength of the cover and the state of the gas-bags.

On the following day, 27th July, a meeting was held at the Air Ministry in Kingsway, London with Sir John Higgins (AMSR) in the chair and Reginald Colmore (DAD) present from Cardington, with Louis Reynolds from Air Ministry (S9) and F. W. Musson, Sir John's Private Secretary.

DAD reported that R.100 had flown for three hours at 70 mph and three hours at 65 mph without undue fluctuation of the cover. No 8 gas-bag was losing a good deal of gas – about 35,700 cubic feet a day – but the holes would be repaired at the mast.

In all the circumstances there would be a good chance of making a start for Canada next morning when there were prospects of a helpful easterly wind. DAD said that he would hope to start with 33 tons of fuel or, possibly a little more.

It was agreed that the aim would be to start for Canada as soon as possible and that the Secretary of State should be so informed. The problem of Major Scott's 'lack of recent experience of airship handling' was raised by DAD and the difficulties which might occur if there was any division of responsibility between Scott and the captain of the ship.

AMSR ruled that the Captains must have full responsibility. Major Scott's position should be that of 'a non-executive Admiral', not the

Captain of the ship. DAD said that he had laid down that Major Scott should give no orders to any member of the crew but should deal only with the Captain. 'His value from an airship point of view was particularly in setting a course because of his experience and meteorological knowledge.' AMSR ruled that Major Scott should give advice to the Captain but that the Captain was not bound to take it. DAD accepted that this had been made clear to the Captain of R.100, Squadon Leader Booth, as was the fact that if he did not agree with a course advised by Major Scott he was to say so and act on his own responsibility. AMSR concluded that Major Scott would not be expected to be on all the flights of R.101 and R.100 between Cardington and Ismailia now to be scheduled for the winter.

Here was a delicate and important issue which had caused unhappiness among the flying staff. Ten years before, Scott had been a skilled and resolute officer who had taken R.34 to New York and back in exemplary manner. Now at forty-two, Scott was clearly not the man he had been; his judgment had frequently been a cause for concern and he resented advice.

Most difficult of all, probably because of mounting ill-health, Scott had begun to take increasingly to unwise lunchtime drink. He had been held in high respect. As Deputy Director of Airship Development (Flying and Training) he was now having difficulty in holding the position he had deservedly won.

But whatever the problems, technical and personal, the flight to Montreal was on. Sir John Higgins reported to the Secretary of State that Colmore and Scott were satisfied with R.100's performance. The airship would be ready to start as soon as it was gassed and fuelled.

Christopher Thomson telephoned Colmore at Cardington that Tuesday evening to wish him God Speed. His one regret, he said, was that because of Parliamentary pressures he could not now be on the flight. But next time – to India in R.101. He asked Colmore for a final assurance that the airship was wholly fit for the Atlantic flights. Colmore replied that, within human and mechanical fallibility, he believed that all would be well.

Thus, both the Air Member for Supply and Research at the Air Ministry and the Director of Airship Development at the Royal Airship Works had assured the Secretary of State that R.100 was now fit and ready for its first, long, overseas flight. But Thomson still had an uncomfortable feeling that still more caution and still more test flying would have been wise.

Staying the following week-end with close friends, Mr and Mrs

Carruthers of Moon Green, Wittersham, Kent,* he confessed that he would be profoundly thankful when R.100 was safely and successfully back at Cardington. He was concerned about its structural integrity after the collapse of the tail-fairing; he was uncertain about the engines, and he was bothered about the outer cover which had not yet been replaced. But his technical advisers, on whom he must rely, had told him that they were satisfied. He would not be justified to act against their assurances. Moreover, a further postponement would be a serious set-back not only to public confidence in the airship programme but also to relations with the Canadian Government which had made extensive preparations for the airship's reception.

Whatever his official view, Colmore was also worried.

On the afternoon of 28th July he had a final consultation with Scott.† He said, 'Suppose the outer cover does fail. What risks are we really running?'

Scott replied, 'Provided we are not caught out in a severe Atlantic storm we should get through all right. Both the *Graf Zeppelin* and R.33 had fabric troubles – and both survived.'

Indeed the 100 per cent record of successful trans-ocean flights by airships inspired confidence.

Nixon recalled:

> On the night of leaving for Canada Wing Commander Colmore was in a very disturbed state of mind. In fact, although I have known him many years and I have often seen him in difficult circumstances, I had never seen him in such a state of nervous tension. Some friends wished to take him out to supper but I persuaded him not to go.

There was a great deal of last minute work to do including the formality of the issue of a Certificate of Airworthiness. On the strength of the inspector's report on the latest flight, the C of A was issued by the Air Ministry and signed by F. G. L. Bertram, Brancker's Deputy, on 28th July.

A pattern of action with R.100 was now set by Colmore and Scott which was repeated with R.101 two months later. A final test flight, followed by an inspection at the mast led up to a flight conference and a decision for a quick departure; all leaving little time for rest

* Letter from Mrs Violet Carruthers, dated 8th November 1930 to Sir John Simon. (AIR 5/916.)

† Recorded by Squadron Leader Sydney Nixon in a note to the Treasury Solicitor, November 1930.

for the crew between the last test flight and the airship's first long overseas journey. Scott, 'the press-on type', was all for rapid action. Colmore weighed every point but yielded to Scott's conviction that there was no reasonable cause for delay. On each occasion there was a schedule to meet. On each occasion Lord Thomson asked for an assurance that the airship was fit and ready to go. On each occasion that assurance was given. Among the airshipmen there was a supreme confidence in airships; an eagerness to demonstrate their qualities.

As Captain George Meager, First Officer of R.100, wrote:*

> As far as feeling any anxiety as to the outcome of the flight, no thought of failure ever entered my head, nor for that matter anyone else's on board except perhaps Colmore who disliked flying at any time; all the more credit to him, therefore, for coming on such a trip as this. What Scott thought of the ship I don't know. One always had the impression that he considered R.101 the better ship but as that ship was partly a child of his own brain that was perhaps understandable. At any rate he certainly never had any qualms about the final success of R.100's flight to Canada and back.

So, only fifty-four hours passed between R.100's return to the mast after its flight round Britain and its departure for Canada in the early hours of 29th July. For R.101, there were fifty-five hours between its last trial flight and its departure for India on the evening of 4th October.

There was, however, one significant difference: the weather.

Gassing, refuelling and provisioning of R.100 began almost as soon as the airship was locked-home at the Cardington mast on the evening of 26th July – and continued up to midnight on 28th July. The gas-bags were brought up to 97 per cent capacity at a gas purity of 96.7 per cent.

The load sheet showed:

10,440 gallons of petrol weighing	34.45 tons
365 gallons of lubricating oil at	1.54 tons
1,220 gallons of water ballast at	5.40 tons
37 crew and seven passengers weighing	3.14 tons
Food, flying gear, etc., at	1.07 tons
480 gallons of drinking and washing water at	2.14 tons
Spares, baggage and miscellaneous items at	0.65 tons

The disposable load on board thus amounted to 48.39 tons, which,

* *The Airship,* Vol 7, No 5, Autumn 1948.

with a fixed weight of 105.52 tons, required a gross lift of 153.91 tons. It was the greatest weight that R.100 ever lifted and, in fact, required the release of 2.1 tons of water ballast to bring the ship into equilibrium at an actual gross lift of 151.8 tons. The ship was thus riding 'heavy' at the mast, as the bow indicator showed.

At the same time, the service-load on board had been cut to the bone. The water ballast was substantially less than the stipulated five per cent of gross lift and a minimum of spares was carried.

'Payload' – seven passengers and their baggage – amounted to 1,365 lb (0.61 tons) and the service load, excluding the 36 tons of fuel and oil, to 13.94 tons.*

The passengers, other than Reginald Colmore, came on board early in the evening and were in bed before midnight. Among them was Sir Dennistoun Burney and Nevil Shute Norway from the Airship Guarantee Company; Archie Wann (former Captain of R.38); Frederick McWade (AID), and Arthur Eldridge (Colmore's official secretary) from the Royal Airship Works, and Lieutenant-Commander R. Prentice, travelling as an Admiralty observer. The crew was made up of six officers, four charge-hand engineers, four coxswains, eight riggers, nine engineers, three wireless operators and three stewards: thirty-seven in all.

A light south-westerly wind was blowing; the night was fine and clear and the forecast fair. At 02.30 hrs on 29th July all was ready for R.100 to leave.

The Tower Officer, Flying Officer Harold F. Luck, at a signal from Ralph Booth in the airship's control car, withdrew the stops which locked R.100's nose-cone into the cup at the mast-head, leaving it secured only by the quick-release mechanism. Nothing happened.

Booth let go half-a-ton of water ballast. Still no move; the nose-cone remained fast in the cup. Booth released another ton of ballast. Still no sign of lift. Yet a further half-ton was discharged, making two tons in all, and Booth gave the order to slip. At last, the nose of the ship lifted gently and R.100 backed away. The time was 02.48 GMT.

Clearly, R.100 had been substantially heavy, loaded with fuel sufficient for some 85 hours of cruising flight with a payload of just 0.61 tons (1,365 lb). It had to do, but it was all quite inadequate for regular Atlantic services. If R.101 had had to be substituted for R.100, with the bags let out, its performance would have been virtually the same.†

* AIR 3/59A. R.100 Flight Log.

† See Appendix 11: 'R.101 for Canada.'

Booth allowed R.100 to drift up to its pressure height of 1,000 feet – now with the three forward engines running at 1,400 rpm. At 02.59, he set course of 288 degrees Magnetic (276 degrees True) to the north-west; now with the pusher engine in the aft car also turning at 1,400 rpm R.100 settled to a true air speed of 47 knots (54 mph) on a four-engine total of 1,200 bhp.

The airship droned through the clear night at 1,200 feet, by way of Wellingborough, Northampton, Rugby, Nantwich, Chester and Liverpool, to cross the English coast at Formby Point, just south of Southport, at 06.10 hrs. R. 100 was over the Point of Ayre, on the northern tip of the Isle of Man, at 08.05 hrs, running into cloud at 1,400 feet. The crew glimpsed the Mull of Galloway, through cloud, at 08.40, flying on a heading of 315 degrees around the northern edge of a depression centred over Ireland, in order to pick up the favourable westerly winds in the anti-clockwise circulation around the 'Low'.

By 10.20 R.100 was over Rathlin Island, off the coast of Antrim. It had covered the first 320 miles in 7 hours 32 minutes at a ground speed of 42.5 miles an hour.

Course was now altered to the west to make good a heading of 250 degrees along the northern coast of Ireland. The last sight of land, five miles to port, was Tory Island, off the coast of Donegal, at 11.25 – 8 hours 26 minutes out from Cardington.

R.100 was now heading into the Atlantic at 1,300 feet and, to conserve fuel, was running on only the three forward engines, all at 1,400 rpm, to give a total of 900 bhp and a true air speed of 45 knots, measured on Michael Rope's Air Log.*

Crew and passengers alike were kept busy in relays pumping fuel by hand from the tanks slung along the corridor on the lower deck, up to the service tanks above the engine cars. As fuel was used – at the rate of about one-third of a ton an hour – so water ballast was recovered from the rain-showers through which the airship constantly passed. Fabric gulleys had been rigged on top of the outer cover to channel rain water into the ballast tanks.

So R.100 set out from the Irish coast on the twenty-sixth non-stop Atlantic crossing by air and on the eleventh by airship. Of these the first two had been by R.34 in 1919 and of the other eight, one by LZ.126 – now the *Los Angeles* – and seven by LZ.127, *Graf Zeppelin*.

By noon on 29th July, after nine hours and one minute on course, R.100 had covered 373 nautical miles and had used 910 gallons of

* The Log of R.100's flights is contained in PRO AIR 3/59A – CAPS.29,261.

fuel at the rate of 101 gallons an hour. The ground speed made good was 41.4 knots or 47.7 mph.

Lunch of tomato soup, lamb chops with peas and potatoes, and a sweet of greengages and custard was served to the passengers with cheese and coffee to the sound of a gramophone in the crew's quarters. The sea, close below, was whipped up into white crests by a wind of 18 knots.

All was smoooth in the air in and out of low stratus cloud at 1,300 feet. The chief problem for those not on watch was boredom; countered by catching up on sleep and by endless games of cards. Meanwhile the airship's position was checked by wireless cross-bearings from liners within range. The day seemed very long because the clock was put back an hour every nine hours or so as the airship flew westwards.

By midnight GMT on 29th July R.100 had covered 840 nautical miles (967 statute miles) in just over 21 hours for a consumption now of 107 gallons an hour or 2.67 gallons to the nautical mile. Calculations showed that the 8,190 gallons of petrol remaining would be sufficient, at this rate for another 76 hours of flying or, if the winds remained constant, for a range of 3,880 miles to dry tanks.

The airship was still running on three engines at a time but ringing the changes to rest each engine in turn for inspection and to change the plugs. The Condors had not been designed for continuous running over so many hours on end, so regular checks were scheduled.

By noon GMT on 30th July R.100 had picked up a 28 knot tail wind – unusual for an east to west crossing – and was making good a ground (or sea) speed of 71 knots. In 33 hours in the air the airship had covered 1,460 n.miles on 3,485 gallons at the rate of 105 gallons an hour for an average ground speed of 44.2 knots.

Now the outer cover was giving trouble again. A couple of foot-long rips opened up on the underside, aft of the passenger accommodation; repaired with needle and thread from inside and reinforced with a doped-on fabric patch. Gas-bag No 7 was also causing concern as a result of an obvious leak.

A further position-check at midnight GMT on 30th July showed that the airship had now flown a total of 2,270 nautical miles since leaving Cardington – and 1,430 miles during the past twenty-four hours at an average of 59.6 knots with the help of the following wind.

R.100 had passed over the Cunard steamer, the SS *Ausonia* at 0510 local time (0710 GMT). There was by now fuel in hand but the wind had backed round to give a head component. So at 2050 hrs GMT, all six engines were run together at 1,400 rpm to give 1,800 bhp and a true air speed of 58 knots. The airship was now, however, making only 40 knots. Even so, at 2042 local time (0042 GMT) the first sight of land came from the lighthouse on Cape Bauld on the northernmost tip of Newfoundland. Thirty-three minutes later, at 0115 GMT on 31st July, R.100 passed over the coast of Belle Isle, in the straits between Labrador and Newfoundland, 48 hours 16 minutes after setting course from Cardington, 2,506 nautical miles away. The average ground speed had been 54.17 knots or 62.4 mph with – for the most part – an unusually helpful wind. From coast to coast 2,200 nautical miles had taken 37 hours 50 minutes.

R.100 was now flying in smooth air at 2,000 feet above a dense layer of fog along the Strait of Belle Isle. Occasional glimpses of the Gulf of St Lawrence, a little later on, gave a check that the ground speed had dropped to about 37 knots against, now, a 21 knot head-wind.

By sunrise on 31st July R.100 was over Anticosti Island, passing at 1,500 feet over the steamships *Empress of Scotland* and *Duchess of Bedford* in the early morning sunshine. A party of riggers under Coxswain Hobbs were hard at work repairing four three-inch holes in No 7 gas-bag and others in No 8 bag while the airship was climbed to 3,000 feet to expand the gas in the bags so that the rents could be more easily reached.

At 1525 local time (1825 GMT) at 1,000 feet over the north-east tip of Green Island (Ile Verte) in the St Lawrence, some 125 miles from Quebec, all six engines were opened up to bring the air speed to 60 knots in an effort to arrive at the Montreal Tower before dusk. Almost at once the ship encountered a cascade of excessively turbulent air where a cold northerly air stream poured down over the hills into the St Lawrence valley.

R.100 rolled rapidly through more than ten degrees and yawed violently, while being borne up to 1,800 feet.

Although R.100 came into calmer air within a minute or so the engine-room telegraph bells from the starboard and aft engine cars were by now ringing for assistance.

The engines were slowed at once and an inspection party under George Meager went to investigate possible difficulties.

What they found showed how susceptible to damage were the fabric covers of airships under the stresses of turbulent air.

First, there were two rents each about three feet long, in the cover of the bottom fin; not serious but flapping violently and beginning to extend. In the starboard horizontal fin the outboard edge was split longitudinally for a length of 12 feet, with ribbons of fabric fluttering in the airstream. Worst of all, the fabric on the underside of the port horizontal fin was in tatters over a length of 15 feet by 12 feet wide.

All the eight riggers in the ship were turned on to work with spare rolls of fabric, needles, thread and dope to make temporary repairs while the engines were cut right back to maintain the airship more or less stationary over Green Island against a 22 knot wind.

Temporary repairs were completed by 2110 GMT (1610 local time) under the most hazardous conditions with the men clinging to the lacing wires within the fins. The airship had been over Green Island for a little more than two hours.

Now R.100 crept ahead against the wind making good about 25 knots – passing over Quebec Bridge at 1905 local time at 24 mph ground speed. Booth now hoped to make the Montreal tower by midnight.

But that was without allowance for the weather.

Halfway between Quebec and Montreal above the small town of Three Rivers, at 1940 hrs local time on 31st July R.100 ran into a line-squall on the edge of a small thunderstorm; precisely the sort of turbulent conditions and vertical currents which had broken the American airship ZR-1 *Shenandoah* in 1925.

The line of the storm clouds, with heavy rain and hail below them, could be seen clearly ahead of the airship right across its path but not extending far to the south. Scott, standing with Booth in the control car, could have ordered R.100 to be flown round the storm to the south. Instead he instructed Booth to fly straight ahead through the storm and a tired Booth, anxious not to precipitate a row at the end of the flight, did not demur. Here was another example of Scott's predilection to 'press-on regardless' which had succeeded with R.34 but had caused so many problems in later years.

As R.100 entered the edge of the storm at 1,200 feet beneath the line of thunder-clouds, a violent up-current tilted the nose up at 25 degrees and then, as the rest of the airship entered the gust wafted the whole craft to 3,000 feet in less than a minute although the nose had now gone down to an angle of 30 degrees.

The ship was now completely out of control and crockery and loose objects, including passengers and crew, all cascaded forwards to come to a halt in a pile against any solid object.

Just as the ship began to level out another, even more violent, gust took it up again in a nose down attitude to more than 4,500 feet where, to those in the control car, a further dimension was added to the confusion by a trickle of blood-red fluid which poured down from above – and (to everyone's relief) turned out to be from a can of red dope which had been opened for the fin repairs. Because of its highly inflammable nature the master switch was tripped and, for ten minutes, the interior of R.100 was plunged into darkness to add to the disarray.

Thanks to adequate gas valves and to the fact that the bags were now well under capacity, the airship survived. There was, however, even further damage to the fabric of the starboard fin which now had two twenty-feet long rips on the underside.

Slowly the ship was brought down to 1,500 feet while the interior was tidied up and additional repairs set in hand by torchlight and lasting until after midnight.

After this further experience of a Canadian summer evening, R.100 crept forward to reach Montreal at 01.10 local time – to be greeted out of the darkness by a fiery cross looming ahead of them in the sky. The tired crew were taken aback until they realised that it was the city's electric symbol on the heights of Mount Royal.

By 0200 local time (0700 GMT), in clear skies the ship nosed over St Hubert aerodrome towards the floodlit mooring tower, a replica of the tower at Cardington. The local (EDST) time was 02.35 hours.

Lieutenant-Commander Arthur R. Pressey of the Royal Canadian Navy, the Tower Officer, trained at Cardington, picked up the airship's nose cable at 04.10 hrs and R.100 was locked-home at 04.37 EST (or 09.37 GMT) on 1st August 1930 – 78 hours and 49 minutes (3¼ days) out from Cardington.

The airship had spent almost exactly 74 hours at various cruising speeds on course, a little more than two hours holding its position while temporary repairs were in hand, and a little more than 2½ hours in the mooring process.

When all was worked out, the log recorded that R.100 had covered 3,364 nautical miles (3,874 statute miles) at a ground speed on course of 45.45 knots (52.3 mph) for a consumption of 8,945 gallons (29.5 tons) of petrol at an average of 119.46 gallons an hour on course – or 2.3 gallons to the statute mile (0.43 mpg).

There were 1,495 gallons of fuel remaining in the airship's tanks on arrival at St Hubert, sufficient for a further 12.5 hours at 119.5 gallons an hour. The ship landed with 9.8 tons of water ballast – a gain of some 6.4 tons through recovery of rain-water off the top of the envelope.

R.100 had been fortunate; first in flying at least part of the Atlantic crossing with favourable winds; secondly, for the smooth air in which most of the flight was accomplished; and thirdly in surviving what might have been serious damage caused by no more than normally turbulent conditions, some of which could have been avoided had Scott shown better judgment.

The mean fuel consumption of 884 lb an hour on course equated to an average of 1,540 horsepower employed and to a mean true air-speed of 52.5 knots – or 60 mph – for the flight. The headwind-component over the whole flight had thus been about seven knots, which was close to the average to be expected throughout the year at the heights flown.

To achieve 'an 85 per cent regularity', however, the airship would have to be capable of completing flights against a mean headwind of 17 knots. In such conditions Cardington to Montreal would take 93 hours and would require 10,920 gallons of petrol excluding reserves. Even with payload limited to only one ton this was beyond the capability of R.100. Like R.101 more lift was essential.

The airship remained at the St Hubert mast for nine days while repairs were completed with the help of Canadian Vickers Ltd at Montreal. Enthusiastic crowds thronged the aerodrome and parties of twenty at a time were allowed up the mast to inspect the ship's quarters.

On Sunday, 3rd August, there was more trouble when No 7 gas-bag split at the top – a tear five feet long. Further repairs were needed not only to this bag but also to the cord netting which restrained the bags laterally. Much of it was found to have rotted and had to be replaced.

By late on 9th August these various repairs were completed in time for the airship to slip from the tower at 1818 local time on 10th August carrying seventeen passengers and a crew of thirty-eight on a long-promised local flight.

The ship was away for 25 hours 57 minutes and covered 805 nautical miles flying at a steady 45 knots westwards by way of Montreal, Hudson Heights, Plantagenet, Cumberland, to Ottawa, then south-west to Smiths Falls and Kingston on Lake Ontario and on to Toronto and along the lake to Hamilton and back eastwards to Niagara Falls at 0800 hrs in the morning. And so across the lake by way of Fair Haven and Kingston to fly up the St Lawrence River to Presscott, Cardinal and Morrisburg back to Montreal, to secure at the mast at 2015 (EDST) on 11th August after dodging local thunder.

There had been a further mishap on the flight. At 07.42 on 11th August, between Hamilton and Niagara Falls, the reduction gear of the forward starboard engine had broken up and smashed the gear-housing.

Pieces of the gearbox flew forward into the propeller, which was not only damaged beyond repair, but also debris from it was thrown into the hull and caused damage to one of the main girders. Shortly afterwards, the magneto of the starboard midship pusher engine in that same car failed and also had to be stopped.

After inspection at St Hubert, it was clear that the forward engine could not be repaired at the mast. So that the engine had to be carried as dead-load on the homeward journey.

The return flight from Montreal to Cardington was begun in fine, clear weather at 20.30 hrs EST on 13th August (01.30 hrs GMT on 14th August), seen off by thousands of people who brought their cars to the airport and hooted continuously, with their headlights on, as the airship moved away.

On slipping from St Hubert mast, R.100 carried a useful load of 47.3 tons, including 9,635 gallons (31.8 tons) of petrol, 300 gallons (1.2 tons) of oil, a ton of food and 1.8 tons of drinking water, together with a crew of 38 and 17 passengers, including nine journalists from leading English and Canadian newspapers, Jacques Cartier, a descendant of the Jacques Cartier who had explored the St Lawrence River in 1535, Group Captain E. W. Stedman, Chief Aeronautical Engineer of Canada's Department of National Defence (and a member of Lord Thompson's Commonwealth delegation to the International Civil Aviation Conference in Washington, DC, in December, 1928) and Wing Commander the Hon. L. J. E. Twistleton-Wykeham-Fiennes, the RAF Air Attaché in Washington, DC.

The payload amounted to 1.6 tons including the seventeen passengers and their baggage and 200 lb of mail and freight Within this load were letters from the Canadian Prime Minister (R. B. Bennett) to Ramsay MacDonald and to Christopher Thomson and a box of cut flowers (peonies) for H M Queen Mary at Buckingham Palace.

R.100 flew southwards for the first few miles from St Hubert and then turned west and north in a wide sweep to come in over Montreal from the west, flying low across its myriad of twinkling lights before heading down the St Lawrence River, pushed along by a fresh following wind.

In rather less than two hours, at 22.46 local time (03.46 GMT on 14th August) the airship was over Quebec, flying at 1,400 feet on what was still a fine, clear night with a temperature of 60 degrees F. Loitering along on only three engines at 1,600 rpm to give 1,230 bhp the Air Log registered a true air speed of 45 knots though so far, for the 118 nautical miles flown from the start, the ground speed had worked out at 67 knots.

By dawn, at 0910 GMT, the airship passed over the West Point tip of Anticosti Island, 500 nautical miles out from St Hubert. The ship had now made good a ground speed of 68 knots with the aid of a 23 knot tailwind. The passengers and crew had their last sight of land before the ocean crossing when, at 1554 GMT, 14 hours 24 minutes from departure, Belle Isle was again abeam to port.

The wind was now beginning to abate but even so the airship had made a fine start; 840 miles at an average ground speed of 60 knots. Shortly afterwards, at noon local time (1600 hrs GMT) R.100, still flying on three Condors, had used 1,425 gallons of petrol at an average of 98.3 gallons an hour. A check showed that there were still 8,210 gallons remaining in the tanks; enough for an additional 83½ hours at the same rate of consumption. That would give an endurance of 98 hours on the fuel uplifted; cruising at 45 knots true air speed. The speed was rather less and the consumption rather more than R.100 had shown on earlier trials probably because of the added drag of the repaired tail surfaces and the fact that the airship was flying tail-heavy, somewhat out of trim.

Ralph Booth now set course over the Atlantic, flying on a heading of 135 degrees magnetic at 1,500 feet, still in fine weather with almost unlimited visibility above a sea dotted with icebergs.

So far the flight had been uneventful but, forty-five minutes later, a sparking plug flew out of the aft, pusher, engine – to the consternation of the engineer on watch in that car. The engine was shut down but restarted with a new plug at 1300 hrs (1700 GMT) some forty minutes before the airship passed the liner *Empress of Australia*, ten miles to port.

The tail wind was now abating and R.100 began to run into slight, and then heavy, rain as it entered a depression. Very soon water began to pour in through the outer cover. It penetrated the crew accommodation and short-circuited the ship's electric heating and also the cooking apparatus so that for the rest of the flight, there was no hot food or drink and the interior temperature of the ship dropped to 45 degrees.* Fortunately, passengers and crew were all kitted out

* Nevil Shute Norway while magnifying R.101's shortcomings in his book *Slide Rule* makes no reference to this and other problems with R.100, except to say 'Everything is streaming water and the whole ship is very wet . . . '

with fleece-lined 'Teddy Bear' flying suits and waterproof sleeping bags. But it was not luxury travel.

By midnight, R.100 was bucking a 15-knot headwind in noticeably colder air, now running on four engines at 1,600 rpm (1640 bhp) to achieve 53½ knots true air speed though making only 38½ knots over the sea.

At 0845 on 15th August R.100's log recorded: 'Ship tail heavy. Air very disturbed and much warmer (63 degrees F). Difficulty in steering.' Shortly afterwards a drift-sight showed that – thanks to a 30 knot tail wind – the airship was now making 82 knots over the water for 52 knots through the air. The difficulty in steering was put down to the stretching of the control cables from the control car to the tail surfaces.

By noon on this second day R.100 had covered 1,805 nautical miles from St Hubert and had used 3,745 gallons of fuel in 38 hours 30 minutes of flying at a rate of consumption of 97.3 gallons an hour. There were now 5,890 gallons remaining, still sufficient for a further 60 hours. The consumption had been pretty well constant, while the mean ground speed made good had dropped to 47 knots.

Later that afternoon, at 1830 local time (20.30 GMT), R.100 passed over one of the few ships spotted on this crossing, the cargo steamer *Beaverbrae* with which wireless messages were exchanged at latitude 52.22N and longitude 19.5W, three-quarters of the way across. The airship was now heading 140 degrees magnetic (115 degrees true) flying at 50 knots TAS and making good a speed of 70 knots over the sea.

Bull Rock, first landfall off the south-west coast of Ireland, was sighted on the port bow in the early hours of 16th August – at 0225 GMT – and R.100 passed over the Fastnet Rocks, south-west of Cape Clear on the southern tip of Ireland, at 0246 GMT. The sea crossing from Anticosti Island had taken 41 hours and 36 minutes.

Booth now set the airship to fly across the Western Approaches at 2,000 feet at 50 knots on four engines at 1,600 rpm making for Lundy Island, at a sea speed now of 60 knots.

After two days and nights in the air the crew were beginning to tire and one of the youngest and newest recruits among the riggers – George (E. R.) Patterson – was put on the elevator control-wheel to relieve the regular coxswains for the final stage of the flight. At 0340 GMT, dozing off, after various oscillations he let R.100 go into an increasingly steep descent from 2,000 feet. Passing 500 feet on the way down Ralph Booth hastily seized the wheel and pulled the airship out of the dive, just in time.

'The silly young mutt almost had us in the drink,' he said to a worried George Meager who had just arrived in the control car.* It was almost the end of R.100. Booth climbed the airship back to 1,600 feet and, now with fuel to spare, called at 0600 hrs for five engines at 1,600 rpm (2,050 bhp) to raise the air speed to 55 knots with a 10 knot tail component.

At 0645 R.100 flew over the north end of Lundy Island, and then passed Ilfracombe to starboard at 0705 flying up the Bristol Channel to cross the English coast between Portishead and Avonmouth at 0800 hrs. Escorted for a short way by two aircraft from the Bristol Aeroplane Company's aerodrome at Filton, R.100 flew on at 2,000 feet in clear weather over Cirencester at 0840, Bicester at 0915, Newport Pagnell at 0935 to Bedford at 0950 to reach Cardington at 0952 hrs GMT – 56 hours and 2 minutes after setting course from St Hubert. The distance flown had been 2,955 nautical miles (3,403 statute miles) at an average ground speed on course of 52.74 knots or 60.73 mph.

The main hauling-in wire was coupled at 1032. The steam winch then puffed steadily for 34 minutes before R.100 was secured back to the Cardington mast at 11.06 on 16th August – in an elapsed time of 57 hours 36 minutes out from the start. Fuel consumption had been 6,115 gallons at an average of 109.13 gallons (807.6 lb) an hour on course, equivalent to an average of 1,400 brake horsepower employed and a true air speed of 50.75 knots (58.4 mph). So, over the whole flight, the headwinds and tailwinds had almost evened out, leaving a small positive component.

R.100 had been away from Cardington for 19 days and had flown 7,124 nautical miles in 154 hours 26 minutes on course for a consumption of 17,730 Imperial gallons. The ground speed had worked out at 46.15 knots and the on course consumption 114.86 gallons an hour or 0.34 miles per gallon at a true air speed which had averaged 51.7 knots.

The hundred per cent success story of Atlantic flights by rigid airship had thus been maintained with this the 12th airship crossing – and R.100's west to east time had been the fastest yet recorded. All this was underlined in the BBC's running commentary of the arrival, broadcast from the gallery around the top of the mooring tower.

* This near disaster was never officially reported. Recorded in the ship's log in Booth's hand in shorthand, however, is the remark 'Patterson nearly crashed the ship into the sea.'

A growing crowd assembled at the foot of the Cardington Tower to greet R.100's return, led by the Secretary of State for Air, Christopher Thomson, and the Director of Civil Aviation, Sefton Brancker.

Noël Atherstone had been detailed to act as Tower Landing Officer. He recorded in his diary for Saturday, 16th August 1930:

> At 0930 GMT the ship was sighted on the western horizon just emerging from the haze. The weather was fine and clear with no clouds and altogether a glorious day. The wind was WSW about 15 mph and quite steady in direction.
>
> I went up the Tower and watched the ship circle round to leeward where she ballasted up and then gradually approached the Tower. The control platform was very full of people, including the S of S, the DCA, the BBC and a few others. It didn't make it easier for me as I had to do the actual mooring and I was continually falling over people and bumping into them.
>
> The ship picked up the wire first time at about 1030 GMT and I had her secured at 1106 GMT. The actual mooring operation took 34 minutes which wasn't so bad for a beginner!
>
> After the Customs Officer – W. Hines – had gone on board and cleared the ship, Colmore, Scott and Booth came off and were officially welcomed by Lord Thomson.

The Secretary of State shook hands warmly with every member of the crew and then, with them all grouped around him at the foot of the mast, congratulated them in an off-the-cuff speech broadcast by the BBC:

'I welcome you home from Canada and congratulate you on having accomplished this first and successful step in the development of our new generation of British airships,' he said. 'This contribution to our Imperial communications will be of incalculable significance. You may look forward with quiet confidence to the successful completion of this great experiment.'

He paid a generous tribute to those who had designed and built R.100 – with a specially friendly reference 'to my friend Barnes Wallis who must be a proud man today; deservedly so. I wish to express my high appreciation of the work of all who have been responsible for the design and construction of R.100 and to pay a special tribute to the officers and staff of both the flying crews and ground organisation.'

Noël Atherstone continued:

> I had been told to go on board with the relief watch and take over

the ship! In a remarkably short space of time all the ship's officers and crew had gone ashore and I found myself alone in the control car while the relief crew was left to get the rollers on. There was much cheering and photographing going on below – but I was rather occupied in working the elevators to take much notice of what was happening.

Irwin relieved me at 1445 BST so that I could get some lunch and told me he would look out till 2000 hrs. No arrangements had been made for letting the relief crew off for meals so I had to let half of them go at a time. A very unsatisfactory arrangement.*

The relief crew – such as it was – consisted of four riggers detailed by Gerrish from R.101's crew and some engineers, also from R.101's crew. There was no chargehand nor assistant cox'n.

I returned to the ship at 1950 hrs BST and found a large hole in the outer cover at the bottom of the ship at Frame 4 with three petrol tanks (of 250 gallons each) hanging out of it. Irwin told me that, shortly before I arrived, while the engineers were refuelling, there was a loud crash and the three tanks appeared through the outer cover!

What had given way was the dural 'spider' on the port end of the aftermost tank on No 4 frame. This dropped and pulled down the next two tanks. All petrol was immediately pumped away from this frame out of the remaining tanks, which were full, while the petrol in the damaged tanks had to be allowed to run on to the ground.

As soon as these tanks were drained they were hauled back into the ship and secured with ropes. Chief Cox'n Hunt of R.101 had come down to R.100 as a volunteer shortly before the tanks collapsed and it was under his direction that they were hauled back into the ship.

Scott paid a visit to the ship and had a look at the damage. The cause of the failure is not at present understood but there is obviously something wrong with the design for such a thing to be possible.†

As weather conditions looked favourable a handling party was ordered for 0300 hrs tomorrow and I turned the ship over to Hunt to see to getting the tower crew together ready for easing the ship to the ground.

Early next morning – Sunday, 17th August 1930 – Squadron Leader

* R.100's own relief crew was, of course, still on its way back from Montreal by sea under the charge of Squadron Leader Wann.

† Colonel Richmond recorded in his diary next day: 'Examined R.100's tanks which appeared to have failed under excessive pressure. This had the effect of damaging the cradle which also allowed other tanks to drop. I understand from Bushfield that these tanks were tested to only four lb pressure.' Had this failure happened when the engines were running there is little doubt that R.100 would have been destroyed.

Sydney Nixon superintended the assembly of the handling party (at the standard rate of 1s-6d ($7\frac{1}{2}$p) an hour) including 400 men from Henlow. R.100 was taken from the tower and put into No 2 shed without difficulty by 0530 hrs in a flat calm.

The airship never flew again. It was in the open only once more – on 1st October – when it was moved into No 1 shed at Cardington, then just vacated by R.101, ready to be cut in half for the insertion of an extra bay and the fitting of a new cover, as had by then been completed with R.101.

Thereafter, R.100 was scheduled to fly a series of return flights to Canada from the early summer of 1931 while R.101 was to ply on the Egypt/India route, 'both ships to build up operating experience while awaiting the new ships of the R.102 class.'

R.100 had made ten flights in its operational life of eight months. The airship had covered 11,134 nautical miles in a total of 296 hours and 57 minutes in the air while it had spent another 800 hours 20 minutes at the mast. During its flying life the airship's Condor engines had consumed 30,574 gallons of petrol and 590 gallons of lubricating oil at an average fuel consumption, on course, of 110.63 gallons (819 lb) an hour or 2.75 gallons a nautical mile, equivalent to 0.42 miles per gallon.

The average ground speed had worked out at 40.29 knots (46.4 miles an hour) for an average true air speed of 51 knots (59 mph) cruising on a mean of 1,420 horsepower employed. Rarely more than four engines had been used at once.

A total of 144 passengers had been carried at an average of 14 on each flight.

In all, up to 17th August, total expenditure on R.100 and its equipment had been £500,113 of which the Air Ministry had paid £392,000 (78 per cent) which was £42,000 in excess of the contract price, while Vickers had contributed £108,113.

A detailed inspection of the ship in the No 2 shed showed that the outer cover and the gas-bags were now only fit for scrapping while later a disconcerting amount of corrosion was found to have occurred in various parts of the main hull structure. A complete refit was essential at the same time as the installation of the new bay. There was also the problem of the continued unreliability of the engine installation.

R.100 had been a lucky ship to survive the various incidents in the course of its ten flights.

Reginald Colmore wrote from Cardington a survey of the situation to Sir Geoffrey Salmond in India on 21st August.* He said:

I have deliberately put off writing to you until we had completed a flight which was something more than just flying around Cardington.

As might be expected we had endless small troubles with both ships during the trials which made it desirable to put off the first overseas flight and, if I had written before my letter would have been full of excuses.

The flight to Montreal has confirmed many views which we have held. Primarily, the meteorological and wireless organisations require developing and we shall have to carry two receiving and transmitting sets in each ship so that the wireless required for meteorological signals will not interfere with direction-finding for navigation purposes; both of which are essential.

No new flying problems are involved in a flight between England and the entrance to the St Lawrence and, given good meteorological and W/T organisations and a ship which can carry sufficient fuel to fly at 70 mph continuously, I think that that part of the route could be flown with reasonable regularity. But, up the St Lawrence and over land in Canada, we met entirely new conditions.

It is very bumpy over land and the movement of the ship is very much quicker than over land in England; and then one also has the thunderstorms to contend with and at certain periods of the year they are frequent and violent.

Aeroplane services are, however, running from Montreal down the Labrador coast and also to Prince Edward Island. I think the solution to this route lies in an intermediate tower somewhere at the entrance of the St Lawrence; mails being picked up by aeroplane for the remainder of the journey.

Aeroplanes are running from Moncton daily, on a four-hour schedule to Montreal and on an eight-hour schedule to New York.

I had a long talk with all the Canadian air authorities, the aeroplane companies and Mr Beatty of the CPR and I am inclined to think that Canada will offer to put up an intermediate tower. As I have said, this would – to my mind – make an Atlantic Service, at any rate during the summer months, a practical proposition with airships which could be built today without any further elaborate experiment. During the winter it is a different proposition but information we get from Canada on weather conditions during the winter will justify us making a careful investigation into conditions on the northern route during the winter months, and we shall start this investigation shortly.

They gave the ship a wonderful welcome in Canada and I think

* AIR 5/906/X.M.08109.

the flight over Ottawa and Toronto was also a success from the Canadian public's point of view.

You do not need to worry too much about the damage to the fins which occurred during the latter part of the outward flight. The structure of the ship is not damaged in any way – only the fin covers. I have realised for some time that the outer cover and the fin covers present a problem in these large ships but it is not one that cannot be got over.

We are now making a spare cover for R.100. The present cover is about two years old and was built at Howden under very bad conditions and also, as a matter of fact, we do not like its design. Until we have completed an examination of the cover I cannot say when the ship will again fly but it is only the outer cover and the gas-bags which are holding up further flying.

In regard to the Indian flight we are working hard to complete R.101 by the end of September and this mainly depends upon the provision of a reversing engine. It would not be wise to undertake the flight with only four engines for forward flight and there would not be sufficient power to avoid bad storms or to ensure having a high enough continuous air speed to deal with bumpy conditions which must be expected over the desert.

Everything that can possibly be done to complete the ship by the end of September is being done and I hope that we shall not disappoint you.

I am afraid that progress during the past twelve months must have appeared very slow to you. It has not been too easy. The two ships are entirely different and there has been an enormous amount of work in getting them through their teething troubles.

I am more confident than ever that the airship can be developed into a useful means of transport but, in the early stages at least, we should confine our operations to over sea work as much as possible and develop our wireless and met: organisations – which are absolutely essential to regular and safe flying.

With the exception of the lift of the ships, both have exceeded my expectations. They handle in the air perfectly, and mooring is now becoming a matter of routine.

We have several developments in hand which will make mooring safer and quicker and will reduce the cost of the equipment and, as I think you know, we hope by the spring of next year to be experimenting with mechanical handling.

I am doing all I can to press on with the mechanical handling and mooring developments – especially the former because, at the moment, it is really not safe to take a ship out of the shed except in a flat calm. With mechanical handling I expect to handle in winds up to 20 mph without any difficulty or danger to the ship. This will mean that, for all practical purposes, there will be no delay in housing the ship or taking her out of the shed as required.

I expect that you also know that Sir John Higgins managed to get approval for us to proceed with the design of a further ship. We are working on a 7½ million cubic feet ship which will be able to carry 50 passengers in considerable luxury with five tons of mail with fuel for 70 hours at 70 miles an hour and a top speed of over 80 mph. This ship will go into the existing sheds when they are lengthened and will not involve any new experimental work of any importance or any radical change in design. Moreover the new ship should carry her full payload and fuel under the worst conditions from Karachi.

I could build the ship within two years from the order to go and could turn out a second new ship eight months after the first. I am hoping that we shall get permission to go ahead after the Imperial Conference.

The development of the Tornado engine is proceeding steadily and, although this may not be the ultimate airship engine, I have every hope that it will give us good service for the next four or five years and enable us to establish the airship firmly as a means of transport.

I am afraid that this is rather a rambling letter but I hope to see you before very long and to be able to explain exactly the lines on which we are working.

Richmond, Scott and Nixon all wish to be remembered to you.

Yours sincerely

R. B. B. Colmore.

Here was a letter of importance. It stated clearly the forward thinking of the Director of Airship Development, the problems of the present and his hopes for the future.

In particular he underlined the fact – reinforced by experience with the *Graf Zeppelin* as well as with R.100 – that airships were very much more suited to operations over water than to flying over land and that safety and regularity was dependent upon better meteorological services through which turbulent conditions could be forecast with some certainty.

Moreover, the outer covers still held pronounced weaknesses until better materials and attachments could be developed.

In order to avoid turbulent conditions and damage to the covers over land in North America, an Atlantic airship service might best be arranged to terminate at a mooring mast in New Brunswick from which passengers and mail could be carried on by aeroplane to Montreal and New York.

Colmore saw and stated plainly that – in the then state of airship development – the one combination of circumstances to be avoided would be flying over land, in bad weather and turbulence with an unsatisfactory outer cover. To fly in such conditions would be asking for trouble.

CHAPTER SIXTEEN

The Way Forward

'Look forward with quiet confidence to the successful completion of the great experiment.'

LORD THOMSON OF CARDINGTON, 16th August 1930

August 1930

SO AIRSHIPS were hitting the headlines in an era of pioneer flights. R.100's North Atlantic success was applauded everywhere; the problems brushed aside. The *Graf Zeppelin* was in the news as well. It had recently flown up the east coast of England, over Hull, Edinburgh and Aberdeen* in the course of a flight from Germany to Iceland and back, after a round trip to Rio-de-Janeiro. The United States was well under way with two new rigid airships for the US Navy each to be of 6,850,000 cubic feet capacity (helium filled) – to be named ZRS-4 *Akron* and ZRS-5 *Macon*.

The Press agreed that, for long non-stop journeys, neither the passenger liner nor the aeroplane would be able to compete with future airships. Soon R.101 would further demonstrate the suitability of airships for Imperial communications during its flight to India.

There were other new stars in the sky. On the showery Bank Holiday Monday of 4th August 1930 – the sixteenth anniversary of the start of the First World War – the Secretary of State for Air flew from Hanworth Air Park to Croydon Airport in a three-seat Desoutter monoplane accompanied by Miss Margaret Bondfield, Minister of Labour and the first woman member of a British Cabinet. At Croydon they were met by Fred Montague, the Under Secretary of State for Air, by Sir Sefton Brancker, the Director of Civil Aviation and by Lord Wakefield of Hythe who had helped to finance the flight. They, with about 50,000 members of the general public had come to welcome Amy Johnson on her arrival back in England from her 19-day solo flight from Croydon to Darwin, Australia, in

* Post-war examination of German archives shows that the *Graf Zeppelin* photographed the Blackburn Aircraft Company's works at Brough, the RAF aerodrome at Turnhouse, Edinburgh, and Dyce aerodrome at Aberdeen, as well as the Royal Navy's anchorages at Rosyth and Scapa Flow, in the course of this flight in the summer of 1930.

her D.H. Moth *Jason*.*

From six o'clock in the evening, when the Armstrong-Whitworth Argosy of Imperial Airways was due to land – after the last leg of Amy's homeward flight from Vienna by way of Brussels – Croydon Airport was besieged by the largest crowd to be seen there since the boisterous welcome to Charles Lindbergh, coming in from Paris, three years before.

Long distance flying, whether by airship or by aeroplane, and especially by a girl, could be relied on then to be a focus of popular interest and enthusiasm.

* Amy Johnson's solo flight from England to Australia in May 1930 had personal links with the First Officer of R.101. Miss Johnson had landed her Gipsy Moth *Jason* at Port Darwin on 24th May, nineteen days out from England, where she had left Croydon Airport on 5th May. She took off from Darwin, for Sydney, on 27th May, escorted by C. W. A. Scott, flying a D.H.50J cabin biplane, chartered by *The Sydney Sun*, and by Harold Owen of Shell in another D.H.60G Gipsy Moth. Scott showed himself to be an ungallant escort to Amy Johnson. He pressed on ahead in his, much faster, aircraft so that Amy could keep up only by flying at full throttle. Scott made no effort to help her at the refuelling stops at Daly Waters, at Alexandra Station and at Cloncurry.

Amy was met at Cloncurry by Ian Grabowsky, Sales Manager of the de Havilland Aircraft Company of Australia, who had flown there from Mascot, Sydney, in a racing Gipsy Moth. He brought her a welcome from Major Hereward de Havilland, Managing Director of D.H. Aircraft, Australia and brother of Geoffrey de Havilland, not then Sir Geoffrey.

Ian Grabowsky found Amy at Cloncurry tearful and exhausted. He was infuriated by the churlish attitude being adopted towards her by Scott and his passenger, W. S. Bird of Wakefield Oil, both of whom insisted that the exhausted girl take off at once for Longreach, where a Wakefield reception had been arranged for her. At the same time, they offered her no assistance with refuelling, nor gave her any opportunity to wash and rest.

Grabowsky did his best to help, but he was held up at Cloncurry, with engine trouble, and he did not catch up with the three aircraft until they reached Brisbane on 29th May, where the distressed Amy severely damaged her Moth in a misjudged landing at Eagle Farm Aerodrome, fortunately without injury to herself.

Subsequently, Hereward de Havilland and Ian Grabowsky flew Amy on a tour of Australia – Sydney, Melbourne, Adelaide, Perth and Fremantle – in a D.H.75 Hawk Moth (VH-UNW). Before Amy Johnson left for Freemantle for England on 7th July, on board the SS *Naldera*, she was subjected to more uncouth treatment by the Wakefield representatives. It would have horrified Lord Wakefield, who had helped to finance the flight. As it was, Ian Grabowsky had to intervene on several occasions on Amy's behalf.

She came to regard Grabowsky as a modern 'knight in shining armour' and remained grateful to him for the rest of her life.

Ian Grabowsky was the younger brother of R.101's First Officer, Noël Grabowsky Atherstone. Ian had been born in Terioki, Finland, on 10th July 1899, five years after Noël. After service in the RNAS and then the RAF in 1919, Ian had gone to Australia with Noël in 1920 to take up sheep and cattle farming in Gippsland, Victoria, under the Soldiers' Settlement Scheme. They were joined by their parents and their two younger brothers, Patrick and Alastair, and their sister, Teresa, in 1921.

When Noël returned to England, to join the Royal Airship Works in 1927,

Three hours late because of strong headwinds against its 85 mph cruising speed, the Argosy eventually landed in the dusk at 21.05 hrs, escorted by three DH Moths.

Beneath the floodlights on the Plough Lane site, a pale Amy, a tired Imperial Airways crew headed by Jimmy Youell, and the long-suffering reception committee, were swamped by the crowd while the scene was broadcast to the nation – including the startling (and fortunately inaccurate) words, 'Here she comes; her aircraft is stalling in to land.'

'Lord Thomson said his usual graceful piece,' reported *The Aeroplane*. 'He spoke glowingly about youth and beauty, airmanship and femininity, courage and determination and leadership iin the air.'

He must have felt that if a girl of twenty-six, flying alone could reach India in record time despite little experience of aviation ('though some of life', according to Brancker) – and then fly on to Australia – how much more successfully and reliably could a professional crew fly a great airship to India and back with all the resources of the State behind them.

Parliament was now in recess. The Prime Minister was in Scotland. The work of the Air Ministry continued unremittingly.

At Cardington in No 1 shed, much was in train to remedy the defects in R.101 brought to light during the ten test flights of the past ten months. On 29th July the framework of the ship had been parted at Frame 8 and the aft portion moved back to make room between the two halves for the assembly of the new bay for which all material had been delivered by 7th August – the longi-

FOOTNOTE—*continued*

Ian, likewise, went back to aviation by joining the newly-formed de Havilland Aircraft Propriety Ltd, in Melbourne, as Sales Manager and Flying Instructor, first at Essendon, Melbourne, and then at Mascot Aerodrome, Sydney, New South Wales.

In 1931, Ian joined New Guinea Airways, of which he became General Manager in 1935, after pioneering air transport services there. He went on to have a distinguished career in Australia and Far East aviation; successively as North Queensland Manager of Airlines of Australia and, then, Australian National Airways. During the Second World War, he joined the Commonwealth Government Department of Aircraft Production and organised the repair and overhaul facilities for the RAAF and the USAAC, before returning to Australian National Airways from 1944 to 1957. During this time, he was also a Director of Cathay-Pacific Airways, and, in 1948 and 1949, on loan as Manager, Air-Ceylon, and, subsequently, in charge of ANA helicopter operations and development. In 1957, he joined Trans-Australia Airlines as Special Projects Manager. After his retirement in 1962, he was commissioned by the Commonwealth Government to write the history of Civil Aviation in New Guinea. Ian Grabowsky died at Vallista, Victoria, on 14th March 1972, aged 73.

tudinals from Boulton and Paul at Norwich and the frame girders from the RAW's own workshops at Cardington. Frame 8A was completed and lifted on 14th August and eight out of the fifteen longitudinals (8A to 9) bolted into place.

Meanwhile a great deal of work had been put into the repair and replacement of the gas-bags. The loss of gas during the flights in June had dismayed everyone at Cardington and a major effort had been made both to rectify the defects and to ensure that, so far as possible, they did not occur again. That required reinforcing of the bags, a greatly improved system of padding, removal of sharp projections in the internal structure and, of course, renewal of the outer cover. Substantial progress was made with this work between the beginning of July and 14th August.

Eleven days later, on 25th August, the structural work was complete. Next day No 9 gas-bag was installed and inflated and, on 28th August, the large new bag (No 8A), to give $15\frac{1}{2}$ tons of extra lift from 510,300 cubic feet of gas, was filled in the new bay.

By this time most of the replacement cover was in place though still awaiting lacing to the framework and it would then have to be doped. So far the job had taken only five weeks from the day on which R.101 had been handed back to the works.

Meanwhile, at the Air Ministry, Major Louis Reynolds, now appointed to be Lord Thomson's Private Secretary, put before the Secretary of State a note, dated 26th August 1930, in answer to his enquiry on 'the latest state of play on R.101'. Following a telephone conversation with Colmore, Reynolds' note read:

1. The date of departure of R.101 for India now depends on five separate factors:

   (i) The completion of the work on inserting the extra bay.
   (ii) The completion and installation of a reversing Tornado engine to enable the airship to have five engines for going ahead.
   (iii) The completion of a reversible airscrew suitable for the engine.*
   (iv) Suitable weather for getting the airship out of the shed when ready for flight.
   (v) The carrying out of a satisfactory trial flight of not less than 24 hours between being taken out of the shed and starting on the flight to India.

* This was somewhat of a misnomer. What was meant was the fitting of a new fixed-pitch wooden airscrew which would give adequate thrust for forwards flight and also some astern thrust when run in the opposite direction.

2. The position with regard to the above five factors is as follows:

(i) The work on the airship at RAW should be completed by 22nd September.

(ii) The reversing engine should be delivered to RAW by 4th September but without the starting gear.

(iii) Airscrews have been ordered to two different designs and due for delivery between 16th and 28th September.

3. If the dates given in (ii) and (iii) above are worked to it will just be possible to have the reverse engine installed with airscrew by the 22nd September. Day and night work is going on on all these items.

4. There would therefore be six days available for items (iv) and (v) – viz, getting the airship out of the shed and carrying out the trial flight. It is clear that this leaves no margin but may just be possible.

On the same day as this note was written at the Air Ministry, Gwydyr House, Colmore held a meeting with Gerrish, Richmond, Rope and Scott to decide on realistic dates on which R.101 could now be expected to be ready to fly to India and on which to schedule the voyage. In the light of current progress a factual programme was needed for the Air Ministry and for transmission to the Air Officers commanding Royal Air Force units in the Middle East, India, Malta, Palestine, Iraq and Aden as well as the Royal Navy's C in C's in Malta and the East Indies.

After a careful analysis of all that remained to be done, Colmore drew up a timetable agreed as practical by all concerned at Cardington and sent it off to the Air Ministry on 27th August. The proposed schedule was:*

1.	Depart Cardington	Midnight	26th September
	Arrive Ismailia	After sunset	28th September (Refuel)
2.	Depart Ismailia	After sunset	29th September
	Arrive Karachi	Before sunrise	1st October (Refuel)
3.	Depart Karachi	After sunset	5th October
	Arrive Ismailia	After sunset	8th October (Refuel)
4.	Depart Ismailia	Before sunrise	9th October
	Arrive Cardington	After sunset	11th October

* PRO AIR 2/364 – 522,085/30 – 6,130.

Thus the scheduled time for the out and return journey would be fifteen days including five days on the outwards flight, four days in Karachi and six days on the flight back.

There was confidence at Cardington that, with the new bay, a new cover and so many detailed improvements suggested by earlier experience, R.101 would be fit and ready not for just one but for a series of flights to India and back during the next few months. Of course payload would still be very limited but there should now be adequate lift for fuel and stores and, for the first outwards journey, Colmore set out his expectations for a complement of fifty-eight made up of five officers, forty-three crewmen, five representatives of the Royal Airship Works and five official passengers including the Secretary of State for Air and the Director of Civil Aviation.

Lord Thomson had inspected the work in progress on R.101 in the shed when he was at Cardington to welcome R.100 home from Canada on 16th August, just after the structure of the new bay had been hoisted into place. Now, on receiving Reynolds' note, followed by Colmore's proposed flight schedule, he called a meeting at Gwydyr House to discuss the flight and to be assured by the Director of Airship Development himself, who was technically responsible, that R.101 would now be ready and adequate for the flight according to the proposed timetable, and then to review the next stage of the airship programme.

The meeting was held in the Secretary of State's room on the first floor of Gwydyr House in Whitehall on Friday, 29th August, with Lord Thomson in the chair.

In addition to the S of S and DAD there were present at this crucial meeting, the about-to-retire Permanent Secretary to the Air Ministry, Sir Walter Nicholson, the Deputy Director of Finance, Henry Self – who conducted all the Air Ministry's financial negotiations on the airship programme with the Treasury – and J. G. Gibson from the Directorate of Civil Aviation, representing Sir Sefton Brancker. Louis Reynolds, took the minutes.*

This was the meeting at which critical decisions were taken for the future – and there were two notable absentees, AMSR and his successor. Sir John Higgins, who had shouldered the final technical responsibility for the airship programme since he had taken over as Air Member for Supply and Research from Sir Geoffrey Salmond on 27th December 1926, was now himself retiring on 31st August 1930 and was away on terminal leave. He was to become Chairman

* AIR 5/987 – 6,357.

of Sir W. G. Armstrong-Whitworth Aircraft Limited at Coventry and of A. V. Roe and Company Ltd at Manchester and to disappear from the airship scene just when his four years of experience there could have been most valuable and, possibly, decisive.

Unfortunately the new AMSR, Hugh Dowding, could not be present either 'to get a feel of the situation' because he did not arrive at the Air Ministry until the following Monday. He had come to the Royal Air Force through the British Regular Army and the Royal Flying Corps. He had recently been Officer Commanding Royal Air Force Trans-Jordan and Palestine with headquarters at Amman and Jerusalem – but though he now assumed responsibility for both the military and civil aircraft Research and Development programmes, he had no knowledge of airships. Because for so many years, Lighter-than-Air developments in Britain had been an offshoot of the Royal Navy and the Royal Naval Air Service, they had come to be regarded by those imbued with the – still strong – RFC traditions as uncouth monsters, descendants of the ponderous Naval 'Battle Bags' of the 1914–18 War. Dowding shared those views – and was not an engineer officer anyway.

The Air Minister's meeting on 29th August began with a review of the prospects for R.101's projected flight to India.

DAD, Reginald Colmore, said that it now looked as if R.101 could not be ready for flight before 1st October on account of delays in the delivery of the starting gear for the new reversing Tornado engine, expected on 11th or 12th September. Three weeks would probably be needed for installation and testing so that would bring the date to the beginning of October – but every effort would be made to beat the time.

He would like to allow four days for 'turning round' at Karachi and the voyage back would take between four and six days.

Lord Thomson asked if the flight could safely start by 4th October? In any event he would need no more than a few days in India and he thought that he could arrange for the air business of the Imperial Conference to be put off until the 20th October if necessary.

Colmore stated that he believed that, subject to one satisfactory test flight, R.101 could be ready to leave by 4th October, possibly earlier.

After discussion, the decision was recorded in the Minutes of the Meeting that, if possible, the flight to India would start in the first week of October, with the idea of remaining in Karachi for between four and six days and bringing the Secretary of State back to England by 18th or 19th October.

The route flown would be to the Mediterranean by way of either the Rhône valley or along the northern foothills of the Pyrenees, and on to Ismailia with the object of arriving between 5 p.m. and 8 p.m. local time and leaving again before 8 a.m. next morning so as to get the best conditions for both mooring at the tower and leaving it again.

On the number of passengers to be carried, and their names, DAD said that he would like to take Colonel Richmond and Major Scott in addition to the officers and crew, and approval had been given to take Major Bishop, Chief Aircraft Inspector of the AID, as far as Ismailia. The S of S said that his party 'would consist of Sir Sefton Brancker; his man, Buck, and an ADC or Secretary who might be an RAF officer.'

DAD summed up the situation by saying that the new bay was now successfully installed and that all systems should be complete by 22nd September and they would then be awaiting only the delivery of the new starting engine and the propeller for the Tornado which was being modified to run both forwards and in reverse. Once these were delivered all that would be needed would be suitably calm weather in which to bring the airship safely out of its shed and on to the mast. Then a trial flight must be completed satisfactorily. It should not be for less than twenty-four hours; preferably more.

Colmore ended with an expression of confidence. The day and night work now in hand should make R.101 fully ready to leave for India during the first few days of October and every effort would be made to be ready sooner if possible. 'The whole atmosphere was of satisfaction and assurance.' *

No mention of gas-bags. No mention of outer cover. Clearly they were technical details with which the S of S should not be concerned. In any case Colmore was satisfied that replacement and rectification work had brought them up to acceptable standards.

Thomson asked: 'Did DAD see any special difficulty in the flight between Cardington and Ismailia?'

DAD replied that the trip was fairly straightforward. They might have to deal with thunderstorms in the Rhône Valley on the way out and perhaps with the Mistral on the return flight. But these were not insurmountable. Later, for regularity of service with a payload, a tower at Malta would be essential.

And that was that.

So, ahead of the trial flight with the new bay – but subject to its

* Discussion with Sir Henry Self on 30th October 1974, and report in PRO AIR.6/904.

success – the decision was taken to start. Contrary to some subsequent allegations there were no financial pressures, no threat of withdrawal of funds, no overriding of technical opinion. There was, however, a general air of confidence which was shared by everyone at the meeting.

They went on to discuss R.100.

DAD said that a new cover was essential. In addition, a number of defects had to be put right, such as leaky gas-bags, trouble with tanks and water ballast bags. With two or three months of steady work R.100 might be made fit for short flights but he would not recommend that. He would much prefer that a new outer cover should be fitted before R.100 flew again.

There was some discussion on whether the Airship Guarantee Company Ltd should be asked to pay for the new cover under their contract but Lord Thomson said that, when the moment came, he would consider waiving such charges even though under the strict terms of the contract the AGC ought to pay. In any case for the next two years there could be no question of offering R.100 back to the AGC for half the sum paid for the airship, as originally proposed. It was essential to progress that the two ships should be operated together by the same organisation – and, so far, there was only one operating base and mooring mast; at Cardington.

DAD went on to say that, because the material for the new cover for R.100 had not yet been made up, a gap of six months would be necessary before the outer cover was on the ship and ready for flight. To put in a new bay would take seven or eight months. He urged most strongly that this should be done while the ship was in the shed awaiting the new outer cover. There was now no reason to believe that a new bay for R.100 would present any more difficulties than had the new bay for R.101 and with the new bay, a new and stronger cover and a number of detailed modifications and repairs R.100 would be a very good airship – especially if the present petrol engines could be replaced by 'dieselised Condors' which now seemed possible.

The meeting agreed that Henry Self should approach the Treasury for authority to spend money on the new bay and other necessary work for R.100.

That led to discussion of the flying programme for R.100 and R.101 during the year 1930/31.

Because of difficulties in over-flying and landing in Italy on the England/Egypt aeroplane service, Lord Thomson said that he was most anxious to mount an airship service on the route, between Cardington and Ismailia, as soon as practical.

DAD said that such an operation should be fairly straightforward so long as a mooring tower was built at Malta to ensure regularity of service.

R.100 could not be available with a new bay before the spring or summer of 1931 but, all being well, R.101 would start a series of services on the Egyptian route on its return from India. There might be the problems of thunderstorms in the Rhône valley, and the Mistral on return flights, but with a good met: service they could be coped with.

At present there were only three spare Tornado engines for R.101, one of which was at Karachi but additional engines could be provided later.

The Minutes recorded that after further discussion, the S of S asked that a programme should be investigated on the possibilities of a regular service between Cardington and Ismailia flown by R.101, starting towards the end of 1930, with a frequency of once a month while, during the summer months, R.100 with its new bay, should operate on the Atlantic route to Canada.

DAD added that he thought that the two airships could earn useful revenue during the next two years. The Royal Airships Works had, for example, received an offer of £20,000 for one airship passage to New York in 1931. He understood that the *Graf Zeppelin* was almost paying her way by offering 'joy rides'. The meeting agreed that, when regular services could be announced, arrangements for passenger and cargo carryings should be made through a suitable travel agency.

The meeting then went on to discuss the projected new airship programme* and the Secretary of State's Paper for the Imperial Conference in October.

Then, having settled the broad lines of the tasks ahead, the S of S went off to tour Royal Air Force stations which had been engaged in the annual Air Exercises within the Defence of Great Britain Command – exercises directed by Air Marshal Sir Edward Ellington and mounted between a hypothetical 'Red Colony', under Air Vice Marshal Hugh C. T. Dowding, and a hypothetical 'Blue Colony' under Air Vice Marshal Sir John Steel.

Colmore returned to Cardington to give instructions there for the agreed programme of work. At the same time, with the dates now provisionally settled, Louis Reynolds drafted a telegram in reply to an enquiry from the Government of India on 13th August, asking

* See Chapter 17 – 'The New Projects – R.102 and R.103.'

Before and After – HM Airship R.101 at the Cardington mooring tower in its original form (below) as first flown in October 1929, and after the extra bay had been inserted as flown in October 1930. The length was increased by 45 feet to 777 feet and the volumetric gas capacity by 510,300 cubic feet – or 10.2 per cent – to 5,509,753 cubic feet. [PLATE XIX]

Before Departure – R.101c at the Cardington mooring tower on the afternoon of 4th October 1930, before departing for India. The portion of the original cover which had not been renewed, just aft of the new nose-cover, is clearly visible. The original photograph shows well, also, the reinforcing strips and patches attached to this part of the cover.

Close-up – R.101c at the mast on 4th October 1930 – a photograph which shows well the control car, the lounge windows and the midships and aft engine cars. [PLATE XX]

for an exact date on which R.101 could be expected to arrive at Karachi.

The reply was attached to a letter, signed by Sir Walter Nicholson, addressed to the India Office under its Secretary of State, the Right Hon William Wedgwood Benn (later the Viscount Stansgate). The letter was dated 30th August:

Proposed Flight of Airship R.101 to Karachi

Sir,

In reply to your letter of 14th instant, forwarding a copy of a telegram from the Government of India, I am commanded by the Air Council to say, for the information of the Secretary of State for India, that their present intention is that R.101 should start on her flight to Karachi at the end of September or during the first week in October. As however, Mr Secretary Wedgwood Benn is aware, an extra section is now being inserted in the airship and, although the work is proceeding satisfactorily, the Council would prefer that no public announcement as to the date of the intended flight should be made until it is more advanced.

If the flight takes place at the end of September or during the first week of October, Lord Thomson proposes to travel in the airship and remain in India for about a week. It is probable that the airship will start on her return flight as soon after her arrival at Karachi as reasonably favourable conditions for such a flight obtain. It is, therefore, not certain whether Lord Thomson will return in R.101 or by aeroplane but he hopes to spend at least three days, and possibly one week, in India. Further details will be sent well before the departure of R.101.

I am, sir, your obedient servant,

W. F. NICHOLSON.

On 3rd September Reynolds added a hand-written note to the new AMSR, Hugh Dowding: 'The question of the dates of the flight was discussed by S of S with DAD on 29th August and S of S agreed that if R.101 is ready by 4th October he could go to India in her, returning by the 18th/19th for the Imperial Conference.'

It was understood in the Air Minister's Private Office that there was still some doubt about whether R.101 would be ready in time and, if it was not, the Secretary of State would remain in London throughout the Imperial Conference and probably postpone his India visit until the Christmas recess.

Though Sir John Higgins and Hugh Dowding had not been able to be present when the arrangements were discussed, because both of them were on leave, they had met earlier to discuss the handing over

30th August, 1930.

522075/30/S.9.a.

Sir,

Proposed Flight of Airship R.101 to Karachi

In reply to your letter of the 14th instant (E.& O.5270/30 forwarding copy of a telegram from the Government of India, I am commanded by the Air Council to say, for the information of the Secretary of State for India, that their present intention is that R.101 should start on her flight to Karachi at the end of September or during the first week in October. As, however, Mr. Secretary Wedgwood Benn is aware, an extra section is now being inserted in the airship, and, although the work is progressing satisfactorily, the Council would prefer that no public announcement as to the date of the intended flight should be made until it is more advanced.

If the flight takes place at the end of September or during the first week of October, Lord Thomson proposes to travel in the airship. It is probable that the airship will start on her return flight in from three days to one week after her arrival at Karachi. It is not certain whether Lord Thomson will return in her or by aeroplane, but he hopes to spend at least three days and possibly one week in India. Further details will be sent well before the departure of R.101.

I am, Sir,

Your obedient Servant

(Sgd) W.F.Nicholson

The Under Secretary of State,
Economic and Overseas Department,
India Office, S.W.

Official Notice: The draft file copy of the letter from Sir Walter Nicholson, Permanent Under Secretary of State at the Air Ministry to the India Office.

of the Department of the Air Member for Supply and Research. In the course of these conversations, Dowding had asked Sir John what he felt about the attitude of mind of DAD and the rest of the Cardington team? Was their judgment good? Could he depend upon them to give him sound advice on the airship programme?

Sir John Higgins had held responsibility for the Royal Airship Works and its people for nearly four years. He told Dowding that – from this intimate knowledge* – the whole Cardington team 'were very enthusiastic over the airship but that he did not think that any advice that they would give would err on the side of rashness at any time. In fact, when it came actually to making a decision their advice would rather be on the cautious side.

'Enthusiastic and confident as the officers of R.101 are they are the last people to leave anything to chance if it can be avoided.'*

That was Sir John Higgins' final contribution to airship affairs. The Cardington team, led by Colmore, were to be relied upon. He reminded Dowding, however, that when R.101 was taken out from the shed the necessary trial flight would have to be made under a temporary Permit to Fly but 'a final and official Certificate of Airworthiness must be provided before she starts for India'; and that would depend upon the results of the trial flight. If problems arose then there could be no question of a C of A.

Sir John Higgins left the Air Ministry for his new career in the aircraft industry. Before that he would have a respite in Canada. He would be back just before R.101 was due to leave.

Meanwhile for policy control – over to Dowding.

* Evidence of Air Vice-Marshal H. C. L. Dowding at the R.101 Inquiry Cmnd 3,825, page 64.

CHAPTER SEVENTEEN

The New Projects - R.102 and 103

'Airship development in this country has already suffered seriously through lack of continuity in policy.... The present is, therefore, a favourable time to consider the question of further development.'

LORD THOMSON OF CARDINGTON,
'Progress of Imperial Communications',
9th September 1930.

August 1930

VINCENT RICHMOND and Michael Rope had been working together since February 1929, on the design of a new rigid airship to succeed R.101. It would be substantially larger than any previous airship and in the course of its development it would incorporate all the lessons which could be learned in the design, the construction and the flying trials of both R.100 and R.101.

By August 1930, the project ('Project H') had been agreed at a capacity of 7½ million cubic feet (36 per cent larger than the lengthened R.101) to be built at Cardington as a development of R.101 and powered with seven improved Tornado engines.

Discussion during 1929 and 1930 had centred on the alternatives of a still larger ship of 9.5 million cubic feet capacity ('Project G'), which would be capable of regular operations carrying a substantial payload non-stop to Egypt and then on, with stops at Karachi, at Rangoon and at Singapore to Australia, and westwards non-stop to Montreal in all weathers – a size of airship which would require much further research and development – or a smaller ship of 7½ million cubic feet ('Project H') which could do the same job once three additional mooring masts had been built for refuelling stops in Malta and at Baghdad on the Indian route and at Monkton, New Brunswick, on the Canadian route.

After looking also at the possibilities of a new ship of 8.38 million cubic feet capacity, the larger ships were ruled out on grounds of cost and complexity and, at a meeting at the Air Ministry on 17th March 1930, chaired by Sir John Higgins and attended by Colmore and Richmond from Cardington, an outline requirement was agreed for a new airship to carry 50 passengers and five tons of cargo and mail (that is a payload of 15 tons, including baggage, food and water) for a still air range of 4,255 nautical miles cruising at 60 knots. It could be achieved with a ship of $7\frac{1}{2}$ million cubic feet if the intermediate masts were available as bad-weather stops.

All the preliminary work in the Cardington Design Office was now concentrated on this design specifically under Michael Rope whose experience and ingenuity now brought him to the front-rank of airship design engineers under Richmond's leadership.

With the Imperial Conference in prospect for October 1930, the work was brought together in a memorandum 'Programme of Airship Development' prepared jointly by Colmore, Richmond and Rope at the Royal Airship Works and submitted through Sir John Higgins to the Secretary of State for Air as the basis for a paper to the Imperial Conference.

This memorandum, dated 23rd August 1930,* set out a three-year programme for the completion of two new airships – R.102 and R.103 – 'as an essential preliminary to commercial services financed by private enterprise'.

The programme would include the lengthening of the existing two sheds at Cardington to accommodate the new ships and the building of one new shed capable of accommodating two airships side by side. A second mooring tower would be built at Cardington on the western side of the Shefford–Bedford road so that both R.100 and R.101 could be operated concurrently from the Cardington base, backed up by use of the mobile transporter tower and supported by a second transporter tower at the Karachi base with its own shed and mooring mast.

By the start of the fourth year of the new programme the two new ships would be ready to embark upon experimental scheduled services in company with the lengthened R.100 and R.101. By then revenue earnings were provisionally estimated to have reached rather more than £20,000 for each round trip to Egypt and Canada while the precise levels of operating and maintenance costs were being established.

In the meantime experience would be gained with the two '1924 Programme' ships. The aim would be to achieve 1,500 flying hours between the two in 1931 and a further 3,000 hours in 1932.

Colmore provided estimates for the cost of the proposed programme including the cost of design, construction and flying trials of R.102

* AIR 5/987/06990: 'Programme of Airship Development'.

at £600,000* and of an identical R.103 at £350,000. In total, spread over four years the costs would be:

		£
1.	Design and construction and trials of R.102 and R.103	950,000
2.	The provision of the necessary ground organisation required for regular flying on the Atlantic and India routes	740,000
3.	Flying and maintenance and spares for R.100 and R.101 and then R.102 and R.103 for a total of 15,000 flying hours over four years	700,000
4.	Personnel costs (flying and ground crews) and base costs over four years	350,000
	GROSS ESTIMATED TOTAL	£2,740,000

From this total of £2,740,000, earnings estimated at £490,000 could be subtracted to arrive at a net total cost for the proposed programme of £2,250,000 – at an average of £562,000 a year† – to set Britain firmly on the airship map by 1934. It seemed a modest price to pay.

The estimates in the fourth year were based on a regular weekly service to Canada for five summer months of the year (21 round trips) and a weekly service to India during the other seven months (31 round trips) carrying an average of two tons of mail and 38 passengers in each direction; an average payload of almost ten tons.

These figures were put up to the Secretary of State for Air on 27th August 1930, with a note from Sir Walter Nicholson that, as a measure of prudence, the estimates of costs ought to be increased by 20 per cent to bring the rounded total to £3,300,000 and a net cost to about £3 million over the four years.

These proposals were, therefore, before the Secretary of State's meeting at Gwydyr House on 29th August.

DAD emphasised that an essential feature of the proposed new programme was that both new airships should be built at the Royal Airship Works to gain the economic advantage of continuity.

By that means the cost of R.103 could be held to £300,000 plus the cost of trials and spares. Whereas the '1924 Programme' had taken

* See Appendix 15. 'Estimated Costs of R.102' (AIR 5/987/06990).

† Equivalent to £8,200,000 a year in 1980 money values.

six years to complete the proposed new programme would cover only four years with consequential savings in overheads.

After Sir Walter Nicholson had made his point about adding 20 per cent to the costs for contingencies and reducing the estimated earnings because of so many unknown factors, Lord Thomson summed up by saying that a further experimental programme costing up to £3 million over four years, with two more airships at the end of it, with R.100 ready to fly to Canada again in the summer of 1931 and R.101 available for a service of flights on the Egypt/India route during the winter of 1930/31,

> appeared to be quite a reasonable programme to put to his colleagues in the Cabinet and then to the Imperial Conference.*
>
> If Canada would pay £100,000 per annum he thought it likely that India would pay something also if flying between Cardington and Ismailia proved to be a solution of the international difficulties confronting air services on the first stage of the route.
>
> The programme would give employment for perhaps 1,000 people or more for four years – an important additional point.
>
> DAD said that he must emphasise that the completion of the programme in four years would depend on starts being made on the construction of the third shed at Cardington and the mooring towers at Malta and in the Maritime Provinces of Canada as soon as the programme was authorised. These points were generally agreed.

The Minutes recorded that, finally, the Secretary of State 'made the following decisions':†

1. The new programme would not be included in the Paper for the Imperial Conference; instead it would be put by him to the Conference for agreement on his return from India in R.101.
2. The form of the Paper for the Conference would be such as would lead logically into a decision by the Conference to go forward with a forward programme on these lines.
3. Orders for new materials for R.102 and R.103 would be planned to be placed in the summer of 1931, by which time further experience would have been gained with R.100 and R.101 so that any necessary amendments could be incorporated in the designs.
4. Steps should be taken at once (by Henry Self) to secure the

* AIR 5/987/6357.
† AIR 5/987/.

agreement of the Treasury to these proposals, as well as to the new programme as a whole, and this should be done before R.101 left for India.

Lord Thomson initialled these decisions on 2nd September as agreed Air Council policy.

So the scene was set for the next step forward in British airship design, construction and operation. It would provide Britain with four airships – two of them substantially bigger than the new American ships *Macon* and *Akron* – and well in the lead over any other nation in the world.

The outline requirement was now translated into Air Ministry Specification A.24/30 drawn up on Wednesday, 3rd September 1930.*

This specification called for two new airships of 7½ million cubic feet to be built to the Airworthiness Requirements of A.P.970. Their lengths were not to exceed 825 feet, the heights not to be more than 147 feet and their diameters not to be more than 140 feet – all to come within new shed dimensions.

The new ships must be designed for operation with compression-ignition engines capable of running on heavy-oil and on gas-fuel for which provision must be made.

The required performance included a specified maximum speed of not less than 70 knots (80.6 mph) and a continuous cruising speed of not less than 60 knots (70 mph). Sufficient fuel must be carried for a minimum still air range of 4,285 nautical miles (4,900 statute miles) cruising at 60 knots for an endurance of not less than 70 hours, carrying a full payload of 15 tons, including fifty passengers in comfortable accommodation, together with five tons of cargo and mail. This load, plus ballast and crew, must be lifted with gas-bags no more than 94 per cent full.

Disposable lift must be available for the required fuel load and payload together with ballast equal to not less than five per cent of the gross lift; that is for not less than 11 tons of ballast excluding fuel.

Vincent Richmond and Michael Rope's design study† to meet this requirement was based on a development of the successful structural layout of R.101, designed, developed and further refined in conjunction with John D. North of Boulton and Paul Aircraft Ltd and T. S. D. Collins of Cardington's Design Staff.

The new design was to be of two feet additional diameter compared

* AIR 5/987–06990 and AIR 2/349–X/M 08109.

† Appendix 4: 'Design Study of R.102 and R.103.'

with R.101 in order to eliminate the problems of gas-bags chafing against the structural members. It would be also one bay – that is 45 feet – longer than the lengthened R.101C, with fuller lines at bows and stern to provide additional gas-bag capacity to carry the weights of mooring gear and of the vertical and horizontal tail surfaces. It would thus provide a better balance and more stability.

Passenger accommodation for fifty would be, as in R.101, on two decks but this time the lower deck, including the smoking room, would be an extension of the control car and navigation and wireless rooms, partly outside and below the hull with windows set to provide a magnificent downwards view. Such arrangements would save weight and would make possible an increase in gas-capacity within the hull, while providing more light and view for passengers at a small sacrifice in aerodynamic drag.

The dimensions of the new ships would be 822 feet long (compared with the 777 feet of R.101C) and 134 feet maximum diameter, to bring the height to 147 feet over the fins and semi-external car.

Seven reversible Tornado heavy-oil engines would be installed; the forward and aft engines, as before, in individual cars but the mid-ships – or 'wing' – engines, aft of the passenger accommodation, would be tandem units as in R.100, with tractor and pusher airscrews.

These developed Tornados were estimated each to be 1,500 lb lighter than the original engines so that – with other weight savings – the total installed power-plant weight of R.102 and R.103 was set at 17.75 tons which would be for seven engines in five cars, only some two per cent more than the weight of the five Tornados of the existing R.101.

The improved Tornados were promised to give a maximum output of 850 bhp each and to have 700 bhp available for continuous cruise. The specified cruising speed of 60 knots for R.102 and R.103 was estimated to be attainable on 3,325 bhp – which would require an output of only 475 horsepower from each engine or 555 horsepower per engine with one engine at a time shut down.

The new airships would thus offer enhanced payload, range and speed compared with the earlier ships, together with many other detailed improvements. They would translate experiment into reliable practice.

It was important to have two of the new ships in order to mount a successful commercial service and the task of arguing that case was put in the capable hands of the Deputy Director of Contracts, A. H. Self.

On Saturday, 6th September 1930, Self – later Sir Henry Self, Permanent Secretary to the Ministry of Civil Aviation – wrote a note* to the Treasury on the Air Ministry's proposed 'Future Airship Programme' (Saturday morning work was still standard practice in 1930). Self wrote:

> For the future programme one new ship would be inadequate to place the operational scheme on a sound footing and, by ordering a second, this deficiency could be overcome at an extra capital cost of £300,000 – as opposed to the cost of one ship alone of £600,000. Such a capital expenditure, added on a gross total of nearly £3 million, is justified on the grounds that it enhances the prospects of successful working in a marked degree and will repay itself rapidly in takings if airships are a success.
>
> There remains the consideration that the Imperial Conference provides a natural opportunity, which will not recur within the development period for airships, for framing an agreed scheme in which all parties have a financial interest. This opportunity, if lost, may well prove expensive in the end and the necessary protection against commitment to a white elephant exists in the fact that the proposed scheme provides a natural 'Break Clause' in that the first year will be devoted to preparatory work alone, and no final commitments will be entered into until the Ministry has had another twelve months of experience with R.101 and R.100.
>
> *To sum up*: the present proposals are submitted as representing sound policy – to produce, with the minimum of cost and time, the next stage towards the setting up of a commercial organisation to take control and to operate an efficient Empire Airship Service. These proposals postulate, of course, the belief that airships are regarded as a practical experiment which must be pursued in the interests of the State.
>
> *Air Ministry 6th September, 1930.* *A. H. Self.*

Henry Self argued the case cogently with the Treasury in the knowledge that he had the full support of the Secretary of State for Air. The upshot was agreement that, subject to endorsement by the Imperial Conference after Lord Thomson's return from India, the money would be forthcoming.

Every informed, dedicated and forward-looking advocate of any branch of technology always seeks to look ahead along the forward path. So the Lighter-than-Air Section at the Air Ministry, hand in hand with the Royal Airship Works, was looking ahead now in developing and refining the new airship programme for the decade

* AIR 5/987.

of the 1930s; encouraged by the Secretary of State for Air.

In the autumn of 1930 the path through the unexplored technical and operational jungle towards a Phase Two of buoyant flight still seemed to lead into a land of promise.

For the Imperial Conference the provisional programme agreed on 27th August 1930, was further revised at Cardington and at the Air Ministry at meetings between Colmore, Self, Reynolds, Richmond and Rope. It was put together in a new Draft Paper dated 12th September 1930.

In this latest proposal the programme was to be spread over five years, 1931 to 1935 inclusive, at an estimated cost revised to a total of £3,210,000 less £310,000 of revenue earnings and rounded to a net of £2,900,000. The average rate of speed was thus seen at almost £600,000 a year,* to include 17,350 airship flying hours.

By 1935 they all hoped that the Royal Airship Works would be embarking on the even wider horizons of a Phase Three for Lighter-than-Air development.

Such was the projected long-range programme worked out between Cardington, Adastral House and Whitehall. Although the Treasury was now favourably inclined the proposals did not, however, command uncritical approval.

On 13th September 1930 Christopher Bullock, by now Permanent Secretary of the Air Ministry in succession to Sir Walter Nicholson, wrote to the new AMSR, Hugh Dowding, at Adastral House, from Gwydyr House in Whitehall. He attached the revised draft Memorandum on Airship Policy, to be forwarded to the Secretary of State. He sought Dowding's views upon this 'comprehensive view of the airship situation' before further commitments were made: †

> But I am quite clear that I myself, if I were Secretary of State, would not consent to further construction at this stage.
>
> My primary objective has been to ensure that, if the Secretary of State insists on a gamble, he does so with his eyes open and realises that he is making a radical departure from the wisely cautious policy he has hitherto pursued.

The note suggested that the proposed programme for building R.102 and R.103 should be postponed and that a further revised and curtailed 'Safety First' programme should be substituted. It should con-

* See Appendix 15 for details of the projected five year forward programme (AIR 5/987–3069).

† AIR 5/987-6357.

centrate upon demonstrating the practicability of intensive regular services with R.100 and R.101 for about a year, flying wholly on the England to Egypt route through one extra mast at Malta. Then, at the end of a year – that is, early in 1932 – the two new airships could be put in hand against a background of much more operating experience.

> It seems to me [wrote Bullock] that a slight setting back of the clock should be faced and really does not matter. In developing a new form of transport (which should last until we let atomic energy loose or achieve some form of auto-projection of the individual) a year's delay, if that should prove inevitable, is after all, a bagatelle.
>
> I suggest, therefore, that we should concentrate as the next stage on what I have called the real crux of the matter; the demonstration of the practicability of regular operations with R.100 and R.101.

'The practicability of regular operations with R.100 and R.101' – that was, indeed, the main purpose of the 1924 Airship Programme to which there had been accorded so much time, effort, money and devotion.

Christopher Bullock had a long talk with his friend and master the Secretary of State, together with Sefton Brancker, later in the month. The upshot was that Christopher Thomson decided that during the five days of the outwards flight to India he and Brancker would together draft a final version of his Paper on the proposed new programme, which he would put to the Conference on his return. Its precise form would depend upon how the flight progressed – but he looked forward to the day, the climax of his policy for Imperial Communications, when – as the next step forward – four British airships would maintain regular scheduled services between England and India and across the Atlantic to Canada.

It was a dream still to be fulfilled.

CHAPTER EIGHTEEN

'To Ride the Storm'

'How may one cull the honey – and yet not pluck the flower?
'And how can man once happy – still keep his happy hour?'
C.B.T.; Posada; 29th August 1915*

August 1930

AND so the scene was set. R.101 was being made ready for the long-awaited flight to India. A new programme had been drawn up to carry forward the vision of Imperial airship routes.

For a short while the Secretary of State could turn from his official life to think of more personal things. Could he fit in one more visit to Paris 'to catch the sunlight' * in Marthe's presence before he left for India?

There was still much to be done. For one thing he was moving house again, to gain a new lease in the block of apartments in Ashley Gardens where he had lived for the past few months.

On Wednesday, 20th August 1930, C. B. wrote to Marthe from 122 Ashley Gardens, Westminster.

Dearest,

As ever, my devotion. As you see, this letter, however late, comes from my new flat at No. 122 which replaces the nearby No 100 which you know. It has several advantages; an extra room for my study, a wider outlook, a better kitchen and – most important – a longer lease.

All that remains is for you to cast your sweet spell upon it – and upon me and on Sambo again. That feline sprite I have now renamed 'Salammbô' † because, like Rosalind, 'he' has turned out to be a she and I have a dark suspicion that there may be some small Sambos and Salammbos about before very long. It is just as well that my quarters are adequately commodious.

* When C. B. and Marthe Bibesco had first met in Roumania in 1915 he had addressed to her some inspired lines:

'How may she catch the sunlight – and wear it in her hair?
'Where is the golden apple – whose core is not despair?
'How may one cull the honey – and yet not pluck the flower?
'And how can man once happy – still keep his happy hour?
C.B.T.; Posada; 29th August, 1915.

† After Gustave Flaubert's *Salammbô*, a tale of Carthage, written in 1862 and read together by C.B. and Marthe in French in 1915 at Posada.

You have been good in writing for a third time while I have not done so once. My spirit has not been unresponsive but I have not been able to make plans as something is always happening in my Department.

Alas, I cannot get away this month but I may be able to manage a day or two in Paris during the first week of September. Will you be there then? I long to see you and especially before I leave.

R.100 gave me pleasure. R.101 will, I hope, give me joy. To ride the storm has always been my ambition and who knows but we may realise it on the way to India but not, I hope, with undue risks to human lives.

Your friend the PM writes in gay spirits from Scotland. He is coming to London at the beginning of September if not earlier. Our unemployment gets worse and worse and it is no consolation that the Germans and Americans suffer even more than we do. There is something wrong with the system which – under Free Trade as under Protection – produces these lamentable results.

No more now – except my fondest love and, of course, my hopes of being with you in Paris in two weeks from now. I cannot tell you how much those glad times with you refresh my spirit.

'God be with you,' as the President of the Irish Free State said to me; (he is a very fervent Catholic).

With all my love

Ever your,
Kit.

For a few more days Christopher Thomson continued to hope that there might still be an opportunity for a flight to Paris early in September for rest and a renewal of inspiration there before he must come back to London to make his final arrangements for the Imperial Conference, for the flight to India and, on his return, for the presentation of his forward-looking paper on Empire Communications to the Conference.

Thereafter, with the next phase of the airship programme under way, to Paris again for Christmas with Marthe.

It was not to be.

On Monday, 25th August, a letter arrived at 122 Ashley Gardens from Marthe in Paris.* She told C.B., sadly, that she must return to Posada to be present at a reception for a delegation of legal professors from universities in Europe and North America who were to attend a conference in Roumania. Marthe had promised, months before, to

* Because, presumably, Thomson took this letter with him in R.101 it has not survived. Its contents is known, however, from the Princess Bibesco's diary and her letters to the Abbé Mugnier.

away this month, but may be able to manage a day or two in Paris during the first week in September. Will you be there then?

R. 100 gave me pleasure. R. 101 will I hope give me joy. To ride the storm has always been my ambition, & who knows but we may realise it on the way to India, but not I hope with undue risk to human lives.

Your friend the P. M.'s writer

Last Letter: Lord Thomson's letter of 20th August 1930 to the Princess Marthe Bibesco.

hold a garden party for them at her 'Castle in the Mountains' at Posada.

There could be no escape from this pleasant but ill-timed obligation. She must go. She grieved to leave him and to miss a re-union before the long voyage so far away.

All her plans had been to offer Kit, by her side, a happy break from his official burdens before he plunged into all that would lie ahead.

She chided him that he had not come to Paris in August when all the world was on holiday – 'all but C.B.' That August was the fifteenth anniversary of the golden days they had passed together at Perros-Guirrec in Brittany – where the 'General de Trestrignel' and the 'Princesse de Kergitzne' had felt that the world was theirs before *le sort en fut jeté* – the die was cast – that he should leave her side and the British Regular Army to enter the hazardous and lonely world of politics.

But now she must leave for Posada. At least there were sweet memories of times together there. But she would not now see him before he set off for a destination so far away.

She felt beaten. She sent her warmest love.

Christopher Thomson wrote back on 25th August a letter addressed to Marthe in Posada:

My dearest love,

I am the beaten one, not you. And I am beaten by force of circumstances just as William the Conqueror might have been beaten by contrary winds which could have changed the course of history.

So far I have had two days holiday since the House rose and on 10th September regular work begins again with committees and other ills to which official flesh is heir.

MacDonald returns to town tomorrow for two nights; he then goes back to Lossiemouth and I to Manchester to attend a function in my special honour.

You write, dearest, as though I did not understand – and had hurt your feelings. This is unjust. There is nothing in the world I should less like to do than cause you any pain. But not only is my work absorbing but it never ceases.

This is partly my own fault since I have identified myself permanently with many things which, as a rule, are left to subordinates and experts.

You, Marthe, however occupy a special place, as you know so well, and even in your absence you are present to my spirit as the kaleidoscope of life unrolls with occasional glimpses of the future.

Ever your

Kit.

Paris in September now had few attractions with Marthe gone to Posada. On Sunday, 31st August, he went down to stay with old friends, Mr and Mrs Carruthers, at Moon Green, Wittersham, Kent, taking with him the half-completed manuscript of an autobiographical book – never to see the light of day.

Of that last visit, Violet Carruthers wrote that he was full of his forthcoming flight to India:

> I can describe his attitude * [she wrote] as being not so much one of confidence as an attitude into which no element of doubt or uncertainty entered. The possibility of risk or danger did not seem to cross his mind. He spoke of a journey in R.101 with as little concern as a journey in the Flying Scotsman.

She added that, at the same time he always emphasised the responsibility for ensuring that no chances were taken with these first passenger airships – that 'every conceivable risk should be foreseen and every precaution taken to make assurance doubly sure'.

Clearly he was well aware of the risks in an experimental, and still largely untried, craft. With that he had implicit confidence in the Cardington team – a confidence which he further underlined when, on Tuesday, 2nd September, back at Gwydyr House, he initialled a copy of the agreed statement of the forward airship programme to go to the Cabinet and, later, to the Imperial Conference.

Two days later he set out again for a few days respite in the West Country before the new Parliamentary session began, taking with him his papers for the Imperial Conference and, for lighter reading, T. F. Powy's new allegorical novel set in Dorset, *Mr Weston's Good Wine*.

He wrote to Ramsay MacDonald from The Bridport Arms in West Bay, between Lyme Regis and Abbotsbury, to extol the beauties of 'the pleasant woods and murmuring streams, the bare rolling downs, the sheltered valleys and the incomparable coastal scenery of this part of Wessex.'

While he rejoiced in the early autumn of the English countryside and looked ahead into the future, his thoughts were far away in Posada.

* In a letter to Sir John Simon dated 8th November 1930 (AIR 5/916–XM 08109).

CHAPTER NINETEEN

Imperial Conference Paper

'The Imperial Conference may again find it convenient to have placed before it a conspectus of the progress which has been made in the past four years on the development of civil aviation in the Empire, and the resulting situation as it appears to us in this country.'

Lord Thomson of Cardington, 9th September 1930.

September 1930

Back at Gwydyr House on Monday, 7th September, ready for the next session of Parliament, the Secretary of State put the final touches to his opening paper for the Imperial Conference, 'The Progress of Imperial Communications'.*

In his foreword, dated 9th September, Lord Thomson looked back to the plans submitted by the Air Ministry to the 1926 Imperial Conference for the further development of Civil Aviation within the Empire.

Though progress had been slower than had been hoped the past four years had been fruitful in practical achievement; for example two new airships had been completed and R.100 had made a successful flight to Canada and back. In addition, the weekly aeroplane service between England and India had now been in regular operation for nearly two years. A weekly service by Imperial Airways to Cape Town would follow shortly.

The first part of the Paper surveyed progress on aircraft development and on aeroplane services throughout the Commonwealth. Part 2 then moved on to airships and began with a review of the 1924 Airship Programme and the situation reached with R.100 and R.101. It went on:

The programme authorised in 1924 has, for a variety of reasons, taken longer to complete than was originally hoped, one of those reasons being that some two years were taken up with research and experiment including full scale trials with R.33. It is perhaps fortunate, however, that the completion of the 1924 Programme should synchronise with the holding of the present Imperial Conference, as it will enable

* AIR 5/908–308.

THE PROGRESS OF IMPERIAL COMMUNICATIONS.

NOTE

BY THE

SECRETARY OF STATE FOR AIR.

In 1926 a memorandum was submitted to the Imperial Conference by the Air Ministry entitled "The Approach towards a System of Imperial Air Communications," in which plans were outlined for the further development of civil aviation in the Empire. If that development has in one or two directions lagged somewhat behind our original hopes in point of time, the past four years have nevertheless been highly fruitful in practical achievement.

We have, for example, completed the construction of the two new airships foreshadowed in 1926, and the R.100 has made a successful flight to Canada and back. The weekly aeroplane service between England and India then projected has now been in regular operation for nearly two years, and has secured very satisfactory results. In co-operation with the South African Government and the other African administrations concerned, a concrete scheme has been formulated for a similar weekly service, also to be operated by Imperial Airways, which should within a very few months link London with Cape Town. In Canada and Australia there has been a most remarkable growth in the air route mileage in regular operation and in the use of air transport generally.

In these circumstances, I have thought that the Conference may again find it convenient to have placed before it a conspectus of the progress which has been made in the last four years and the resulting situation as it appears to us in this country. The present memorandum has been prepared for this purpose.

9th September, 1930. Thomson.

Foreword: Lord Thomson's Introduction to his Paper to the Imperial Conference headed 'The Progress of Imperial Communications' surveying achievements since the previous Imperial Conference in 1926 and proposals for the future.

the various Governments of the Empire to consider the question of further airship development at a convenient moment.

The main items of the 1924 Programme included the building of two airships of 5 million cubic feet gas capacity, or about 150 tons gross lift – nearly twice as large as any airship previously constructed in this country and the carrying out by each airship of exhaustive flying trials in this country and of a final trial flight to India.

The construction of two airships of such a size was an engineering essay almost unprecedented in any other field. The greatest difficulty in all aircraft construction is to reconcile the conflicting claims of speed, safety and useful load and it is seldom, if ever, that the first experimental type succeeds in satisfying all three when it first emerges from the maker's works. It is a high tribute to the skill of the two designers that, so far as the structure of the airship is concerned, the exacting requirements have been met with the exception – admittedly an important one – that both R.100 and R.101 have a much smaller proportion of useful load to gross lift than was originally hoped.

'Adequate and comfortable living and sleeping accommodation' was required for 100 passengers 'including provision of heating, lighting, cooking, ventilation and sanitary facilities'. This condition was complied with and in R.101, in addition to a dining room, a large lounge with a total area of 2,700 square feet, stretching from side to side of the airship with promenades and windows at each end, together with a separate smoking room of about 220 square feet, specially protected against any risk of fire.

In both ships, however, the inclusion of passenger accommodation on this scale together with many other experimental features (especially in R.101, the heavy-oil compression ignition engines) has brought the structure weights up to a figure appreciably in excess of the 90 tons originally aimed at.

As with all aircraft the excess of structure weight has to be borne by the lift available for payload and consequently, on stages of 1,500 miles or so, R. 100 and R.101 as they are at present, will be able to carry only some twenty-five passengers. It is in fact clear that the requirements as to passengers was put too high in 1924 not only having regard to the fixed weight of the passenger quarters but also to the weight allowances which must be made for luggage, fuel and water. Experience has shown that in calculating passenger-load a figure of about five to the ton (450 lb each) must be taken.

The organisation of an airship route to India and a flight to India were, from the beginning, essential features of the present programme of development and it was always intended that, with this end in view, both airships should use some fuel other than petrol in order to avoid the risk of fire in sub-tropical regions.

The Paper went on to explain that because the hydrogen-kerosene engines originally planned for R.100 were not sufficiently advanced when power-plants were required for that airship, the Airship Guarantee Company installed R.R. Condor petrol engines. With them R.100 would not fly beyond Egypt at any time and not to Egypt during the summer.

The Paper then went on to explain the fact that the lift of an airship is decreased by high temperatures and/or a low barometric pressure to the extent that, during the summer months at Karachi R.101's lift could be reduced by as much as 11 tons. When allowance is made for some 25 tons of fuel needed for the sectors Karachi–Ismailia and Ismailia–Cardington, an airship with a gas capacity of about 9½ million cubic feet would be needed to provide a regular operation with a commercial load of fifty passengers on the India route or from Cardington to Montreal non-stop.

If however additional mooring towers were erected for refuelling en route – at Baghdad and Malta for the Indian route and on the east coast of Canada for the Atlantic route, though many flights might be made non-stop, an additional refuelling stop would make regular services possible even against continuous headwinds. In these circumstances commercial services would be possible with airships of about 7½ million cubic feet gas capacity and such new airships could be built more quickly and at lower costs than could the larger size, which would need a good deal further research and development.

The Paper then discussed the importance of a mechanical handling system using a 'mobile stub mast' on caterpillar tracks so that airships could be moved from mooring tower to shed even in winds of up to 18 or 20 miles an hour.

During the next six months, R.101 would be scheduled to make a series of flights between Cardington and Ismailia.

> It is not to be expected that R.101 with her smaller capacity and useful lift (compared with a projected airship of 7½ million cubic feet) will be able, without intermediate mooring towers, to fly during the coming winter with the regularity of, say, the cross-Channel aeroplane service. The flight from Ismailia to Cardington will probably take an average of about forty-eight hours. There may be occasions on which, although the airship might have no difficulty in reaching Cardington, the weather would be such as to make a landing, even to the tower, impossible for several hours. To deal with this difficulty it may be possible to make temporary arrangements with foreign governments for facilities for emergency landings. But it seems clear that on an

airship route, intermediate towers must be provided just as emergency landing grounds are provided for aeroplanes and ports of call for seagoing vessels.

When the present programme was authorised in 1924, the essential object in view was to produce two airships strong enough and big enough to demonstrate the practicability of airships as a means of improving Imperial communications.

Satisfactory as were the trials of R.100 and R.101 it would be premature to say that either they or *Graf Zeppelin* have established the capacity of airships to operate regular services. It is hoped that the development of a mechanical handling system, improved power plants and a period of systematic operation will, in due course, place matters in a different light.

By the end of the financial year 1930, the United Kingdom will have spent some £2,400,000 on the airship development programme. In addition, Canada has provided a mooring base at St Hubert Airport while India has provided the land at Karachi for the airship base and has paid for the mooring tower there. The exact cost of these contributions made by Canada and India is not known but it is probably not less than £120,000 in the case of Canada and £100,000 in the case of India. The assets produced by this expenditure consist of the two largest airships in the world, four large airship sheds, and four mooring towers; a large amount of valuable knowledge and experience in the design, construction and operation of airships and of the meteorological conditions on the India and Atlantic routes and the technical and works staff of the AGC and the administrative, technical and works staff at the Royal Airship Works.

These assets have been acquired by considerable expenditure of money and effort in an attempt to achieve for the British Empire a leading position in the development and use of a new system of transport.

As a vehicle of transport the airship has certain undoubted advantages over the aeroplane and over any form of land or sea transport. Of approximately 530 persons who have crossed the Atlantic once in each direction by air, only 36 have travelled in aeroplanes.

There is no reason to doubt that an airship of 7½ million cubic feet – which can be constructed without any radical departure from the design of R.100 or R.101 – would be able to fly the Indian and Canadian routes with the assistance of additional mooring towers, with adequate regularity under average conditions. The time of passage to Montreal would be about 3½ to 4 days while the passage back might often be done in 48 to 60 hours. On the Indian route, using intermediate mooring towers, Karachi should normally be reached in 4 to 4½ days; the return passage taking between 5 and 7 days.

From the point of view of comfort the main attraction of airship travel is its smoothness, cleanliness and quietness. R.100 and R.101

have shown that accommodation can be on a far more ample scale than that of the best express train even if it does not approach the elaborate luxury of a first-class liner.

With the meteorological organisation which has been built up, airships of the type of R.100 and R.101 should be safe in any weather they are likely to encounter and, from the point of view of actual flying should give regular service on stages of 1,500 miles or so. The real question which is still unsolved is whether regularity of operation can be secured, having regard to the necessity of re-housing the airship for inspection and re-fit at the end of so many hundred flying hours. The answer to this question must depend on the development of the mobile stub mast for mechanical handling. If, as is hoped, this proves a success, and if a regular service, say between Cardington and Ismailia can be carried out, then it may well be claimed that the practicability of the airship has been established. Some data should also be available then on the economic side of airship operation. Further capital expenditure on route organisation and construction will be necessary before regular commercial services can begin; but there seems reason for thinking that at that stage, the earnings ought certainly to be enough to cover operational costs though not, perhaps for some time, interest on capital as well.

In the meantime, His Majesty's Government in the United Kingdom will have to make a decision as to their future policy and, in so doing, great weight will necessarily attach to the views expressed on this subject by members of the Imperial Conference.

As regards the future policy to be pursued there would seem to be three possible courses:

1. *To continue* a policy of development at the sole expense of the Governments concerned;
2. *To hand over* the task of development to a commercial concern with such assistance as may be necessary from the Governments concerned.
3. *To abandon* the enterprise altogether, so far as Government action is concerned.

'The possibility of adopting either of the first two alternatives will, of course, depend on the extent to which the Governments concerned are willing to contribute jointly towards the expenditure involved in an enterprise which must be regarded as of interest to the whole British Commonwealth of Nations.'

Here, for the Imperial Conference, was a summary of the whole airship situation on the eve of R.101's flight to India. The airships of the 1924 Programme were now seen to be much too small to be able to provide adequate payload/range over Empire routes; and even new airships, of fifty per cent more volumetric capacity, would require additional mooring towers as refuelling points in order to achieve regular services with half the passenger loads originally planned. The extent of the optimism of the 1924 Specification could be measured by the fact that airships of 9½ million cubic feet capacity were now seen to be required to carry fifty passengers and five tons of cargo and mail non-stop to Egypt and on to India and on the route from England to Canada.

The original, optimistic, technical assessment had been prepared by Sir Geoffrey Salmond, Reginald Colmore, Vincent Richmond and Herbert Scott and endorsed by Dennistoun Burney and Barnes Wallis of Vickers.

Now, the Cardington team – Colmore, Richmond, Scott and Rope, had six more years of practical experience – and could judge the prospects much more accurately.

The objective of Imperial communications and the determination to achieve it remained the same.

Nothing daunted, Christopher Thomson began to draft, with Reynolds in his office and Colmore at Cardington, the follow up paper for the Imperial Conference to be presented on his return from India in R.101.

The initial Paper would be circulated with the Conference Agenda on the opening day. The second paper would be printed immediately he returned to London for circulation on Monday, 20th October. Of the three options there was no doubt at all which the Secretary of State would propose to the Conference, provided that R.101 acquitted itself satisfactorily as everyone was confident it would do.

It would be 'to continue the policy of airship development' financed primarily by the Government of the United Kingdom with some support (perhaps to the extent of £600,000 over five years) from Canada and India.

With Treasury agreement already gained by Henry Self, subject to the endorsement by the Imperial Conference, this 'appeared to be quite a reasonable programme' to put forward to the Conference.

CHAPTER TWENTY

Final Trial

> 'According to the programme to which DAD is working, R.101 should be completed with the new bay by 22nd September. One trial flight will be necessary before the ship leaves for India.'
>
> SIR JOHN HIGGINS, AMSR Note to S of S, 30th June 1930.

September 1930

ON MONDAY, 8th September 1930, Vincent Richmond delivered a Paper on 'The Development of Rigid Airship Construction' to the Annual Meeting of the British Association at Bristol.*

Richmond re-stated the requirements to which he continued to work – the development of the commercial airship for long-distance transport.

> Such airships must provide adequate safety and comfort for the passengers and they must be capable of perpetually running to a scheduled time-table over long distances with the necessary margin of fuel to meet adverse weather. Great altitude is not essential and although inflation with helium is a safety precaution which it is hoped will ultimately be adopted, the risk of using hydrogren for commercial vessels has now been proved to be extraordinarily small. On the other hand, the use of petrol as a fuel, while introducing no more than a justifiable risk in a military vessel is hardly to be contemplated in a commercial vessel, especially for operation in tropical temperatures.

He went on to justify the use of the heavy-oil engines the installed weight of which had been such an absorber of the available lift of R.101.

> The Diesel is the engine *par excellence* for airship work; its fuel is much safer than petrol and the consumption of fuel per horse-power is low. The ultimate criterion of performance must be the weight of fuel, in any form, consumed per horse-power hour and the possibility of burning heavy oil with a light gas in a diesel engine appears attractive provided the engine weight is reasonable.

* The Paper to the British Association by Lieutenant-Colonel V. C. Richmond OBE, BSc., was printed in full in *Engineering* for 12th September 1930.

So far diesel engines for airships had come out a good deal heavier and of rather less power than originally planned but there were good prospects that weight would be reduced and power increased shortly.

> From the point of view of safety there is no doubt that diesel oil is at the top of the scale and petrol at the bottom. Regarding a helium-filled airship as the ultimate end to be attained in the interests of safety, petrol is likely to disappear. The use of gas with or without diesel oil would offer an attractive method of maintaining static equilibrium and thus conserving the helium without resorting to the inefficient expedient of exhaust water recovery.

He went on to discuss the methods of construction used in R.101, how a future airship could be built with no less strength but substantially lighter as a result of experience with the present ship, and then to consider R.101's streamlined form and the layout of its passenger accommodation.

Then he turned to the one really unsatisfactory feature of both R.100 and R.101 : their outer covers. What could be done for the future?

The problem was that doped cotton fabric, though relatively light, deteriorated with time, was liable to flap and reacted to changing humidity. Yet a metal covering would be too heavy, subject to corrosion and – even in very thin gauges – unnecessarily strong.

For R.101's 27,500 square yards of outer cover, doped cotton fabric – at 0.369 lb per square yard and a strength of 700 lb per foot run – weighed, with attachments, 8.1 tons.

Duralumin sheet, at 1.2 lb per square yard for 0.0098 of an inch thickness, would weigh 14.6 tons; stainless steel sheet, at 2.95 lb per square yard at 0.0086 of an inch thick, would weigh 36.2 tons for a strength of 1,560 lb per foot run.

Phosphor-bronze gauze was a possibility, at 0.46 lb per square yard and an attached weight for R.101 of 9.23 tons.

There seems no doubt that Richmond had in mind a note from Arthur Gerrish, Shed Manager at Cardington, written on 30th June after the Hendon flights, that

> R.101's outer cover forward of Frame 13 [that is, just in front of the fins] flaps continuously and will not have a long life. A cover 'wave', trailing aft, meets a return wave at this point. A short length of metal cover might, perhaps, be added at this section of the cover?*

* AIR 5/911.

The problem was the old enemy – weight. For the present, reinforcing bands of fabric, doped on, seemed to be the best compromise.

Richmond's Paper was followed by a Paper by Barnes Wallis from Vickers Ltd on R.100 and then by Dr Ludwig Dürr on the *Graf Zeppelin* which, he reminded the audience, had had to be limited in size by the dimensions of the only construction shed available in Germany in 1927/28.

Commenting on Richmond's Paper and on R.101, Dr Dürr said that, in his opinion, R.101 was the safest form of transport on land, sea or in the air which had yet been produced anywhere in the world.

Returning to Cardington from Bristol, encouraged by this *cachet* from the foremost German airship designer, Richmond was further pleased to find that the first Tornado reversible engine had been delivered from Beardmores, and that its starting engine had arrived too, ahead of schedule, on 10th September. A satisfactory test was run on the reversing engine on 12th September while on Monday, 15th September a second reversing Tornado was unexpectedly promised for 17th September – and actually arrived on the 16th. Things were really looking up.

Installation of this second engine would add usefully to the manoeuvrability of the ship so Richmond gave authority for it to be fitted although this work would put back the date at which R.101 could leave the shed until Thursday night, 25th September.

A week before this, on 18th September, the new Air Member for Supply and Research, Hugh Dowding, had made his first visit to Cardington. Richmond recorded in his diary:

> He gave us a brief outline of his policy for the next twelve months. As this was rather contrary to what had been expected, DAD and I told him that we should like time to consider the matter.

There is no direct evidence of what Dowding said but there can be little doubt that it would be in character for 'Stuffy' Dowding to accept Christopher Bullock's proposition for a delay in starting construction of R.102 and R.103 until both R.100 and R.101 had provided further practical experience from intensive route operations. All this would be subject to the views of the S of S on his return from India.

Noël Atherstone returned from two weeks' leave on 15th September ready for R.101's third commission and noted, with satisfaction, in his diary on Saturday, the 20th:

I found the ship nearly completed. Very good progress has been made during the past fortnight and now there only remain the cleaning up jobs to finish off. There has been great success with the reversible Tornado and it has functioned perfectly on the gantry.

The most unheard of feature of this engine is that it is actually 150 lb *lighter* than the non-reversing engines! It would almost appear that Beardmores have actually grasped the idea that these engines are for use in the air and not to push railcars about in Canada.

Many of the new parts are of cast electron which is considerably lighter than aluminium.

The ship was to have been handed over to the flying staff today but, owing to the almost miraculous delivery of a second reversible Tornado last Tuesday it has been decided to fit this one also – which is very much to our taste. The second engine-car complete is due for delivery to the shed on Tuesday, 23rd, and it will also have the new Beverley compression-ignition starting engine instead of the petrol Ricardo.

Both forward engine cars will now be reversible and that with the Beverley starter will also be a generator car as the Beverley can be run continuously and will start the main engine while driving the generator, without any difficulty. At least, so they say, but have to prove this yet.

With the two reversing engines the ship will have nearly as much astern thrust available as she had before with one engine fitted with an astern airscrew.*

I have heard that by next March an improved type of Tornado engine will be available. It will have an available cruising HP of 700 instead of the present 585 and it will weigh 1,200 lb less than the present engines.

This sounds almost too good to be true but if it is it will mean an increase in cruising range of about 600 miles which is a considerable gain.

At this encouraging stage in R.101's progress, Sir Sefton Brancker visited Cardington on 22nd September to inspect the work. He was impressed by the confidence of Colmore, Richmond and Scott over a lunchtime talk.

Next day, on Tuesday, 23rd September, Atherstone added a further entry to his diary:

Today, at noon, the second reversible engine car was delivered to the shed. It has done a little running on the gantry and, as everything appeared to function well, they called it a day and shot the thing down

* The efficiency of a fixed pitch airscrew for normal forward thrust, when run backwards, was low; hence two engines achieving almost the same thrust as one with an astern airscrew.

to the shed. This car will go on forward on the starboard side and will be the only one with the CI starting engine.

From what I have heard from high sources, the ship is supposed to be ready for handing over to the flying staff on Thursday evening, the 25th, and if weather conditions are suitable, we will be expected to take the ship out the following a.m.; very much a.m.! it is rather extraordinary that, on every occasion so far, the wretched flying staff are expected to take over the ship and crash into full flight within a few hours of taking over but everything in the reverse direction is never done unless there has been due deliberation and much 'cagging'.

However, we can survive incidents like these as long as we are given an airworthy ship with a decent performance. As far as I can see, a real effort has been made this time to put right most of the glaring defects and I think that, with ordinary luck, we should be able to keep the ship in commission for the best part of a year.

R.101 is to do a 24-hour flight at the first opportunity after leaving the shed and, if everything is all right then, she is to start for India about thirty-six hours later. Every effort is to be made to leave for India on or before 4th October.

No one was more pro-R.101 than 'Grabby' Atherstone but his note underlines the gap in communication between Scott, in charge of Flying Policy, and the flying staff for whom he was responsible. There was little consultation.

There is no doubt that, after twelve weeks of concentrated work in the shed since 30th June, R.101 was now in a much better state than had been either of the two airships at any time before. Not only had the lift been substantially increased by the addition of the new bay but the gas-bags had been completely overhauled and an improved form of padding with fabric bandages was now in place wherever bolts or projections might rub on the gas-bags. McWade passed them as satisfactory. A completely new cover had been fitted and doped after it was in place; completely new that is except for the sections at nose and stern which had previously been doped after installation. All the pre-doped portions of the cover were renewed.

The forward portion of the outer cover at the airship's nose and for the first twenty-six feet of double-curvature back to Frame 1 had been renewed and replaced by a brand new cover, doped after fitting when the front reefing girders were removed to save weight at the nose.

Then, for the next eighty-six feet, between Frames 1 and 3, the original cover was left in position. The reason for this was that this section of the outer cover had not been pre-doped but doped after it had been placed in position, chiefly because it also had much double-curvature which made the fitment of the pre-doped panels difficult, and it was this pre-doping which was thought to have caused the deterioration of the original cover. The reefing girders to which the original cover between Frames 1 and 3 was laced were not removed and this was an added reason for leaving it in place.

For the main 474 feet of the airship's length, between Frames 3 and 12, the original pre-doped cover was replaced by a new linen cover doped in position. This was the main section of the outer cover.

Aft of this, between Frames 12 and 13, a further fifty-three feet length of the original cover, where the hull tapered ahead of the fins, was left in place with its lacing to the reefing girders. This was another portion of the old cover which had been doped after fitment and was deemed to be airworthy, well aft of No 5 Power Car.

Behind this section, in the area between the fins, the original cover and the reefing girders had been removed to save weight where it was most needed to be saved, at the tail, and a new section of cover fitted and then doped. It extended for seventy-five feet between Frames 13 and 15.

Aft of the fins, sixty-three feet of the original tail-cover remained. Like the other portions of the original cover not renewed it had been doped in position and so was held to be sound.

Thus, out of the new length of 777 feet of R.101 with its extra bay, there was 575 feet length of brand new cover, doped after placing in position, together with 202 feet of the original cover in three sections, just aft of the nose, just forward of the fins and at the tail – all of which had also been doped after fitment.

Two sections of reefing girders had been removed – at the nose and in the portion of the hull adjacent to the fins where weight saving was important to offset the weight of mooring gear and of the fins, rudders and elevators. In all, this scrapping of reefing girders over 101 feet length of the hull had saved 1,168 lb (0.52 tons) of weight. The old cover was retained over those parts where double curvature had made pre-doping impossible.

So 74 per cent of the length of the outer cover had been renewed and 26 per cent length of the original cover remained.

Significantly, the portion of the original front cover from Frames 1 to 3, eighty-six feet long, was that from which most of the dynamic lift would be generated when the airship was flying pitched nose-up to counteract heaviness. For that reason the reefing girders were not

* The fact that substantial portions of the old cover were not renewed especially at the nose, was not specifically mentioned at the Official Inquiry.

removed from this portion of the forward part of the ship where the fabric of the cover could be laced to them to reduce the unsupported areas.*

In all, the outer cover of the lengthened R.101C had an area of 22,993 square metres (247,494 square feet) equivalent to 5.68 acres and weighed 8.11 tons at 1.174 ounces per square foot. In his lecture to the British Association, Richmond admitted that 'the design of the outer cover presents one of the most serious difficulties especially at the higher speeds which are being demanded of future airships'. He did not think that a metal cover would be a practical solution because of its weight, its liability to corrosion and the difficulties of maintenance and repair.

A further problem cropped up on 22nd September just when R.101 was almost ready to be taken out of the shed. The new Beverley starting engine installed in the starboard forward (No 1) car to start the main Tornado, itself stubbornly refused to start in the shed. The trouble was eventually traced to wrong connections of the fuel pipes – and all was well again.

Meanwhile, Professor Leonard Bairstow, on whom responsibility rested for an independent assessment of the airworthiness of the airship, asked for a special analysis to be made of the control reactions of R.101C pitched down by the nose with one gas-bag deflated. What elevator power was available to raise the nose in such conditions?

Richmond and Rope worked out the stability and control characteristics in a special memorandum and sent them off to Bairstow on 24th September. They showed that the ship had ample control response even with a forward gas-bag deflated, provided power was maintained to keep normal cruising speed.†

But yet another – and potentially more difficult – issue had come to light on the previous day when Frederick McWade, Chief Inspector of the AID at Cardington was engaged in a final survey of R.101 before it was cleared and signed out to leave the shed.

In the two main sections where the original outer cover remained

* A summary of the final outer cover of R.101C is as follows; from nose aft: 1. Frame OA to Frame 1. Forward nose cover 26 ft long. Reefing girders removed. New cover. 2. Frame 1 to Frame 3. Front cover 86 ft long. Reefing girders retained. *Original cover*. 3. Frame 3 to Frame 12. Main cover 474 ft long. Reefing girders retained. New cover. 4 Frame 12 to Frame 13. Aft cover 53 ft long. Reefing girders retained. *Original cover*. 5. Frame 13 to Frame 15. Fin section 75 ft long. Reefing girders removed. New cover. 6. Frame 15 to Frame 16A. Tail cover 63 ft long. No reefing girders. *Original cover*.

† Lieutenant-Colonel V. C. Richmond's diary 22/24 September, 1930 (RAF Museum, Hendon).

in place, for 86 feet just aft of the nose and for 53 feet around the fins at the tail, rubber solution had been used to attach a number of circular patches and some lengthwise reinforcing strips. The patches were added to the cover at places where tabs or loops attached 'catenary wires' (wires curved to the contour of the hull) used for the tensioning of the cover. Reinforcing strips were glued over any suspected weak places where the original cover was attached by lacing to the fore and aft reefing girders.

Now McWade found that wherever the rubber solution had been applied to the original cover and was not covered by the patches or reinforcing strips, there had been a serious deterioration of the fabric which was so rotten that 'a finger could be pushed through the cover.'*

McWade at once reported these facts to Colmore and to Richmond who noted in his diary that he 'consequently gave orders for additional reinforcing strips to be stuck over all places where rubber solution had been used'.

Richmond's note to Arthur Gerrish the Shed Manager, dated 24th September, read as follows:

R.101. Rubber solution on portions of the outer cover between Frames 1 & 3 and 12 and 13

It having been discovered that rubber solution on covers doped in place has had a tendering effect on the cloth, will you please note that reinforcing strips must be stuck on the outside of the cover along all lines of rubber solution in the areas mentioned. These strips must extend for a distance of two inches beyond the boundary line of the solution.

The material used for the strips is to be 'B' cotton which may be stuck down with red dope.

In the case of the reefing girders, the existing cover strip may be stripped off if it is found not adequately to cover the line of the solution.

Cardington's records show that 'The work of placing these reinforcing strips in position on all suspected weak positions was completed to the satisfaction of the RAW Inspection Section and also the AID.'†

* Only a casual reference was made to this important fact at the Official Inquiry (Note on Page 63 of Cmd 3825) from a comment in Colonel Richmond's diary. A more detailed analysis was made after the Inquiry had closed. (AIR5–1380–8292–2nd April, 1932).

† Cardington Inspection Sheet No 2026.

(TOP RIGHT)
'wan Song – Early in the
norning on Wednesday, 1st
)ctober 1930, after R.101 had
een walked out of No 1 Shed at
Cardington and put on to the
nooring tower in preparation for
s final trial flight before leaving
or India on the Saturday
vening, R.100 was taken from
he No 2 Shed and moved in to
No 1 Shed in which it was to be
arted for the installation of a
ew bay to increase its lift – as
ad just been completed with
R.101. This was the only
ccasion on which both R.100
nd R.101 were seen in the open
ogether and this is the only
hotograph of both airships at
he same time. It was also the
ast appearance of R.100 which
ever came out again and was
roken up in December 1931.

'he Largest Vessel in the World –
R.101 at the Cardington mooring
nast on the afternoon of
Saturday, 4th October 1930,
aking aboard fuel, gas and stores
or the 2,235 nautical miles from
Cardington to Ismailia in Egypt
expected to be flown in about 48
ours.

HM Airship R.101 – at the
mooring mast at Cardington,
Bedfordshire on the afternoon of
Saturday, 4th October 1930
before leaving for India.

[PLATE XXI]

[PLATE XXII]

(LEFT) *The Bridge* – Looking forward in R.101's control-car slung beneath the hull some 26 feet from the bows. In front is th wheel through which the Steering Coxswain moves the rudders 440 feet aft. On the righ is the elevator control wheel through which the Height Coxswain regulates pitch and height. Above is the single large needle of the altimeter, the air speed indicator above (graduate in knots) with a clock to the righ and a manometer to measure ga pressure on the left. In front is a binnacle compass. Speaking tubes to the front hatch and to a and midships positions hang to the left. Engine telegraph indicators are on the left wall.

(BELOW) *A Room With a View* – One of the two 'promenade decks' of R.101 which flanked the lounge on both sides. Measuring 7 feet 6 inches wide by 32 feet long with a ceiling height of 6 feet 6 inches the viewing balcony was lit by full length windows 8 feet deep let i to the outer cover.

Had time permitted the right course would clearly have been to renew all those sections of the cover where the original, pre-doped, fabric remained; some 33,000 square feet of it in all. But there was no prospect of doing that and maintaining the schedule which had been given to the Air Ministry and sent to all overseas Commands of the Royal Air Force along the route. Unless serious problems showed up during the final test flight the departure date must be kept. In any case, Richmond who had long experience of airship fabric, together with the Manager of the Fabric Section at the Royal Airship Works, J. W. W. Dyer, was confident that, with the reinforcements doped in place, those sections of the original cover which had been doped after fitment would be perfectly sound and ought not to have deteriorated seriously in no more than a year of service. A sharp watch would be kept on the cover when R.101 took the air. There the matter rested.

Flight Instruction No 31 for the trial flight after refit, included the instruction that at least five minutes should be run at full power from the five engines which, for the first time, would be available for forward flight. With, now, 3,250 bhp available for short periods at 935 rpm the maximum level speed of R.101c, trimmed neutral, was estimated at 62 knots, or 71 miles an hour. Up till now R.101 had never flown faster than 55 knots on four engines at their maximum continuous 890 rpm yielding a total of 2,340 bhp – and that for just a few minutes on the fourth flight on 2nd November 1929, when the original cover was relatively new.

The loads on the outer cover increase as the square of the speed; double the speed, four times the stress. A speed of 62 knots would have almost doubled the loads on the cover compared with the 50 knots or so at which R.101 had done most of its flying on three engines at 450 hp each.

It could well have been that, under the stress of a full-speed test, the vulnerable front portion of the original cover still in place would had split. In good weather and in daylight such a split would have been serious but not necessarily disastrous.

In the event, the failure of an oil cooler on 1st October meant that a full-speed test was never run. Consequently the most severe test of the cover came only on the first stages of the flight towards India when R.101 was flown at maximum cruising speed through some of the most turbulent conditions any airship had encountered and in the face of heavy rain.

No one could foresee all this when R.101 was handed over from the

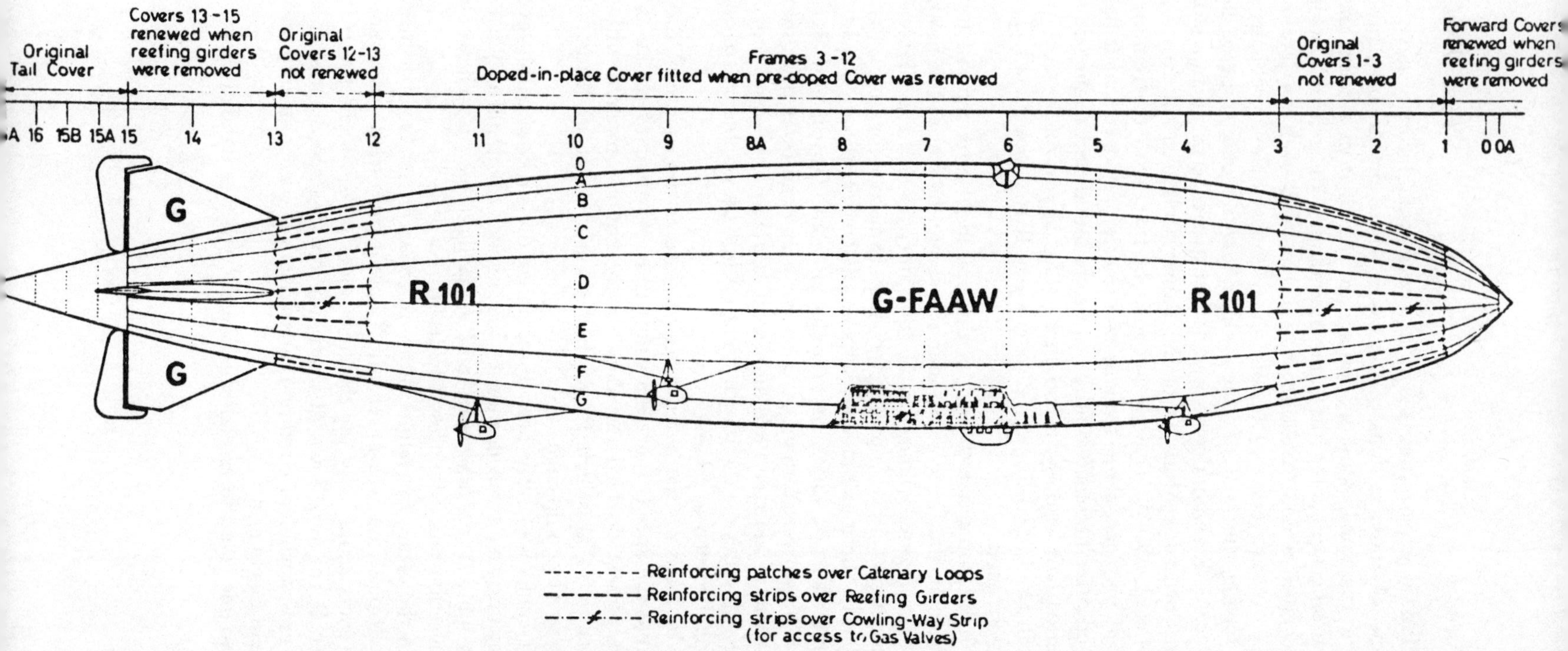

The Outer Cover: The Cardington drawing of the portion of R.101's outer cover renewed during the refit of

Works to the Flying Staff on Saturday, 27th September, on the completion of its comprehensive three months' refit.

During that Saturday afternoon the ship was gassed to 100 per cent capacity and the lift and trim measurements taken. As Noël Atherstone noted in his diary: 'And so started the Third Commission. I hope that it will be a really good one.'

The results of the lift and trim calculations certainly seemed to justify that hope. The volumetric capacity with the new bay was now found to be 5,509,700 cubic feet, to give a standard gross lift of 167.259 tons. The fixed empty weight was established at 117.9 tons. So the disposable lift was 49.359 tons (or 110,564 lb); an improvement of 14.4 tons compared with its original disposable lift, and 8.919 tons (19,978 lb) more than that of R.101B; the ship lightened for flying in June with the gas-bags let out. The new bay had thus contributed a net increase of lift of almost nine tons.

Compared with R.100, the lengthened R.101C had 1.642 tons (3,678 lb) less disposable lift but this figure included in R.101's weights the fact that it was carrying around rather more than seven tons of additional power-plant weight.* For a flight of 60 hours R.101 would require almost four tons less fuel, cruising at 55 knots, so under such conditions R.101C would have 2.3 tons more lift available than R.100 for useful load. Even so, neither airship had adequate disposable lift for long flights.

On Sunday, 28th September, Atherstone noted:

> Handling party was cancelled yesterday morning but the crew all turned up this morning complete in uniforms and had their photos taken. As there is still a possibility of getting the ship out early in the morning the gear for holding the ship while running engines was not rigged. The wind is persistently in the north-east – which is exactly in the wrong direction as it would mean taking the ship out with her tail to the wind and then having to swing her 180 degrees which would bring her tail over the main Bedford–London road. What we are waiting for is a very light wind in any direction or a wind of up to 12 mph from the SW.

* The comparative figures were:

	*R.101*C	*R.100*
Gross lift	167.259 tons	156.521 tons
Fixed empty weight	117.900 tons	105.520 tons
including:		
Power plant installation	17.386 tons	10.189 tons
Disposable lift	49.359 tons	51.001 tons
Fuel required for 60 hrs cruising at 55 knots	26.116 tons	30.033 tons
Remaining useful lift for 60 hours	23.243 tons	20.963 tons

But the wind remained too high and from the north-east. So, next day, the bow restraining-gear was rigged and the day was spent in running engine tests in the shed. All went well except that, once again, the Beverley compression-ignition (diesel) starting engine in the starboard forward (No 1) car could not be persuaded to start until 2200 hrs in the evening when, at length, the No 1 engine was given a short run with, successively, forwards and astern thrust.

Colmore had thus been as good as his word to the Secretary of State on 29th August. R.101 had been made ready and able to leave the shed by 27th September with all systems complete and all inspections made. R.101 had been signed out by the AID as fully serviceable with all modifications and rectifications complete.

Now, frustratingly, time was being lost because of the impossibility of taking the ship out of the shed safely until calmer weather arrived.

That Sunday, 28th September, while the handling party remained at readiness, C.B. dined with an old friend from Roumanian days, Vladimir I. Einstein. Of that dinner, Einstein later wrote to the Prime Minister:*

> Lord Thomson was one of my dearest friends. When we dined together at the Hungaria Restaurant he told me that he had hoped to leave for India on the following, Monday, evening had R.101 been ready; in time to be back for the Imperial Conference's meeting on Civil Aviation on Monday, 20th October. But the airship was still held up in the shed by gales and inclement weather.
>
> He said that he wanted to be on board R.101 when it went to India because Imperial Communications by airship was his idea and he would not ask anyone to do anything he would not do himself. He thought also that his presence would give more confidence in airships as a psychological factor with the British public and that the gesture of the S of S being aboard would help to create a more friendly impression towards Britain in India.

But still the winds blew and R.101 remained ready but bottled up in its shed at Cardington.

On Tuesday, 30th September, after three days of waiting, Colmore telephoned Maurice Dean† in AMSR's (Hugh Dowding's) office to

* Letter from V. I. Einstein to J. Ramsay MacDonald from 3 Central Buildings, Westminster, London, S.W.1., dated 8th October 1930. (PRO.30 69/6/114–06,130).

† Later Sir Maurice Dean, Permanent Under Secretary of State at the Air Ministry, 1955–63.

ask if AMSR 'would agree to the trial flight being reduced from 24 hours if all went well and Major Scott was satisfied with the ship?'* In a later conversation, on 1st October, DAD explained that: 'provided Major Scott was satisfied with the behaviour of the airship there would be no object in continuing the flight to complete 24 hours which was a purely arbitrary figure. The object of the suggestion was to do everything possible to ensure that the airship should leave for India on the evening of 3rd October'.

After some discussion, Dowding accepted Colmore's proposal for a shorter test flight, in the belief that it could be 'taken as expressing the opinion of everyone who had any experience of airships and of the crew who were to do the flight'.

All day on 30th September the handling party had been standing by in the hope that the wind would drop, to be finally dismissed at 1930 hrs, only to be recalled in the early hours when the north-easterly fell to very light.

Then at last, R.101 was taken out of the shed at 0630 hrs on 1st October and put on the tower after some difficulties with the hauling down wire which 'shook up the airship somewhat'.

As soon as R.101 was on the tower, the handling party moved over to No 2 shed and took out R.100 and moved that ship to the larger No 1 workshop shed, ready to be parted for the installation of an extra bay. This was the only occasion on which R.100 and R.101 were seen in the open together at Cardington.

During the morning orders were given to have everything ready for flight by 1500 hrs GMT. Flight Instruction No 31, signed by Sydney Nixon for DAD issued from Cardington to Carmichael Irwin, Captain of R.101, was now amended so that the time of landing was brought forward from 0800 hrs BST on 2nd October to a 'time of landing which should be previous to 0600 hrs GMT at which time the ship is to be secure at the Tower'.

There was only one reason for such a precise time and that was so that AMSR could accompany the trial flight and yet keep an appointment with the Controller of the Admiralty on the early morning of 2nd October – an appointment which had been put off several times on account of the delayed flight – this in spite of the subsequent statement that AMSR 'made it clear to Colmore that on no account was the duration of the flight to be altered on his account'.

On the morning of 1st October telegrams were sent to the Air

* AIR 5/904/311.

Officers Commanding in India, Palestine, Mediterranean, Iraq and Aden. They read: 'R.101 left shed this morning for mooring tower and will start this afternoon on trial flight. Subject to trials being satisfactory earliest possible date for departure would be evening of 3rd October.'

The ship at the mooring tower was, in all significant respects a new and untried airship, some six per cent longer and of ten per cent more volumetric capacity compared with the original. Moreover, now for the first time all five engines were available for forwards flight. The new and bigger airship had, therefore, a total of 3,300 bhp available for forwards flight compared with only 2,600 bhp before.

With the increased length and capacity and the additional power, all the control characteristics were changed. A new 'Permit to Fly' was required after all the modifications and changes had been considered and approved by the Air Ministry's two independent consultants on airworthiness, Professors Leonard Bairstow and Sutton Pippard. This necessary 'Permit to Fly' was gained on 1st October accompanied by a letter from Leonard Bairstow which included the sentences:

> We have satisfied ourselves that R.101, as now existing with its additional bay, complies with the specified requirements of the Airworthiness of Airships Panel.
>
> The difference between the conditions of loading of R.101 now submitted and those of the original design on which our previous report was based, surprised us by their magnitude; the differences are not primarily a consequence of the addition of the new bay.
>
> A good deal of general thinking and comparison on limited information has been required in reaching our conclusion and we have not had time since essential information from the RAW to prepare a sufficiently considered final report.

Enough had been secured, however, to justify the issue of a 'Permit to Fly' for the trial flight on 1st October. According to the airworthiness standards of the day the procedure to be adopted was wholly in keeping with established practice and all requirements were being met.

When the final report was completed, Professor Bairstow noted: 'The most important considerations before us – although due in part to the addition of the new bay – arise chiefly from the difference between the fixed weights of R.101 as completed and flown and the weights assumed in design.' In fact the increased weights of the power cars, passenger accommodation and the structure at the bow and tail were 'such as to produce in the light condition of the ship an excessive tail moment'. In other words, when flying light, R.101 would be tail heavy – and this has been borne out in practice but rectified by the removal of the servo gear at the tail and deletion of reefing girders.

Detailed examination showed, however, that the airworthiness and control characteristics were all adequate and that the lengthening of the ship had in some respects 'the effect of reducing rather than increasing the stresses' on the ship.

The fact remains that R.101C – as lengthened and modified – was virtually a new airship and that logically, a complete series of test flights ought to have been flown before a Certificate of Airworthiness was granted.

Indeed, in 1929, Flight Lieutenant Irwin had drawn up a schedule of recommended test flights for R.101 which had included a requirement for:

> a flight of 24 hours duration under reasonably adverse weather conditions, to be followed by a flight of 48 hours duration under adverse weather conditions to windward of base. Ship to be flown for at least six hours continuous full-speed through bumpy conditions and the rest of the flight at cruising speed. Ship to be berthed in shed as soon as possible after landing to the mast and a complete bow to stern inspection carried out.

None of this had been followed. The longest R.101 had flown at a stretch, and in its original condition, were the 29 hours on course of its endurance flight on 18th/19th November 1929, during 30 hours 41 minutes away from the mast. It had not flown in really adverse weather though it had ridden out a storm at the mast. It had not flown for even one hour at full speed in any conditions. Yet, of R.100's 54 hour 52 minutes endurance flight in January 1930, Ralph Booth subsequently pointed out that 'certain defects – and very important defects – occurred after flying about 45 hours which would have seriously hampered the ship if they had occurred over the middle of the Atlantic'.

Yet, here was R.101 in a new form about to be launched on a 10,000 mile flight to India and back carrying the Secretary of State for Air, a mission which would require four flights in widely different weather, each of about 48 hours duration, and all to be based on one additional test flight.

As long ago as 30th June 1930, Sir John Higgins had said in a note to the S of S* that 'R.101 should be completed with the new bay by 22nd September. *One trial flight* will be necessary before the airship leaves for India' . . .

The fact that one trial flight would be insufficient even if it was completely successful was not raised at that stage – or subsequently – from Cardington, although had the point been made it would have been quite out of character for the Secretary of State to have done other than to agree; the more so with his insistence on 'Safety First' and that 'there is going to be no risk, while I am in charge, of the thing being rushed or of any lives being sacrificed through lack of foresight'.†

Now Sir John Higgins had gone and a new AMSR had been installed who had never been in an airship in his life. He must rely wholly on advice from Cardington. The onus lay squarely upon the Director of Airship Development, Reginald Colmore, and upon his Chief of Flight Operations, Herbert Scott.

What was their advice? Only that the duration of the impending trial flight should be reduced to less than 24 hours if the ship behaved well.

Examination of all the surviving records, and later discussion with some of those concerned, leads to the clear conclusion that there was an absolute confidence that all would be well and that no unjustifiable risks were being taken. The ship had been thoroughly overhauled after ten flights and 102 hours 56 minutes of flying; all known defects had been put right. There was more lift and more power than ever before. Other airships – the LZ.126 *Los Angeles*, the LZ.127 *Graf Zeppelin* and R.100 – had all set out on their first, long overseas flights with less background than had R.101.‡ All three had succeeded. R.101 would too.

So, at 1536 GMT (1636 BST) on Wednesday, 1st October 1930, R.101 slipped from the Cardington mast carrying four officers, 38 crew and 15 passengers (57 in all) together with 6,037 gallons (22.8 tons) of fuel and ten tons of water ballast which, together with stores and spares, brought the disposable load to a total of 42.222 tons. It was calculated to be sufficient for a flight duration of 68½ hours at normal cruising speed.

* AIR 2-364/523024/30-06,130.

† House of Lords, 3rd June 1930. Hansard Column 1360.

‡ *Los Angeles*: 5 flights: 51 hrs 12 mins.
Graf Zeppelin: 6 flights: 64 hrs 55 mins.
R.100: 7 flights: 134 hrs 35 mins.

The chief passengers were Air Vice-Marshal Hugh Dowding (AMSR), Air Vice-Marshal Sir John Steel of Headquarters, Wessex Bombing Area and Viscount Gage, Parliamentary Secretary at the India Office, together with Colmore, Richmond, Rope and Collins from the Royal Airship Works.

Atherstone noted in his diary for 1st and 2nd October:

> At 1530 BST the passengers came on board and the ship slipped from the tower shortly after 1630. The ship flew over London and then down the Thames and over Southend. The night was spent off the East Coast and in the morning we came in just North of Yarmouth and straight back to Cardington.
>
> I turned out at 0400 hrs to keep the Morning Watch with the ship in the vicinity of the Newarp Light Vessel off Yarmouth. There was very little wind and as we had to be back at Cardington at 0700 hrs BST I set course for home and got the ship about five miles to leeward of the tower by 0645 hrs BST. The landing was rather a long affair and Irwin did not get the ship on to the wire until 0830. I think the ship was rather too light and there was also a small inversion. We did not pick up the wire until the third attempt. The mooring also took too long as a yaw guy was lowered too soon, on the advice of Scott, with the result that it fouled the tower. We were secured to the tower by 0930 BST.
>
> The trial was very successful except for a burst oil cooler in the starboard forward engine car which put that engine and the Beverley out of action for the rest of the flight. The ship appeared to be much better in the air than before and the cover was really good.

All this was encouraging even though the flight had had to be cut short to get Hugh Dowding back in time for his Admiralty meeting.

Richmond wrote:

> Impossible to carry out full speed test owing to the early failure of the oil cooler in the forward starboard engine. Flying conditions were very perfect and under these conditions all other items in the ship behaved admirably.

The airship had been away from the mast for 16 hours and 51 minutes on this, its eleventh flight. It had been on course under power for 14 hours 24 minutes and had covered a distance of 533 nautical miles at a speed on course of 38.4 knots (44.2 mph) flying most of the time on four engines. Fuel used amounted to 802 gallons at 55.7 gallons (471 lb) an hour – a very low rate of consumption which equated to cruising on four engines at only 700 rpm each for a total of 1,180 bhp and a true air speed of 44 knots (51 mph). At that speed and consumption R.101C would have a still air range on the 6,037 gallons of fuel lifted of 4,770 nautical miles (5,490 statute miles) at a duration of 108 hours.

It was all satisfactory – except that the cover had not been tested at higher speeds nor in any but good, clear weather with no turbulence.

Richmond recorded, on 2nd October, that he had investigated the oil cooler failure with Forsyth and Cave-Browne-Cave 'and decided that the matter was not serious. Certain simple precautions and modifications were made which were further tested on Friday with completely satisfactory results'.

In fact, a leak in one of the gills of the oil cooler had caused the trouble. A spare had been fitted in the air but, in the dark, the engineers in the car could not achieve a complete sealing of the pipe connections so that the engine was shut down for the rest of the flight. In addition, because both the forward, reversible, engines had shown higher oil pressures than the other engines – later diagnosed as an over reading of the pressure gauges – the decision was made not to run the port forward engine at other than low speeds until the high oil pressure was sorted out.

G. D. Raisbeck, Chief Examiner AID at the Royal Airship Works, flew on the trial to make a special report of an inspection of the ship during the flight.

His report* on the flight gave the ship a clear bill of health. He noted that:

> The final inspection of HMA R.101 had been completed in the shed and every system and assembly had been checked so that no definite item required any special criticism. As a result of this I conferred with the Captain of R.101, Flight Lieutenant Irwin, previous to flying and agreed to give an independent opinion of certain things which I thought were important under flying conditions including the Cover, Gas Valves, Water Recovery, Gas Bags, Tanks, Controls.

He then reported on each item.†

On the covers, Raisbeck reported:

> Practically no movement was noted of the covers and no 'tugging' at lacing edges. All sealing strips appeared securely held. The new bay covers were steady and sound.

* AIR 5/14–X/M. 08109.
† AIR 5–14–X/M. 08109.

On the gas valves:

No leaks were observed. No surging or currents of air were in evidence. Previous to flight the covers cleared the valves by approximately three inches and it was noticed during flight that at least four inches clearance was obtained by the effect of pressures.

The only slight movement of gas-bags was as the ship possibly rolled or elevated for height. This was so slight that I took quite ten minutes to observe it and it had no effect on the valves.

Gas-Bags: During the tour various bags were examined with their bulkhead and lifting wiring systems especially at the positions between Frames 5 and 8A. All padding observed was secure and no movement was noticed except the slight effect mentioned in connection with Gas Valves.

The Water Recovery System was installed previous to this flight and I examined it to see that the fabric funnels acting as inlets for the water, were not dragging. All proved satisfactory.

All other items inspected were likewise found to be in good order and functioning satisfactorily.

Raisbeck concluded:

I met Captain Irwin several times during the night and reported progress and handed to him some final notes on the morning of 2nd October which stated that the agreed programme had been carried out and every item was entirely satisfactory. He said that he was satisfied with the independent inspection which agreed with the reports of his crew.

Later Dowding mentioned that: 'During the trial flight I talked to all the officers – Colmore, Scott, Irwin and Johnstone – and the atmosphere was quite serene.'

One more report was made on the trial flight in addition to the Ship's Log which remained on R.101. This report was by Charles T. Wright, Charge-Hand Rigger at Cardington.

He noted:*

I was on the trial flight of R.101 on 1st–2nd October 1930, attached to the crew. I had no special duty but I was available for any emergency.

This was my first flight. I thought the ship very steady. I was particularly interested in the gas-bag wiring and the covers of the ship as I had been working on them.

* AIR S/905–XM 08109.

The gas-bag wiring seemed completely satisfactory and so was the cover. The quilted mattresses were functioning as they should and were preventing any flapping.

Captain Irwin spoke to me on the trip and I told him that the behaviour of the ship was splendid and I would go anywhere in it.

He said it was a splendid and stable ship.

*

Meanwhile, writing to his old friend, Mrs T. G. Conrad of San Francisco, C.B. had outlined his hopes and expectations in a letter of Sunday, 28th September. He wrote:

I have to go to India this coming week in the airship R.101, the sister ship of R.100 which went to Canada last August.

I am taking only a few experts with me as this is the first trial flight in tropical climates. We expect to be four or five days from England to Karachi and from there I go into Simla to stay with the Viceroy for two days.

I then return by airship to London, where I have promised to be on October 20th as MacDonald wants me during the concluding stages of the Dominion Conference.

The voyage will be my first real holiday since June 1929 as my work in this Cabinet has included much outside my own Department. I am on seven Cabinet Committees and Chairman of three of them.

The Government still survives, not so much because of its popularity as because there seems no alternatiive to it. The Tories are divided and the Liberals do not want a General Election.

We may be in office for another year – but, in the meantime, there is talk of my going to India, next May, as Viceroy. I am in two minds about this matter. It is, of course, a great position but full of difficulties and could easily lose such little reputation as one might possess – and to please even a small majority is next door to impossible. There are too many views about India and the whole business is in a tangle.

We expect to finish the Dominion Conference early in November and the India Round Table Conference will then begin.

I have never felt better in my life, fortunately. But how long this good state of health will last is always uncertain.

CHAPTER TWENTY-ONE

Outward Bound

'Sail fast, sail fast
'Ark of my hopes, Ark of my dreams,
'Sweep lordly o'er the drowned Past,
'Fly glittering through the sun's strange beams;
'Sail fast, sail fast
'Breaths of new buds from off some drying lea,
'With news about the Future scent the sea.'

SIDNEY LANIER, 1878.

On Tuesday, 30th September – while still awaiting anxiously news that the weather at Cardington was calm enough for R.101 to be brought out of its shed – Christopher Thomson accompanied the Prime Minister to the Foreign Office for a preliminary gathering of delegates to the Imperial Conference for 'an informal exchange of views and discussion of procedure.'

One of the items on the agenda was agreement on the date when the Secretary of State for Air should plan to present to his colleagues on the Conference's Civil Aviation Committee, his Paper 'Progress of Imperial Communications', together with his recommendations for the future. After some discussion, Monday, 20th October, was fixed for the start of the air transport sessions. It would give the Air Minister one day short of three weeks to fly to India and back – that is, if he went at all? It would depend upon how soon R.101 could be brought out of the shed and, even more, upon whether the trial flight revealed any significant defects.

Unless everything was satisfactory then the flight to India would have to be put off until after the end of the Conference. Instead, delegates would be invited to visit Cardington and, if all was well, to take a short flight in R.101 themselves. But that would be very much a second best compared with a demonstration flight to India and back while the Conference was in session.

That very day, 30th September, encouraged by his visit to Cardington the previous week, Sefton Brancker wrote cheerfully to his old friend, J. A. Wilson, Director of Civil Aviation in Canada:

The Secretary of State and I are by way of starting to India this week. The S of S hopes to get back in time to dispose of various air problems

> which may arise at the Imperial Conference but, as I am going as far as India, I must take the opportunity of doing some work there – and in Persia and Greece on the way back.

For long an enthusiast for airships, Brancker recalled his own remarks on an earlier occasion following a lecture by Vincent Richmond before the Royal Aeronautical Society:*

> I feel that our airship development is one of the most important things which is taking place in Aviation today and there are no bounds to what the success of these airships will mean to the British Empire if the experimental services work out all right. It is the greatest adventure in construction engineering of our time.

Next morning, 1st October, at 10 o'clock – with the news that R.101 was at the Cardington mast – the Prime Minister officially opened the Imperial Conference at the Reception Room of the Foreign Office, welcoming on behalf of His Majesty's Government the representatives from all the Dominions and Colonies.

The Conference would, he said, probably continue for about six weeks † and, as agreed on the previous day, its business would be conducted largely in Committees. Among them, the Civil Aviation Committee, chaired by Hugh Guthrie, Minister of Justice for Canada, would begin its work on 20th October 'immediately after Lord Thomson's return from India".

By next morning, Thursday, 2nd October, there was news that R.101 had successfully completed its trial flight and, though there had been the minor trouble with one cooler, Colmore, Scott, Richmond and Irwin were all satisfied that the airship was sound and airworthy and that the flight to India could now be mounted with confidence, leaving at the end of the week.

Now there was repeated with R.101 the pattern which had been followed with R.100 after its final test flight before its Atlantic crossing. On both occasions the recommendation to go was made by Colmore as Director of Airship Development and Scott in charge of flying – the one cautious, the other all for pressing on. Together they were more experienced than anyone else in Britain on the operation of airships.

After the ship had been inspected again at the mast and a conference

* The Royal Aeronautical Society: V. C. Richmond on Airship Research and Development, 29th April 1926.

† In fact it was concluded on 13th November 1930.

had been held on the oil-cooler problems, Colmore telephoned F. W. Musson, Private Secretary to Air Vice-Marshal Dowding, to say that 'nothing we can foresee should prevent the ship leaving on Saturday evening.'*

The Secretary of State was anxious, however, to receive a first hand report and to be satisfied himself that everything was really in order before a final decision on if and when to start was taken. Accordingly, he called a meeting at Gwydyr House with AMSR, Hugh Dowding and DAD, Reginald Colmore, to receive their impressions of the trial flight and to discuss the forward arrangements. The meeting was held on the evening of 2nd October with Louis Reynolds taking a note.†

Before going in to the Secretary of State's room – while awaiting his return from a Cabinet meeting at 1900 hrs – Dowding and Colmore had a brief talk about the trial flight and the prospective journey to India.

Dowding reported later:‡ 'I said to Colmore that he was my adviser and that I would back him up whatever line he took and if he were pressed to anything against his better judgment.'

When the meeting began, Lord Thomson asked how the trial had gone. Colmore said that it was quite satisfactory except that one engine had had to be shut down for what was now established was a defect in one of the fins of an oil cooler. It had now been repaired and had turned out to be a problem which was easily put right. Everything else was in good order. The behaviour of the centre bay and of the outer cover had been admirable. There were still, however, a number of odds and ends to be done. He wanted to put an additional reserve oil tank on board and review the list of spares to be carried. Two of the gas-bags were a little down on gas purity and they would need bringing up to standard. All in all, the airship would be ready to leave on Saturday evening. Scott was satisfied. The airship looked 'very good'.

Dowding spoke of the test flight 'as a layman'. He had enjoyed his first experience of an airship. The only thing was that because of the failure of the oil cooler there had been no opportunity to do the full-power, maximum speed, test which had been scheduled. That ought to be run near home during the flight to Egypt.¶ He had proposed that to Colmore.

* AIR 5/903–XM.08109 and AIR 5/14–XM 08109.
† AIR 5/14–07640 and AIR 5/912–06990.
‡ AIR 5/903–XM.08109 and AIR 5/14–XM 08109.
¶ Nothing more stupid can be imagined than to run a full-speed test during a long flight on which every gallon of fuel was precious and no undue strain ought to be placed on engines or airframe with so many hours of flying, out and back, ahead of them. It was not done.

'That all sounds fine,' said the S of S. 'So what about the possibility of starting on Friday night or Saturday morning to give ourselves as much time in hand as possible?'

'Well, sir,' said Colmore, 'That is not absolutely impossible but it would be very difficult. The ship has to be gassed and fuelled. And there wouldn't be time for the crew to have the rest they need before they start on so long a journey. Then, an early start on Saturday morning would be no good because it would bring us to Ismailia at just the wrong time of day for landing. We want to get in a little after sunset to gain the advantage of extra lift in the cool of the evening.'

'Well, of course I understand,' said Thomson. 'As I've always said, I don't want anything to be rushed. I rely entirely on your advice and judgement.'

Colmore added that the best timing would be to cross Northern France during the night so as to be in daylight when they came to the higher ground on the edge of the Massif. Then they'd have a pleasant second night cruising down the Mediterranean. 'So, if it suits you, S of S, we should aim to start on Saturday evening; weather permitting of course. Then all the times should work out right.'

'Of course that's all right,' said Thomson. 'Naturally I am anxious to get off as soon as I can. But you must not allow my natural impatience, or anxiety to start, to influence you in any way. I rely on your judgement.'

Reynolds added a further, abbreviated, note to his record of the meeting:

> *Colmore*: Leave Saturday, 5 or 6 o'clock.
> *Thomson*: No rush on my account.

They then went on to discuss the final passenger list. The S of S, Sefton Brancker, Percy Bishop of the AID, and the S of S's valet, James Buck, were already on the list, together with Colmore himself, Major Herbert Scott, Colonel Vincent Richmond and Squadron Leader Michael Rope from the Royal Airship Works.

Colmore now confirmed that with the extra bay there would at last be lift to spare – at least on the outwards journey. They could take, say, four more passengers (to make a dozen in all) if there were priority requirements? Sefton Brancker had already put forward the name of Squadron Leader William H. L. O'Neill, whose departure for India to take up an appointment as Deputy Director of Civil Aviation in Delhi had been delayed because of the illness of his wife. Colmore suggested that two more, experienced, men from the Royal Airship Works would be useful to have on board, Alexander Bushfield of the AID, who had stood down to allow Frederick McWade to fly to Canada in R.100, and Harry Leech, Foreman Engineer, who was well versed in all aspects of R.101's structure and equipment.

These names were agreed and, at Dowding's suggestion, Squadron Leader William Palstra, MC, of the Royal Australian Air Force and now Australia's Liaison Officer at the Air Ministry was added as an Air Aide to the S of S and to represent the Australian Government.

Dowding added:

> We then discussed various alternative methods of return for the S of S in the event of the ship being delayed at any stage of the journey.
>
> At the end of the interview, Lord Thomson appeared to be absolutely confident in his own mind that he would be back in London on the 20th. I thought it wise to point out to him that he must not be too certain on this point and that there was a definite chance that the ship might be delayed at some stage of the journey. Weather conditions were principally in my mind when I spoke.*

Mention was then made of the State Dinner Party which the Secretary of State proposed to hold on board R.101 on the evening of arrival at the Ismailia mooring mast. Depending upon the local time of arrival the dinner would be provisionally set at 2130 hrs local time. Detailed arrangements would be made by the Officer-in-Charge at Ismailia (Squadron Leader Claude Cox). Those present would include the S of S, DCA, DAD (Colmore), A/D (F) Scott, A. D. (T) Richmond, Captain of R.101 (Irwin) and about eight or ten guests including the High Commissioner, the AOC (Air Vice-Marshal Francis Scarlett) and Sefton Brancker's son, John, who was with Imperial Airways in Cairo.

All these matters being sorted out and no further problems having been raised by his professional advisors, the Secretary of State wound up the discussion at 2000 hrs by saying, 'Well, that's settled then. We leave on Saturday evening.' He turned to Colmore. 'Let me know if there's any change. I can't tell you how much I'm looking forward to this voyage.'

Nothing had been said at this meeting about the Certificate of Airworthiness which was essential if R.101 was legally to fly outside United Kingdom's airspace. In the event the C of A — 'Number L.A.6 in Special Category (Research and Development)' – was signed, later that evening, by Major F. G. L. Bertram, Deputy Director of

* Air Vice Marshal H. C. T. Dowding. (AIR 5/903–X/M 08109).

Civil Aviation who signed all such certificates. It was handed to Reginald Colmore next day on the strength of the satisfactory report of the trial flight and a final inspection of R.101 at the mast by the meticulous Frederick McWade who passed the ship as fully airworthy remarking that 'In fact I need not have gone on the ship at all because Mr Bushfield had already signed it out and Squadron Leader Rope had made an additional check of the gas-bag and valves'.*

The meeting did not end until after 8 o'clock that evening and C. B. hurried straight from Gwydyr House to the Cafe Royal in Regent Street where there was awaiting him for dinner an old family friend, Mrs Margaret N. Taylor, the wife of a cousin on his maternal grandmother's side.

It was obviously a congenial dinner and Margaret Taylor was not only good-looking and elegant but also well read in French literature on which C.B. likewise prided himself. They talked family affairs and the coming flight to India.†

Next morning, 3rd October, Christopher Thomson went to see the Prime Minister at No 10, Downing Street.

Ramsay MacDonald asked him if he *really* had to go?‡ What were the risks? The Imperial Conference was just getting going. Why not stay and help all the way through?

> C.B. brushed aside any risk [wrote MacDonald]. He had set his heart on the flight, he said. It was the right thing for him to do. He believed

* AIR 5/904.

† On 29th October 1930, Mrs Taylor wrote to Sir William Jowitt, the Attorney General, from the Hotel Scribe in Paris:

'So far I have refrained from offering any assistance in the R.101 Inquiry as I felt that my testimony would be of such a personal nature – that is, a conversation with Lord Thomson concerning the flight on the eve of his departure – that it might not be of as great importance as the technical evidence.

'However, I gather from the newspapers that there is some feeling that perhaps Lord Thomson was over-anxious for the flight and determined that it should take place at that particular time. As I know that this was not the case I am writing to say that I shall be very glad if my conversation with him will be any help in clearing up the matter in regard to his attitude, to come over to London at my own expense to testify.

'It was my privilege to be a personal friend of Lord Thomson and, as such, to know the great consideration he always had for others.

'On Thursday evening, October second, I had dinner with Lord Thomson. His remarks to me then about the flight did not in any way indicate a particular anxiety on his part for it, certainly they were not indicative of a desire to push it at any cost. I sincerely hope that the public will not get the false impression and am anxious to do anything I can to prevent it.

'Sincerely yours

MARGARET N. TAYLOR.'

‡ Ramsay MacDonald's account of the meeting set out in his Introduction to Lord Thomson's book *Smaranda* (Jonathan Cape, London, 1931).

...égrammes
...TEL - PARIS 96
...INE Nº 65.476

Téléphone:
RICHELIEU 92 70 & LA SUITE
INTER SPÉCIAL 12-61
12-62.

SOCIÉTÉ DES HÔTELS RÉUNIS

Hôtel Scribe

...ection
...OYAL
... HOTEL
... PALACE
... HOTEL
...E PROVENCE

1, RUE SCRIBE
PARIS

Même Direction
MONTE-CARLO: HÔTEL DE PARIS
• CAFÉ DE PARIS
• (RESTAURANTS)
• HÔTEL DE L'HERMITAGE

29th October 1930.

Sir William Jowitt,
Hall of the Institute of Civil Engineers,
Great George Street, Westminster.
London.

R. 101 Inquiry.

Dear Sir William,

So far I have refrained from offering any ...istance in the R. 101 Inquiry as I felt that my testimony would ...f such a personal nature, that is, a conversation with Lord ...son concerning the flight on the eve of his departure, that ...ight not be of as great importance as the technical evidence.

However, I gather from the newspapers and from ...ers brought to my attention that there is some feeling that ...haps Lord Thomson was over anxious for the flight and determined ... it should take place at that particular time. As I know this ... not the case and cannot conceive that any blame should be attached ...Lord Thomson personally, in however roundabout a manner, I am ...ting to say that I shall be very glad, if my conversation with ... will be of any help in clearing up the matter with regard to ... attitude, to come over to London at my own expense and testify.

It was my privilege to be a personal friend ...Lord Thomson and as such to know the great consideration he ...ays had for others.

On Thursday evening, October second, I had dinner ...h Lord Thomson and was undoubtedly therefore one of the last to ... him before the fatal trip. His remarks to me then about the ...ght did not in any way indicate a particular anxiety on his part ... it, certainly they were not indicative of a desire to push it ...any cost. I sincerely hope that the public will not get this ...se impression and am anxious to do anything I can to prevent it.

Sincerely yours,

...t.:
...Margaret N.Taylor.

Margaret N. Taylor

Personal Account: The letter written to the Attorney General, Sir William Jowitt by Mrs Margaret Taylor concerning Lord Thomson's attitude to the flight to India as seen by her during dinner with Lord Thomson at the Café Royal on 2nd October 1930.

in the ship. It was his child. He had watched it grow. How could he stay at home whilst it went on its way to attainment?

I asked if recent alterations made to the ship had been good? He was confident. Feeling I was doubtful he chaffed me about my earlier desire to cross the Atlantic in R.100 and said that I was the last man on earth to hold him back. Then, more seriously, 'You know I have never let you down. I shall not do it now. I shall meet you here on the 20th.'

Later, as he descended the stairs and I leant over the balustrade at the top to see the last of him, he stopped and called up in lightsome words that, if the worse came it would soon be over and that the Fate of all of us was written.

The Prime Minister left for an audience of the King, 'to consider various constitutional aspects of inter-Imperial relations'.* C.B. went back to his desk at the Air Ministry in Gwydyr House.

Then he did what might be thought to have been an uncharacteristic thing. He wrote out his Will.

Certainly he had complete faith in his professional advisors from Cardington. At the same time, as he had said exactly a year before: †

No one would claim that the present airships are the last word in airship design or other than experimental. I see no occasion at present for the smallest misgivings on technical grounds . . . but I take it that no one would suggest that the operation of experimental aircraft is free from risk. It is easy, however, to exaggerate that risk – and, such as it is it should, in my opinion, undoubtedly be faced in the interests of aeronautical progress.

Perhaps his talk with the Prime Minister had reminded him of what he then had said. Now he wrote out, on Gwydyr House notepaper:

In the event of my death during the flight of R.101 to and from India – or as a result thereof – I bequeath everything of which I die

* At a session of the Imperial Conference, four days later, the Prime Minister was to say: 'Lord Thomson has been the dearest of colleagues. He has passed in and out of No. 10 as a member of the family. He has been a bulwark in political matters – always ready with advice when approached. His door has always been open His table has always been spread. His fireside has always been ready with a welcome for me. There were times when we would not talk of political things but of all the delightful matters of the World – of its beauty, its energy, of the affections of life.

'He has enabled me to get over many harsh circumstances and to bear many heavy burdens – a distinguished colleague, a well-loved friend.'

† In a letter of 3rd October 1929, from Gwydyr House to Colonel L'Estrange Malone, MP. (See pages 116 and 117.)

possessed (cash, shares, chattels and papers) to my brother, Colonel Roger Gordon Thomson, CMG, DSO, of Springhill, Widdington, near Newport, Essex'.*

Nothing for Marthe – but then she already possessed lands and wealth and there was little he could add; except just one more telegram of his devotion.

Later that afternoon at 1500 hours, Sefton Brancker came up to the S of S's room from his own on the ground floor, to discuss with him the serious financial straits of National Flying Services, the Hanworth Air Park Company which had promoted five new flying clubs and the Hanworth School and, in doing so, had over extended itself. They agreed that, somehow, something must be done to set it back on its feet when they returned from India.

Brancker had lunched that day at the Army and Navy Club with his Deputy, the phlegmatic Irishman, Francis Bertram, (who had signed R.101's C of A the previous evening) with Reginald Colmore and with Francis Shelmerdine, the Director of Civil Aviation in India since 1927 and now in London for the Imperial Conference. The lunch was so that they might discuss with Shelmerdine the state of Indian Aviation against the arrival of R.101 in Karachi. Brancker reported to the S of S that Colmore had said to them over lunch that he would bring with him in R.101 'Document 16' which contained proposals and calculations upon the setting up of the additional mooring masts at Malta and Baghdad needed for regular commercial airship services. 'We have heaps of time to go over it all with the S of S on the way to India,' he had said.†

They discussed what should be done if the atrocious weather of the past ten days should, after all, prevent R.101 leaving the next day – as seemed quite possible.

Brancker suggested that if the flight should be seriously delayed it might be better to take the delegates to the Imperial Conference for a flight in R.101 after all so that they could see for themselves the spaciousness, quietness and comfort of airship flight. It might even be a worthwhile thing to do before the Indian flight and it would offer opportunities for another trial flight or two, including a full speed test.

Thomson agreed that such a course had its merits but it would be

* Lord Thomson left £1,737 with a net personality of £1,647 worth about £24,000 in 1980 values. He had already arranged with his brother, Colonel Roger Thomson, that in the event that he should be appointed next Viceroy of India, Colonel Thomson should accompany him to Delhi as ADC.

† AIR 5/911.

no more than a last resort if the India flight was seriously postponed. He did not want to admit that airships were only fair weather craft.

'To ride the storm has always been my ambition . . .'

'We have already made the first flight to Canada,' he had said. 'Now we must do the first airship flight to India so that when we approach the Dominions for further co-operation in the Programme I can say that we have done Canada and we have done India – and now you see what airships can do.

'Anyway let us see what tomorrow's weather brings.'

With that Thomson and Brancker went off together to call on Mr J. H. Scullin, Prime Minister of Australia at Australia House. Mr Scullin wrote of this meeting:*

> I met Lord Thomson for the first time when, with Sir Sefton Brancker, he called to talk over the development of civil air communications to Australia. They were both full of enthusiasm on the subject of linking up Great Britain with Australia by airways. They undertook on their return from their journey to bring me their impressions so that we could discuss the matter further. I was deeply impressed by Lord Thomson's strong character and personal charm.

Tomorrow evening they would be on their way to India. This evening Brancker had tickets again for Noël Coward's popular, melodious and nostalgic operetta *Bitter Sweet* – and though both C.B. and Brancker had seen it before, C.B. with Marthe and Brancker with his devoted Auriol Lee, here was an opportunity to enjoy it once more before the weeks of hard endeavour which seemed to be ahead.

Marthe was in Roumania and Auriol in New York and there is no record of who made up the party that evening. But there can be no doubt that the old friends, C.B. and Brancker, would have taken much delight in congenial company and at an after-threatre supper at their favourite Cafe Royal.

And no doubt too – there echoed in their ears that farewell melody which brought with it memories of those two who were far away:

> This sweet memory
> Across the years will come to me
> Though the end is drawing nigh
> I shall love you till I die
> In my heart will ever lie
> Just the echo of a sigh
> Good-bye.

* *The Times*, 6th October 1930.

PART TWO

THE INDIA TRAIL

4th/5th October 1930

CHAPTER TWENTY-TWO

Departure

'It is never any good dwelling on good-byes. It is not the being together that they prolong. it is the parting.'

ELIZABETH ASQUITH BIBESCO.

Saturday, 4th October 1930. The last day of Summer Time – and of more besides.

All the chief characters in the airship saga were to come together at Cardington that evening. All of them knew that the critical moment had arrived at last. A successful flight to India and back at the time of the Imperial Conference, carrying the Secretary of State for Air and the Director of Civil Aviation, would set the seal on the forward programme and on the future of airships for long-range travel.

Failure was unthinkable.

With confidence in the outcome went keyed-up feelings about the importance of the flight. At the same time there were certainly no illusions. This was going to be substantially the most difficult flight which any British airship had ever been called on to undertake. Flying over water was one thing. Flying over land; above mountains and across deserts was something quite different.

As with all pioneering, long-distance, flights in new aircraft, there was an understanding among the best informed, of the uncertainties and of a measure of risk. But that had to be accepted as part of the business of aviation. Everything possible had been done to prepare for all foreseeable eventualities. Now the time had come to put it all to test. They did so cheerfully.

R.101 on its first flight outside British air space was now to be set the task of four, long, non-stop journeys to and from mooring towers in Egypt and at Karachi, both hitherto untried.

The first sector across France and down the Mediterranean to Ismailia – some 2,235 nautical miles – would take perhaps forty-eight hours. The second sector by way of Baghdad and the Persian Gulf – another 2,125 nautical miles – might take another forty-six hours. Then, with lift reduced by the high temperatures at Karachi, the return journey would take somewhat longer against the prevailing winds. But, by then, the airship and crew would have found themselves and everyone would be looking for a triumphant arrival back at Cardington.

There was no doubt in anyone's mind that R.101 was now better prepared and flying to higher standards than any British airship ever before. Not only had R.101 now more gross lift than any other airship in the world but the four main defects which had shown up on the earlier flights had all been studiously rectified.

There was now sufficient disposable lift – 49.36 tons – to accommodate, even out of Karachi, a full service load of 17 tons together with sufficient fuel – 31 tons – to make possible the flight with adequate reserves in even the most adverse weather likely to be encountered between October and January on the critical Karachi–Ismailia leg.

The power-plant installation – 'the machinery' as it was called in airship terms – previously limited to only four engines available for forwards flight, was now back to the original specification with five engines for running ahead and, now, two for reverse. In addition, the aluminium coolant-pipes, which had caused problems early on, were now all replaced by copper pipes and no further trouble was expected.

The gas-bags, which had been such a worry after the restraining wires were let out, had now all been repaired and reinforced, while a new and more effective form of padding, with cotton bandages around all projecting points of the structure, was now confidently believed to have solved the chafing problem. And, after all, every German rigid ship had used padding for years without difficulties and the new padding for R.101 was of the highest standard yet devised.

Evidence that the work on the gas-bags had been effective was shown by the fact that, in the sixty-four days between 24th July, when the work was completed, and 26th September, when the ship was ready for service, the loss of gas in the shed had amounted to an average of only 14,272 cubic feet a day which was almost exactly the theoretical normal seepage rate to be expected.

This low rate of loss of gas compared with the average of 79,982 cubic feet a day lost by R.100 during the same period, and with the 51,071 cubic feet a day which had been experienced by R.101A in the shed during October 1929.

Work on the other main problem area, the outer cover, was thought to be equally satisfactory. The whole of the original, predoped, portion of the cover had been renewed – that is some 214,000 square feet out of the total of 247,494 square feet. The remaining two major portions of the original outer cover – portions which had been doped *in situ* – had been reinforced with doped-on bands and some additional patches over fixing points. Air-inflated mattresses were located inside wherever there had been areas of flapping. On the trial flight the cover had been carefully inspected and reported to be 'very good'.

Thus, everyone believed that their confidence was well-founded. Of course, still more time would have been useful – but then one could go on for ever and now, with so much accomplished and the Imperial Conference in session, now indeed was the time to show the flag.

So, to join together in doing just that on the way to India, the three main streams of the Airship Service now converged on the mooring tower at Cardington on that Saturday afternoon; each of them with its own atmosphere and ambience, its own background and ambitions, its own ideas of what it was all about.

There were, first, the originators of the whole affair: the Air Ministry team from Whitehall and Kingsway, led by the Secretary of State for Air, and, among them, Sir John Higgins now back from Canada ready to take up his new duties in industry.

The men from the Ministry saw the airships as worthy instruments of a long-term policy towards greatly improved Empire communications, leading on into an expansive future for Lighter-than-Air as part of a national plan for an 'All-Red-Route' linking British territories around the world. So, the flight to India was seen as 'the end of the beginning' of a new era for British airships and, as Maurice Dean from AMSR's office at the Air Ministry recorded: *

> No one connected with airships had the slightest doubt that the final flight would be a tremendous success. No one, that is to say, from the Secretary of State downwards. I, for example, closely connected as I was with these events, had no doubts, fears or hesitations whatsoever. I knew all too well that the weights had come out too high and, therefore, the chances of my being offered the trip as a passenger were non-existent. But, had I been offered the trip I would have gone with no more doubts or hesitations than troubled me when I climbed on board the R.100 for its endurance flight of 54 hours.

At Cardington, the team of designers, works staff, craftsmen and administrators of the Royal Airship Works – 755 of them in all –

* Sir Maurice Dean, KCB, KCMG, writing in *The Air Force Department Society Journal of May 1966* on 'The Loss of the Airship R.101'.

looked on R.101 as the present centre of their world; the symbol of future progress and security. They were all dedicated to airships, earnest, hard-working believers in Lighter-than-Air, immersed in detail, eager to achieve the success on which the future depended.

Third, and apart from the other two groups as a separate breed, were the eight officers and ninety-eight men who made up the flying staffs of the two airships – all of them reared in the Royal Navy or from the Merchant Service. As yet there had not been enough flying to create the camaraderie of good ships' companies but none doubted that that would come for R.101 on the voyage to India and back.

True, there were some hard cases among the crew and recent evidence of this had surfaced just after R.100 returned from Canada when a relief watch from among R.101's crew had gone aboard to take charge of R.100 at the mast. Noël Atherstone noted, on 17th August, somewhat frigidly, in stout Royal Navy style:

> I was informed by Booth that three of the relief crew had been discovered lying drunk in the crew quarters on board and that several bottles of liquor were missing as well as private articles from officers' cabins. Apparently some of the relief crew had broken open various lockers and had gone on the jag during the night. There is going to be an inquiry about this.
>
> While making no excuse for what took place,* I can't help feeling that the initial mistake lay with Booth in allowing all his officers and crew to leave the ship the moment she was moored. There was no necessity for it and I was very surprised that he should have allowed it to happen. Not my idea of a ship's captain.

Next day he added:

> The enquiry has commenced under the chairmanship of Cave and, judging from what I've heard it will take about a week and produce nothing. In the meantime, the three men concerned have been stood off, one with a very sore face the result of being knocked down in the shed yesterday morning by Johnston for insolence. Very Merchant Service but rather unnecessary, especially ashore!

Both Atherstone and Johnston were strict disciplinarians in their different ways and there is no doubt that Atherstone was much liked and respected by the crew and by his superiors alike. As the up and

* Later, during the R.101 Inquiry, a number of letters were received by Sir John Simon from people in Bedford alleging drunkenness among the crew. There is, however, no evidence that – with one possible exception – this was so on 4th October.

coming man on the flying side, he had already been designated at the Air Ministry as the future captain of R.101 when Irwin took over the new ship R.102.

Like all professional flying staff, before and since, the officers and crewmen of R.100 and R.101 regarded their craft as wholly their own and their handling as entirely their professional prerogative once the ships had been handed over to them from the Works. In their views everyone else was ancillary.

Then – as now – communications between the different groups, administrative headquarters, design and manufacturing staffs, and the flying staff were beset with misunderstandings, united as they were by a common purpose. These problems were partly because of the remoteness brought about by geographical separation, and partly because of the widely different backgrounds, outlook and responsibilities of the different groups.

Whitehall was remote from Cardington in mind as well as in presence. Design staffs, works staffs and flying staffs have, traditionally, each found something of a mental gulf between each of them. So it was with the Airship Service.

Colmore and Scott could have done a good deal to bridge these gaps and to achieve a coalition; the more so because so relatively few people were involved. Colmore, for all his responsibilities and talents, was shy and diffident. Scott was dogmatic, aloof and ill, reluctant to delegate to the Captains of the ships. So, in the time available, a closer accord did not happen.

In the course of time, and in the close association which a series of long flights would have brought about, there can be no doubt that many of these human problems would have been resolved. But, on 4th October 1930, less than a year had passed since R.101's first flight and out of that time the airship had spent nine months in the shed and only fourteen days in the air.

Now, all three disparate groups were to come together to join in the flight to India.

Reginald Colmore telephoned Louis Reynolds on Saturday morning that, in the light of the weather forecast and after consultation with Major Scott the start for India had been fixed for 7 p.m. BST that evening. He asked Reynolds to arrange for the Secretary of State to be at Cardington at 6 p.m. Reynolds suggested that 'an hour was a long time to keep the S of S waiting, and Colmore agreed that 6.15 would do'.*

* AIR 5/904-311.

That morning Squadron Leader Sydney Nixon, Assistant Director (Organisation) at the Royal Airship Works, circulated Flight Instruction No 32* with orders to the 'Captain R.101' for 'a demonstration flight to India via Ismailia and return'.

The instruction stated that position signals must be sent every six hours and listed the French and Mediterranean 'Prohibited Areas' to be avoided. 'The French are very touchy about their various Prohibited Areas and it is essential to conform with the French regulations which, in the neighbourhood of Toulon are particularly stringent.'

A camera would be carried to be operated by Richard Blake, an engineer allocated to No 4 Engine Car, but no photographs must be taken while flying over European countries; suggested subjects included the approach to and departure from Ismailia and Karachi.

Luggage would be limited to 30 lb for each passenger and 15 lb for each member of the crew.

The Royal Air Force Ensign would be flown at the stern.

At midday, Scott, Johnston and Steff went off for a convivial lunch at the Bridge Hotel† in Bedford. They returned in a cheerful state, in time for Scott to conduct Sir John Higgins over the ship at the mast – Sir John was making a private visit to see R.101 in its lengthened condition.

> I took the opportunity to ask Scott how the trial flight had gone off [wrote Higgins]‡ and whether everything was satisfactory? He told me that everyone had been greatly pleased with the ship as altered and that the trials were, in fact, better than those it had done before being lengthened. She answered her helm better.
>
> I afterwards saw Wing Commander Colmore who also told me that everyone was delighted with the behaviour of the ship at the trial flight and that nothing had gone wrong excepting for the failure of the oil cooler on one of the forward engines.

While Sir John was with Colmore in the Administrative Block, Scott joined Irwin in a final check over the ship.

'Nice to see old Josh again,' he remarked. 'We are going to miss him here. Dowding really knows damn-all about airships – and he's not much interested I think.'¶

Stores were coming on board. Arthur Finch, the RAW's Chief Storekeeper, was checking the final items.

* AIR 5/904–X/M.03109.
† Now The Angler's Rest.
‡ AIR 5/921–X.M 08109.
¶ Reported by Arthur Finch, Chief Storekeeper, Cardington.

On the lower deck, in the chart room above the control cabin, Irwin and Atherstone had their heads together working on the airship's lift and load capacity against the complex variables of temperature, atmospheric density, gas purity and the percentage fullness of the gas-bags, upon all of which Lighter-than-Air craft were so dependent.

Noël Atherstone had recorded in his diary on the previous evening that all Thursday and Friday had been needed 'to get everything on the top line'.

> A reserve lubricating oil tank was put in today to hold 112 gallons of lubricating oil and a spare oil cooler is also to be carried! One of the emergency ballast bags was found to be defective and had to be renewed and all the others carefully examined. Also the gas-bags with low purities were purged through and altogether the ship was given a proper look over.
>
> The weather conditions appear to be pretty good with not much wind about. I think we should be able to get away with about 28 tons of fuel on board which should give us nearly 100 per cent reserve.

Now, on the following afternoon a detailed Load Sheet was being drawn up in the light of the latest facts.

The Service Load (equipment, food and water for fifty-four people in all; the weight of officers and crew and their baggage; plus lubricating oil for the engines, but excluding water ballast) had been calculated at 8.6 tons. The dozen passengers and their baggage – estimated at 190 lb each – added a shade more than another ton over and above the Service Load.

Cruising at 50 knots on five engines, each at 750 rpm, to deliver a total of 1,800 bhp for a consumption of 760 lb of fuel an hour, Atherstone calculated that 22 tons of fuel would be needed for the voyage to Egypt against an average headwind of 15 knots. That was the worst condition contemplated; a headwind which would extend the time of flight to almost 64 hours. A further three tons of 'contingency fuel' in reserve would provide for another 12 hours duration at 45 knots and provide an additional 540 nautical miles of still air range. That 25 tons was three tons short of Atherstone's expectation on the previous evening but it would give an adequate margin in any likely wind.

The forecast meteorological conditions at take-off were for a temperature of 55 degrees F, a barometric pressure of 1,012 millibars and a density of 1,222 grammes per cubic metre. At a gas purity of

96.5 per cent, these conditions would give a specific lift of 68.5 lb per 1,000 cubic feet of hydrogen – as read off from the Lift Charts at Cardington. The bags were inflated to 96 per cent of their capacity – equivalent to a pressure height of 1,200 feet before gas would need to be valved. Thus the estimated gross lift came to 161.752 tons or 362,324 lb.

That would leave some 8.75 tons of lift available for water ballast to reach equilibrium at the mast at a disposable lift of 45.84 tons – which would be about 2½ tons less than that of R.100 on its departure for Canada.

The disposable lift included the 2,630 square feet of blue Axminster carpet laid along the entrance gangway and in the lounge, at the request of the Secretary of State. That fitted carpet weighed 1,052 lb together with 8,076 lb of furniture, fittings and movable equipment.*

Scott, Irwin and Atherstone went over the details together. Compared with the trial flight of three days before the airship would be carrying an additional 2.92 tons of disposable load including 2.2 more tons of fuel. But Scott was not satisfied.

'I'm looking for another four tons,' he said to Finch as the Chief Storekeeper checked the food coming on board. 'We need four tons more disposable: three tons for fuel and one for ballast.'

Just then two of the three stewards, Hodnett and Graham, passed, loaded down with tins of biscuits.

Scott stopped them.

'We don't need those tins,' he said. 'Turn out the biscuits. Put them in paper bags and get the tins off the ship. I want to save every pound of weight I can.'†

He looked into the lounge. The sky-blue carpet on the floor, the aviation pictures on the walls and the fabric-covered wicker chairs and settees now arranged for thirty people at a time, gave it an air of solidity and comfort not seen in any aircraft before.

Scott walked up the carpeted walkway to the bows, off the ship and down the lift from the mast-head to the ground to join the small group of staff, relations and friends who were assembling there. Among them was an old friend, F. A. de V. Robertson of the editorial staff of *Flight*. They shook hands and Robertson asked him how things were looking for the flight?

'Fine,' said Scott. 'The ship is really good now. I'm sure we'll have

* Appendix 5B, 'R.101 Equipment weights' (Pages 477–8).

† AIR 5/903.

Ready for off – R.101C swings at the Cardington mast with the lift half-way between the ground and the boarding gallery just before dusk fell on the evening of 4th October 1930. All five engines are ready to start and fuel and stores are now all stowed – the 12 passengers and 42 crew members going aboard. [PLATE XXIII]

The Great Ship – R.101c floating at the mooring tower at Cardington on the afternoon of Saturday, 4th October 1930, taking aboard fuel, gas and stores before the airship's departure for India. The additional bay, 45 feet long, can just be distinguished aft of the passenger windows and ahead of the midships engine cars. The original sections of the outer cover not renewed when the rest of the cover was replaced in August 1930, are those containing the letters 'R.101'.

Dining Aloft – A section of R.101's dining saloon laid with Royal Airship Works crested cutlery and table-ware and set for 24 passengers to dine at a sitting, cooked by E.A. Graham and served by Stewards Savidge and Hodnett. [PLATE XXIV]

a splendid flight – and we'll see you back here just two weeks from now.'*

Robertson observed that Scott was in uniform.

Scott, defensively, was at pains to emphasise that he was not a passenger on R.101, nor had he been a passenger on R.100 during its voyage to Canada and back.

'I am officer-in-command of the flight,' he said, 'and that is why I am wearing uniform. I decide such matters as when the ship shall depart, her course, her speed and the height at which she will fly. The captain of the ship commands the crew and he is responsible for discipline – and responsible, too, for carrying out my orders.'

This was not what had been laid down – and Major Robertson was concerned to notice that Scott's speech was not clear. Sadly he reflected that his friend was not the man he used to be. Tom Cave-Browne-Cave commented:† 'Scott's failure to deal with problems adequately showed how greatly his effective understanding and determination had deteriorated.'

At 0930 that morning Scott had been present, with Irwin and Giblett, the Chief Meteorological Officer, at a weather conference in the Forecast Room in the Meteorological Office at Cardington. They went over the weather charts together. A shallow depression with its centre on the North-East Coast near Tynemouth was receding into the North Sea but an associated cold front was moving eastwards over France.

By night, the winds over Northern France were thought likely to be of moderate strength from the west to south-west with fair weather extending over Southern France and the Mediterranean. It looked all right.

Now Scott drove up back to the Administrative Block with Robertson. At the door they were met by Colmore, Nixon and Giblett, who had the latest weather report covering the period up to 0100 hrs on Sunday morning. It was rather less favourable.

A trough of low pressure off Western Ireland was spreading in quickly with winds from the south-west of 10 to 15 miles an hour freshening later to 20 to 30 miles an hour at 2,000 feet over Northern France, lighter in Southern France. Weather would be fair at first with rain later.

* Major F. A. de V. Robertson in *Flight*, Page 1107, 10th October 1930, and in conversation with the author, June 1938.

† In a private memorandum held by Mrs Cave-Browne-Cave.

Scott read the sheet. 'The glass is falling,' he said. 'I want to get all stores and passengers on board as early as possible so that we can get away before the rain starts and in case the wind does get up a bit.'

He went into his office to pick up his bags.

Robertson turned to another old friend, Reginald Colmore. 'As always,' wrote Robertson, 'DAD was imperturbably charming, calmly summing up the situation.'

He said to *Flight*'s representative that he was thoroughly satisfied with the forecast conditions along the route on the outwards flight; the weather seemed 'reasonably favourable' and there was a large anti-cyclone spread over the Mediterranean. As to the airship, everything was well with it.

He thought it probable that they would pass over the French coast east of Rouen and R.101's course would then lie between Rouen and Paris before turning south-west towards Tours to miss the Rhône Valley.

> He was in an entirely different mood from the Canadian flight [wrote Nixon] happy and contented, with not the slightest trace of anxiety.
>
> I can say with every confidence that, had he thought that there was any risk in starting out for India after such a short trial, he would not have advised leaving before further trials were carried out.*

By now, the first of the passengers were beginning to arrive; driving in through the main gates in front of the Administrative Block and then on down the hill, past the two great airships' sheds to the mooring tower beyond.

That morning, Sefton Brancker had driven from the Royal Air Force Club in Piccadilly, up the Edgware Road to the de Havilland aerodrome at Stag Lane, Burnt Oak. There he was met by Winifred Spooner, the leading British airwoman and the first professional lady pilot of her day. He climbed iinto the front seat of a D.H. Gipsy Moth to be flown by Miss Spooner for lunch with the Norfolk and Norwich Aero Club at Mousehold Aerodrome, two miles north-east of Norwich.

After a convivial Aero Club lunch, rounded off by one of Sefton Brancker's off-the-cuff speeches, in the course of which he promised to come to the Club's annual dinner and dance at Norwich in December, he and Miss Spooner climbed back into the Moth to fit in yet

* Statement by Squadron Leader Sidney Nixon to the Treasury Solicitor; 21st October 1930.

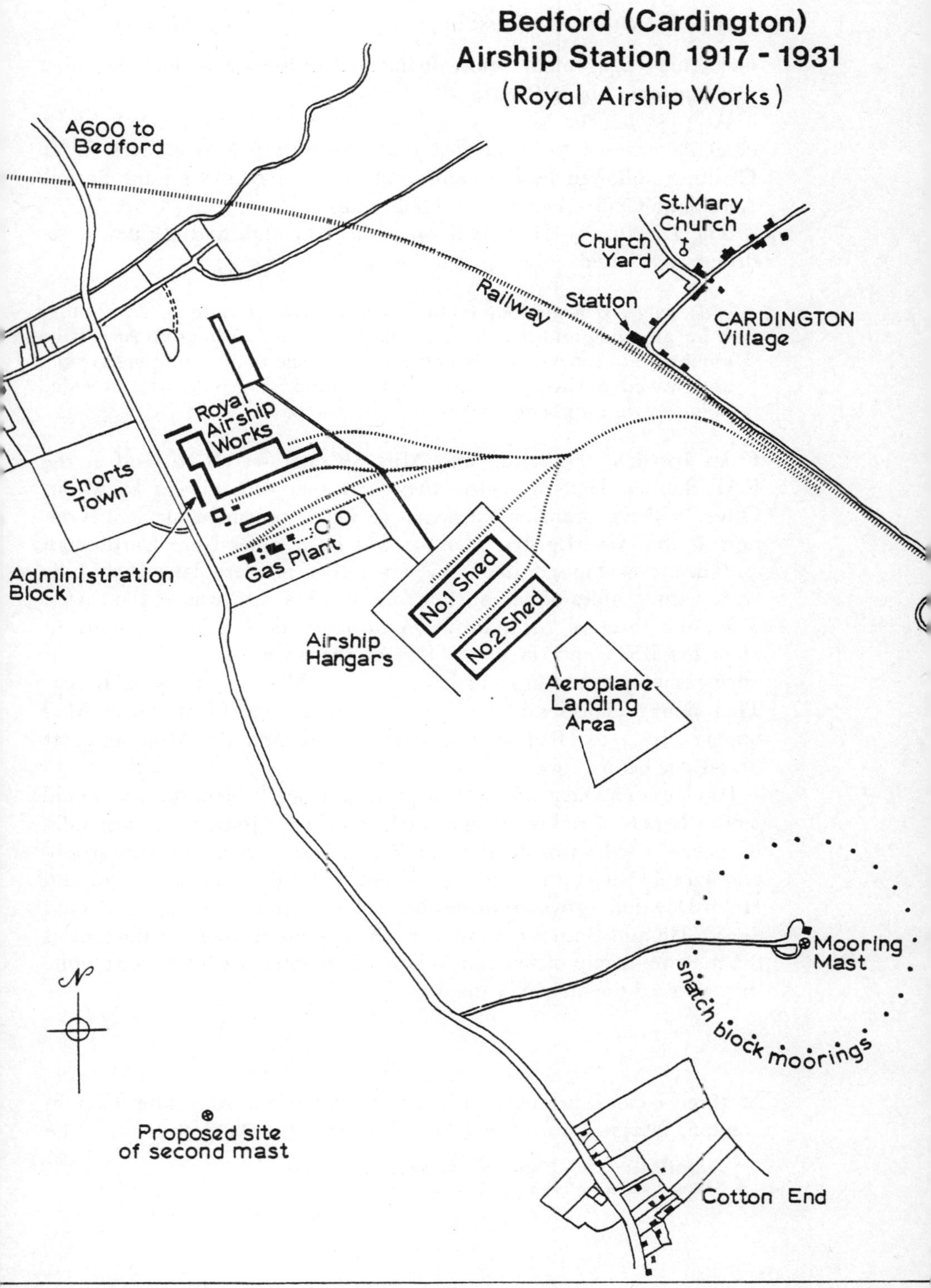

The Royal Airship Works: The layout of the airship construction works, the two great sheds and the mooring mast at Cardington.

one other engagement before flying to Cardington to join the other passengers on board R.101.

At 1445 hrs the Moth landed at the new 144-acre Ipswich Municipal Aerodrome to be opened that day by the Mayor, Mrs. A. L. Clouting, followed by the naming of two new Moths for the Suffolk Aeroplane Club: 'Pegasus' and 'Chrysalis'.

The Director of Civil Aviation wished the club and the new aerodrome well – and

> spoke eagerly of the long flight to India saying that he would not miss it for anything and that he must not be late at Cardington. At 3.35 he climbed into the cockpit, removed his monocle to put on flying goggles and waved a cheery farewell to the hundred people who crowded about to wish him *bon voyage* to India and back.*

From Ipswich, Sir Sefton and Miss Spooner set off to land at the RAF Station, Henlow, where they were met with his car by Major Oliver Villiers, Brancker's assistant in the Department of Civil Aviation at the Air Ministry. Henlow was eight miles from Cardington and Brancker himself had laid down that no aeroplane should fly within three miles of the airship base while a ship was at the tower.

So, the three of them drove in through the Cardington gates at 1630 hrs BST, and on to the foot of the tower. There Brancker at once gained permission from Irwin to show Miss Spooner over R.101. That done, Villiers said his good-byes to his chief and drove Miss Spooner back to Henlow so that she could land the Moth back at Stag Lane before dark.

Having seen them off, with a promise from Villiers that he would send a bunch of red roses with his love to Amy Johnson in hospital,† Brancker posed with members of R.101's crew for a cheerful group-photograph before they went on board. Now the time was 1730 and Hugh Dowding drove up to see the airship off and to bid them all God Speed. He and Brancker went in to the operations room at the foot of the mast for a cup of tea and to look at the latest weather chart while they awaited the arrival of the Secretary of State.

*

At three o'clock precisely on that fine, grey Saturday afternoon in London, Major Louis Reynolds, Principal Private Secretary to the

* *East Anglian Daily Times*, 6th October 1930 (Page 7).

† She was suffering from nervous exhaustion.

S of S had driven up to Block 9, Ashley Gardens in the Air Minister's blue Daimler accompanied by a green Trojan van for the luggage.

Reynolds – at 6 feet 6 inches the tallest man in the Air Ministry – climbed out of the car and, carrying a red Ministerial box, went in through the glass porch and up in the venerable lift to Flat 122 on the third floor.

Lord Thomson himself answered the ring at the solid, white-painted door – elegant in his dark suit and the grey tie which matched the hair at his temples; the one indication of his fifty-five years.

'Come in, Reynolds, come in,' he said. 'Delighted to see you. Everything's ready but we'll need all hands to get the luggage down.' *

One by one, with the help of the two drivers, James Buck, Lord Thomson's young valet, Louis Reynolds and the Secretary of State himself, the two green cabin trunks, four suitcases, two cases of champagne for the State Dinner at Ismailia, and a dress sword, were all piled into the lift and into the waiting van.

Last of all came a mackintosh-covered roll, fully ten feet long.

'My precious carpet from Sulimaniya,' said C.B., by way of explanation. 'We'll have it down for the dinners at Ismailia and Karachi to do the thing in style.'

In fact the carpet roll was too long to fit in the lift. In the end it had to be manhandled down the stairs and into the van.

Reynolds, without his bowler hat, displaying the completely bald dome which had earned him in Whitehall the affectionate name of 'Curley', handed the red box to the S of S while the last items were being stowed.

'You'll find the latest weather report there, Minister,' he said. 'Not too bad, I think. Improving over Southern France and clear over the Mediterranean. It should be a good flight.'

'Fine,' said Thomson. 'What else is in the box? We'll have a look at it in the car.'

'There's a typewritten draft of your follow-up Paper for the Conference which you want to work on,' said Reynolds. 'And your speech for Karachi – oh, and a cable from Cairo about the dinner on Monday evening. It confirms that all ten guests can come; including the High Commissioner.'

The Air Minister made his farewells to his small staff in his flat, Rose the housekeeper, Betty his house-parlour maid and 'Salammbô'

* AIR 5/905–X/M–08,109 and discussion with Major Reynolds in November 1943, at Gwydyr House.

the black cat, and climbed into the car with James Buck in front and Reynolds in the back beside him.

There was one more stop to be made before they took the road to the North. The Daimler drew in to the Post Office on the north side of Victoria Street within sight of Westminster Hall and still open on Saturday afternoon. Lord Thomson went in himself and handed the message across the counter.

It was written out in his own hand.

> S.A. The Princess Bibesco, Posada, Par Comarmic Prahova, Roumania.
> Off to India at last in R.101. Stop. My thoughts as ever with you. Stop. Our talisman flies too. Stop. All my love and blessings.
> Kit.

Back in the car the little procession set off again; past Westminster Abbey, up Whitehall and past Gwydyr House. On through Trafalgar Square, into Piccadilly and up Regent Street to Portland Place and Regents Park. Thence past Lord's Cricket Ground and Swiss Cottage, through Whetstone and Barnet to Potters Bar.

The Secretary of State was deep in his Red Box. There was a message from the Prime Minister. Reynolds made notes for a reply. There was a Good Luck note from Laurence Pritchard, Secretary of the Royal Aeronautical Society.

They spread out the weather report. It had been received at the Air Ministry, Kingsway, at 1220 that morning. It read:

> *Inference: Cardington to Ismailia*
> Anticyclone over the Balkans. Depression Eastern Atlantic spreading east. Occluded front running from shallow Low centred near Tynemouth to South of France moving east. Winds tonight over France are likely to be from west or south-west of moderate strength over Northern France, light over Northern France. Weather mainly cloudy.
>
> In Western Mediterranean light and variable or westerly winds are possible tomorrow with fair weather.
>
> Central Mediterranean light easterly or variable winds; weather fair.
>
> Eastern Mediterranean winds northerly or north-easterly, Athens to Crete, lighter towards Egypt. Weather mainly fair, perhaps local thunderstorms.

'That seems all right,' said Thomson with, perhaps, a touch of relief. He looked out. 'Certainly, the weather here seems to me to be just about perfect for airships; practically no wind and the sky overcast so I suppose the temperature will be stable. Ah well; this time tomorrow, Reynolds, we ought to be coming up to Malta, I suppose. I'm sorry you're not coming too, old chap.'

'I'm sorry, too,' replied Reynolds. 'But someone has to stay behind to mind the shop – and, of course, I'll keep you posted with any matters of moment from day to day.' *

The Ministerial car and its accompanying van sped on through Welwyn, Hitchen and Henlow to Shefford. As they came into the village, Lord Thomson glanced at his watch.

'Five twenty-five,' he said. 'We're a bit ahead of time. We don't want to get there too early or we'll be in the way. Just time for a cup of tea, I think. I could do with one too.'

The Daimler drew up at the little White Hart Inn at Shefford, eight miles from Cardington. A large sign stood out even in the fading light. 'Teas with Hovis,' it read.

'Come in, all of you,' called the Secretary of State. 'Bring the van driver, Buck,' he added. 'I'm sure he'll be glad of a hot cup of tea like the rest of us. And I can say "thank you" for all your help.'

They all trooped in.

Forty-five minutes later as the car breasted the top of Hammer Hill and the road to Bedford wound downwards again, Christopher Thomson called for the car to stop. He alighted and walked across the road and up the bank on the right to gaze down on the level plain, 200 feet below.

There at the mooring tower – and seeming enormous even at this two miles distance – the silver form of R.101, the largest vessel in the world, floated almost broadside-on in the rising, south-westerly wind.

Here was the vast and elegant outcome of six years of vision and endeavour; a magic carpet to link this green countryside with Egypt and India and with all which might lie ahead.

*

So we come to Cardington at just before six o'clock on that Saturday afternoon.

R.101 rides at the mast; nose to the increasingly blustery wind from the south-west. There is rain about. The occasional flurry portends a downpour. The grey and overcast afternoon is fading into an early dusk. The calm air and clear skies of the trial flight three days before, have gone.

* Major L. G. S. Reynolds made a detailed note of the journey from Westminster to Cardington.

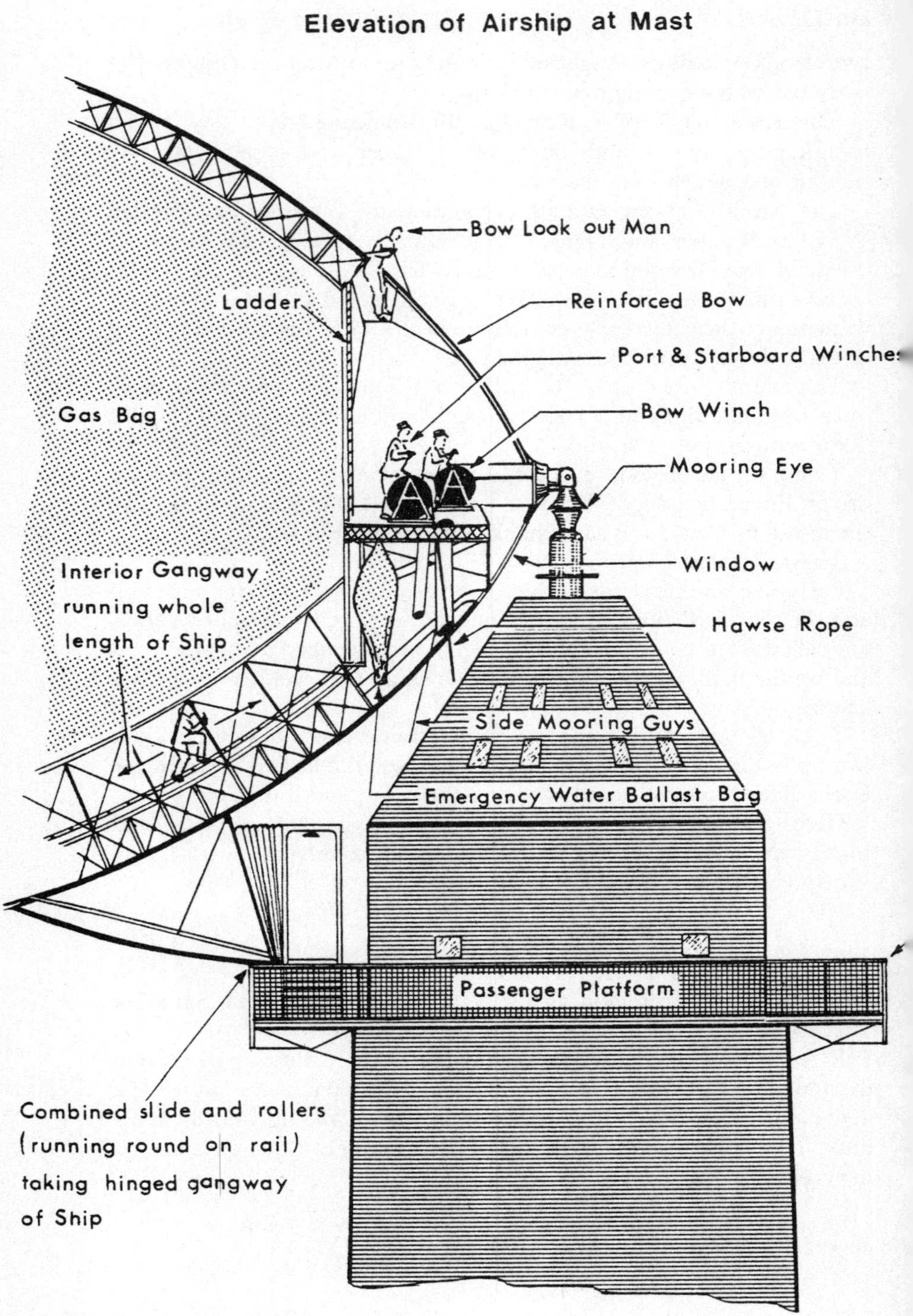

Mooring System: A diametric drawing of the nose of R.101 secured to the mooring eye at the top of the mooring tower at Cardington.

Everything at Cardington on this autumnal evening is dominated by the airship. At close range it seems vast. Locked by the nose to the top of the tower, 170 feet above the ground, it moves restlessly with every gust, swinging like a massive weather-vane from the lighted mast-head. With its length of 777 feet, its girth of 420 feet and with a surface area of almost six acres, the airship now looms dark against the sky while a cheerful radiance shines from the windows in the hull and from the little control-car underneath.

A red light from the top of the tower casts a curious ruddy glow over a section of the nose.

All is bustle and activity around the tower as the last loads go up in the lift. In twos and threes those who will fly – passengers and crew – have been ascending to the gallery at the mast-head and down the carpeted walkway into the ship. Bag and baggage they have gone after their last farewells, helped by little Herbert 'Joe' Mann, the RAW's lift attendant, always ready to give a hand with a bag or to bring a letter down from the ship.

Soon they must be all aboard; five officers, thirty-seven crew members, six officials from the Royal Airship Works with Reginald Colmore at their head, and the six passengers led by Lord Thomson himself. Fifty-four in all; as eager and cheerful a band as ever embarked upon a pioneering voyage.

Joe Mann has come up to the mast-head for what must be the hundredth time that day. With him now in the lift are Michael Rope and Tom Cave-Browne-Cave. Michael Rope to fly on the voyage – a man who, in the view of his colleague, T. S. D. Collins, 'knew more about R.101 and about the mechanics of handling an airship than anyone else in Britain.* Cave-Browne-Cave would stay behind to press on with modifications to the spare Tornado engine at Cardington ready for R.101's return. Then it would be installed to run, experimentally, on hydrogen-gas as well as on oil. If successful that would be a break-through in the saving in fuel weight with tremendous prospects for the future.

Now, at just before 6.15, the crowd around the mast, growing every minute, looks expectantly towards the sheds. Past them and across the aerodrome can be seen the stately approach of the angular Daimler of the Secretary of State for Air, followed by the, even more upright, Air Ministry Trojan van. They come to a stop alongside the wooden buildings at the foot of the tower.

Ten minutes earlier, the Ministerial car had stopped at the entrance

* AIR 5–902–86990 Page 967.

to the track from the Shefford–Cardington road leading to the tower and Reynolds had enquired of the gatekeeper where Reginald Colmore was to be found? The gateman directed them to the Administration Block but, arrived there, Sydney Nixon reported that Colmore had just left to go on board the ship. 'Jump in then,' called the Secretary of State and, on the way down, past the sheds, enquired the latest news of the weather, the ship and the flight.

They were met at the tower by Higgins and Dowding. Beside them were Brancker, Richmond and Johnston with Herbert Mann, the RAW liftman, waiting to take them up to board the ship.

Christopher Thomson shook hands all round and the three who were flying with him posed for a final photograph; Louis Reynolds towering in the middle. Hugh Dowding remarked 'Everyone seemed very cheerful.' *

Meanwhile the Ministerial luggage was being loaded into the lift.

As the flying party turned to go, a note was handed to Lord Thomson. It was from Sir Dennistoun Burney, wishing him good luck on the momentous flight to India which had been the central point of 'The Burney Plan' first mooted eight years before. It was a generous gesture, written from hospital where Burney was recovering from a minor operation. Typically Thomson responded. He went into the operations office and scribbled a hasty, friendly, note to Burney, thanking him, wishing him well and agreeing that this was indeed a great day at last when, for the first time ever, an airship would fly 'down the trail of Marco Polo to the East'.

It was delivered on the following Monday.

Ten minutes earlier, at 6.16 p.m. local time, Johnston had telephoned Eric Mayers, in the Cardington Met Office, from the foot of the mast to ask for a report on the present wind conditions at various relevant heights, available and recorded from the release of small met balloons.

The wind was clearly getting up.

Mayers reported : †

At 150 feet	24 mph from 235 degrees.
At 1,000 feet	23 mph from 233 degrees.
At 2,000 feet	25 mph from 243 degrees.

Certainly the wind was strengthening from the south-west. It would present R.101 with a substantial head component on the first stages of

* AIR 5/903–X/M.08,109.
† AIR 5/903.

the flight along its projected track of some 160 degrees true.

Now the Secretary of State came out of the operations room.

'What's our route this evening?' he asked of the Chief Navigator.

'As a start, Minister,' said Johnston, 'I've plotted a course to London, then across Kent to Hastings, over the Channel to pass just north of Paris and then west of the Rhône Valley to Toulouse and on to the sea at Narbonne. After that we'll see but it ought to be plain sailing down the Mediterranean past Malta to Ismailia. We'll have a bit of a head wind to begin with but it'll die down as we get south.'

'Well, Reynolds,' said the Secretary of State, 'It's time I said good-bye and went up on board. Thank you for coming out with me.'

They shook hands.

'Good luck. *Bon voyage*, Minister,' said Reynolds.

'Thank you,' replied Thomson. 'As you know, this is the voyage for which I have been waiting all my life.'

He stepped into the lift with Brancker.

A smile and a wave and they were gone.

*

Irwin and Atherstone had now completed the final Load Sheet and had passed a Ballast and Fuel Report back to the ground.

It read:*

	Gallons	*Tons*
Fuel oil	6,724	25.3951
Lubricating oil	250	1.0045
Petrol (starting motors)	65	0.2147
Drinking water	250	1.1161
Water ballast	2,072	9.2500
Total	9,361	36.9804

The fuel load was carried in thirty out of thirty-four tanks, each with a nominal capacity of 224 gallons. Of the total, ten tons could be jettisoned in an emergency; two tons from Frame 3 in the nose; two tons from Frame 5 forward and two tons from Frame 10 aft – all quickly releasable from the control car, while four tons could be jettisoned in emergency from the new Frame 8A by local control.

The weighed weight of baggage, checked aboard by Arthur Finch

* AIR 5/903 XM–08,109.

for officers, crew and passengers was reported to Irwin at 1,207 lb (or 0.5388 tons). Within it the Secretary of State's seven items amounted to 254 lb of which the carpet-roll weighed 129 lb and the sword precisely three pounds.*

The nominal payload of 12 passengers and their bags was 2,508 lb (1.12 tons).

In all, as set down on the Load Sheet,† the final calculation of Disposable Load amounted to 44.58 tons which would require a gross lift of 162.11 tons to bring the ship to equilibrium. Some ballast would have to be dropped to achieve a pressure height of 1,200 feet.

Now the lift ascends to the mast-head for the last time with the Captain of the ship, 'Bird' Irwin, and the Chief Coxswain, 'Sky' Hunt, who have been down briefly to say their farewells and for Irwin to collect and sign the final Load Sheet. At the top gallery Irwin turned to shake Mann's hand.

'Well done, Joe,' he said. 'We'll see you in about two weeks time, DV.'

A cheerful wave and a 'Cheerio; keep smiling' from Hunt and the gangway of the ship is hauled shut from inside.

'What a contrast,' observed Mann to his friend, Bill Williams, later.‡ 'The bright and cheery manner of all those who joined the ship compared with the glum faces of the spare watch who had to stay behind. You could see which was which,' he said, 'even if you didn't know them all so well.'

There were, of course, a few who had grumbled that it was not the best of evenings to be setting out. Rain is clearly blowing up and the wind is rising too. Even so, R.100 had set out for Canada in quite a breeze. And how could the flight be put off now, with all those distinguished passengers on board? What's more, airships had to show that they were not just good weather craft. Everyone at Cardington and in Shortstown were agreed that the time was overdue when R.101 should show what it could do. No doubt it would turn in a splendid flight.

*

Now, a floodlight lit up the silver underside of the ship at the mast. Red and green navigation lights glowed at the tips of the port and starboard lateral fins. The inside of the control car beneath the body was

* Appendix 5. Baggage weights taken on board R.101, 4th October 1930. (Page 479).

† See Table Page 349.

‡ Letter of 31st July 1974, from W. Williams of Bedford to the author.

R.101c LOAD SHEET

Details of Lift on Leaving the Ground; 4th October, 1930

Temperature 55°F; Pressure 1012 Millibars; Density 1,222 grammes per cubic metre.
Gas Purity 96 per cent; bags filled to 96.2 per cent.
Specific Lift 68.5 lb per 1,000 cu ft; Gross lift 161.75 tons.

		Imperial Gallons	lb.	Tons	
1.	Fuel Oil (30 tanks filled)	6,724	56,885	25.3951	1.
2.	Petrol (for four starting motors)	65	481	0.2147	2.
3.	Lubricating Oil	250	2,250	1.0045	3.
4.	Water Ballast	2,072	20,720	9.2500	4.
5.	Drinking Water	250	2,500	1.1161	5.
	Sub-Total	9,361	82,836	36.9804	
6.	Officers and Crew (42)		6,720	3.0000	6.
7.	Passengers (12)		1,920	0.8571	7.
8.	Mails and Freight		12	0.0054	8.
9.	Food		896	0.4000	9.
10.	Baggage		1,207	0.5388	10.
11.	Crews Kit and Equipment		1,840	0.8214	11.
12.	Loose Furnishings and Equipment		2,832	1.2643	12.
13.	Red Cross Box		15	0.0067	13.
14.	Miscellaneous Items (mooring, etc.)		767	0.3424	14.
15.	**Total Disposable Load**		99,854	44.5774	15.
16.	Airship Fixed Weight (on 4th October 1930)		263,276	117.5340	16.
17.	Total Lift Required		363,130	162.1114	17.
18.	Theoretical Lift Available		377,414	168.4885	18.
19.	Lift Available at 96 per cent filled for Pressure Height of 1,200 ft		362,317	161.7488	19.
20.	Estimated Ballast Release to achieve One Ton Light at Slip from Mast		3.253	1.3626	20.

brilliant with white light while a golden yellow glow shone from the long window panels in the hull.

A narrow spot-light-beam began to shine from the control car, first to one engine car then to another to check as each two-blade propeller began to turn. Soon, all but that of the forward starboard (No 1) engine car were in motion, each propeller a bright and shining disk in the floodlight, turning over at 600 rpm to warm up the engines.

For a time, the new Beverley heavy-oil starting motor, in the starboard forward engine car itself refused to start under the impulse of compressed air. Then, with it running on its generator the No 1 engine was turned over. After some eight or ten attempts, covering four or five minutes, eventually it caught with a shower of sparks. By 1820 BST all five Beardmore Tornadoes had settled down to a steady throb and growl; the two forward engines running in reverse to give backwards thrust.

The wind had now freshened from the south-west to, perhaps, 20 miles an hour; rather stronger than forecast and, unhelpfully, from a direction which would present a substantial headwind-component on the first leg to London. Scattered drops of rain were blowing in on sudden gusts while the temperature fell a degree or so.

In the ship Irwin was now clearly 'ballasting up' ready to slip. Several times a fine mist of water fell from the forward ballast-tanks – 'the drops as they cascaded down were illuminated by the floodlight from the tower and gave a fairy-like touch to the scene.*

Ralph Booth, watching from the gallery of the tower estimated that four ½-ton (112 gallon) ballast bags were emptied at Frame 1 in the bows, together with two more at Frame 6 just before the ship left the tower – a total release of three tons.† It suggested that the boarded load had come out rather more than had been expected, leaving now 6¼ tons of water ballast on board.

At 18.36 GMT, a cheer from the crowd told that the mooring cone had swung free. The nose of the airship lifted and floated clear of the mast as another ton of water ballast was released from Frame 6.‡

R.101 now drifted slowly backwards while the hull lifted above the head of the tower. Then, as the engines were opened up to 700 rpm – about half power – the airship drew away, turning and climbing

* F. A. de V. Robertson. *Flight*, 10th October 1930. Page 1109.

† AIR 5/903. X/M. 08,109.

‡ R.101C thus had 5.25 tons of water-ballast remaining at the moment of departure for India. That compares with the 3.4 tons of water-ballast remaining on board R.100 after it had slipped from the mast on its departure for Canada. R.101 was thus 1.85 tons or 4,144 lb better off.

gently into the darkness, now nearly complete. Families on the ground flashed back their acknowledgements to last signals with torches from their men.

'It struck me as extremely good handling,' wrote Robertson. 'As she rose, the shape of the hull was lost in the darkness and only the lights could be seen as a drizzling rain began to fall and, as I drove back to town as the rain got heavier so the wind got more violent.'

The deep throb of the airship's engines still sounded out of the darkness as R.101 flew low over the Shefford–Bedford road, lined with cars whose headlights made a continuous belt of light and added their horns in a last salute.

R.101 turned north over Shortstown for the traditional circuit of its home town; a routine for Bedford's ships before they set off on any voyage. Aboard in the saloon the lights had been switched off so that passengers could have a better view of the ground.

For minutes more for those around the mast the low drumming of the engines told of the start of the ship's voyage towards India and into history, bearing fifty-four men and the hopes and plaudits of a nation.

At Cardington, Noël Atherstone's diary lay open on his desk where he had made the last entry:

> Everyone is rather keyed up now as we all feel that the future of airships very largely depends on what sort of show we put up.
>
> There are very many unknown factors and I feel that that thing called 'Luck' will figure rather conspicuously in our flight.
>
> Let's hope for good luck and do our best.

On the floor above, Miss Foster, Colonel Richmond's secretary, wrote in his Design Office Diary a final sentence: 'R.101 left for India this evening at 7.36 BST.'

CHAPTER TWENTY-THREE

With Hindsight

'Hindsight is always One Hundred Per Cent'
(*Beverley Shenstone 1906–1979*)

4the October 1930

As R.101 droned out into the darkness, into the freshening wind and into the heavier rain, what were the real prospects for the voyage?

So deep and lasting a shadow has been cast, by the disaster, over all R.101's affairs that – in retrospect – the confident and unworried atmosphere of the departure and of the time which led up to it (so often misreported since) is difficult to recapture.

As already quoted, Sir Maurice Dean recalled that 'No one connected with airships had the slightest doubt that the flight would be a tremendous success.'

Sir Walter Nicholson, Permanent Secretary of the Air Ministry from 1922 to 1930 was of the same opinion. He said:

> The atmosphere was one of complete confidence. It was simply taken as an obvious matter of straightforward commonsense that this ship would, if possible, be got ready. And, of course, the Secretary of State, Lord Thomson, had all along intended to go to India in her – and, of course, he would go. It was just accepted as natural and obvious.

Sir John Simon finally set the scene:*

> For four years [he wrote] they had been preparing for this moment and, notwithstanding that the date of the Imperial Conference tended to condition the hour of their start, they believed themselves to be prepared for it. What is so impressive, on studying the details of those, last, hours before R.101 started for India, is the high courage and genuine confidence of all concerned.

The fact is that, at the time, this confidence seemed to be well-founded. Rigid airships had established an impressive record for successful long-range flights since the War. In succession, R.34, the *Los Angeles*, and

* *Report of the R.101 Inquiry*. Page 67, para. 76.

the *Graf Zeppelin* had set out with less flying hours than R.101 now had in its log books. They had all got through; they would be followed by R.101 as well.

Those three airships, and R.100 too, on its flights to Canada and back, had all had their share of good luck; more than their share, some would say. But, significantly, all their first long flights had been across oceans and during summer months. Did the Cardington team make sufficient allowances for autumn weather and for the heavy turbulence so often found over land, compared with the, smoother, conditions over the sea?

In practical terms, the letting out of R.101's gas-bag wiring and the installation of the extra bay had given the airship sufficient lift to be able to start the flight to India with a fuel load which was considered to be ample for the task. With five engines set at their 'normal cruise' regime of 810 rpm to give 450 bhp each – a total of 2,250 horse-power – the rate of consumption had been carefully established by Cave-Browne-Cave at 950 lb an hour at a true air speed of 54.2 knots with the ship in equilibrium; neither pitched up nor pitched down. With 25 tons of fuel, R.101c's endurance would then be 59 hours and the still air range just on 3,200 nautical miles. Against continuous 15 knot headwinds it would be reduced to 2,310 nautical miles.

The distance from Cardington to Ismailia by way of Paris, Toulouse, Narbonne and Malta was 2,235 nautical miles.

If, by an outside chance, a continuous headwind-component should turn out to be stronger still – say, an almost unheard-of average of 20 knots – then, by slowing down to the long-range-cruise regime, and taking a long time about the flight, the rate of fuel consumption could be cut to no more than 610 lb an hour at 750 rpm and 360 bhp from each of only four engines. The cruising speed would then be 46 knots on 1,440 horse-power. At that speed and rate of consumption the endurance would be 93 hours on the 25.4 tons of fuel uplifted. Then, against a 20 knot headwind, the ship would have a range of 2,420 nautical miles; sufficient for the Cardington–Ismailia sector with some 185 miles to spare.

That would be the straightforward and obvious view on the subject of range and duration on the amount of fuel lifted. It was certainly the view at the time.

A more sophisticated approach, based upon the experience of later years of air transport operations, suggests that a number of additional factors must be taken into account in any such assessment. Modern experience dictates that allowances need to be made for :

(i) *Reductions* in true air speed as a result of having to fly out of trim. A loss of three knots might represent a fair and prudent figure for an airship such as R.101.
(ii) *An increase* in sector-distance flown, to allow for navigational deviations en route. An addition of five per cent might be expected to be about right at the relatively low cruising speeds of airships.
(iii) *Reserve fuel* required on arrival to cover delays in mooring. The equivalent of three hours additional fuel at 'long-range-cruise' powers ought to cover all such contingencies. Such an additional duration could be extended substantially by a further cut-back in power near the mast;
(iv) *An additional* fuel allowance for higher-than-normal fuel consumption to cover engine maladjustments. A 2½ per cent increase in the rate of consumption is prudent for such an allowance;
(v) *Provision for* 'unusable fuel' trapped in tanks and fuel lines. A figure of two per cent is reasonable.

Adding all these allowances together, the additional fuel consequently required for R.101 on the Cardington–Ismailia sector, compared with the basic calculation under still-air conditions and in the long-range-cruise regime, would amount to some 3.3 tons. Under these conditions, a sector fuel load of 16.5 tons would have been needed for the flight in still air, compared with the 25.4 tons actually uplifted.

What then would have been the limiting headwind-component for R.101 on the Cardington–Ismailia sector?

The available fuel, reduced by two per cent for 'unusable fuel', would have amounted to 24.9 tons.

An allowance for a 2½ per cent excess fuel consumption in long-range-cruise would increase fuel used to 625 lb an hour.

Provision for three hours mooring-reserve in the Ismailia area, at 625 lb an hour, would reduce the available-sector-fuel en route to 24.1 tons.

A five per cent addition to the sector-distance would bring the 2,235 nautical miles to 2,347 nautical miles to be flown.

A reduction of three knots in cruising speed would bring it to 43 knots at the 1,440 bhp long-range-cruise power on four engines.

With 24.1 tons of en-route fuel available, consumption at the rate of 625 lb an hour, would result in a flight duration of 86.3 hours. To cover the required distance of 2,347 nautical miles in 86.3 hours, would demand a ground speed-made-good of 27.2 knots while flying at 43 knots through the air.

The limiting headwind-component would, thus, work out at 15.8 knots.

Analysis of the meteorological records on the UK Egypt route, for a height of 1,500 feet, shows that a headwind-component of about 15 knots is likely to be encountered on about 15 per cent of occasions – that is on one day out of seven.

On a flight which was expected to take two days, the chance of a *continuous* headwind of 15 knots or more was, thus, sufficiently remote to justify the start with 25.4 tons of fuel; even had modern allowances been applied.

The forecasts on 4th October were, however, for light winds after France – and for some time – light, helping winds. Although they had hoped for 28 tons of fuel at the start, to give increased contingency reserves, clearly the officers of R.101 on 4th October had no real concern about fuel or range on the outwards flight.

But what about gas?

On the Hendon flights in June after the gas-bag wiring had been let out, R.101B had been losing gas in flight in bumpy air at the rate of about 33,000 cubic feet – or one ton of lift – an hour. That was when the bags had been badly holed by chafing on the structure. Since then the bags had been repaired and reinforced and the padding of all points of likely contact between them and the structure had been studiously improved with stronger, thicker and softer, linen-bandages.

There is no doubt that everyone concerned believed that this new padding was effective after the nasty shock they had all had in June. Evidence that the bags appeared to be in first-rate condition at the start of the India flight had come from the very low rate of gas loss measured in the shed during September and the fact that after the final trial flight, not only did inspection of the bags show them to be in perfect condition but also, for almost the first time R.101 was substantially light when it returned to Cardington on 2nd October after some 15 hours away from the mast.

Though obviously the bags would have to be watched on the long flights ahead, there seemed to be good reason to believe that the earlier problems had now been resolved.

Thirdly, the outer cover. Although the cover of R.101 had generally behaved better than the cover of R.100 (R.101's cover was designed to billow out with air pressure, whereas R.100's cover was pulled in with tapes) the pre-doped sections had deteriorated badly. They had all been replaced by a new cotton and linen fabric, doped after lacing into position on the structure. There were, however, the two quite substantial sections of the original cover which had not been renewed because they had been doped in place before the first flight. Instead, reinforcing bands had been passed around the envelope at these sections, as well as on the fins, together with strengthening strips and patches where deterioration had been discovered in places where rubber solution had been used. On the last trial flight in good weather all parts of the outer cover had been closely inspected and reported to be completely satisfactory. But the maximum speed attained on that flight had been 47 knots on four engines; maintained for four hours.

So, compared with R.100 which had left for Canada with what had seemed to be more uncertainties about the cover, the gas-bags and the engines, the three months' of work on R.101 were believed to have brought it to a new standard of reliability and safety. A relevant fact, however, was that on both its Atlantic crossings R.100's air speed had been held well down.

Taking all this into account there can be no doubt that the Cardington team believed that they were taking no undue risks with R.101 in clearing it for India or in carrying the Secretary of State for Air and the Director of Civil Aviation.

But there was one further point : weather.

The forecast on the Saturday morning had promised tolerable conditions across France which would improve steadily beyond Narbonne. Colmore had remarked to Nixon that his chief concern was with strong vertical movements of air over the desert and along the Persian Gulf but that, after his experience with R.100 over the St Lawrence, he believed that the large gas-valves of R.101 would be adequate to cope.

So, those at the Royal Airship Works assured themselves that they had done everything possible to prepare R.101 for the long-planned flight and that they tried to foresee every eventuality.

These endeavours had included the setting up by the Air Ministry, over the course of years, of a special Meteorological Section at Cardington and the establishment of a series of weather-reporting stations down the route to Karachi; all co-ordinated by one of the Air Ministry's leading meteorological officers of the day, Maurice Giblett, who was to add his expertise to the knowledge available to the officers of R.101 aboard the airship for the flight to India.

With all these preparations and with, certainly, one of the best meteorological services in the world, weather forecasting was still an inexact science. R.101 'started with a promise of winds over the next twelve hours of 20 to 30 miles an hour at 2,000 feet; yet an hour-and-a-half after leaving she is informed that she is to meet winds of 40 to

50 mph – as she actually did – an increase of roughly 100 per cent. The fact must be brought out clearly that never, with or without the extra bay, had R.101 ever experienced weather approaching in violence anything met with during this journey'.*

Significantly, within two hours of starting for Egypt, R.101 was flying into the worst weather that it – or R.100 – had ever encountered, other than the storm ridden-out by R.101 at the mast in November 1929. The combination of high, gusting, winds, heavy rain, and low-level turbulence, all maintained for hours on end for 250 miles from Bedford to Beauvais, was possibly never exceeded in airship experience, even taking account of the substantial total of some 61,000 hours flown by German rigid airships (Zeppelins and Schutte-Lanz), compared with the less than 4,000 hours flown by British rigids and the almost 90,000 flying hours of the British non-rigids between 1914 and 1918.

On the day, and despite the weather, Herbert Scott was true to his 'press-on' reputation. In his own words, he 'decided the course, speed and height to be flown', and, setting off for India and with five engines available to R.101 for the first time for forward flight, he called for the airship to be driven at fast-cruise with each engine set at 825 rpm, to give a total of 2,375 horsepower. The resulting level speed, in trim, was 55.4 knots (63.8 mph), a good deal faster than R.101 – or R.100 – had ever flown before for any continuous period.

That relatively high cruising speed would, of course, reduce the time during which the airship would have to be exposed to the very bad weather, but, at the same time, it would impose heavier loads than had ever been borne before by the outer cover and, by way of accelerations in turbulence, by the gas-bags and their restraining wires.

Certainly, Scott did not expect the engines to let him down. One of the merits of the heavy oil engine is that it will keep plugging along at relatively low revs and at a low-rate of fuel consumption for days on end. Moreover, no doubt Scott had in mind that should there be any renewed trouble with the gas-bags, the faster the cruising speed, the shorter the time on the journey.

On 2,375 bhp at 55.2 knots R.101 would be consuming a ton of fuel every $2\frac{1}{4}$ hours and, on the forecast weather, the flight to Ismalia could be expected to take about 50 hours at an average ten knot headwind. Thus some 22 tons of fuel would be expected to be consumed on the journey. Normal seepage of gas would reduce lift by about one ton during that time. The loss of a further 21 tons of lift through gas-bag

* *Report of the R.101 Inquiry.* Cmnd, Page 106.

leakage could thus be tolerated without difficulties, because it would be wholly offset by the consumption of fuel. The officers of R.101 could thus contemplate without concern a gas-bag leakage of some 332,000 cubic feet a day (equivalent to five square inches of holes in the bags) and that would be much in excess of any expectations.

There was, however, the possibility of the bellows-type gas-valves opening if the airship rolled excessively in turbulent air. T. S. D. Collins, of the Royal Airship Works, who was fully informed on the subject, calculated that, in the worst conditions, the maximum loss of lift through valve oscillations in very rough air would be at the rate of 0.35 tons (785 lb) an hour or 2.63 tons in 7½ hours of continuous rough flight. It was not, therefore, of outstanding significance.

With hindsight, we can now see that, apart from the inherent problems of Lighter-than-Air operations – slow speed because of great bulk, variable lift from the effects of changes of temperature and atmospheric pressure and density, the absolute need to carry large quantities of water ballast, the necessity to fly low to conserve gas and the vulnerability of any very light structure of enormous surface area – both R.100 and R.101 were sound within their limitations, but with two notable points of weakness: the gas-bags and the outer covers. For R.101 the structure was sound and strong. The engines were heavy and now believed to be dependable. But how dependable were the cover and the bags? Already they had almost brought R.100 to ruin. Indeed, one may speculate on whether R.100 might have fared any better than R.101 had it been despatched instead of R.101 to India on the night of 4th October 1930?

In the light of all the circumstances, especially the storm conditions of high, gusting winds and heavy rain, the known bad state of R.100's cover and the need to fly faster to make headway against headwinds than R.100 had flown continuously before, there seems to be a strong probability that there would have been serious failures in the saturated cover and bags. How things would have transpired then would have depended upon how the airship was handled.

Nor were the German and the American airships immune. The cover of the *Graf Zeppelin* split on several occasions though it had smaller unsupported areas than had the British ships. Gas-bag leakage was a primary cause of the loss of the *Hindenburg* in 1937.

In 1930, however, the success of the *Graf Zeppelin* since 1928, in spite of some narrow squeaks, and the largely erroneous impression of successful wartime Zeppelin operations, induced a climate of con-

* AIR 5/910. 12th December 1930.

fidence about airship flying which belied the facts and was certainly unjustified.

The loss of the helium-inflated *Akron* in 1933, of the similar *Macon* in 1935 and of the hydrogen-filled *Hindenburg* in 1937 – three new airships lost out of three new airships built – provides a measure of the odds.

A major contribution to the vulnerability of the large rigids was the need to fly low, often in turbulent air. For every hundred feet of gain in height for R.101 there had to be a sacrifice of half a ton of lift. It was not a question of whether it was prudent or safe to fly low. It was accepted that it was a normal necessity of Lighter-than-Air navigation; inevitable, a peculiarity of the breed. The fact that, in rough air, all airships were liable to 'hunt' up and down some 250 to 300 feet on each side of the designated flying height, despite little airspace in which to manoeuvre, was also an accepted fact.

Add to this the further facts that for an airship of the size of R.101 every increase of one degree Fahrenheit, in temperature meant the loss of one-third of a ton of lift; that light rain added up to three tons in weight on the cover and that heavy rain could add up to seven tons – and some of the problems come into perspective.

So, with one hundred per cent of hindsight, what should those at Cardington have done on 4th October 1930?

The answer must be that there were many options but that only one of them seemed open then. It was accepted that the flight could be accomplished safely. That being so, any alternative would undermine confidence in the whole philosophy of airship travel and add unbudgeted, and so unacceptable, costs.

But, still with hindsight, in the face of the imperfect weather forecast there could have been a case – difficult to justify but nevertheless a case – to postpone departure for twelve hours until the early morning of Sunday, 5th October; a morning which dawned clear and bright after the trough had gone through, with no more than the lightest of breezes. There was fair weather then all the way from Cardington to Paris and on to Marseille where there were supplies of gas had they been needed.

Instead the crew of R.101 decided to face up to the problems of weather and the problems of flying low in the dark. They were all in the night's work.

Earlier than that, hindsight makes clear that just one trial flight at moderate speeds – and that in good weather – after all the troubles in June and the extensive refit which followed, was quite inadequate.

Prudence should have dictated that the cover and the gas-bags, the known weak spots, ought to have been subjected to more stringent tests in rough weather and in rain while still in reach of base. A long and searching second trial flight followed by any needed modifications and then a demonstration flight of a few hours duration for the delegates to the Imperial Conference, might have met all essential requirements leading to the projected new airship programme.

Thirdly, and more fundamental still, the possibility which had been lightly discussed three months before by Richmond and Rope, that the gas-bag wiring should be pulled in again to provide ample clearance between bags and structure, and that the loss of lift should be more than restored by the addition of a second extra bay – if combined with a wholly new cover, might have provided the sought after long-term answer provided that the extreme further delay and the cost could have been afforded. But the time and money were not available, the resultant R.101D would have been, at 822 feet long, only just able to fit into the Cardington sheds and extensive further trials would have been needed. It was not on.

Hindsight shows that, taking all human attitudes as well as technical factors into account, there really was no alternative to the decision to depart for India on 4th October 1930, following the successful last trial flight.

But there could have been a decision, difficult but possible, to turn back when on the early stages of the flight to India the wind and rain clearly built up to an altogether different weather pattern than had been forecast. That decision was not taken.

Those in charge of the flight – Scott with Colmore – took a risk which, not for the first time (nor for the last) for men facing great issues, they persuaded themselves was but little risk at all. Scott's motto always had been, 'Nothing venture; nothing gain'; the antithesis of the Secretary of State's admonition that ' there is going to be no risk, while I am in charge, of the thing being rushed or of any lives being sacrificed through lack of foresight'.

Once on board, the Secretary of State was a passenger and not in charge.

There could be no turning back when all was set and everyone keyed up 'like batsmen walking out to face fast bowling in a vital match'.

So R.101 was set to do the best that could be achieved in 1930 with an imaginative concept of 1925.

CHAPTER TWENTY-FOUR

India Flight

'There's a long, long trail awinding
'Into the land of my dreams.'
(*Stoddard King*)

4th October 1930

R.101's flight towards India contained, within its concept, seeds of one of the epic pioneering flights of aviation's history.

No airship had ever flown that route before. No aircraft carrying passengers had ever made the eastwards flight with less than a dozen stops. The long, long trail to India had, for ten years now, been the centre of dreams and ambitions.

Fuelled, equipped and victualled for the journey, and just sufficiently light at the moment of lift-off from the mast, R.101 moved forwards, past and above the doors of the No 1 Shed in which it had been assembled, and out over Shortstown, the home of so many members of its crew.

At 1840 hrs GMT, four minutes after slipping from the Cardington tower, all five of R.101's engines were up to their fast-cruising regime of 825 rpm and the 160 tons of the ship – the greatest load so far lifted by any aircraft – were being worked up to a steady air speed of 55 knots.

Three minutes from slip, the wireless room in R.101 received the customary 'Departure Data' from the Met Office, Cardington.

> *Barometer* (MSL) 1009.8 millibars. Surface temperature (dry bulb; 4 feet) 56.0°F; 150 feet, 56.5°F; Character of Thermagram : steady.

That, and the rain on the cover, would account for the need to discharge more water-ballast than expected. The drop in temperature a few minutes before departure had increased lift. Then, just before slipping from the mast the temperature had gone up so that temperature and the atmospheric pressure were now both higher than forecast. Hence the available lift was down by a little more than a ton. Even so the ship had got away light.

Six minutes later at 18.45, just north of Elstow, R.101 was turned ninety degrees to starboard, over South End, to fly north-eastwards, down wind, for its ritual tour of the Bedford area: over St Paul's Church, across the River Ouse, to come low over Bedford School – where Sefton Brancker had been a boarder, forty years before – and on above the Bedford-to-Kimbolton road (B.660) past Alex Bushfield's house at 20 Goldington Road, and at 1848 hrs to sweep low over Carmichael Irwin's home at Longacre, in Putroe Lane.*

The airship was still very low indeed; barely 600 feet above the houses. But Bedford was familiar with low flying airships; objects of civic pride, and everyone knew that this was R.101's farewell, outward bound for India.

For a couple of minutes, with the wind behind it, R.101 made a ground speed of more than 70 knots – 80 miles an hour – until, some three miles north of the town and just west of Ravensden, Irwin turned the airship 180 degrees to port in a wide curve to head back into wind to cross Wood End and Twin Wood Farm to fly back on a parallel course and, at 1854, 18 minutes after slipping from the mast to pass above Clapham High Street and the River Ouse west of Bedford.

Certainly Irwin and Atherstone would both be in the control car at this time with Steff, the Officer of the Watch; Atherstone probably at the wheel and recognised as the most skilful holder of both heading and height among all the coxswains of either R.101 or R.100.

The airship was now 20 minutes and 12½ miles from the mast along the track flown. Crossing the river the ship was turned through 60 degrees to port to fly over Biddenham, past Maurice Steff's home at 65 Ashburnham Road and then, on crossing the river again to turn sharply through 130 degrees to starboard to pass over 28 Biddenham Turn, the home of R.100's Captain, Ralph Booth – a last farewell.

Port helm again at less than 700 feet above the ground and R.101 flew up the shallow valley which runs south-west from Kempston Church-End, beside the river, towards Wootton and Cranfield.

Already, in the control car, and the chart room above, the discussion would certainly be of how the ship was handling in this, already the worst weather it had ever encountered, and whether the outward voyage ought to be postponed?

* The detailed track followed by R.101, and the airship's times of passing specific check-points, have been assembled from wireless reports from the airship to Cardington during the course of the flight, together with more than 500 letters written by members of the public who lived under the route and kindly wrote in reply to requests in local newspapers between 1972 and 1978. (See Appendix 13, pages 512–16.)

A fair impression of the controllability of the lengthened R.101C, at continuous 'fast cruise' on five engines in rough air, would by now have been established in two sharp turns to starboard, together with, first a U-turn to port, and then two more, adding up to another change of heading of 180 degrees.

At this stage it would be in character for Irwin and Atherstone to be giving consideration to the wisdom of a temporary and precautionary return to base to await better weather – but for Scott to be all for pressing on.

The fact that no airship had ever moored to a mast in such weather and at night, must have weighed heavily in the minds of those who had to take the decision and, for another twenty minutes the option was kept open while R.101 continued to be flown in a wide circuit of Cardington, up-wind to the south-west.

At 1900 hrs, over Kempston Church End – 24 minutes and 15½ miles from the start on the track flown, (and still only 4½ miles from the Cardington mast) – the first of the 'on the hour' entries in the standard Journey Log would have been made according to routine:

FIRST ENTRY: JOURNEY LOG BOOK
4th October 1930

Evening Watch: (1600 to 2000 hrs)
Time: 1900 hrs GMT
Altitude: 1,000 ft
Temperature: 54°F
Wind Direction: 230° True
Through Air Heading: 220° True
True Air Speed: 55.2 knots
Position: Three miles S.W. of Bedford
Fuel Used: 400 lb
Ballast Remaining: 5.25 tons

Officer of the Watch: F/O M. H. Steff
Duration from Start: 24 mins
Pressure Height: 1,200 ft
Track: 215° True
Wind Force: 30 mph
Engines: Five at 825 rpm
Ground Speed Made Good: 33.75 knots
Distance Flown: 13.5 n. miles
Fuel Remaining: 25.2165 tons
Cloud Base: 1,500 ft
Weather: Moderate rain

Now the airship was flying on a south-westerly heading of 220° degrees making good a track of 215° degrees almost into wind, flying some 700 feet above the surface at a ground speed of about 35 miles an hour.

Mr Samuel Evans, of Walnut Tree Farm, Lower Shelton, remembered seeing the ship passing low to the west of Wootton Wood on that rough and stormy night. Standing in the driving rain he watched R.101, with all its lights on, fly over Roxhill Manor and then swing round in a 90 degree turn over Wootton Thrift Wood, to pass just south of Millbrook Station at 1915 hrs GMT flying crabwise across the wind towards Hall End.

'The airship was very low', he said, 'on an absolutely horrible night.'*

Skilfully in the dark, but over familiar ground Johnston kept the airship to the north of the ridge which rises steeply to 400 feet from Ridgmont to Ampthill, flying at almost right angles to the direction of London; now making 70 miles an hour over the ground with some help from the wind.

All this time, now for more than half-an-hour from lift-off, the ship had been making a sweep to the north-west, the west and now the south-west of Bedford while keeping within seven miles of the base. At 1916 hrs, forty minutes from the start, crossing above the main line railway from Luton to Bedford, R.101 had covered 24½ miles at a ground speed of 36.75 miles an hour in rough air.

'I remarked that the ship was pitching and rolling a good deal', recalled Harry J. Leech,† R.101's Foreman Engineer, of a conversation with Michael Rope as they climbed around the ship on a detailed tour of inspection. 'Squadron Leader Rope didn't seem to be worried about the behaviour of the ship', he added, nor was Arthur Disley, the Chief Electrician, who likewise completed a general inspection from one end of the airship to the other and reported everything in order 'except one lamp at No 4 Frame, and this I changed.'*

Certainly, thus early in the flight, everyone seemed well satisfied with the manner in which R.101 was riding the turbulent conditions. The rain was driving down and, if anything, the wind seemed to be increasing in force and in its sudden gusts. The high cruising speed of 55 knots was, however, being kept up with each of the engines delivering 475 horse-power to its propeller.

There seems no doubt that during this time while R.101 was flying at right angles to the route to London, on a heading of 134 degrees to make good a track of 110 degrees at a ground speed of 53 knots – neither setting out southwards nor returning to Cardington – a final conference must have been held in the chart-room above the control-car between Scott, Irwin, Atherstone, Johnston and Giblett with

* In conversation with the author, 3rd August 1974.

† AIR 5/903. Both Leech and Disley survived the catastrophe.

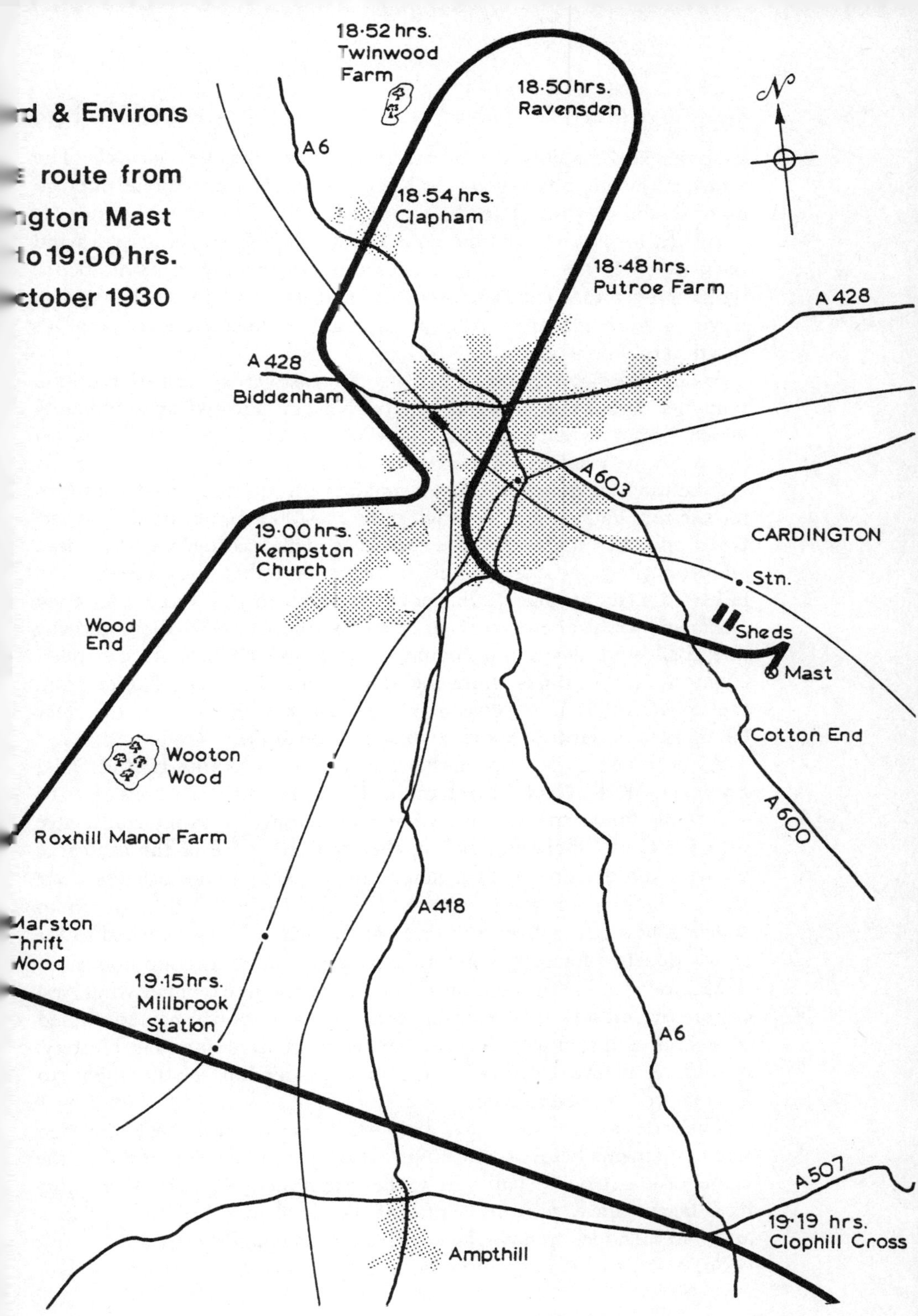

Testing the Wind: The course flown by R.101 during the first half hour of its voyage towards India on the evening of 4th October 1930 – circling Bedford and keeping within easy distance for a return to the Cardington mast while assessing the airship's performance in the turbulent flying conditions.

Colmore and possibly the Secretary of State consulted as well. The weather was already a good deal worse than forecast. Should they go on or should they turn back?

A turn to port at any time now would bring the airship down wind back over Cardington in a few minutes to moor and await better weather over the first four hundred miles or so of the long flight to Egypt. At present, apart from a reasonable temperature, the weather could hardly be worse.

While the airship continued to fly eastwards, and the option remained open, the situation was further complicated by a difficulty which began to emerge with No 5 engine in the aft car, attached below Frame 11 ahead of the ventral fin.

Like the four others, the aft engine had run normally at its 825 rpm for the first half hour since leaving the mast. Then, at about 1906 hrs GMT the duty engineer in the car, A. V. Bell, recorded that the main oil pressure gauge had begun to flicker and that the pressure had fallen from its normal 34 lb per square inch to 27 lb and continued on down. So, at about 19.16 hrs, crossing the Luton/Bedford railway line, Bell shut down the engine – action which brought his relief engineer, J. H. Binks, hurrying down the windswept ladder from the crew's quarters to give help. Bell asked him to fetch the duty Charge Hand, George Short, to diagnose the fault.

As he reached the top of the ladder Binks was met by the Chief Engineer, W. R. Gent – a veteran of R.34's Atlantic crossings of 1919 – coming down to find out what was wrong. A rapid conference between Gent, Short and Bell in the confined space of the car came up with the decision to fit a new main pressure gauge because there seemed nothing obviously wrong with the engine. Bell went up to collect a new gauge from the ship's spares while Binks returned to the crew's quarters for supper before coming on watch at 2000 hours.

Meanwhile, R.101 continued on the cross-wind course while, no doubt, the officers in the chart-room and the control car continued to feel out and to discuss R.101's behaviour in the unfamiliar, blustery, conditions before electing to commit the airship to the flight to Egypt – or to return to base.

Their decision cannot have been made easier with the knowledge that – moments before – one engine had been shut down and that the cause of the trouble had still to be established, though there must have been a message from Gent that the problem did not look serious and that he hoped to have the engine running again shortly.

So, at 19.19 hrs – 43 minutes and 29½ miles out from base – over the Clophill crossroads (A6/A507) quite positively and clearly the decision was made, certainly with Scott as the final arbiter. The order given, R.101 turned away from Cardington and home and on to a southerly heading of 185 degrees to make good a track of 157 degrees, ploughing on towards London and the coast through the rain and into the teeth of the wind, now making rather less than 40 knots ground speed.

Here now was a fulfilment of C.B.'s ambition 'to ride the storm', to echo the hope that 'R.101 will give me joy'. He must have been assured that all would be well, that – even with one engine out, at least for the moment – the ship was behaving admirably and that there was 'no undue risk to human lives'.

When R.101 turned south for London it was still only eight miles in a direct line from the Cardington tower although, at 1923 hrs over Hanscombe End, it had covered 33 miles 'as the airship flew', and it was now 47 minutes from the start at an average ground speed, so far, of 48 miles an hour.

Now Johnston set a straight-line course for the traditional British airship check-point in London; the unmistakable landmark from the air of the white tower and glass-covered square of the 'Metropolitan Cattle Market', Islington, between the Caledonian Road and York Way. Leading to it was the navigational guide of the main line of the Great Northern Railway.

By now, except for those in the aft engine car, the crew were settling down to its flight routine. The Evening Watch of fifteen, on duty till 2000 hrs, was made up of two coxswains, seven engineers, four riggers and one W/T operator – all headed by Flying Officer Steff, as the Officer of the Watch.

Passengers, and the officers who were not required for duty had, by then, assembled in the lounge and promptly at 1930 moved into the brightly-lit dining room where places had been laid for twenty-four at six tables with white linen and the ship's silver plate, presented by the town of Bedford. Already R.101 was demonstrating the comforts of airship travel even in rough air; the spaciousness, the low noise levels, the ability to walk and talk not enjoyed to the same extent by any other means of transport except the ocean liner.

A delectable cold dinner, preceded by hot soup, had been prepared by E. A. Graham, the ship's cook, for the first evening out; served by A. H. Savidge, the Chief Steward with Assistant Steward F. Hodnett and by J. W. Megginson, the Galley Boy. In the crew's quarters there was bread-and-cheese with pickles washed down by mugs of steaming cocoa. Whatever the weather outside, all was warm and comfortable within.

One hour out from Cardington, at 1936 GMT with a ground distance of 41 miles covered, R.101 passed over the little hamlet of Bendish between Luton and Knebworth, 350 feet above sea level. So low did the ship appear that Mrs Shane Leslie, wife of the author, claimed in dudgeon that the airship 'came up her garden path and was about to enter the living room window when, fortunately, it just cleared the roof'.

The bulk of the airship flying at less than its own length above the ground, with four engines at 475 horse-power each, would indeed appear menacing. A more rational view of it was, however, provided by another observer who had an aeronautical background, Richard M. Clarkson, then an apprentice at the de Havilland Aircraft Company's works at Hatfield and later to become D.H. Chief Designer responsible for the Comet and Trident airliners.

Clarkson and his father watched R.101 pass at 1938 hrs precisely – and later calculated, with some precision – on a track 430 yards west of their house at Hoo End, near Whitwell.

Clarkson wrote down* that they heard and then watched R.101 approach from the north, 'flying steadily on a compass course and an even keel, the engines running very strongly. She had some drift but kept her nose straight on her course.'

Not every one of those in the airship's lounge had sat down promptly at 7.30 for the evening meal, because Clarkson notes that: 'Two persons were definitely distinguished looking out of the lounge windows on the port side and silhouetted against the light within.'

He estimated, (and then checked with other witnesses and by painstaking trigonometry) that the airship was flying almost exactly 600 feet above the point of observation which would make it just about 1,000 feet above sea level.

So the airship passed on its way, closely observed, making a ground speed of 37 miles an hour against the, by now, 22 mph head-wind component.

Meanwhile, in the aft engine car with the engine shut down, Bell had returned carrying the spare pressure-gauge to replace the suspected faulty gauge which had been removed by Binks before he went up to finish his supper ready to return on watch.

The airship flew on to pass just west of Bernard Shaw's house, 'Shaws Corner' at Ayot St Lawrence and, at 1944, to cross above the

* AIR 5/906–X.M. 08,109.

Fifty Years Later – The Bois de Coutumes between Allonne and Bongenoult near Beauvais photographed from 600 feet at the heading on which R.101 was flying a few moments before the disaster on 5th October 1930. The nose of the airship struck the ground a few yards into the wood leaving the major portion of the hull on the open ground. The Beauvais Ridge is beyond. (Photograph by the author on the 5th October 1980.)

Catastrophe – R.101's burned-out skeleton lying with its nose in the trees of the Bois de Coutumes in the early sunshine of the bright Sunday morning on 5th October 1930 – sunshine which followed the storm of the previous night. The collapsed centre-section contained the wreckage of the passenger accommodation, where the fire had burned fiercest. The bodies recovered from the wreck lie on the edge of the wood on the right side of the air to ground photograph.

[PLATE XXV]

Gaunt Skeleton – The lightly damaged tail section of R.101 on the morning of 5th October 1930, the fabric of the outer cover burned off except for that on the underside of the port elevator. The Royal Air Force ensign still flies from the look-out position at the extreme tail.

Return Journey – Departure from Allonne on the morning of Tuesday, 7th October 1930, of the carriages carrying the coffins of the 48 victims of the R.101 disaster. Behind one of the carriages, escorted by French soldiers, walk three survivors of R.101's crew, Harry Leech (Foreman Engineer), and A.V. Bell and J.H. Binks, Engineers from No. 5 (Aft) engine car.

[PLATE XXVI]

railway and the River Lea at Waterend, a little east of Wheathampsted and – while dinner was in full swing – to fly over the eastern corner of de Havilland's new aerodrome at Hatfield, to pass west of ancient Hatfield House at 19.50 and so on close above the main LMS railway line to fly over Potters Bar Station at two minutes before eight o'clock; all clocks in the airship by now had been put back to GMT.

Now the second 'on the hour' entry in the Journey Log would be due; to be completed by the oncoming Officer of the Watch, Noël Atherstone, who would have just returned to the control car from the dining-room above where, surely, he had savoured the presence on board of the Air Minister and the Director of Civil Aviation with the other passengers. He would have enjoyed the easy flow of their conversation and their pleasure in the delights of Lighter-than-Air travel – looking ahead to next day's cruise down the Mediterranean under blue skies.

By now No 5 engine, its new pressure gauge installed, had been re-started at 19.56 hrs over Brookmans Park. The new entry in the log at 20.00 hrs could again record that R.101 was flying on all five engines and was back at a true air speed of 55 knots or 63.3 miles an hour.

SECOND ENTRY: JOURNEY LOG BOOK*

First Night Watch: (20.00 to 23.00 hrs)
Time: 20.00 hrs
Altitude: 1,000 ft
Temperature: 52° F
Wind Direction: 235° True
Through Air Heading: 190° True
True Air Speed: 55 knots
Position: Potters Bar
Fuel Used: 1,300 lb (0.58 tons)
Ballast Remaining: 5.25 tons
Weather: Moderate rain

Officer of the Watch: N. G. Atherstone
Duration from Start: 1h 24m
Pressure Height: 1,200 ft
Track: 157° True
Wind Force: 28 miles an hour
Engines: Five at 825 rpm
Ground Speed Made Good: 34.5 knots
Distance Flown: 48 n. miles
Fuel Remaining: 24.8 tons
(Water recovery in progress)
Cloud Base: 1,500 feet

During the past hour since 19.00 hrs GMT R.101 had flown 40 statute miles against a headwind component which had averaged

* Compiled from wireless messages from R.101 to Cardington on the airship's standard flying log entry form.

about 21 miles an hour; the wind blowing at nearly 80 degrees to the airship's course.

R.101's route now took the airship over the north-eastern outskirts of London over Hadley Wood and Hadley Common* over New Barnet and East Barnet to fly across Brunswick Park at 20.11 hrs and two minutes later over Friern Hospital (then Colney Hatch) to pass overhead the Alexandra Palace Band-Stand at 2016, and so over the railway stations at Crouch End, Hornsey and Holloway Road in quick succession.

Across London, the passage of the airship – so large and so low, with all its lights on, flying with its nose pointing almost 80 degrees to starboard of the track being followed, in order to offset the force of the gale – brought people pouring out into the streets, braving the driving rain.

The noise of the engines resounded from house to house even though at 2016 over Crouch End, just south of Alexandra Palace and one hour and 40 minutes from the start, No 5 engine had again to be shut down because of a renewed drop in oil pressure.

To those on board the airship the 'aiming point', the unmistakable roofs of the Metropolitan Cattle Market at Islington, must just then have loomed out of the rain a quarter of a mile away on the starboard bow with Pentonville Prison just ahead. So many years later, Ernest Johnston's quick calculation and order to the rudder coxswain for a change of heading, remains clear and precise from a plot on the map of the passage of the airship.† Johnston had his Cattle Market check-point and at 2020 hrs R.101 was turned five degrees to port over Bride Street on to a new heading of 195 degrees to make good a track of 132 degrees, true, towards Greenwich and the Channel coast.

Fifty years later those who had witnessed it, still remembered the tumultuous passage of the airship overhead and the wildness of the night below. Among those who never forgot was Mr G. R. Lewiston, who watched the airship pass directly over his home at Offord Road, beside the Highbury and Islington Station.

He checked that the time was precisely 2020 hrs GMT. The BBC's nine o'clock News had just ended; telling of R.101's departure for India. Here now was the airship, its cabin windows brilliant with

* By an odd co-incidence, at 20.05 hrs on 4th October 1930, R.101 passed directly over the house at Hadley Common where, a generation later, Noël Atherstone's son, Dr Richard G. Atherstone, came to live with his wife and family – and was there fifty years exactly after R.101's passage.

† Recorded from a succession of witnesses on the ground who wrote to the author in reply to published requests for information.

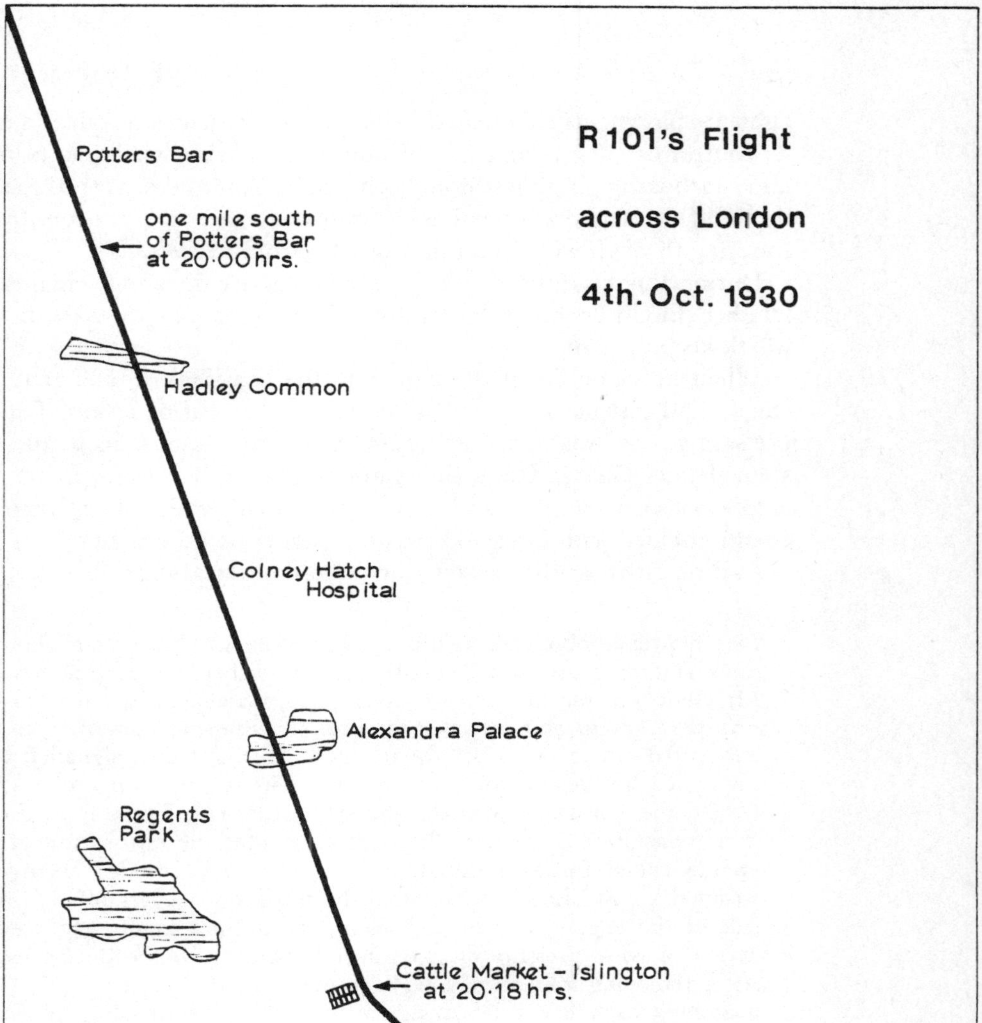

Across London: R.101's passage across London, from Potters Bar over the Alexandra Palace and Islington to cross the River Thames at Shadwell and Millbank bound for the Channel coast just east of Hastings.

light, its passengers silhouetted beside the pink shades of table lamps, its control car aglow and its red and green navigation lights to port and starboard, glinting through the rain. Amid the deep-throated throb of its engines the airship seemed to young Lewiston to be hurrying through wind and rain eager for tropical skies.

During this time there was continued activity in the aft engine-car whither Engineer Binks had returned at 20.00 hrs to take up his watch-keeping duty.

When he came down the ladder to the car, Binks found that the engine had just been restarted after 40 minutes out of action. The oil pressure-gauge was showing 32 to 33 lb per square inch and all seemed fine. Charge-Hand Short and Engineer Bell went off up the ladder for their supper, and to rest off-watch, while Chief Engineer Gent remained with Binks to keep an expert eye on the engine.

In Binks' own words, written down shortly thereafter:*

In a matter of about ten minutes, oil pressure was once again showing signs of falling and did, actually, fall to 27 lb per square inch when Mr Gent told me to reduce to half speed, bringing it down from 825 revs per min to 700 revs per minute . . . Pressure continued to fall and got down to 22 lb per square inch. Mr Gent then signalled with his hands for me to stop (shut down). This second stop brought Mr Leech, the Foreman Engineer, down into the car. These two, after a short consultation came to the conclusion that the oil conduit of the engine called for a thorough investigation. I proceeded (being instructed by Mr Leech) to remove the inspection covers off the Port-side of the engine, first of all locking engine in a manner to prevent action of wind pressure on the airscrew turning engine during exam. After removing these covers, eight in number, Mr Leech and Mr Gent examined each big-end bearing, main bearings, connecting rods and oiling system in crankcase and agreed that the defect was *not* in that part of the engine.

I was instructed to replace inspection covers while they inspected relief valve, which valve is situated on starboard side of engine and would not, therefore, cause any delay or impede me whilst I replaced covers etc: During the operation of replacing covers the Charge-Hand Engineer of my Watch, S. E. Scott, came down to assist me. In the meantime relief valve was examined, additional packing discs inserted and engine generally was prepared for starting and was started by myself and Charge Hand S. E. Scott, Mr Gent and Mr Leech still being in car.

It had now come to 11 p.m. Down came Engineer Bell and relieved

* AIR 5/9871.

> me – we had, on this flight, been put on three-hour watches unlike our former flights which had all been carried out in four hour watches.
>
> By this time I was ready for a rest after having spent the largest portion of my watch in a very cramped position. And so back up to our quarters and I turned in to the bed Bell had a short time previously occupied.*

Binks' simple and direct account, shows well how quite major work could be done on an airship engine in flight.

In all, No 5 engine was out of action for 3 hours and 20 minutes; first, for 40 minutes between 19.16 and 19.56 hrs between Millbrook and Brookmans Park and then again for 2 hours and 40 minutes betweeen 20.16 (overhead the City of London) and 22.56 hrs when the airship passed mid-Channel. During this time, while flying on four engines, R.101's air speed fell from the 55.2 knots attained on 2,375 bhp on five engines to 51 knots (58.7 mph) on 1,900 bhp on four engines. During the same period the rate of fuel consumption came down by some 200 lb an hour, from 1,000 to 800 lb an hour.

Curiously, no mention was ever made in any of R.101's wireless messages that night, of the fact that, for more than 3¼ hours and for 113 nautical miles, between Millbrook and mid-Channel, the airship had been running with one engine stopped. The intention may well have been to avoid causing any undue alarm at Cardington about R.101's power-plant, or to raise any question about the decision to press on in bad weather. No doubt the plan would be to mention the subject later when the engine was back in service and, the weather had improved, at a time when no concern would be caused.

The interest caused on the ground that Saturday evening by the low and thunderous passage of the great airship across London, so vividly remembered by everyone below its path†, was no doubt echoed by those in R.101 itself, observing from no more than 800 feet above them the brightly lighted rain-washed streets and houses with familiar landmarks appearing and disappearing one by one.

In the control car, below the hull, the handling of a great rigid airship was different from that of any other form of craft. One man, the Steering Coxswain, stood facing forwards with the spoked wheel, like that of a ship, mounted vertically in front of him, its motion linked through cables to the rudders some 400 feet behind. The heading to be steered, set in degrees magnetic, was inscribed on a

* AIR 5/9871.

† The author is grateful to the very large number of people who wrote to recall the airship's passage and to help to pin-point its route.

board above his head and read off from a compass binnacle in front of the wheel.

A second wheel, mounted sideways to starboard, in the control car, controlled the elevators and hence the height and the pitch, nose up or nose down, of the airship. This fore-and-aft wheel, at right angles to the steering wheel, was manned by the Height Coxswain who required not only skill and experience but also continuous concentration and intelligent anticipation to keep the airship in relatively level flight, especially in bumpy air.

An altimeter, an air speed indicator, calculated in knots, and fore-and-aft spirit levels, all grouped on an overhead panel, were visible to both duty coxswains.

In turbulence, though steering in yaw was fairly precise, all airships – and R.101 not least – tended to hunt up and down in pitch. To keep within 200 feet of a set height was regarded as good work – and, flying at less than 1,000 feet above the ground, there was not much manoeuvring space below.

A recognised fact was that, at every change of watch, about ten minutes would be needed for the relieving coxswains to 'get the feel of the ship' – especially if it was flying 'heavy' or 'light', pitched nose-up or nose-down to maintain a constant height under power. Until the Height Coxswain, in particular, settled down, the flight path was liable to be erratic.

On this watch, from 20.00 hrs to 23.00 hrs; Assistant Coxswain E. G. Rudd had relieved Assistant Coxswain C. H. Mason at the rudders while Assistant Coxswain L. F. Oughton had taken over from Flight Sergeant W. A. Potter on the elevators; standing together in the control car with, no doubt, Johnston beside them to check visually the course he had plotted on his map.

Just above the control car and beside the chart room, Wireless Operator F. Elliot was transmitting to Cardington a message from R.101:

> *Serial No 2.* – 20.21 hrs. Over London. All well. Moderate rain. Base of low cloud 1,500 feet. Wind 240 degrees. 25 mph. Course now set for Paris. Intend to proceed via Paris, Tours, Toulouse and Narbonne. Ends.

No mention of the trouble with No 5 engine. Weather was the primary preoccupation. But clearly there was no concern about progress. No doubt all those aboard, the Secretary of State included, were keenly looking forward to the blue skies and smooth air over the Mediterranean next day and, ahead of that, to the completion of the first flight of an airship between the centre of London and the centre of Paris, en route southwards.

So R.101 passed on, flying over Shoreditch, just north of Liverpool Street Station and Spitalfields Market, over Whitechapel High Street, a little west of the London Hospital to come to the River Thames at Shadwell over King Edward VII Memorial Park at 20.25 hrs GMT.

The airship thundered close over the ships lying in the Lower Pool, across Rotherhithe, the Surrey Commercial Docks and the Limehouse Reach to the Isle of Dogs, Millwall and Tower Hamlets, to cross the river again at the Greenwich Reach to fly at 20.28 hrs directly over the Royal Naval College and the Painted Hall at Greenwich; a final tribute to the Service to which all the airship's officers had belonged.

Thence across Greenwich Park and over the Royal Observatory 73 statute miles and one hour and 52 minutes from casting off at Cardington.

Even in the dark and the rain those landmarks would show up as precise navigational aids; the U-shaped bend in the river and the rain-washed domes at Greenwich. Johnston would be able to re-calcuate wind speed and direction and concentrate upon the most precise piece of navigation required of this stage of the flight; to guide the airship through gaps in the North Downs where they rose ahead to nearly 800 feet above sea level.

So far, since turning on to course for London at 19.19 hrs, above the Clophill cross-roads, R.101 had covered 45 miles in 71 minutes to be overhead Blackheath Park Station* at 20.30 hrs for a ground speed from Clophill of 38.2 miles an hour or 33.2 knots.

The airship was climbing slowly now to its 'pressure height' of 1,200 feet in order to gain a little more air space between it and the rising ground ahead; not any higher though, so as to conserve every cubic foot of gas. To fly higher would, also, increase the headwind component.

Johnston was indeed a master navigator. In the wind, the rain and the dark of this October night, with R.101 flying in and out of wispy cloud, he brought it along a precise and careful track over Eltham and New Eltham, between Sidcup and Chislehurst, past Crockenhill, to turn through 48 degrees to starboard over Lullingstone Park where the waters and weirs of the River Darent came into view, with its lighted bank near Eynesford beyond, and to fly south down the Darent valley along the Shoreham/Otford gap

* Now Kidbrooke Station.

through the Downs. The hills now rose to within a few hundred feet of the airship on the port and starboard bows.

Another turn of almost 90 degrees to port, just north of Sevenoaks, brought the airship from its southerly heading to fly now due east past Kemsing on the high ground to port, to turn again and to fly through the second gap in the hills, between Ightham and Borough Green over Plaxtol – and on, due south again, towards Tonbridge.

It could not have been better done on a fine, clear day. Years afterwards each turn of the airship could be plotted faithfully from the many independent memories of those on the ground. There could be no mistaking the passage of so great a vessel, flying so low; to be seen and never forgotten.*

Now, above Higham Wood with the River Medway ahead R.101 was clear of the North Downs and the ship was turned firmly and precisely over Hadlow Stair to leave Tonbridge to starboard and to fly on a new heading of 195 degrees across Kent towards the coast.

During the past half hour additional weather information had been coming in from the Met Office at Cardington to the Wireless Cabin of R.101.

At 20.34 hrs when the airship was overhead New Eltham Station, a transmission from Cardington began in morse of a long message made up of 76 Groups of the meteorological code – to provide detailed synoptic data from which Maurice Giblett would be able to draw a chart of the weather actuals at 18.00 hrs GMT along the whole area to the Mediterranean. The message was acknowledged from R.101 at 20.39; the ship then coming up to Crockenhill and Lullingstone Park.

One minute later, Cardington began to transmit, also in morse, the 18.00 hrs GMT Forecast for the route from Cardington to the South of France for the twelve hours ahead. Ominously, it predicted steadily worsening conditions up to and beyond Central France, to such an extent that, had this forecast come through three hours before, the start would almost certainly have been postponed. The message read: †

Serial No 3. To R.101. From Meteor, Airships, Bedford.

18.00 GMT situation. Trough of low pressure along coasts of British Isles, moving East. Ridge of high pressure over Southern France. Forecast for next 12 hours of flight. SE England, Channel and Northern

* Curiously, the Official Inquiry of October 1930/March 1931 made no checks of the airship's passage between Cardington and the Channel coast and it was left to the researches of the author to establish the precise track flown.

† AIR 5/903.

France. Wind at 2,000 feet from about 240 degrees, 40 to 50 miles an hour. Much low cloud with rain. Central France; upper wind about 250 degrees 30 to 40 miles an hour. Cloudy and local rain. Southern France wind westerly to variable, mainly light, variable amount of cloud. Local morning mist and fog. Ends. (20.39 GMT).

R.101 acknowledged the message at 2045 from between Hulberry and Shoreham, Kent.

This new forecast of low cloud with rain, and of a wind speed getting up to 50 mph from 240 degrees, would mean a headwind component of about 35 mph for at least the first part of the journey across France; slowing down the ground speed of the airship to no more than 25 mph. If it lasted for another twelve hours they would certainly have to reduce speed to conserve fuel, even though the duration of flight would be substantially increased. No need for such a decision yet, however.

For another fifteen minutes, up to a little after 2100 hrs GMT, Johnston and Atherstone would be too occupied with their low-level visual navigation through the North Downs to give attention to the new forecast; Irwin would be resting before coming on watch at 2300 hrs and Steff, no doubt, in his cabin bunk ready for his stint as Officer of the Watch from 0200 to 0500 hrs. Meanwhile the airship was brought carefully and precisely past Plaxtol at 20.56 hrs, over Hadlow Stair at 21.02 hrs and set on its course for the coast over Tudeley Hale at 21.04 hrs.

Now, with the airship flying at 1,200 feet, making a ground speed of 50 mph over the wooded Kentish countryside which rose in places to about 300 feet above sea level, Atherstone and Johnston would be able to turn from low level navigation to examine the forecast.

Certainly it was worrying because the next leg of the planned route – from Paris to Tours – would lie almost dead into the forecast wind. Even if the wind became light and variable later on, R.101 would be bound to take between eleven and twelve hours from Paris to Narbonne instead of the six hours planned.

The concern in R.101's Chart Room was clear from a wireless message to the Met Office, Cardington, transmitted from the airship at 2119 hrs when it was just south of Hawkhurst and coming up to Bodiam. It read:

Serial No 3. 21.19 GMT. *To Meteor, Cardington from R.101.* Request forecast Paris, Dijon, Lyons, Marseilles with special reference to wind and cloud.

Obviously, Johnston and Atherstone were looking at the prospects of replotting an alternative route to avoid the worst of the headwind after Paris. A course further to the south-east would substantially reduce the adverse wind component. On the latest met evidence, a quick calculation would show Johnston that flying by way of Lyons and the Rhône Valley would be likely to cut the flying time to the Mediterranean coast at Marseille to less than seven hours, with Malta ten hours further on. That would compare with eleven hours to Malta from Narbonne.

Meanwhile the third hourly entry in the Journey Log had been due at 2100 hrs. It would show that, so far, R.101 had averaged a ground speed of just over 41 mph along the route flown, including a brief four minute spurt of 82 mph down wind between Otford and Borough Green from 2052 to 2054 hrs.

THIRD ENTRY: JOURNEY LOG BOOK

First Night Watch: (20.00 to 23.00 hrs)
Time: 21.00 hrs
Altitude: 1,200 ft
Temperature: 51°F
Wind Direction: 230° True
Through Air Heading: 208° True
True Air Speed: 51 knots
Position: 1½ miles N.E. of Tonbridge
Fuel Used: 2,100 lb (0.94 tons)
Ballast Remaining: 5.75 tons
Weather: Heavy rain

Officer of the Watch: N. G. Atherstone
Duration from Start: 2 hrs 24 m
Pressure Height: 1,200 ft
Track: 180° True
Wind Force: 37 miles an hour
Engines: Four at 825 rpm
Ground Speed Made Good: 41.3 knots
Distance Flown: 85 n. miles
Fuel Remaining: 24.46 tons
(Water recovery in progress)
Cloud Base: 1,500 feet

So, during the hour between 2000 and 2100 hrs, R.101 had covered 37 nautical miles across London and into Kent. The average headwind component was fairly steady at 22 mph. The airship's height was being maintained at 1,200 feet above sea level – between 200 feet and 300 feet less above the ground.

Historic Bodiam Castle was right on track at 2120 GMT. The airship passed just west of Ewhurst Green to fly precisely over the Broad Oak cross-roads (A.28/B.2089) at 2128, thence across the little River Brede with the ground falling away to pass east of Guestling Green and over Pett at 2133 and so out to sea at Haddock Gap; a farewell

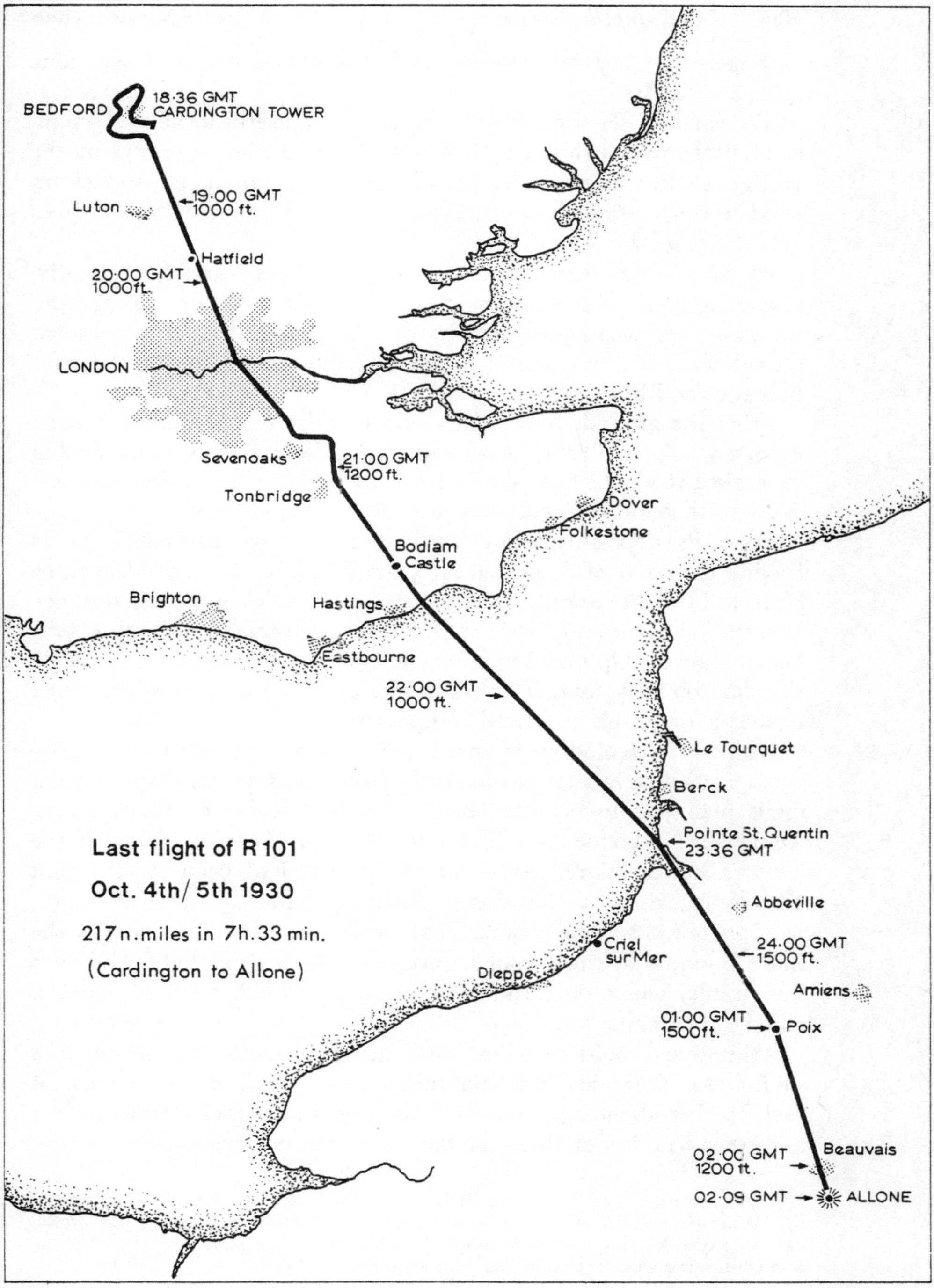

Final Flight: The course flown by R.101 on its voyage towards India on the night of 4th/5th October 1930, crossing the English coast just east of Hastings at 21.35 hrs GMT and arriving over France at the estuary of the Rover Somme one minute more than two hours later.

to England at a point between Cliff End and Fairlight Cove; four miles east of Hastings.

To watchers on the cliffs, as R.101 swept overhead at 10.35 p.m. local time, the airship was still appearing to slide sideways in the gusting south-westerly gale, its nose pointing almost 40 degrees up wind to the south-west of the direction of flight, holding 141 degrees over the ground.

As the airship flew out to sea, the watchers could hear clearly, above the drone of the engines, the sound of dance music coming from the passenger accommodation. Large loudspeakers had been mounted on the walls of the lounge and dining-room, with a wireless receiver tuned to the BBC.*

From the ground, passengers could be clearly seen behind R.101's lighted windows. From the airship the lights of Hastings and Bexhill to starboard would fade slowly into the mist and rain while to port, likewise, those of Rye and Dungeness disappeared behind.

From its departure from Cardington at 1836 hrs GMT to its leaving the coast of Sussex at 2135 hrs GMT R.101 had taken three hours, all but one minute, for the 127 statute miles of ground distance covered. The true speed through the air, had averaged 60.3 mph (52.4 knots). The airship had flown for one hour on five engines; the rest of the time on only four. The ground speed made good was 42.4 mph against a 17.9 mph headwind component.

No doubt calculations of speed and range were made in the chart-room immediately after passing this precise check-point. If the original flight plan was to be maintained – to fly by way of Paris, Tours, Toulouse, Narbonne and Malta to Moascar, Ismailia – and if the average headwind remained the same as it had been for the past three hours, then the duration to Ismailia, at the same true air speed, would work out at 60½ hours. That would make them too late for the State Banquet at the Ismailia tower. The forecast of light, and even favourable, winds over the Mediterranean would, however, offer a chance of catching up.

Atherstone would calculate, with his usual meticulous detail, that so far the Tornados had consumed about 2,600 lb (1.16 tons) of fuel, leaving about 23.7 tons of usable fuel in the tanks; sufficient for a further 61¼ hrs of flying at the same rate of consumption as that

* The Archive's Section of the BBC record that, at 10.30 p.m. BST that evening, Ambrose and His Orchestra had begun their regular Saturday night broadcast from the Mayfair Hotel, London – starting with 'My baby just cares for me' and continuing with 'Free and Easy' at 1035. (2135 GMT.)

during the past three hours. So, there ought to be just enough fuel. If the wind speed increased still further and then fell light and variable as promised over the Mediterranean, then there ought to be a comfortable margin. But it needed watching and the latest forecast was still awaited.

However that might come out, for the present the decision was made to press on at the same power settings – and to hope that the aft engine would be back in business shortly. Fast-cruise conditions were obviously right for the present to make way against the strong winds. A decision to eke out fuel in long-range-cruise at lower powers could wait until the winds dropped.

Generally, there must have been complete confidence in the behaviour of the airship, its cover and its gas-bags. Clearly the bags were not losing gas because, despite the weight of rain on the cover (by now probably three tons or more) water ballast was being recovered through the rain-water channels from the top of the envelope. The speed being maintained showed, too, that the ship was nicely in equilibrium.

All this was confirmed by the wireless report transmitted back to Cardington from R.101, five minutes after moving out over the English Channel. It read:

Code CROW. To Cardington. From R.101. Series No 4.

At 2135 GMT. Crossing coast in vicinity of Hastings. It is raining hard and there is a strong south-westerly wind. Cloud base is at 1,500 feet. After a good get away from the mooring tower at 1830 hrs ship circled Bedford before setting course.

Course was set for London at 1854. Engines running well at cruising speed giving 54.2 knots. Reached London at 2000 hrs and then set course for Paris gradually increasing height so as to avoid high land. Ship is behaving well generally and we have already begun to recover water ballast. Ends. (2140 hrs)

This was an odd message, obviously intended for Press consumption and probably originated by Scott, less particular on detail than were the ship's officers. No mention that R.101 had been flying on four engines for the past eighty-four minutes. Vague, and not wholly accurate, timings for the start and over 'London'. 'Course set for London at 1854' while the ship was still circling Bedford. Interesting, however, that the air speed was quoted at a precise 54.2 knots which may perhaps have been a transposition for 52.4 knots – the exact average of one hour on five engines at 55.2 knots and one hour and 59 minutes on four engines at 51.0 knots.*

However that might be, Johnston with Atherstone beside him had kept R.101 on an immaculate course, bringing the ship to the English coast on a course which had threaded its way through the North Downs and had then been laid to avoid high ground around Wadhurst, Rotherfield and Heathfield, which were on the direct route to Paris.

The direct line of flight from Haddock Gap to overhead the (then) French military aerodrome at Orly, south of Paris, would take R.101 on a Channel crossing of 57 nautical miles to make a landfall at Criel-sur-Mer, just west of Le Tréport. If the wind speed and direction remained the same, the French coast would be reached in about an hour and a half, at just after 2300 hrs GMT and R.101 should then be over Argenteuil, the Champs-Elysées and the west centre of Paris by one o'clock in the morning before turning over Orly, either south-west to Tours or south-east to Dijon.

By arrangement with the French Air Ministry, gas, fuel and a handling party were available at Orly in case of emergency; but, surely, none of those facilities would be needed. It looked as though the flight from over the centre of London to overhead the centre of Paris would be accomplished in just about five hours; not bad in the atrocious conditions.

Johnston set a new course for Criel-sur-Mer on a ship's heading of 190 degrees – 11 degrees to starboard – to make good a track of 139 degrees, assuming a constant wind speed and direction of 36 miles an hour from 230 degrees.

Fifteen minutes out from the English coast a new message began to come through from Cardington.

It was from the Met Office and gave a general inference for the more distant part of the route beyond the south of France; an encouraging pattern once the airship had ploughed through the bad weather still ahead. It read:

Serial No 4. To R.101 from Meteor, Airships, Bedford.

Anticyclone centred over Balkans and Greece extends over Western Mediterranean. Winds over Western and Central Mediterranean light and variable with fair weather. Eastern Mediterranean, winds northerly. Athens to Crete fresh at times. Crete to Ismailia weather mainly fair with local showers.

* A speed of 54.2 knots was some three knots faster than R.101C could achieve on four engines at 825 rpm each.

This message was acknowledged from R.101 at 2152 hrs. It must have been greeted with pleasure. There should be plain sailing once the south of France was reached.

Two minutes later the awaited forecast for the weather beyond Paris was transmitted from Cardington.

Serial No 5. To R.101 from Meteor, Airships, Bedford.
Forecast Paris—Dijon—Lyons. Wind Westerly, 25 to 30 mph at 2,000 feet. Lyons to Marseille, wind variable and light. Cloudy with cloud base 1,000 to 2,000 feet.

It was acknowledged by R.101 at 2156.

That would seem to settle it. There would be a tail wind component from Paris to Dijon and light winds beyond. There would however be a problem of low cloud over the comparatively high ground of the Côte d'Or, north-west of Dijon; cloud which might well be down on the wooded hills which rise almost to 2,000 feet on both sides of the Burgundy Canal, north-west of Dijon. From Dijon, southwards, the route over the Bresse Plain would be easier down the River Saône to Mâcon, and thereafter on to Lyons. From there the narrow valley of the Rhône would be followed, in better weather, to the sea, a route often traversed by the *Graf Zeppelin*.

So the choice would be strong headwinds over low ground by way of Tours and Toulouse to Narbonne, or tail winds but high ground and low cloud via Sens to Dijon. After that things would be easier – and there would be no concern about sufficient fuel. It would mean, however, a need to rise well above pressure height to clear the higher ground.

No doubt a final decision was left for another weather forecast from Le Bourget just before reaching the decision-point at Orly some hours ahead.

Now the time had come for the 2200 hrs entry in the Journey Log.

FOURTH ENTRY: JOURNEY LOG BOOK

First Night Watch: (20.00 to 23.00 hrs)
Officer of the Watch: N. G. Atherstone
Time: 22.00 hrs GMT
Duration from Start: 3h 24m
Altitude: 1,000 ft
Pressure Height: 1,200 ft
Temperature: 51°F
Track: 139° True
Wind Direction: 230° True
Wind Force: 36 miles an hour

Through Air Heading: 190° True
True Air Speed: 51 knots
Position: 18 n. miles South East of Hastings
Fuel Used: 2,900 lb (1.29 tons)
Weather: Heavy rain
Engines: Four at 825 rpm
Ground Sped Made Good: 37 knots
Distance Flown: 126 n. miles
Fuel Remaining: 24.11 tons
Cloud Base: 1,500 ft

Over the sea the position could be calculated only by dead reckoning but this showed that during the past hour, R.101 had covered 41 nautical miles and that on the track flown the average headwind component was now around 17 mph.

In fact the wind was veering towards the south and increasing still further in velocity. This was difficult to establish over the sea in the dark and rain but Johnston was doing his best with the navigational aids available to him. But, as Sir John Simon emphasised later,* 'This fact must be brought out clearly, that never, with or without the extra bay, had R.101 ever experienced weather approaching in violence anything met with during this journey.'

During the next hour Johnston dropped seven calcium flares into the Channel from the aft end of the control car to determine drift as the brilliant light on the surface of the water receded into the darkness. By mid-Channel, out of the shelter of the land, the wind had risen to some 44 miles an hour at 1,000 feet, with fierce gusts, blowing from 221 degrees; a gale from almost due south-west.

The result was that, while the airship beat its way across the Channel on a heading of 190 degrees (ten degrees west of south) the track made good was some 139 degrees (41 degrees east of south) at a speed over the surface which averaged only 28.3 miles an hour (24.5 knots).

Arthur Disley, the ship's leading electrician, reported: †

> At about 2230 hours I turned in but for some time before that, possibly half an hour, I happened to be in the chart room and heard Commander Atherstone tell the Flight Coxswain (I think it was L. F. Oughton) not to let the ship go below 1,000 feet. I saw him take the elevator wheel and pull the ship up to this height, which change of height I actually read off the altimeter. At that time Major Johnston was taking drift readings using calcium flares. While I was there Major Scott came down into the room and passed on down into the control

* *Report of The R.101 Inquiry*, Cmd 3825, March 1931.
† AIR 5/903–XM 08,109, 9th October 1930.

car. I did not leave the electrical switch room again after I had turned in at about 2230 hours.

Harry Leech, the Foreman Engineer, added a further comment about this stage of the flight.

> The weather certainly got worse over the Channel and particularly so towards the French side. Over the Channel I judged our height to be not more than 800 feet.

Arthur Cook, in charge of the port mid engine, observed that the white spume on the waves seemed very close, even in the dark and the rain.

As always over the sea the turbulence in the air was less than over land and possibly Ernest Johnston allowed the Height Coxswain to come low to help in establishing his drift sights from the calcium flares.

In any case, Noël Atherstone was well aware of the dangers of flying too low and he had remonstrated with Oughton on earlier flights for allowing the ship to come below 1,000 feet.

By 2236 hours R.101 was four hours out from base and close to mid-Channel with 155 statute miles behind it and about 130 statute miles to go to reach Paris. By now Gent, Leech, Short and Bell in the aft engine car had, at last, fixed the recalcitrant oil-pressure relief-valve. The engine itself was fully serviceable. At 2256 the engine was re-started and brought up to its cruising 825 rpm and 475 horse-power. R.101 resumed its former five engine gait of 55 knots – and so put a still heavier strain on its, by now saturated, outer cover.

For the rest of the flight No 5 engine ran steadily at its full cruising power.

In the smoking room on the lower deck, just aft of the chart-room and wireless cabin, the passengers were enjoying their final cigars with their ports and brandies The coast of France ought soon to be in sight from the deck above and, of course, the conversation must have run on the weather, the progress of the flight, the performance and accommodation of the airship and the prospects for the morrow,* satisfaction combined with an edge of excitement and the flavour of the unknown!

Now the time was coming up to 2300 hours, to the change of

* This, from the author's first hand experience of a series of long-range flights in prototype aircraft on trials, over the course of the years since 1944. The pattern of thought and discussion among those on board during such flights is invariably the same.

watch, to Noël Atherstone handing over to 'Bird' Irwin, and to the 'on the hour' entry in the Journey Log. Still over the sea, the routine figures would differ little from those of an hour before except that R.101 was now on five engines again, the air speed was up but the ground speed was clearly down, though by how much must be guesswork until landfall was made.

Atherstone, satisfied with the airship's behaviour, would go to his cabin after a cup of hot cocoa with the crew members of his watch. They had all been on duty for an arduous three hours since the ship was coming up to Potters Bar on its way to London. Rest was needed before the next watch took over again at 0200 hrs. Atherstone and Johnston were due to come on watch again at 5 a.m. for another critical stint.

Johnston would almost certainly stay up to check the place of crossing the French coast in case the freshening winds had caused an unexpected amount of drift over the sea. Atherstone, dog-tired from more than fifteen hours of almost continuous duty, handed over and turned in.

During the next fifteen minutes R.101 was in contact by wireless with Le Bourget to get cross bearings from Le Bourget and Valenciennes on the airship's position off the English coast.

The smoking room was alongside the wireless cabin on the lower deck and it would be natural for the Chief Wireless Operator, S. T. Keeley, to put his head round the door. There was now a pause in the wireless activities so that (as the evidence shows) Keeley asked whether there were any brief personal messages to send? Nearing midnight the thoughts of many aboard were with those dear ones who were far away.

Brancker and O'Neil scribbled brief notes and handed them to Keeley. They were transmitted to Cardington in succession.*

Serial No 5. 2320 hrs GMT. *From R.101*

To Mrs O'Neill; West London Hospital, Hammersmith.
Good wishes for further progress. All well here.

W.H.L.O'N.

* AIR 5/903.

rior Layout: The ngement of the two s of R.101 in which upper deck comprised lounge, 60 feet by eet, and the dining n 32 feet by 20 feet ther with the enger sleeping cabins. lower deck housed crew's quarters ther with the king room and the rt room, wireless in and galley with ess to the control car, ow, from the chart m.

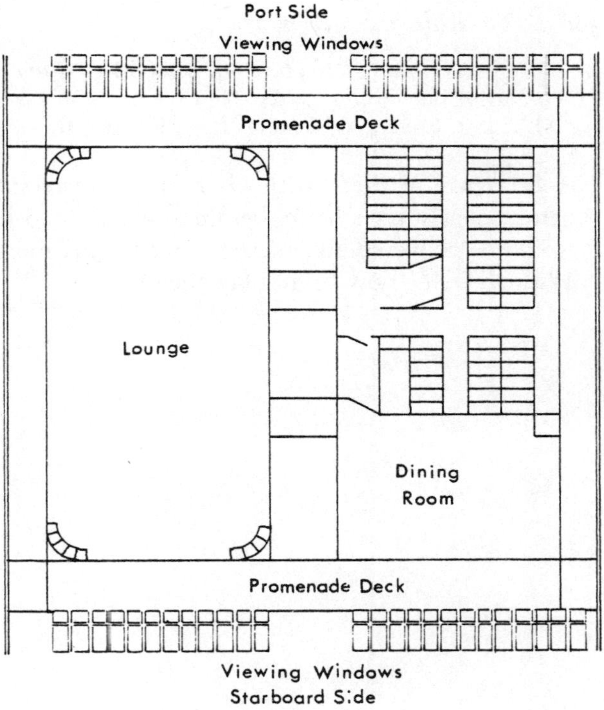

PLAN OF UPPER DECK

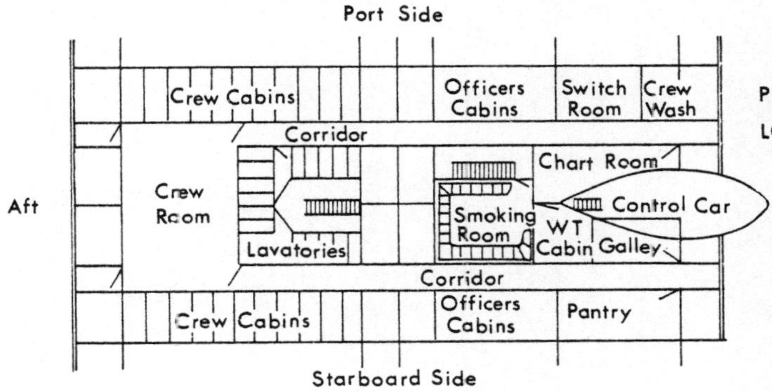

Forward

PLAN OF LOWER DECK

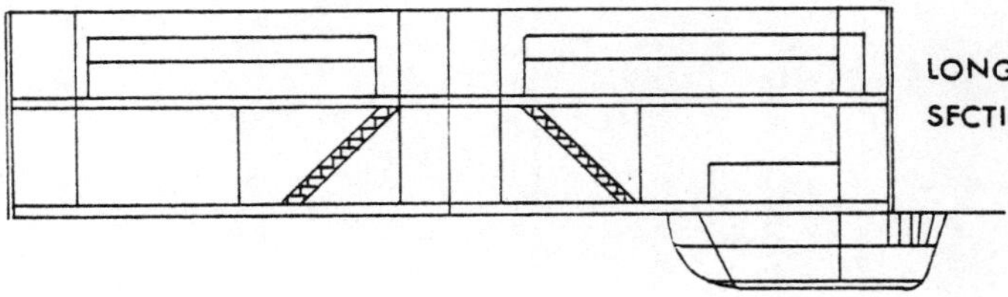

LONGITUDINAL SECTION

Serial No 6. 2325 hrs GMT. *From R.101*
To Auriol Lee, Hotel Elysée, New York.
Off at last. Love and blessings from R.101 – B.

The Secretary of State with his habitual foresight had sent his own telegram nearly nine hours before – from Victoria Street to Posada. It would be delivered, like the rest, on Monday morning.

With all well – now France lay ahead.

CHAPTER TWENTY-FIVE

Allonne

'Hic situs est Phaeton; currus auriga paterni;
'Quem si non tenuit, magnis tamen excidit ausis.'*
Ovid: Metamorphoses Book ii: 327.

4th October 1930

'COAST OF FRANCE in sight.' The first glimpse of the lights on the Pointe de St Quentin, on the north side of the Baie de Somme, would have been at a little after 2300 hrs GMT from the control car beneath R.101's hull where Johnston was peering forward to confirm the point of landfall. Irwin had just taken over from Atherstone as Officer of the Watch – from 2300 hrs to 0200 hrs.

France, at the Pointe de St Quentin, was below at 2336 GMT, exactly five hours from casting off from Cardington; 160 nautical miles away as the airship had flown. The ground speed had thus worked out at 32 knots (36.7 mph) precisely, against a rising head-wind. Three hours and twenty minutes out of the five hours had been on four engines.

The airship's Channel crossing had taken half an hour longer than expected. It was evidence of the change in direction and still greater strength of the winds. When, a few moments later, the flashing beacon on the starboard bow had been identified as that at Neu Brighton by Cayeux-sur-Mer, it was clear that R.101 had been drifted some twenty miles to port of its intended track.

The 50 nautical miles of the Channel crossing had been flown at a surface speed of only 24.8 knots (28.3 miles an hour).

Before going off to bed after his long spell on duty, Johnston plotted a new course on a heading of 200 degrees to bring the airship overhead Orly. The track would pass about nine miles south-west of Abbeville and four miles west of Beauvais. Johnston gave a new estimate of 39 mph for the ground speed on the Orly sector, making allowance for strengthened wind now estimated to have risen to 35 mph from 245 degrees.†

That was, in fact, a miscalculation by an understandably tired

* 'Here at this place lies Phaethon; charioteer of the sun across the heavens;
'Who, if he did not succeed, at least fell in a greatly daring endeavour.'

† Reported to Cardington by wireless from R.101.

navigator. The wind was in fact blowing at about 220 degrees – further round to the south – at 41 mph so that, under its influence, the airship drifted further to the east during its progress southwards.

By this time all the passengers had come up from the smoking room to the viewing balconies on each side of the lounge. From behind the large cellon windows they could now catch their own first sight of France; the lights of Le Crotoy to port and those of St Valéry almost below as R.101 flew up the mouth of the River Somme, of such tragic memories for the British race.

As, two by two, the twelve passengers trooped up the stairs, they passed the dining room where breakfast had already been laid for 0730 a.m. in the morning.

They would all be there together; tall, urbane Christopher Thomson at their head, with ebullient Sefton Brancker; quiet, reflective Colmore; stolid Vincent Richmond and retiring, brilliant Michael Rope; while from the AID there was Alexander Bushfield with Major Percy Bishop of the Air Ministry and with them the two squadron leaders, O'Neill of India and Palstra of Australia. No doubt Major Herbert Scott would have joined them again at this time, now that the first stage of the journey had been accomplished and course had been set from the French coast to Paris.

Now, at not much short of 1 a.m. in the morning by the clock time at which all the passengers had risen (midnight GMT) 'good-nights' would be said and the passengers would troop off to their cabins to sleep before the expected excitements of the day ahead.

The Secretary of State had been allocated two adjoining outer cabins, fore and aft, on the starboard side of the ship against the inside bulkhead of the dining room so that he might use the forward cabin for sleeping and the other as a sitting room and as an office,* both curtained and with rush mats on the floor.

There remained on duty watch 'Bird' Irwin as Officer of the Watch until 0200 hrs, with C. H. Mason as Rudder Coxswain and Flight

* AIR 11/23–07444. 'The two outer cabins on the starboard side, one forward and one aft of Frame 6½ – that is those against the dining room inside bulkhead – are for the exclusive use of the S of S. In the forward cabin, which will be used as the sleeping apartment, the upper berth is to be removed and the fabric in way of berth supports made good. In the after cabin, intended as a sitting room, the upper and lower berths are to be removed; the fabric again being made good. The fabric panel under the lower berth to be covered with 3 m/m plywood secured to cross batten. A table of poplar balsa to be fitted in the bare space. In the passage way, a rail to carry a curtain to be fitted. This means that the two curtains at the entrances to the two cabins are not to be fitted.

'One folding stool to be fitted in each cabin, the seat of Air Force blue. Hinged shades to be fitted to all bulkhead lights in all 28 cabins.'

Sergeant W. A. Potter on the elevators, and G. W. Short as Charge-Hand Engineer.

Further confirmation of the ship's position was sought in a wireless signal to Le Bourget just before midnight. At 2359 GMT Le Bourget reported back to R.101 that the bearing of its signals from Le Bourget was 343 degrees true – which placed the airship on a line between Le Crotoy and Moyenneville, to the west of Abbeville and rather to the east of the desired track to Orly. There seems little doubt that, by now, Johnston had gone off to get some rest. Certainly no action was taken to correct the heading to allow for increasing wind velocity at this stage.

In any case, as midnight GMT came up on the ship's clock, Irwin would be occupied with making the required sixth 'on the hour' entry in the Journey Log. With all five engines running well a true air speed of 54 knots was showing on the Air Speed Indicator.

R.101 was now flying at 1,500 feet in order to give reasonable clearance to the higher ground ahead where the ground west of Abbeville went up to almost 800 feet. The airship's climb to some 300 feet above its former pressure height would have spilled off around 49,000 cubic feet of gas, worth about 1.5 tons of lift, while calculations would have been made for the effects of rain on the cover, for temperature changes and for normal gas seepage. At this time, therefore, after $5\frac{1}{2}$ flying hours, the weight and lift equation could be expected to look approximately thus:

Weight and lift for equilibrium, at pressure height, at start:	157.75 tons
Increase in weight from rain on cover and recovery of water ballast:	5.25 tons
Reduction in weight from consumption of fuel, oil and water:	2.25 tons
Thus: net increase in weight to be lifted at 2400 hrs GMT:	3.00 tons
Reduction in available lift from 300 feet increase in pressure height:	1.5 tons
Reduction in lift from normal gas wastage in $5\frac{1}{2}$ hrs flight time:	0.25 tons
Gain in lift from three degrees drop in temperature:	1.00 ton
Thus: net reduction in lift available at 2400 hrs GMT	0.75 tons
So: At 2400 hrs GMT airship flying approximately:	3.75 tons *heavy*

Experience showed that after some hours of flying in rain the cover would have absorbed about four tons weight of water. The remaining one-and-a-quarter tons of weight would be in the form of water ballast recovered down the fabric funnels into the water ballast tanks. The available water ballast on board would, therefore, now amount to about 6.5 tons within the flying weight of some 161.5 tons. By the discharge of 3.75 tons of this water ballast the ship could be brought into equilibrium but, clearly, Irwin was content to fly R.101 at the couple of degrees of nose-up trim which would give sufficient dynamic lift to offset that extra weight, in the knowledge that it was being reduced all the time by the consumption of fuel.

The air speed to carry 3.75 tons of dynamic lift at a constant 2,375 bhp would be reduced by about one knot, to 54 knots. Such then would be the situation when at 0010 hrs GMT on the morning of 5th October, Irwin handed a signal to wireless operator F. Elliot to transmit to Cardington for repetition also to RAF Malta, Ismailia, Cairo, Karachi and Simla.

The message, based upon the entry in the Journey Log at midnight, read:*

From *R.101* To *Airships, Bedford.*

Code 'Cent' 2400hrs. *Position* 15 miles SW of Abbeville

Course and speed made good from 1830 hrs. Various headings and 33 knots; *Wind* 245 degrees, 35 mph.
Altimeter height: 1,500 feet.
Air temperature: 51 degrees F. *Weather*: intermittent rain.
Cloud: Nimbus at 50 feet. *Conditions* since departure: Similar. *Temperature*: uniform.

Items of interest: After an excellent supper our distinguished passengers smoked a final cigar and having sighted the French coast have now gone to bed to rest after the excitement of their leave takings. All essential services are functioning satisfactorily. The crew have settled down to watch keeping routine. Ends. 2400 hrs.

The message was acknowledged by Cardington at 0018 hrs and was repeated to the AOCs, RAF Malta; Ismailia; Cairo; Hinaidi, Baghdad; Karachi, and Simla at 0050 hrs.

Immediately after sending this signal R.101 followed up – at 0013 – with a further message to Cardington, which was also received direct in the Control Tower at Croydon Airport: 'If you 'phone "G.E.D."

* AIR 5/903.

Croydon please tell them we will not work Direction Finding any more tonight.'

Croydon replied to Cardington at 0023 hrs: 'Am remaining on watch.'

Irwin had now been on duty since before six o'clock on the previous morning, for eighteen hours, with very little rest while shouldering the chief responsibility for the preparations for departure and for the VIP passengers on board. It must have been a heavy load. In principle, the ship's captains should not keep watch but, with only three watch-keeping officers on board, there was no alternative but for Irwin, Atherstone and Steff each to take his turn in three-hour stints.

Irwin was certainly very tired – and this showed up in his midnight signal. The airship had passed over Pointe de Quentin at 2336 and over St Valéry-sur-Somme at 2350 hrs, making only 28 mph over the ground. Ten minutes later, at midnight, R.101 would be 4.6 miles further to the south-west, two miles north of Quesnoy, almost exactly six miles north-west of Abbeville; not 'fifteen miles SW of Abbeville' which would have put the airship between Gamaches and Blangy on the River Bresle. To achieve that R.101 would have had to cover the twenty-one miles from Pointe de St Quentin to near Blangy in twenty-four minutes at a grand speed of 52.5 mph.

R.101 could have been there at midnight only if the wind had sunk to a virtual calm.

A near-exhausted Irwin, highly stressed by his responsibilities and the people on board, must have placed a ruler quickly on the map, and then forgotten to make any allowance for wind though, almost at the same time, he had reported a wind estimate of '35 mph from 243 degrees'.

Over this rolling, wooded, countryside between the valleys of the Somme and the Bresle, there were no obvious landmarks or check-points on this dark, wet, night. R.101 rumbled on with all but a dozen of the fifty-four on board, asleep. On duty in the control car was Irwin with duty Coxswain W. A. Potter and, on the wheels, M. G. Rampton and H. E. Ford. Above, in the radio cabin F. Elliot was listening out. One engineer was on watch in each of the five engine cars under Charge Hand G. W. Short while Cardington's Foreman Engineer Harry J. Leech was making a tour of inspection of the engine cars, and the Chief Electrician, Arthur Disley, was dozing in the switch room.

Just before 0100 hrs, after a brief hour's rest there seems no doubt that Herbert Johnston was back to check on the navigation.

Fortunately, there must have loomed up out of the rainswept blackness below and ahead a useful check-point on the port bow, the lights of the familiar little town of Poix, perched on a 500 foot hill at the intersection of a railway and six roads, each with at least some street lights and readily recognisable from no more than 900 feet above them. Johnston had flown regularly on the Croydon-Le Bourget air route and Poix, with its landing ground to the north of the town, would be well known to him, day or night.

Here then was a valuable navigational fix. It showed that, during the past seventy minutes from St Valéry, R.101 had drifted about fifteen miles to the east of the designated track from the mouth of the Somme to Orly. Johnston gave a new heading to steer to bring the ship further round into wind so as to recover the leeway; 209 degrees to make good a new track of 168 degrees to Orly.* The wind had now risen still further to a mean speed of all but 50 mph from 226 degrees; almost due south-west.

Now was the time for the next entry in the Journey Log to record the position and situation at 0100 hrs GMT.

SEVENTH ENTRY: JOURNEY LOG BOOK

Middle Watch: (23.00 to 02.00 hrs)
Time: 01.00 hrs GMT
Altitude: 1,500 ft
Temperature: 50°F
Wind Direction: 226 degrees
Wind Direction: 226°
Through Air Heading: 209° True
True Air Speed: 54 knots
Position: One mile north-west of Poix
Fuel Used: 5,740 lb (2.56 tons)
Ballast Remaining: 6.5 tons
Weather: Moderate rain

Officer of the Watch: H. C. Irwin
Duration from Start: 6 hrs 24 m
Pressure Height: 1,500 ft
Track: 168° True
Wind Force: 49 miles an hour
Engines: Five at 825 rpm
Ground Speed Made Good: 30.15 knots
Distance Flown: 193 n. miles
Usable Fuel Remaining: 22.33 tons
Cloud Base: Scattered 100 ft

All five engines were running well and giving no trouble. The airship was handling well also, though flying in continuous low-level turbulence in the fiercely gusting wind. Since crossing the French coast

* The alteration in R.101's heading over Poix can be seen clearly from a plot of the airship's course as shown by the observed points and times at which it passed over fixes at Pointe St Quentin, at Poix and at Beauvais.

R.101 had covered 33 nautical miles in 84 minutes at a ground speed of 23.6 knots or 27 mph; even less than on the Channel crossing. The airship was bucking into the teeth of the storm and, on the new heading, making only 22 mph over the ground, its nose pointing to the south west to travel south-south-east. But beyond Paris, en route to Dijon, they could hope for a tail wind.

Instead of passing some ten miles west of Beauvais (twenty-four miles south of Poix), R.101 would now fly just to the east of the city centre and across the famous Ridge which, Johnston would remember, was renowned for its low-level turbulence athwart the London–Paris route. Flying at 1,500 feet R.101 would have a little more than 730 feet of clearance above the Ridge's highest point; compared with 900 feet between R.101 and the ground just south of Poix.

Had Sefton Brancker been awake and following the airship's course he would have, no doubt, remembered a day – two years before, in February 1928 – when he had passed the 774 feet Ridge on board an Imperial Airways Armstrong-Whitworth Argosy en route from Le Bourget to Croydon.

Afterwards he had addressed a note to Captain Freddie Entwistle of the Air Ministry Met Service:

> At about 12.30 p.m. on Saturday, February 11th, I was flying from Paris to London on an Argosy belonging to Imperial Airways when we encountered a most exceptional down current. I estimate that the aeroplane fell like a stone for about 200 feet. The cabin roof was damaged by the impact of passengers flying upwards, one emergency exit was torn and blown off, chairs were broken and two passengers were slightly injured.
>
> During the rest of the voyage, we had a very heavy headwind and the bumps were from time to time unpleasant but, generally speaking, the conditions were not very bad. The weather was thick and I am not sure of our location when this incident occurred but it must have been in the neighbourhood of Beauvais. Could you give me any meteorological explanation of this particular disturbance? I have never experienced anything quite so bad on the London–Paris service though in the East, when flying over mountains, similar incidents are not infrequent. The sky was overcast and we were close under the clouds at about 2,000 feet.
>
> *W.S.B. (D.C.A.) 15-2-28.*

Captain Entwistle replied:

> Strong westerly winds prevailed throughout Saturday along the whole of the Paris route reaching the force of a whole gale over the

Channel. The winds were of polar origin sweeping down the western side of the British Isles and then turning west along the Channel. The air was, in consequence, very unstable, particularly between Paris and Abbeville where the wind had previously crossed a relatively wide stretch of sea. Severe local squalls moved eastwards in this westerly current and, doubtless, gave rise to considerable vertical disturbances in their neighbourhood.

It seems probable that the Argosy was caught in the rear of one of these squalls where there would be a decided downward current. If, as seems likely, the machine was at the time on the leeward side of the Beauvais Ridge, the downward current would have been intensified, the resultant downward current being sufficiently strong to account for the unusual effect on the aeroplane. A sudden, temporary, change of wind direction to a direction from behind the machine – which may have co-incided with the squall – would, of course, have helped matters. – *F. Entwistle*. M.O.6. 17-2-28.

What was certain is that, in high winds, at low levels around Beauvais the air was inclined to be very rough.

On 5th October 1930, at 0124 hrs, near Sommereux, Johnston was manifestly still doubtful about his position. He asked Wireless Operator Elliot to seek another bearing and this time, at 0145 Le Bourget reported the bearing of R.101 from Le Bourget was 347 degrees true and from Valenciennes, 228 degrees true. Johnston therefore had a cross-bearing 'fix' which set the ship just south of Crèvecoeur-le-Grand above a ground level of 550 feet. It was right on course and Johnston could go off to rest again until the ship reached Orly at about 0345 in the morning.

The time was coming up for the change-over to the Morning Watch from 0200 to 0500 hrs and for Irwin, dog-tired, to hand over to Maurice Steff with Chief Coxswain 'Sky' Hunt at his elbow.

R.101 was now coming up towards Beauvais, almost head to wind through bursts of heavy rain with the ground speed down to less than 20 mph, the wind soughing iin fierce gusts.

To those on board, however, by now acclimatised to the motion of the ship and reassured by the steady drone of the engines, all seemed well. R.101 was, indeed, riding the storm. There was an exhilaration in the progress through the night.

The routine of a long flight was now well established, the Middle Watch to take over from the First Watch in the traditions of the Royal Navy from which the Airship Service had sprung.

Maurice Steff, the Second Officer, came down from his cabin into the control car to relieve his captain. 'Bird' Irwin went straight up to take off his clothes, fling on his pyjamas* and to climb into his bunk and, doubtless, to fall instantly asleep. Clearly he felt no concern about the progress of the flight or of the integrity of R.101. The ship was still flying in trim, a little nose-up to carry around 3½ tons of dynamic lift but no more than that. There had been no significant leakage of gas and the airship was maintaining its air speed satisfactorily.

On the crew deck, Joe Binks was still 'out to the world' in his bunk when, at a minute or two before 2 a.m. George Short, the Engineer of the Watch, roused him with a steaming mug of cocoa with the news that his No 5 engine was running well.

'A hurried drink of the hot cocoa,' wrote Binks,† 'a last look around and remark to G. Short and proceeded along the keel through Frames 8, 8a, 9, 10 and into Frame 11 – a distance of approximately 140 or 150 feet.'

On the way, along the dimly lit passage, Binks recognised a well-wrapped figure, the popular and conscientious Squadron Leader Michael Rope, moving forward and then hoisting himself into the framework around the dark surging gas-bags, restrained within the complex cradle of their 'parachute wires'.

Rope, acting as an official observer and passenger on this flight and not a member of the crew, was nonetheless up and busy even at this late hour, making a personal inspection of the bags and outer cover. If there were weak features of the airship these were they and, after more than seven hours of fast-cruise in turbulence, Rope was seeing for himself that all was still secure.

Binks went on: 'I descended the ladder into the Aft Car, the time now being three minutes past two a.m. by engine chronometer (Bill drawing my attention to this in cheerful manner) and some small talk and general information re temps and pressures.'

This, then, was the time in the early hours when human initiative is at its lowest ebb; that time between yesterday and tomorrow when the world seems less than real – and anything can happen.

Aboard R.101 – manfully breasting the roaring gale – the Middle Watch, from 2 to 5 a.m., was to be shared by Steff and Hunt in the control car, by C. H. Mason on the elevators and P. A. Foster on the rudders with L. F. Oughton standing by. These coxswains, 'good lads', would take ten to twelve minutes, once more to gain the feel of the

* Subsequent evidence.

† AIR 5/911–9871.

ship before they could begin to settle down to steady control in turbulent conditions. Mason had taken over on the elevator control instead of Oughton because Oughton had already done a three-hour trick on the wheel between 2000 and 2300 hrs.

Mason, like Oughton, had been liable to let the airship gradually lose height; there was no Atherstone in the car now to pull the ship up to safe clearances over land.

Yet all seemed well and there was no reason for the Old Watch to wait around. The airship was 7½ hours out on its voyage with a long day ahead and everyone needed all the rest he could get. Clearly there was confidence that R.101 would continue ahead through the next hours as it had done since crossing the French coast.

In almost 7½ hours R.101 had been able to put only 248 statute miles behind it – an average of 33 mph, at a consumption of 3.23 gallons of fuel to the mile. They were running about 2½ hours (or 130 miles) behind schedule and things must improve if they were to make the Moascar mast at Ismailia in time for dinner on Monday night. There were about 2,280 statute miles to go by way of Dijon, Lyons, Marseille and Malta and 42 hours to do it in. An average of 54 mph would still just be possible in the winds forecast after Paris. At the present rate of fuel consumption there would be about three tons of fuel remaining on reaching Ismailia – sufficient there for 8½ hours further duration at long-range-cruise.

Steff in the control car would be busy entering up the eighth 'on the hour' Journey Log. Maybe he reflected that there would still be forty-one more such entries to come.

EIGHTH ENTRY: JOURNEY LOG BOOK R.101

Morning Watch: (02.00 to 05.00 hrs)
Time: 02.00 hrs GMT
Altitude: 1,200 ft
Temperature: 50°F
Wind Direction: 225° True
Through Air Heading: 209° True
True Air Speed: 53 knots
Position: Beauvais
Fuel Used: 6,740 lb (3.0 tons)
Weather: Moderate rain
Cloud Base: 1,200 ft

Officer of the Watch: M. H. Steff
Duration from Start: 07 hr 24 m
Pressure Height: 1,500 feet
Track: 168° True
Wind Force: 50 mph
Engines: Five at 825 rpm
Ground Speed Made Good: 29.0 knots
Distance Flown: 215 n. miles
Usable Fuel Remaining: 21.88 tons

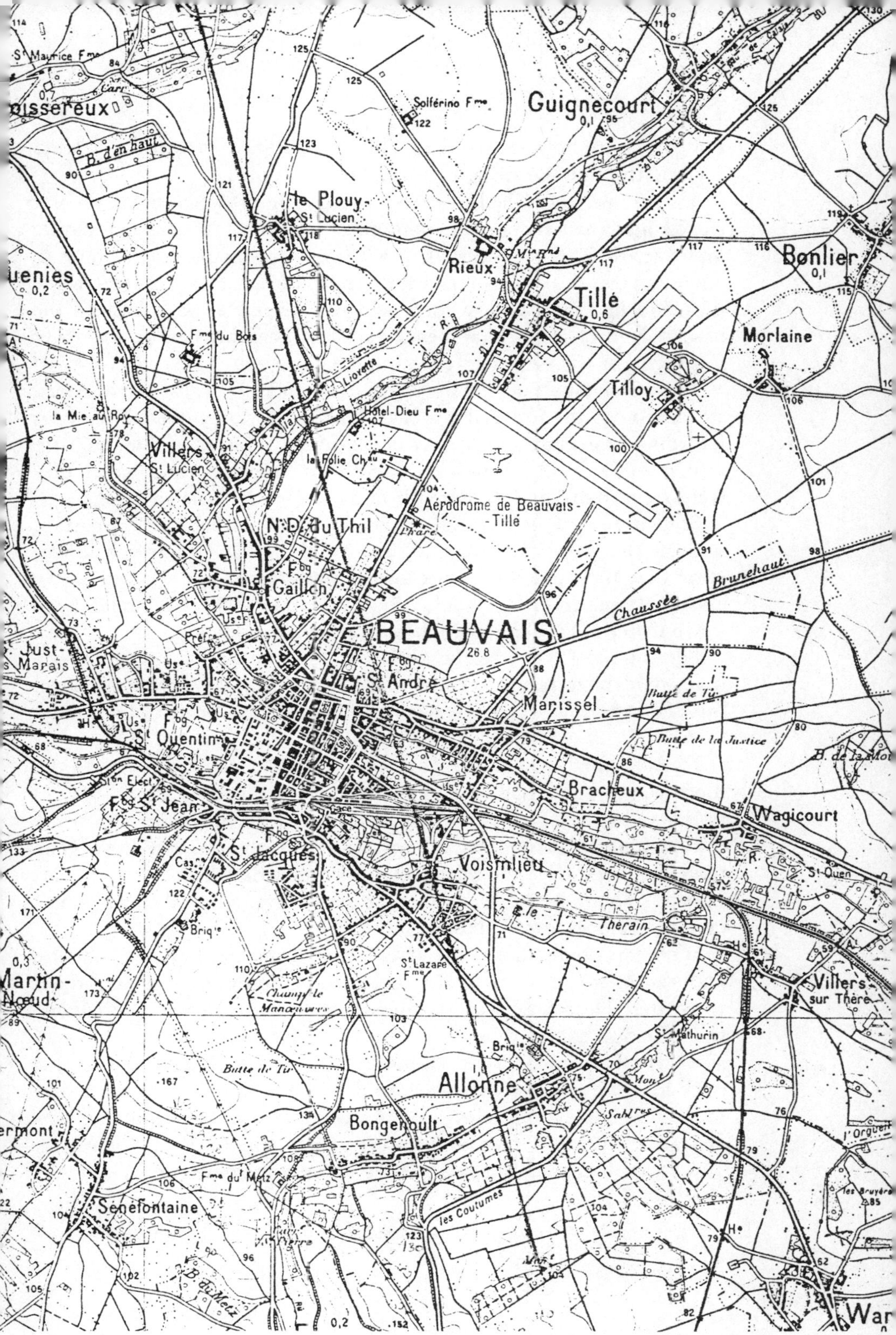

Last Lap: The final stage of R.101's flight in the early hours of 5th October 1930 across Beauvais to come to ground in the Bois du Coutumes just beyond Allonne and Bougenoult.

For $7\frac{1}{2}$ hours now, R.101 had battered its way forward at around 60 mph through the air in a gusting Force 10 gale.

The distance travelled in each successive hour between the regular checks had varied from a maximum of 41 nautical miles (47 statute mph) across Kent and Sussex – with one downwind leg – down to only 22.5 nautical miles (26 statute mph) during the past hour when the wind had increased in force and had veered to be almost on the nose. For most of the time heavy rain had been beating on the cover in stinging shafts – especially around the nose.

The cover was not wholly waterproof and rain must have seeped steadily on to the forward gas-bags. The 'goldbeaters skin' of the bags had the unfortunate propensity of absorbing moisture. They built up weight – the gas-bags weighed nine tons already – and when they were saturated they lost strength and were susceptible to rents.

Still R.101 ploughed ahead. Scott, Irwin, Atherstone, Johnston and all the passengers were asleep. To the twelve on watch it was clear that those off duty must be left undisturbed for the heavy day ahead – and morning would now soon dawn. R.101 would have flown through as wild a night as any of them had known.

For the passengers, as for the crew, the ship had done everything that could have been expected of it. It flew on, comfortable, warm, full of purpose as – surely it would continue to be for almost two days to come.

Dined, wined, at peace with the world and looking forward to the morrow, they would all be glad to sleep.

For Christopher Thomson, instigator of it all, as he lay down for this first night in an airship's berth, his thoughts must have gone tumbling back over the years; over so much that had led up to this voyage. Thoughts of R.101 – certainly. Thoughts of people and events; of hopes and ambitions.

Who can doubt that C.B., with Sefton Brancker, had been the life and soul of the party at dinner that evening and in the smoking-room thereafter. Here now was so much that he had dreamed of coming true. In due course those other dreams might come to pass as well.

So, for C.B. there were – surely – thoughts of Marthe. Thoughts back into the past – to the Rue de Rivoli in Paris, soon to be below; of a summer's day there in 1902, 'of a moment of enchantment, of a girl all in white, a radiant creature under a parasol in a carriage. Just an exchange of glances; a smile – and a memory for a lifetime'.*

* Christopher Thomson's Journal, Paris, July 1902.

Cardington – The burial service in the churchyard of the parish church of St Mary, Cardington, on Saturday, 11th October 1930, when 48 dead of R.101 were laid to rest exactly a week from the day they had left with such high hopes on the voyage to India. In the photograph trumpeters of the Royal Air Force sound the Last Post.

Cardington – The memorial tomb in the churchyard of St Mary, Cardington.

Ensign – The Royal Air Force ensign flown from the tail of R.101 and, tattered and scorched, a survivor of the disaster. The ensign is now mounted on a pillar in the north transept of the parish church of St Mary, Cardington, above a memorial tablet inscribed with the names of those who were lost.

[PLATE XXVIII]

And then, eleven years later in Bucharest, at a soirée held by the English-born Queen Marie, at the Cotrocini Palace, there had appeared 'a vision of delight; a lady on the arm of the King, slim, elegant in a black velvet dress, a mass of chestnut hair piled high on her head and then falling in ripples over a white and slender neck; proudly held.'*

She was the girl of the carriage.

Later that evening she had been introduced to the British Military Attaché, standing in his red and gold full dress uniform of the Royal Corps of Engineers.

'The Princess Marthe Bibesco.'

That had been fifteen years ago. Through all that had happened since, his devotion to Marthe had remained constant – always in the forefront of his mind. And now, the fine Persian carpet which held such deep memories for them both, was on board R.101 – a talisman from the East, soon to grace the floor at the State Dinners to be held at the masts at Ismailia and at Karachi.

Maybe as he was lulled to sleep with the motion of the airship he thought, too, of another sentimental token which had accompanied him on all his travels throughout the fourteen years. It was a shoe which Marthe had left in his car one evening long ago in 1916 in the flurry of their escape from capture by a German flying column of Uhlans during the invasion of Roumania.†

That was a moment when he had posed again to Marthe those questions from a moonlit night in Posada:

> How may one cull the honey and yet not pluck the flower?
> And how can man, once happy, still keep his happy hour?

At long last, to fly to India in R.101 must surely rank high among any reckoning of happy hours.

But now – 'to sleep – perchance to dream'.

And there, we may believe, sleep possessed him.

*

In R.101's smoking room on the centre line of the ship, aft of the chart room and wireless cabin and beneath the deck where lay the sleeping passengers, Harry Leech had gone for a last cigarette before turning in.

* Christopher Thomson's Journal, 24th March 1915, Bucharest.

† Recovered from the wreckage of R.101 and the subject of a special report by Sir Bernard Spilsbury (AIR 5/14).

He had been there for about fifteen minutes, sitting drowsily on a settee against the wall on the starboard side, when he felt the ship plunge downwards in a sudden dive.

> Tables and all loose articles, including myself, slipped towards the forward bulkhead [he later dictated].* Sitting on the fixed settee with a cushion under me, I slipped from about the centre to the front end. I think we must have dived quite a considerable distance before the ship resumed on an even keel; the engines still running at cruising speed.
>
> She remained on an even keel for perhaps a minute, possibly less. I picked up some of the glasses and a soda-water siphon and put them on the table again.
>
> Then the ship dived again at about the same angle as before – and the crash came a moment later. The shock of the impact was certainly not great; it was more of a crunch than a violent blow. I was not even shaken by the impact.
>
> Within two seconds of striking, a blinding flash of fire occurred and this appeared to me to originate immediately above the control car. The door of the smoking room was burst open by the impact and it was through this opening that I saw the mass of flame.
>
> The next thing I knew was that the upper passenger deck had collapsed downwards on to a level with the tops of the backs of the settees in the smoking room, leaving me a space about three feet high. This space was filled with choking fumes but no flames.
>
> I heard people screaming and moaning in the crew quarters and also from the upper passenger deck which was then blazing.
>
> I tore one of the settees away from the bulkhead and managed to scramble through the opening and found myself inside the hull on the starboard side. By that time the outer cover had been completely burned away from that section except for the cellon windows which were still blazing and through which I had to force my way.

Within a short distance of Leech, at 0207 hrs when the first dive began, was Rigger S. Church who had been acting as a messenger from the control car during the previous watch. In hospital, later, his words were taken down :†

> The new watch had just come on and I was walking back to the crew's quarters when the ship took on a steep diving attitude. At this moment I received an order to release half a ton of emergency forward water ballast. But before I could get to it the crash came.

* AIR 5/903.X.M. 08107. Statement by J. H. Leech in Beauvais Hospital, 6th October 1930.

† AIR 5/903. S. Church died in Beauvais Hospital on 8th October.

Arthur J. Cook, the engineer on duty in the No 4 (port midship) engine car, said from a hospital bed:

> I had taken over from R. Blake at 0200 hrs. He reported everything OK and I had a general look round myself. I did not notice any marked change in the trim of the ship until after I had received an order by the engine telegraph to 'slow' (450 rpm).
>
> I was actually reducing engine speed when the ship took up a diving attitude and because of this change I happened to look out of the door of the nacelle towards the ground. I was looking out when the ship struck. After the first impact the ship seemed to rebound and then crashed again and, immediately after this, became enveloped in flames.
>
> My car sank on to the ground quite an appreciable time after the ship struck.* In fact I had time to stop the engine. I remember pieces of structure falling on to and round the car.
>
> My first attempt to get out was unsuccessful and, finally, after a struggle I managed to get clear of the burning wreckage.

In the aft car, Joe Binks was still talking to Bell, when:

> The ship puts her nose down, responds to high engine speeds and quickly loses height. She then returns to a more even keel but, almost immediately, gets into another dive and during this latter change, the pointer of the engine telegraph revolves to that section of dial which indicates to change from present speed to slow speed, namely from 825 to 450 revs.†
>
> Bell instantly performs this operation which was completely carried out.
>
> A crash! Engine is stopped immediately. This also was completed. Explosion and fire.
>
> Aft car does not strike ground heavily but, in the ship's effort to telescope the car is bumped along the ground causing the bottom to cave in; a blaze coming through the opening.
>
> 'Looks as though all is up, Bell!'
>
> But no. Although by this time our car light must have gone out it was not missed owing to the fact of receiving plenty of illumination from the fire which was by this time all around the car.
>
> We endeavoured to quench the flames coming through the bottom of the car; getting too close to the petrol tank and further menacing our chance of escape; this tank for the starting motor now contained something like twelve gallons of petrol and should it become hot it might burst.

* Nos 3 and 4 midships engine cars were mounted well up on the sides of the airship.

† AIR 5/911–9871.

However, it did not do any such thing even though it must have got fairly warm. The heat was certainly affecting both Bell and myself.

At this stage there was no chance or hope of escape through the exit of the car for we could see nothing but fire. There was much smoke also in the car and we begin to think we should be suffocated.

Far better, Bell, than being roasted.

Thought of many things now; family, friends; even discussed what a blow to Cardington Works and the general idea that R.101 could hardly take fire. What a fallacy!

Now, a sudden change and also our saviour. A deluge of water just when it was most needed, in our personal case; Bell and myself getting a goodly share, but I imagine the majority ran off the upper portion of the car and so subdued the flames in the immediate vicinity of the exits.

We could now, thanks to this splendid and merciful water, put our heads out of the car-exit and could see that much structure had either fallen on, or around, it or, possibly, the action of striking the ground had forced us up and the strut girders from which the car was suspended were now resting on the car. One of the girders was athwart ships and inclined downwards just clear of the exit.

I feel sure that my colleague, Bell, like myself, must have been amazed at the transformation of *our* ship. Such pride and confidence we had in her and although not exactly being our home, our lives were very much wrapped up in her in many ways, apart from the fact that it was the means of earning our liiving.

The exit being on the port side we could see straight across and through that side of the ship but we could not see what was beyond – and, personally, I didn't care, for now we were presented with a chance of getting clear of the fire so long as we kept our heads. And so, with wet rags held to our faces we left the car; Bell in the lead and, in spite of various obstructions, we quickly arrived at the ship's side and, jumping or falling clear we landed on the fringe of a plantation of small trees.

A few paces left and in the direction of the port quarter of the ship and we were now in rough grass, it was raining heavily and half a gale blowing diagonally across the ship from the starboard bow and bringing towards us thick smoke and flame from the accommodation parts of the ship.

We were now shouting 'Hello, Hello – anybody out; where are you?' This is soon answered by a voice coming from the starboard side of the ship, replying, 'Here's one – Mr Leech'.

'Where are you and how many of you?'

We gave him our position and we met somewhere about the stern of the ship.

We continued shouting and we are answered by Disley and quickly by V. Savory from No 3, starboard midships car. How glad we all are to see each other; how eager to hear more voices. But none is coming.

Arthur Disley, the Chief Electrician, had been half asleep in the switch room, alongside the chart room on the lower deck. He remembered that:

> I became conscious of the ship dipping a little; not more than it had done before, and this change of altitude seemed to be corrected immediately.*
>
> At that moment, George Hunt, the Chief Coxswain, came into the switch room and said, '*We're down, lads*'. And then, just as I was trying to get up, the ship went into a steep dive which threw me back on to the bed. I then heard the engine telegraph bell ring – and very shortly afterwards the crash came.
>
> I would say the crash was more of a crunch than a violent or sudden impact. I have a faint recollection of 'tripping' one of the electric field switches before the crash came.
>
> The first explosion followed immediately after the crash and was very violent. The two which followed were far less violent. I think the final dive was very steep.

One more significant fact was added by Joe Binks' friend, A. V. Bell. He recalled for the record from Beauvais Hospital on 6th October:

> At about seven minutes past 2 a.m. I noticed the ship rather suddenly change its altitude and take a nose down angle. This dive was quickly rectified and the ship resumed horizontal flight. But at the next moment the nose went down again to about the same angle as before. I then received an order over the engine-room telegraph indicator to go to 'slow' and while I was taking action the crash came. I was thrown forward but not very violently. I am certainly of the opinion that the nose of the ship struck before my, aft, engine car touched.
>
> The first thing I knew after the car hit the ground was that there were flames all round us and, owing to the heat and fumes, we gave ourselves up for lost. A little later water started to run from the ballast tank above the engine nacelle and this doubtless saved our lives. We saw gaps in the flames and managed somehow to scramble out and run clear of the fire. Mr Leech and others joined us as we got clear.
>
> By that time the passenger accommodation was blazing but the bow and stern of the ship only showed patches of fire.

What had happened so suddenly?

At 0207 hrs R.101 had been flying at more than 50 knots through turbulent air exactly as it had been flying for hours past.

At 0209 the airship was on the ground, a blazing wreck. In less

* AIR 5–903. X/M. 08109. A. Disley at Beauvais Hospital, 9th October 1930.

than two minutes six years of high endeavour had been brought to disaster.

At 0207 the new watch would have barely settled down to its three hour tasks. Steff would still be occupied with his writing up of the 0200 hrs Journey Log. The height and rudder coxswains would still be getting the feel of the ship. R.101 was flying at, perhaps, a little below the mean height of 1,200 feet in heavy turbulence above ground varying in height between 200 and 300 feet.

Inevitably – and quite normally in the rough air and with fresh hands on the wheel – the nose of the ship would be hunting up and down, the height of the ship varying by two or three hundred feet around the mean. For a combination of reasons, one of these downwards surges took on a steeper tilt and then, it seems probable, entered a downwards gust in the lee of the Beauvais Ridge, which depressed the nose still further. The power of the engines, inclined downwards, would continue to drive the airship towards the ground.

In such conditions, the Height Coxswain probably took several seconds to grasp the situation. When he did, the evidence shows that he reacted with urgency and spun the wheel to wind the elevators to their full-up position. That would take about twenty-five seconds. The time would be about 0208.

'Full-up elevator' would have one immediate effect on an airship flying at cruising speed. It would depress the tail and so bring the whole ship nearer to the ground, squashing downwards to come gradually into a horizontal position after losing some 700 feet of height. Then, with up-elevator and under power, in normal conditions the ship would begin to climb away from a position uncomfortably close to the ground.

But there was, certainly, something else going on as well during these critical sixty seconds of time.

Arthur Disley, dozing in the switch room, recalled that at the moment at which the airship levelled out, 'Sky' Hunt had come in from the chart room, moving aft towards the crew quarters, calling out *'We're down, lads'*.

Why?

For Hunt, one of the most experienced of British airshipmen, to have had time to size up what he must have judged to be a sudden and desperate situation, to have decided to leave the control-car and climb up to the lower deck to bring a warning to the crew, something definite, drastic and critical must have happened. Further, it must have become known and obvious to those in the control-car before the airship entered its first dive at about half-a-minute or so after 0207 hrs. Otherwise the known times do not fit.

This supposition is further confirmed by the despatch of Rigger S. Church (almost certainly by Hunt) to release half-a-ton of water ballast from the nose. At the same time, *in extremis*, and in order to save weight, there is no doubt that some fuel would have been dumped from the quick-release 'tin-opener' mechanism which connected the control-car to some of the fuel tanks. The contents would fall beneath the struggling airship.

There seems only one explanation which would account for a decision to take such sudden and positive action from the control-car remote from the rest of the ship and at the moment when the R.101 was levelling-out from what, without a specific warning, would have appeared to have been no more than a severe oscillation.

A significant fact was that, at that hour of two o'clock in the morning, only one man was out and about, other than those who were on watch in the control-car, in the wireless cabin, in the engine cars and as stand-by riggers – a dozen in all.* That man was Michael Rope, the best airship 'all-rounder' of them all and the most knowledgeable about the structure, the fabrics and the flying characteristics of R.101 and, indeed, of airships in general. As Assistant Chief Designer at the Royal Airship Works for the past six years (and, earlier than that, responsible for the successful SS Zero non-rigids) a pilot and a practical engineer, he knew and was familiar with every aspect of the business.

When Binks walked down the companion-way to his engine car at a little after 0202 hrs he had met Rope coming forward, bent on inspection of bags and cover, well aware of the potential weaknesses after so rough a ride through heavy rain.

In all the circumstances, virtually the only way in which those in the control car could have learned of a potentially catastrophic failure in the remote forward part of the ship, would have been if this dire condition had been discovered, and instantly weighed up, by Rope. Then, post haste, he would either have hurried back to the chart room or, possibly, he might have spoken to Steff in the control-car through the speaking-tube system running aft from the nose gangway.

What could he have found?

The evidence (and not least, the established slowing of the ship's forward speed south of Beauvais) adds up to the probability that, at that moment, a split in the outer cover was just beginning to develop

* Evidence from discussions with the survivors.

at the most highly-stressed place of all, on the upper surface just aft of the nose. Through a sad mischance that was the section where the original cover had not been renewed but merely reinforced with stuck-on fabric bands. After 7½ hours of saturation and battering by an airstream of more than 50 knots, the fabric bands and the section of the original cover could well have been starting to come apart.*

Rope would recognise, at once, the hazards of any such damage; the disastrous effect of the airstream on the saturated and weakened forward gas-bags and the likelihood of a rapid loss of lift at the forward end of the ship.

Certainly only an urgent report from such a knowledgeable and authoritative source as Michael Rope about a failure of this sort, could have produced such a rapid and positive reaction in the control car.

Established facts from survivors are that the experienced Chief Coxswain George Hunt, who had been in the control-car with Steff, came up the ladder into the chart-room where he found Rigger Church and sent him forward to release water-ballast by hand. Hunt then ran aft to warn the crew. While this was going on the nose of the airship had inclined downwards and the first dive had begun.

The action of sending Rigger Church forward from the chart-room to release water ballast, and the decision of Hunt to leave the control-car to give a warning, must have been initiated from a knowledge of a serious situation before the first dive began.

Rope's presence among the bags and Rope's detailed knowledge and grasp of what would be critical to R.101's survival was, thus, by far the most obvious and probable link between the spreading danger and those in the control-car.

Hence, a recapitulation of all the facts (including important evidence which has been established since the Official Inquiry) suggests that the only sequence of events which would seem to fit all the known circumstances would have been that at about 0206 hrs Rope dis-

* An analysis made at Cardington in April 1931, after the publication of the *Report of the Official Inquiry*, made the point that the steadiness of R.101's outer cover depended upon: 'a positive internal pressure supplied through comparatively small orifices at bow and tail. Any damage to the outer cover would release this pressure and perhaps cause flapping. Further failure would considerably increase the fabric tension in the bows. The dynamic effect of heavy rain would also increase the tension in the bow fabric. At speed, this loading on the bow fabric might considerably exceed that to which it had ever been subjected since the removal of the support girders in the bows. If failure of the bow fabric occurred it is quite possible that the inrush of air striking the forward gas-bags would cause comparatively rapid deflation of bags exposed to the airstream.' (AIR 5/380 Pt II X./R.08109.)

covered that a failure was beginning to develop in the original outer cover between Frames 2 and 3,* where the inrush of air would certainly lead to a collapse of the front two gas-bags. Rope would have made haste to report this situation while the airship was still in more-or-less level flight. The height coxswain might, indeed, have already begun to wind on some 'up-elevator' to keep the nose from beginning to drop.

If this was so, it is probable that the combination of some loss of dynamic lift, of a leak beginning to develop in one of the forward bags, together with a down-gust on the nose in the turbulent air, would have begun to over-ride the amount of elevator applied and so result in the beginnings of the first dive – at a little after 0207 hrs and 30 seconds. A close analysis of the time scale suggests that a simple explanation of damage to the nose, followed by the gradual running out of up-elevator to keep the nose from dropping – which would, of course, cause desperation in the control-car – was not the likely explanation for the catastrophe. There must have been some prior warning of damage in order for Hunt to sum up the situation and reach the switch-room from the control-car at the moment he did.

What is certain is that at about 0207 and 35 seconds the nose of the airship began to go down and a dive to begin.

Even so, things were still not desperate. In analysing the stability and control of R.101 after the accident, the National Physical Laboratory commented upon the high degree of dynamic stability of the airship in fast-cruise regime, with a powerful elevator capable of exerting positive control even when two forward gas-bags were fully deflated.†

* This part of the outer cover (where the fabric had not been renewed) is the area at which the dynamic lift would be exerting its maximum force and where, following Arthur Gerrish's ideas, a short section of metal skin would have brought security, at a penalty of rather more than half a ton of weight. A semi-circular band, 90 feet long and of about 25,000 square feet of surface area would have weighed some 3,250 lb in duralumin sheet, 0.0098 of an inch thick, compared with the weight of 1,835 lb for the doped cotton fabric with its attachments which it would replace. The increase in weight for a metal skin over this upper section of the hull would, thus, have been 0.63 of a ton.

† AIR 5/901. The NPL report added that there must have been a positive reason for R.101's first prolonged dive, 'not just an unfamiliar coxswain. Something catastrophic must have occurred to upset the trim of a stable ship' (R & M. No 1401 (Ac. 522) of May 1931). R.101 had smaller tail surfaces than R.100; deliberately designed to achieve the best stability and control at normal cruising speeds of between 50 and 55 knots. Because the ship spent most of its trial flights at slower speeds on three or four engines, it gained an uninformed reputation for a somewhat sluggish response to its controls. At cruising speed the handling reports show that R.101 handled in a precise and satisfactory manner.

What is clearly established is that the application of full-up elevator did pull R.101 out of the dive and held the airship level for perhaps 20 seconds at a height of, probably by then, no more than 500 feet above the ground.

At that moment there was just one chance of survival, one chance to pluck safety out of disaster. Sadly, but understandably, it was not taken.

Full engine power with full up elevators would not only have continued to check the dive but would also have driven R.101 upwards even in its dire condition.* The airship would have climbed slowly away and time would have been won for the jettisoning of more ballast and some of the remaining 22 tons of fuel-oil to lighten and to trim the ship. The situation was desperate but not catastrophic. In somewhat similar circumstances on 27th March 1929, the vastly experienced Hugo Eckener had called for full power and had saved the *Graf Zeppelin* from ploughing into the ground in the Danube valley.

With such action R.101 could have regained height and then limped home, down-wind, flying on low engine powers to reduce the strain on the cover and to conserve the remaining fuel.

But with an urgent report of damage forward there would, no doubt seem – as a snap judgement – to be an over-riding need to reduce speed.

For Steff, relatively inexperienced, in the control-car of R.101 the instinctive and obvious reaction was to shut down power. He did just that. It was the fatal step.

So, tragically, but understandably, Steff took exactly the opposite action to that which would have saved the ship. He rang the engine telegraphs for a reduction in rpm from fast-cruise to slow.

Bereft of power, bereft of forwards way, bereft of dynamic lift, R.101's nose fell away again slowly and then more steeply. At 300 feet above the ground the airship was less than twenty seconds from disaster.†

At 0209 hrs, at a forward speed of no more than ten miles an hour over the ground, the under-side of R.101's nose drove hard into

* A total of 3,250 horse-power was available from the five engines at 935 rpm for emergency use compared with the 2,375 bhp then being developed at 825 rpm. The analysis by the NPL showed conclusively that at full power R.101 had sufficient elevator effectiveness to enable the ship to climb away with two forward gasbags gone.

† See Pages 427 and 428 for an analysis of the sequence of events.

the thick undergrowth of the Bois de Coutumes on the 200 feet contour-line in a little valley just south of the hamlets of Allonne and Bongenoult.

The control-car beneath the hull was crushed up into the main structure and with it a row of calcium flares, slung along the sides, were split open against the soaking ground and brushwood. In such conditions they would instantly ignite to start the fierce fire below the passenger accommodation so clearly remarked upon by all the survivors from positions further aft.

At the same time the two forward engine cars, attached below the hull, were swung round and pushed up into the envelope where their hot exhaust pipes would ignite the mixture of hydrogen and air around the ruptured bags.

Christopher Thomson, with forty-seven more besides, had come to die in France.

In a moment, in the high wind and the driving rain, the whole countryside was lit by the glare of flames.

Eight survivors, four of them from the three rearward engine cars, one from the fire-proofed smoking room, scrambled somehow out of the wreck and stumbled across the wet grass beside the wood; shocked, scorched, bleeding; two of them fatally injured.*

They watched helplessly as the holocaust consumed R.101, consumed the future of British airships, consumed the work, the hopes, the ambitions, the wit and the wisdom, the achievements and the miscalculations too, of a gallant company of men.

In two minutes of disaster a door had closed on a chapter of aviation's history.

*

On Christmas Eve of 1930 there arrived at the Rue de l'Université in Paris a basket filled with red roses : 'Variety : General Jacqueminot'.

The envelope with them was addressed, in a bold, a flowing and a familiar hand, to '*S. A. La Princesse Bibesco*'.

The card inside was headed '122 Ashley Gardens' and the date 28th August 1930; an anniversary of eleven years before; 'for delivery on Christmas Eve'.

* The survivors were: A. V. Bell (Engineer: No 5 aft, car); J. H. Binks (Engineer: No 5, aft, car); A. J. Cook (Engineer: No 4, port mid, car); A. Disley (Electrician-Wireless Operator); H. J. Leech (Foreman Engineer); V. Savory (Engineer: No 3, stbd mid, car). S. Church (Rigger) died in Beauvais Hospital on 8th October and W. G. Radcliffe (Rigger) died in Beauvais Hospital on 6th October 1930.

The note read :

If not September, nor even October, then for Christmas and for always.

Here, with all my love now and for ever, some more of our roses of yesterday and for tomorrow. They come for one who has brightened fifteen long years since 1915 – so long ago and now so far away. They come from one who cheerfully then began to suffer that incurable malady which some call love –and I call *Marthe*.

A Very Happy Christmas, dearest, though we be apart.

From my heart and mind, with my devotion eternally :

Ever your –

Kit

CHAPTER TWENTY-SIX

Swansong

7th–10th October 1930

THEY brought them back to London on Tuesday, 7th October 1930, from Boulogne to Dover on the sea they had overflown. This time they were on board the destroyer HMS *Tempest*, aptly named.

On Friday, 10th October, the forty-eight coffins, each draped in a Union Jack, lay in state in Westminster Hall, C.B.'s 'most beautiful Gothic room in the world' where, on 28th July, he and Marthe had said good-bye.

Thousands filed past all day; from eight o'clock in the morning until half an hour past midnight. The quiet queue stretched all the way past Big Ben and the Boadicea statue to the Royal Air Force Memorial and Cleopatra's Needle from Egypt, along the Thames Embankment. The whole nation had been stirred by the drama of the disaster, just when another aviation triumph had been expected to add to the air speed and distance records already held by British aircraft.

On the same day, at noon, a Memorial Service was held in St Paul's Cathedral in the presence of a great congregation led by HRH The Prince of Wales accompanied by the Lord Mayor of London, the Prime Minister and all the members of the Cabinet. They were joined by *mon cher enemie* Laurant Eynac, by General Balbo and a legion of friends and relations, among them Bell, Binks and Leech, the three survivors well enough to be present.

R.101's scorched and tattered flag which, as Scott had said it would be, had flown proudly from the stern on the ship's first flight abroad, was now draped over the Cathedral's altar.*

At the same time, a Requiem Mass was said in Westminster Cathedral, a stone's throw from 122 Ashley Gardens. The service was conducted by Father Harry Rope, brother of R.101's Michael Rope.

*

A heartbroken and prostrate Princess Marthe Bibesco remained in seclusion at Posada. On Sunday afternoon, the 5th October, before the news reached Roumania, she had written:

* R.101's Royal Air Force ensign now hangs on a pillar in the north transept of Cardington's Parish Church of St Mary where it was installed at a service on 28th September 1931.

Early this morning, a little after 2 a.m. I awoke with a pain in my heart, so violent, so sudden, so inexplicable and yet so persistent that, after half an hour's struggling with myself, I had to awaken the servants and get them to send for a doctor. He came, nobly, still in the early hours.

His report was, perhaps, a heart attack?

But he tells me that he can find nothing obviously wrong.

In spite of sedatives I still cannot sleep.

With daylight, however, the pain continues. Now, writing this Sunday afternoon, I feel weak and depressed and I have stayed in bed.

Perhaps another night I shall be all right again and I can begin to think of my next journey back to Paris; and then of Kit's presence there again after his return from his long voyage.

What should I do without those times together to tide me through those long, lonely, months apart?

*

In London, on Saturday, 11th October, the dead of R.101 under their Union Jacks were lifted on to twenty-four artillery-waggons, each drawn by four horses, to start the long journey through London – down Whitehall, past Gwydyr House, through Trafalgar Square and on to Euston.

The two-mile long procession was made up, behind the cortège, of detachments from the Royal Air Force, the Royal Navy, the Grenadier and the Welsh Guards and with them the spare watch of R.101 (those who had been dismayed to be left behind) and the officers and crew of R.100. Following the cars with relatives of the dead were cars carrying the Prime Minister, the members of the Cabinet and representatives of the Dominions, of India, of foreign governments, the Air Council, the Board of Admiralty, the Army Council and of airshipmen from around the world.

That afternoon the forty-eight dead of R.101, only twenty-six of the coffins named, (among them those of Noël Atherstone, Sefton Brancker, Reginald Colmore, Carmichael Irwin, Vincent Richmond and Michael Rope – but not that of Christopher Thomson)* were

* Four of those who were killed in the crash of R.101 were identified at Beauvais. They were: Richard Blake (Engineer: No 4 car); W. A. Potter (Assistant Coxswain), Sydney Ernest Scott (Charge-Hand Engineer) and E. G. Rudd (Rigger). All except S. E. Scott were off duty after taking their watch up till 02.00 hrs. In addition: Samuel Church (Rigger) and W. G. Radcliffe (Rigger) died in Beauvais Hospital. After the bodies had been brought back to England a further twenty were identified through dental records and personal effects. They were: G. H. Atkins (Wireless Operator); Lieutenant Commander Noël Atherstone (First Officer), Sir Sefton Brancker (DCA); James Buck (Lord Thomson's valet); C. H.

taken by special train from Euston to Bedford, carried on gun carriages back to Cardington and laid to rest in a single grave in the country churchyard and within sight of the Royal Airship Works and of the empty mooring tower whence they had departed, so full of hopes and confidence exactly a week before.

*

A noticeable fact is that those who were not identified included most of the passengers who were in the sleeping quarters on the upper deck immediately above the control car, together with all the stewards who had accommodation alongside the Chart Room above the control car.

Because of the calcium flares, and the igniting water on the rain-soaked ground, the control car became the seat of the initial, and the fiercest, fire.

The wreckage of R.101 lay on the hillside beyond Allonne and Bongenoult between Frocourt and Warluis.

Early in December 1930 Marthe, dressed all in black and accompanied only by the Abbé Mugnier, friend also of C.B. since 1919, paid a tearful visit to Allonne where still lay the gaunt skeleton of R.101.

> Voilà le dénouement – Here is the denouement, said the Abbé. Here is the end of a long, long journey. Yet, in the end, all will be well. We shall be reunited with all our former mortal attachments.
>
> Alors, chacun, sera libre d'aimer.*
> Then, each will be free to love.

*

Burton (Engineer); Alexander Bushfield (AID), W/C R. B. B. Colmore (DAD), Frank Elliot (Wireless Operator); W. R. Gent (First Engineer); M. A. Giblett (Chief Met. Officer); Flight-Lieutenant H. C. Irwin (Captain); W. H. King (Engineer No 2 car); Wilfred Moule (Engineer No 2 car), Squadron Leader W. H. L. O'Neill; L. F. Oughton (Assistant Coxswain); M. G. Rowston (Rigger), A. J. Richardson (Rigger), Lieutenant-Colonel Vincent C. Richmond (ADAD Tech.), Squadron Leader F. M. Rope (Asst. to ADT), C. E. Taylor (Rigger).

Not identified were: Major Percy Bishop (AID), C. J. Fergusson (Engineer), H. E. Ford (Rigger); P. A. Foster (Assistant Coxswain); E. A. Graham (Steward), A. C. Hastings (Engineer); F. Hodnett (Steward), G. W. Hunt (Chief Coxswain), Squadron Leader E. L. Johnston (Nav Officer); S. T. Keeley (Chief Wireless Operator); T. Key (Charge-Hand Engineer), M. F. Littlewick (Engineer), C. H. Mason (Assistant Coxswain), J. W. Megginson (Galley Boy); A. W. J. Norcott (Rigger), Squadron Leader W. Palstra, A. H. Savidge (Chief Steward); Major G. H. Scott (ADFT); I. G. W. Short (Charge-Hand Engineer), Flying Officer M. F. Steff (Second Officer), Lord Thomson, A. H. Watkins (Engineer).

* Recorded by the Princess Bibesco. *La Vie d'une Amitie* Volume 2. Page 389.

An official Public Inquiry to investigate the accident was begun in London on 22nd October 1930, under the authority of the new Secretary of State for Air, Lord Amulree.

Viscount Simon (then Sir John Simon, KC) was appointed to preside over the Inquiry with two Assessors; Lord Brabazon of Tara (then, Lieutenant-Colonel J. T. C. Moore-Brabazon, MC) and Professor Sir Charles Inglis, OBE, FRS, of Cambridge University.

The Court of Inquiry held thirteen public sessions at the Institution of Civil Engineers starting on 28th October and ending on 5th December 1930. The Court heard evidence from forty-two witnesses including the six survivors of the disaster. It produced a 129-page Report, published on 27th March 1931 (Cmd 3,825). That report, elegant in its language, but incomplete in its researches and imprecise in its conclusions, found that the disaster was caused 'by a substantial loss of gas in very bumpy conditions' and that no one was to blame.

By modern standards of accident investigation the Inquiry was perfunctory and went into little detail or analysis of the events and considerations which led up to R.101's departure for India. It made virtually no study either of the course of the last flight itself or of the experience, the qualifications or the attitudes of those concerned.

A Minister had died. The Prime Minister's chief confidant had gone. Leaders of British civil aviation had perished with him, notably the popular and hard-working Sefton Brancker, and alongside them – with a few notable exceptions – all the skill and experience of the British airship service.

Who had the knowledge, or the heart, to look more closely at the actions and the motives which had led up to what was, up till then, the world's second greatest air disaster? *

A final tribute was paid in a letter to *The Times* on 19th October 1930 from Professor R. V. Southwell. He wrote:

> May I as one who has watched in hope and admiration the design, construction, and early trials of R.101 add my word of tribute to friends I have lost?
>
> Barely a week ago Richmond sat by my side at the Air Ministry discussing with other members of a small Committee how best to apply experience gained in the design of R.100 and R.101. It was characteristic of the man that, expecting every moment his recall for the flight on which his thoughts had centred during six most arduous

* The worst air accident, up till 1930, had been the wreck of the ex-German Zeppelin IZ.114, flown by the French Navy as the *Dixmude* and destroyed by lightning off Sicily with the loss of fifty lives on 21st December 1923.

years, he could bend his whole attention to problems more remote; planning investigations which might lead to improvements in future designs.

To him, as to Colmore, Scott, Rope and others who I was privileged to know at Cardington, this first spectacular voyage to Egypt and to India was not in itself the goal but a step on the road to their vision of safe and rapid travel throughout the Empire.

Now they are dead and with them has perished all but the whole of that accumulation of skill and knowledge which seemed at long last to have repaired our losses in the R.38 disaster. Whether in our time such store will again become available I do not dare to guess; for it must come, if it does, not by a mere association of persons skilled in various ways but by their fusion into one common fellowship; single of purpose as these men were.

No one can truly gauge their work who has not watched them at their work at Cardington. By those that have, the words of Sir Samuel Hoare, (close friend, though political rival, of Lord Thomson), will be seen as simple truth : 'They were as a religious community intent upon a high ideal. Neither disappointment nor success deflected them from their life's work.'

Of this I am certain; that in so far as man's thought and labour could have ensured success, the best of man's thought and labour was given without stint; that no chances were taken where certainty could be attained. The spirit in which that work was conceived and brought to completion is known to few who are alive today.

To those who were so fortunate as to have some share in the great adventure one possession will still remain – the memory of a gallant company bound together by selfless devotion to the cause they had at heart.

*

There survived, and survives still, the memory; the tattered ensign – and the flying carpet.

END

EPILOGUE

HOW IT HAPPENED

I

The Loss of R.101

AFTER YEARS of research and consultation, after detailed analyses of all available records and information and applying modern techniques of accident investigation against a background of aeronautical experience, the true causes of the disaster to R.101 can now, I believe, be established with reasonable certainty. They differ materially from the suppositions arrived at in the Report of the Official Inquiry published in March 1931.

Few major accidents are caused by one single or simple failure or deficiency. Almost always a series of misadventures and miscalculations lead to a situation from which there is no return. So it was with R. 101.

Had an accurate weather forecast been available at Cardington two hours earlier, there can be no doubt that a decision would have been taken to postpone the departure of the airship from 18.36 GMT on 4th October 1930, until the belt of high winds and rain had passed across Southern England and Northern France. It did that by shortly after dawn on 5th October.

Had there not been a delay in slipping from the Cardington mast because of difficulties with the Beverley starting-motor in No 1 engine car, R.101 would have been on its way twenty-five minutes earlier, before the rain began and the wind started to rise. Less time would have been spent before setting course and the airship would have reached the critical Beauvais area well before the change of watch at 02.00 hrs.

The more experienced Irwin, as Officer of the Watch as well as Captain of the ship, together with two coxswains by now fully conversant with the feel of the ship, would have been able to cope better with the crisis when it arose after 7½ hours of flight.

Had the oil-cooler on No 1 engine not failed on the last trial flight on 1st October (it was thought by someone stepping on it), a full-speed trial on all five engines would have been run and R.101 worked up to a speed of some 62 knots (71 mph) on 3,250 bhp. Under such a test there is a possibility that the weak portion of the cover would have split either then, or at an early part of the flight towards India, when such a failiure could almost certainly have been managed without disastrous consequences.

In any event, there had been a cardinal misjudgement in the decision not to renew the portions of the original cover which had been doped in situ when the remainder of the, pre-doped, cover was replaced. A few hundred pounds saved there possibly cost the future of R.101 and all aboard.

An added factor was, certainly that the officers of R.101 were suffering from severe fatigue at the time of departure. Most of them had been on continuous duty, with only brief snatches of sleep, since the ship had been brought out of its Cardington shed in the early hours of 1st October. Consequently the judgement and performance of officers and crew were, inevitably, below par when the flight began.

In addition, just after 02.00 hrs on 5th October, R.101 had been allowed to drift down below a reasonable height above rising ground in a region of heavy turbulence. There was inadequate room to manoeuvre when things went wrong following the change of watch.

All of these were contributory factors but none of them was catastrophic in itself. Together they built up – as in a Greek tragedy – to a situation in which only the final straws were needed to bring about irrevocable disaster.

For R.101 the final straw – the ultimate, but avoidable, single cause of the catastrophe – was the error of judgement in throttling back the engines.

For many years the clue to this fact has lain unheeded among the wealth of abstruse and detailed Reports and Memoranda from the Royal Aircraft Establishment at South Farnborough, Hants, and the National Physical Laboratory at Teddington.

Buried among fifty-six other official papers in 'The Technical Report of the Aeronautical Council' for the year 1931/32, dated June 1932, and addressed to the Secretary of State for Air,* and labelled 'For Official Use,' are two mathematical and wind-tunnel analyses of the final flight path of R.101 : '*R & M 1400* by Dr R. Jones and A. H. Bell' and '*R & M 1401* by Dan H. Williams and A. R. Collar'. They contain within them the answers to the questions of why R.101 crashed. Embedded deep in the complex equations of lift and drag, of thrust and pitching moments and among reports of wind-tunnel tests, lies the solution to the whole tragic affair.

Because by 1933 the Report of the Official Inquiry was over and done with, because the case was closed and British rigid airships were

* *Aeronautics*. Technical Report of the Aeronautical Research Committee for the Year 1931/32 (with Appendices) Vol 1, Aerodynamics, Airships and Meteorology. Crown Copyright Reserved. HMSO 1933.

dead, these Reports and Memoranda have never previously been critically examined nor their conclusions remarked upon.

A few quotations, from among the mathematical data and the alternative cases examined, will suffice:

First, the background. From the Aerodynamics Sub-Committee (Page 50) the statement is made that, as a result of the calculations and research 'on possible paths of the airship during the last moments of its flight it seems clear that the airship with its extra bay had satisfactory stability and control characteristics'.

Then, (*from R & M 1400* (Ac 521) Paragraph 37 Page 584):

> All the above analyses tend to show that the controls provided on the airship were more than adequate to keep the head of the airship well up even after losing lift to a considerable extent. It definitely indicates that some additional assumptions other than loss of gas should be sought in order to explain the final motion of the ship before she struck the ground.

Finally, (*from R & M 1401* (Ac 522) Pages 607 and 608):

> Heaviness in the nose alone would not give a dive comparable with that experienced by the ship. The ship could always be brought out of the dive by putting the elevator up.

But – 'shutting down the engines to zero thrust (halving the revolutions would produce the effect) would result in the second dive'.

And further:

> The conditions of the final situation would be justified by the following considerations:
>
> The ship was flying in gusty weather and would undergo changes of height of 200 feet on each side of the main flight path.
>
> The Height Coxswain would make elevator movements of anything up to eight to ten degrees (at the rate of one degree a second) to counteract this.
>
> A possible condition is that the ship would be climbing nose-up and the elevators would be put down to bring her horizontal. In the initial conditions chosen the ship was assumed to be five degrees nose up with the elevators seven degrees down.

This would induce an initial dive accentuated by a downwards gust.

Analysis of the effect on the ship 'with various amounts of gross heaviness through loss of gas shows that the ship had ample control surfaces for keeping the axis pointing upwards at the nose even after a considerable gas leakage and a possible damage to the outer cover had caused an increase in drag; the second dive being caused by the shutting down of the engines.'

(Page 615): 'Shutting down the engines and dropping ballast causes the ship to lose speed and therefore aerodynamic moment and, after being horizontal, the nose drops reaching an angle of 14 degrees.'

And, sadly, that was that.

But there were additional factors in the tide of events which came about with such dramatic and tragic suddenness after the change of watch.

The detailed evidence from survivors and French witnesses on the ground, together with the stopped watches and the chronometers in the engine cars, all confirm that the crash came at almost precisely 0209 hrs.

Analysis of the times sets the start of the first dive at about 0207 and 35 seconds. Engineer Bell, in the aft engine-car, put the start of the dive 'at about seven minutes past 2 a.m. or a little after', by the engine chronometer. Binks had come down into that car at between 0203 and 0205 hrs and Bell and Binks had exchanged shouted remarks for, they both thought, about two or three minutes before the first dive began.

At about the same time Rigger Church was walking back down the companionway to the crew's quarters aft of the chartroom (above the control-car) following the end of his watch, when he received an urgent order to go right forward – a distance of about 300 feet – to release emergency ballast at Frame O. Just as this order was given to him the ship began its first dive. It hit the ground before he could get to the ballast bags. Church had, however, reached a position well ahead of the control car when the airship struck and so escaped being engulfed in the instant violent fire from its calcium flares.

The significant fact which comes from this is that a decision to drop ballast from the nose had been taken from the control car *before* the first dive began. Moreover the only person who could have been in a position, then, to give such an order would have been the Chief Coxswain 'Sky' Hunt.

In seeking to establish the relevant facts and the sequence of events, specific statements of the survivors offer vital clues:

1. *Church* (who died shortly afterwards):

I was walking back to the crew's quarters when the ship took on a sudden steep diving attitude. *At this moment* I received an order to

> release half-a-ton of emergency forward ballast. But before I could get to it the crash came.

The order could only have originated from the control car and must have been initiated before the first dive began. It would appear that it was given to Church, therefore, some time after 0205 and before 0207–40 hrs.

2. *Cook* (In the No 4, port midships, engine car):

> I did not notice any marked change in the trim of the ship *until after* I had received an order by the engine telegraph to slow. I was actually reducing the engine speed when the ship took up a diving attitude.

This suggests that the order to cut back engine speed reached the No 4 car *before* the first dive. Because the telegraphs had to be rung individually and one at a time, they would probably be rung in sequence, No 1 to No 5. Thus engines No 1 and 2 (from which there were no survivors) might well have been slowed down before the first dive and at the time that the order was given to Church to release ballast.

The order to slow the No 5 engine in the aft car certainly came later.

3. *Binks* (Aft, No 5, engine car).

> The ship puts her nose down, responds to high engine speed and quickly loses height. She then returns to a more even keel but, almost immediately, gets into another dive and, during this latter change, the pointer of the engine telegraph revolves to that portion of the dial which indicates to change from present speed to slow speed, namely from 825 to 450 revs. Bell instantly performs this operation which is completely carried out.

Binks thus states that the order to shut down the aft engine came during the second dive; that is after 0208–30.

4. *Disley* (asleep in the switch-room aft of the chart-room):

> I became conscious of the ship dipping a little; not more than it had done before and this change of attitude seemed to be corrected instantly.
>
> *At that moment* George Hunt, the Chief Coxswain, came into the switch-room and said *'We're down, lads'*, and then, just as I was trying to get up the ship went into a steep dive which threw me on to the bed. I then heard the engine telegraph bell ring – and very shortly afterwards the crash came.

From this, Hunt came into the switch-room (aft of the chart-room) at the moment the airship levelled off during which time the telegraph bell could be heard ringing to slow the aft engine.

5. *Bell* (No 5, aft engine car):

> *At about seven minutes past 2 a.m.* I noticed the ship rather suddenly change its attitude and take a nose down attitude. This dive was quickly rectified and the ship resumed horizontal flight. But, at the next moment, the nose went down again to about the same angle as before. I *then* received an order over the engine-room telegraph indicator to go to 'slow' and while I was taking action the crash came.

Bell's evidence confirming that the first dive did not begin before 0207 and he confirms also that the order to slow the *aft* engine came during the second dive; that is after 0208–30.

6. *Leech* (in the smoking room ahead of the chart room):

> I felt the ship take up a sudden dive. Tables and all loose articles, including myself, slipped towards the forward bulk-head. Sitting on the fixed settee with a cushion under me, I slipped from about the centre to the front end. I think we must have dived quite a considerable distance before the ship resumed on an even keel. The engines were still running at cruising speed. She remained on an even keel for perhaps a minute, possibly less. I picked up some of the glasses and a soda-water siphon and put them on the table again. Then the ship dived again at about the same angle as before – and the crash came a moment later.

In cross-examination, Leech said that the first dive took somewhere between forty-five and twenty seconds. When the second dive began he heard the engine telegraph ring. Both Bell and Binks estimated the first dive as about thirty seconds 'or a bit more', and the time levelled out 'not more than ten seconds'. But experiment shows that to get up, pick up two or three objects off the floor and put them on a table takes fifteen to twenty seconds.

Analysing the survivors' statements, the discussions recorded in the Minutes of the Inquiry and the NPL calculation, one arrives at a time period of eighty-five seconds from the start of the first dive to the moment of striking the ground.

If we work back from the time of 0209, we therefore arrive at 0207–35 as the time of the start of the first dive.

By that time it appears that the telegraphs had been rung in No 1 and No 2 engine cars to reduce engine speed and that the period of calling for this reduction in speed for all five engines extended for about sixty-five seconds; from about 0207–25 for Engine No 1, to about 0207–30 for Engine No 4, to about 0208–30 for Engine No 5. According to the survivors there seems to have been a gap of sixty seconds between the order for No 4 engine to be cut back and the order to reduce speed for No 5.

One might think that Cook, in Engine Car No 4, did not notice the first dive and was speaking only of the second dive when he remarked 'I was actually reducing the engine speed when the ship took up a diving attitude', but he then went on to say that, having reduced the engine speed from 825 rpm to 450 rpm, he went 'to look out of the door of the nacelle towards the ground. I was still looking out when the ship struck'.

To reduce the engine speed and to move over to the door would, almost certainly, have taken longer than the brief duration of the second dive.

So there seems no doubt that he received the order just before the first dive at about the same time as Church was ordered to go forward to release ballast.

This would be consistent with a crisis having been perceived in the control car *before* the first dive. It is consistent also with the fact that Hunt had time to leave the control car, climb into the chart-room above and then walk back past the switch-room aft, towards the crew's quarters still further aft, and to pass the chart-room (according to Disley) at the moment the ship levelled out – that is at about 0208–10.

When we add this together we come to a time scale, as follows:

1. 0200 hrs GMT	Change of Watch. R.101 at 1,200 feet at 52 knots TAS at 825 rpm on five engines. In substantially level flight with 200 feet variations around the mean.
2. 0205 hrs (approx.)	A split starts to develop in the forward upper part of the outer cover. Ship at 1,100 feet at 52 knots at 825 rpm. Drag increases.
3. 0206 hrs (approx.)	Michael Rope discovers the developing failure of the outer cover and realises that it will spread rapidly and that it is likely to cause extensive damage to the forward gas-bags.
4. 0206–45 hrs (approx.)	Rope reports to control-car serious forward damage.

5. About 0207–15	An emergency situation is recognised in the control-car. Airship flying at about 1,100 feet. The reaction in the control-car was to slow down to avoid further damage.
6. 0207–25	Telegraph rung from control-car to slow No 2 engine. (Some seconds would lapse before action was taken.)
7. 0207–27 to 30	Telegraphs rung from control-car to slow engines 2, 3 and 4 while Hunt begins to climb up from the control-car to warn crew.
8. 0207–35	With engine power now being reduced the ship's nose begins to drop and the first dive to start. Hunt finds Church in the corridor aft of the chart-room and sends him forward with instructions to release ballast.
9. 0208–10	After thirty-five seconds of the first dive, with the No 5 engine still at cruising power and, by now full-up elevator, the ship begins to level off at about 500 feet. Hunt moves aft and calls out '*We're down, lads*', as he comes into crew's quarters, aft.
10. 0208–35	Telegraph rung by Steff in control-car to No 5 engine to reduce power. Nose falls away again and second – more shallow – dive begins at 450 feet with elevator still full up.
11. 0209–00	Airship hits ground at 200 feet at 54 knots through the air but at no more than ten knots over the ground at impact. Takes fire immediately from calcium flares in the crushed control-car.

That seems to fit all the known facts.

It does, of course, bring us back to the emergency at about 0207–15.

The NPL analysis in the R & M 1400 and 1401 shows that even with a split in the cover causing extra drag, and even with two deflated bags, there was enough elevator power at cruising speed to keep the nose up – but that it would drop if power were significantly reduced.

The slowing of the ground speed apparent during the last five minutes of flight, making allowance for the wind speed, could be accounted for by the increased drag of a cover split.

The Report comments (Page 92) 'If the fore part of the cover had become torn and the wind entered the envelope, serious damage to gas-bags would most likely occur with startling suddenness.'

If we add to that a downwards gust (R.101's luck was clearly out) and then the throttling back of some of the engines, then the first dive becomes inevitable.

But why were the engines throttled back? And how would those in the control-car know of a sudden emergency in a remote forward part of the ship where, certainly, all had been well up to some minutes after 0200 when the watch was changed?

It is, in fact, clear that all was thought to be well at 0205 because the Middle Watch had all had time to get clear of their stations and, Irwin, for instance, had changed into pyjamas and had got into his bunk.

This brings us back to the knowledge – almost certainly from Rope – that serious damage had developed in the forward cover.

Why did Hunt think 'We're down, lads' – his shout to the crew members who had come off watch a few minutes before and were settling to sleep as Hunt moved aft towards their quarters?

The only explanation is that Hunt had been warned of a serious emergency before he left the control-car at, probably, just after 0207. A message from Rope, or the arrival of Rope helter-skelter into the control-car, seems by far the most logical explanation. He would have brought news of a fabric failure in the forward part of the ship.

This probability is supported by the fact that five incinerated bodies were found in the crushed remains of the control-car which was the main seat of the flash-fire of high intensity. They could not be identified but they were, almost certainly, those of Steff, the two coxswains, Oughton and Michael Rope. No bodies were found in the forward part of the ship, from which Church escaped with fatal injuries.

News of the developing failure of the forward cover would cause a natural reaction to slow down to limit the damage. Possibly Steff started to cut power immediately without waiting to discuss the situation with Rope and Hunt or to consult on whether this would be the right action. The experienced Hunt and Rope would know that with reduced power the nose must drop, especially because it was carrying dynamic lift. They may have all felt that, as an instant judgement, it was the right thing to do.

A grounding would not, necessarily, have been considered catastrophic. R.34, in which Hunt had served, had hit the ground without any fire or loss of life. The German airship *Bodensee* – also inflated with hydrogen – had likewise run into high ground without serious consequences.

But for the fire in R.101 most on board would have been safe. The tragedy was compounded by the fact that the control-car with its calcium flares was immediately below the passenger accommodation.

Once the engines were shut down the grounding was inevitable. Recording the calculation of the situation at the moment when the airship levelled out, R & M 1401 states categorically:

> If the engines were not shut down the nose would continue to rise. Shutting down the engines and dropping ballast (fuel or water) causes the ship will lose speed and, therefore, aerodynamic moment and the nose will not rise above the horizontal. Shutting down the engines (halving the revolutions would produce this effect) would result in the second dive.

So the critical events were, almost certainly:

1. *The discovery* of a serious failure forward.
2. *The reporting* of this discovery to the control-car.
3. *A decision* to reduce the speed of the ship by cutting the power – a natural reaction but, in the event, a fatal error of judgement.
4. *The nose drops* – possibly helped by a turbulent down-gust.
5. *The airship levels* out with full up-elevator but with the continued reduction of power, the nose drops again.
6. *R.101 hits the ground* at low forward speed into the very strong wind.

R.101 could have been saved.

If the reaction of those in the control-car had been to go for height rather than to reduce speed and if full power had been applied, there is no doubt that – damaged as it was – R.101 would have pulled away under full-up elevator; slowly and ponderously but positively. The ship could then have been trimmed and lightened by the dumping of some fuel. The crisis would have been surmounted and the airship could have turned back to reach Cardington by dawn with the wind behind it; damaged, disappointed but safe.

There can be no criticism of the snap decision. But the most poignant factor of all was that the act of throttling back – however understandable – inevitably 'parked' the airship on the ground. Its luck was out – and fire consumed it and almost all on board.

II

Factors in Disaster

The trail of events which led up to the loss of R.101

Few of the unfortunate series of catastrophies to various means of transport and travel which have marked the progress of human endeavour can be laid at the door of a single cause.

Carefully thought-out safety measures and, eventually, 'fail-safe' techniques, have increasingly been able to ward off disaster from any single failure or even from two or three compounded together. Catastrophies have happened when, against long odds, failures – both mechanical and human – multiply beyond any foreseen possibilities.

So it was with the SS *Titanic* in 1912. And so it was with R.101 in 1930.

For the *Titanic* on its maiden voyage, the latest and most advanced form of ship architecture at that date, the loss of most of the 1,502 lives did not come from just the simple fact that the ship struck an iceberg. There was a compounding of circumstances. The crew was new; the look-out inexperienced. If the iceberg had been seen earlier – or later – the damage might not have been fatal. As it was, there was just enough time to put over the helm; hence the ship scraped along the side of the iceberg which ripped out the sides of a series of compartments below the water-line.

Even then, had there been enough lifeboats most of the passengers could have been saved. There were not enough to go round. Few lives would have been lost also, had the *Titanic*'s distress flares been recognised by ships which were within sight. A response could have come in time.

At least half a dozen successive failures together resulted in the tragedy of the *Titanic*.

So with R.101.

Over the course of sixteen months, from June 1929 to October 1930, thirteen separate and cumulative factors can be recognised as adding together to build up to disaster like the inexorable tide of a Greek tragedy. Without any one of them – right to the end – the disaster to R.101 would not have happened.

Those factors which led to the tragedy were :

1. *The India Flight as a Prototype Trial* (16th June 1929)

The intention to fly R.101 to India and back as the culmination of the airship's proving trials was originally stated in 1924 and, at that time, scheduled for January 1927. By 16th June 1929 on advice from the Air Ministry and from Cardington, the Secretary of State for Air and the Director of Civil Aviation had stated their resolve to accompany the flight, then re-scheduled for December 1929.

Consequently, all sights were set on the preparation of R.101 for the India flight at the earliest possible date.

2. *Weak Outer Cover* (2nd June 1930)

Michael Rope's note to Vincent Richmond and Reginald Colmore, dated 2nd June 1930, called attention to the serious weaknesses, by then apparent, in the pre-doped outer covers of both R.100 and R.101; facts backed up by Noël Atherstone's investigations. Although the outer cover was largely renewed by a stronger, doped-in-position, cover when the extra bay was installed between July and September 1930, two areas of the original cover were left unrenewed, at nose and stern because these portions had been doped in situ.

Consequently there remained two portions of the outer cover which were not new; one of them at the area of maximum stress just aft of the nose.

3. *No Full Speed Trial* (1st October 1930)

The failure of an oil cooler, though a small defect in one of its fins (it had been stepped on) during the final trial flight on 1st October 1930, meant that the projected full speed trial on all five engines was never performed.

Consequently the outer cover was not exposed to its full stress, especially at the un-renewed nose portion. If it had been, the evidence suggests that the forward original section of the cover would probably have split and the India flight would have been postponed until a replacement forward section had been installed.

4. *Over Confidence and Inadequate Trials* (2nd October 1930)

R.101 was cleared as airworthy to depart for India after one, sixteen-hour, trial flight following its refit and the installation of the extra bay. Previous deficiencies – a weak outer cover, chafing gas-bags, inadequate padding, only four engines for forward flight and inadequate lift to counteract the heavy power-plant – had all been remedied to the satisfaction of those responsible for design standards, works action and AID inspection. The ship was signed out as up to requirements and sound and satisfactory for service. Those responsible were confident that work had been well done – and in most respects it had been. The credibility of Cardington was at stake. A year from first flight, after 120 flying hours and 603,400 man-hours of rectifications and modifications, (162,400 design and 441,000 direct labour man hours) those best qualified to know believed R.101 to be ready. By the standards of the day, in which previous airships had set out on long flights with less preparation and had succeeded triumphantly, they were right. By today's standards there was over-confidence and an inadequate test programme.

Consequently – with hindsight – R.101 left for India, with the Secretary of State on board, unproven in a number of important respects. There could have been no argument or recriminations, in the face of Lord Thomson's avowed 'Safety First' and 'no rush' policy, had Colmore, Richmond, Scott and Irwin declared R.101 not yet ready for its first long overseas flight on 2nd October. In the words of the Official Inquiry:

> What is so impressive, on studying the details of those last hours before R.101 started for India is the high courage and genuine confidence of all concerned. For four years they had been preparing for this moment and, notwithstanding that the date of the Imperial Conference tended to condition the hour of their start, they believed themselves to be prepared for it.

5. *Decision to Go* (4th October 1930)

The decision to depart for India on 4th October was left by Air Vice-Marshal Hugh Dowding, AMSR – first, on general airworthiness grounds – to Reginald Colmore, the Director of Airship Development and – second – on operational grounds – to Major Herbert Scott, Deputy Director of Airship Development (Flying and Training). After examining the weather forecasts during the morning and afternoon of Saturday, 4th October, Scott confirmed the decision to depart that evening – rather than to postpone in the face of an unfavourable weather report. By delaying the flight for twenty-four hours the expected good weather and light winds would have arrived behind the occluded front.

Consequently R.101 departed as rain began to fall and the wind to rise into conditions which turned out to be significantly worse than had been forecast.

6. *Scott's Judgement* (4th October 1930)

Successive senior officers at the Air Ministry and those at Cardington turned a blind eye to the known and obvious deterioration in the judgement and competence of Major G. H. Scott to whom was delegated all decisions upon the flying of R.101. Always recognised – and admired – for his predilection 'to press on regardless', Scott had shown during recent months that he was no longer the man he had been. In addition he refused to admit that airships were not 'all-weather craft'.

Consequently R.101 departed for India when prudence would have dictated an operational postponement for better weather.

7. *Engine Starting Delay* (4th October 1930)

Once the decision had been taken to hasten departure ahead of the advancing front, there was a twenty-five minute delay in starting the starboard forward engine because the diesel starting motor would not itself start. This delay brought the crisis – when it came to 0207 hrs on 5th October – immediately after the change of watch and over difficult ground.

Consequently, the highly competent Noël Atherstone and Carmichael Irwin, Captain of R.101, had both just gone off watch when their skill and experience would otherwise have been available in the control car to meet the emergency and cope with it.

8. *Weather* (4th/5th October 1930)

An inaccurate weather forecast before departure, although predicting rain and high winds which would have justified delay, failed to forecast the storm conditions which engulfed R.101 by the time it reached the English coast.

Consequently R.101 found itself having to plough into the teeth of a gale accompanied by heavy rain and low-level turbulence; certainly the worst conditions ever encountered by any British rigid airship in free flight. The tensile strengths of both the cover and the gas-bags would be seriously reduced by that prolonged saturation at faster speeds than those at which R.101 had ever flown continuously before.

9. *Fast-Cruise-Regime* (4th/5th October 1930)

Scott's decision to drive R.101 for the whole of the $7\frac{1}{4}$ hours of its flight at the 'fast cruise regime' in order to make headway against the fierce headwinds, certainly exposed the airship to more extreme conditions than either R.101 or R.100 had ever encountered for such a long period before. All engines were run at a continuous 825 rpm

(475 bhp each) to give a cruising speed of 55 knots (63 mph) – and 50 knots (58 mph) while one engine was shut down. The objective was to reach Ismailia in as near forty-eight hours flying time as possible.

Consequently, both the outer cover and the forward gas-bags were exposed to greater stresses than they had ever been subjected to before.

10. *Change of Watch* (5th October 1930 0200 hrs)

In the belief that all was well – as it was then – the watch was changed on schedule at 0200 hrs GMT and while Irwin and Atherstone retired to rest for the strenuous day ahead, the least experienced officer, Maurice Steff, took over as Officer of the Watch at the hour of the lowest ebb in human endeavour.

Consequently at a critical moment, the Officer of the Watch was faced with a situation outside his experience while, at the same time, the Height Coxswain had not had sufficient time at his elevator controls to gauge the 'feel' of the ship and to react quickly enough to take effective action in an emergency.

11. *Crew Fatigue* (4th/5th October 1930)

There was, undoubtedly, a state of moderate to severe fatigue among both officers and crew at the time of R.101's departure. Many of them had been on continuous duty, with only brief snatches of sleep, since the airship was walked out of the shed at Cardington at 0600 hrs on 1st October.

Consequently by 0200 hrs on 5th October, this fatigue must have become extreme among those on duty, and to have affected their decision to leave the control of the ship to the least qualified officer, backed up by the highly competent Chief Coxswain Hunt. Both Steff and Hunt must have been very tired. Crew fatigue in 1930 was not recognised as the problem it is known to be today. At the time of the crash Scott, Irwin and Atherstone had all retired to their bunks to sleep as confirmed by identification the following morning.

12. *Engines Throttled Back* (5th October 1930, 0208–20)

The decision to signal from the control car to the power cars, one-by-one, to throttle back the engines to idle-speed – an instinctive and fatal reaction to a situation of not-fully-appreciated danger (almost certainly a forward cover split) – removed dynamic lift gained from flying speed.

Consequently the nose of the airship dropped with the loss of flying speed, the second dive was precipitated and the airship nosed into the ground.

13. *Calcium Flares* (4th/5th October 1930)

Squadron Leader Ernest Johnston had arranged for a rack of calcium flares to be installed in the control car. He used some of these to obtain drift sights over the Channel during the crossing from England to France. The calcium flares ignited instantly on contact with the water and burned with a bright light. When the control car hit the ground in the soaked undergrowth of the Bois de Coutumes near Allonne they were broken open and ignited.

Consequently a fierce fire was initiated from the aft part of the control car, immediately below the passenger accommodation, and spread from the centre of the airship to the gas-bags above – and then fore and aft until the whole airship was engulfed.

*

These thirteen successive and cumulative factors all built up to cause the tragedy of R.101. Without any one of them the chain of events would have been broken and R.101 and those on board would – almost certainly – have been safe.

In the final analysis, the weather, the cover failure, the decision to throttle back and the calcium flares were the predominant factors.

A Cardington Memorandum of 21st November 1930 in Air Ministry File S.23,843 (now AIR 5/380) reads:

> Any damage to the outer cover would release the positive internal pressure and cause flapping and further fractures and considerably increase the internal fabric tensions in the bow. The dynamic effect of heavy rain would also increase the tension in the bow fabric.
>
> If the ship was becoming heavy and if full speed was used to obtain dynamic lift, the loading of the bow fabric might considerably exceed that to which it had ever been subjected since the removal of the support girders in the bow. The reason for the local deterioration of the fabric in the forward part of the ship does not appear to be fully explained. It is possible that there was a loss of strength in places besides those which had been discovered and rectified.
>
> If a failure of the bow fabric occurred, it is quite possible that the inrush of air striking the forward gas-bags would cause comparatively rapid deflation of the bags exposed to the air stream. Flapping of the outer cover may have also disturbed the valves of the forward bags. Any loss of gas would pass through the circumferential gills round the centre of the ship but some might flow to the tail and, mixed with air, would create an explosive mixture.

We know now that the forward part of the cover was not renewed. The tear in the cover could have been survived but for the throttling back of the engines and with it, the loss of dynamic lift. And the low-speed hitting of the ground could have been survived but for the calcium flares.

R.101's luck was out.

APPENDICES

APPENDIX I

AIRSHIP DEVELOPMENT

1900–1924

Reference page 51

February 1924

BUOYANT FLIGHT with Lighter-than-Air craft provided the means for Man's first successful ventures towards air travel.

In Britain, experiments with Lighter-than-Air (LTA) vessels were a primary part of aeronautical thought and, to a lesser extent, action from the time when the English Channel was first crossed by balloon on 7th January 1785, only fourteen months after Man's first successful free flight – in a hot air balloon from the Château la Meutte, in the Bois de Boulogne in Paris on 21st November 1783.

By the time that the Boer War had started in South Africa, rather more than a hundred years later in October 1899, sporting ascents and military ballooning for observation purposes were all well established in Britain. The sporting events were under the auspices of The Aero Club of the United Kingdom (not to receive the title 'Royal' until 1910) while the military side was under the British Army's Corps of Royal Engineers based at Chatham.

When, on Friday, 29th February 1924, the 'Airship Development Committee' of the first British Labour Government met for the first time* the new Secretary of State for Air, Brigadier-General the Right Hon the Lord Thomson of Cardington, presented to the Committee a Paper which reviewed the history of Lighter-than-Air developments as a background to his proposals for a radically new airship programme. From a personal point of view, Lord Thomson remarked to his colleagues that he had formed his own first link with aviation during the Boer War when as a Captain, Royal Engineers, he had been detailed to assist the 2nd Balloon Section, RE, outside Kimberley.

'In 1900,' he said, 'military balloons were the eggs from which future airships were to be hatched.'

The evolution from the spherical balloon to the powered and controllable rigid airship had, by then, already arrived by way of 'sausage balloons', or non-rigids, mostly of French derivation and in

* AIR5/1062 and AIR 8/74 (C.P. 104(24) and AD(24)2 of February 1924.

the form of the first, unsuccessful, rigid airship, built by an Hungarian, David Schwarz, in St Petersburg, in 1880. The real advance came, however, with the invention, by a German cavalry officer, Count Ferdinand von Zeppelin,* of the rigid structural frame which held a number of separate gas-bags, or ballonettes, all contained within a cigar-shaped fabric envelope, irreverently described as 'a bunch of balloons in a birdcage'.

On 2nd July 1900, at the age of sixty-two, Count Zeppelin flew his first dirigible from a floating dock at Manzell, on Lake Constance, near Friedrichshafen. The airship was LZ-1 (*Luftschiff-Zeppelin 1*), of 399,054 cubic feet capacity, powered with two 14 horse-power Daimler engines. Though LZ-1 was not a success, Zeppelin went doggedly on to develop and improve his airships and, by 1924, one hundred and fifteen Zeppelins had been flown.

Between 1910 and 1914, a commercial Zeppelin Company, DELAG (Deutsche Luftfahrts Aktien Gessellschaft), carried 34,028 passengers – 10,197 fare-paying – in a series of 1,588 summer-season pleasure flights inside Germany. They were followed by a further seventy-eight flights carrying 2,253 passengers after the War, all without fatalities. They seemed to point the way to successful commercial airship services, until, at the end of 1919, they were closed down by the Inter-Allied Commission of Control.

Following this forced suspension of activity, by the beginning of 1924 a total of 138 German rigid airships had flown some 32,000 hours and had gained for their crews a substantial store of operating experience, albeit with heavy wartime casualties.

The possibilities of future peacetime uses had been underlined by an outstanding, non-stop, flight of 4,230 miles in 95 hours, when the German Army Zeppelin L.59 (LZ-104) had flown from Jamboli, in Bulgaria, almost to Khartoum in the Sudan – and then back – in an attempt to bring supplies to a German garrison. That failed, but the out and return flight, between 21st and 25th November 1917, was held to show that tropical flying by airships over long ranges was practical and could be of great value in peace-time.

British involvement with rigid airships began in 1908, and, at that time, owed little except the broad concept, to Count Zeppelin, who,

* Count Ferdinand von Zeppelin was born at Konstanz, in Wurttemberg, on 3rd July 1838, of French descent on his mother's side. In 1916, he turned from airships to the design and construction of large, four-engine, biplane bombers and remarked that the future lay more with them than with Lighter-than-Air craft. Zeppelin died in 1917 age 79.

by then, had produced only three, not very successful, airships. HM Rigid Airship No 1 was built at Barrow-in-Furness, by a combined Admiralty-Vickers team – and piquantly named *Mayfly*. That turned out to be optimistic, because the airship broke its back on 24th September 1911, when it was taken out of its floating shed at Cavendish Dock, Walney Island, after its keel had, unwisely, been removed to lighten its excessive weight. It had been designed on the only experience available – as an ultra-light submarine – which made it an ultra-heavy airship.

The misfortune to *Mayfly* halted British airship development for five years, until the use of Zeppelins by the German Navy brought about another essay into rigid airship construction in Britain, again at Barrow, by Vickers and the Admiralty in conjunction. HM Airship No 9 (the intervening numbers were non-rigids) was flown for the first time on 27th November 1916, under the command of a respected airship man, Wing Captain E. A. D. Masterman, RN.

As the first successful British rigid airship, No 9 flew a total of 198 hours and 16 minutes, largely on crew training, up to the end of 1917. It was dismantled in June 1918.

Between 1916 and 1921, fourteen British rigid airships, out of twenty-four planned, were completed and flew a total of 3,302 hours at a cost of rather more than £6 million; an average of about £1,820 for each hour flown. This part of the story ended, however, on 24th August 1921, when the newest British rigid, R.38, broke up in the air over the Humber, on a trial flight on the eve of being handed over to the United States Navy. Thereafter, the four remaining British airships were grounded.

Little had been gained in the six years. The first seven British rigid airships to fly – No 9 to R.29 – were all designed on rather sketchy first principles allied to information gathered from photographs of early Zeppelins and from data gleaned by 25-year-old Hartley Pratt of Vickers Ltd, when he had snooped around the Zeppelin Works at Friedrichshafen, on a holiday, just before the War. He measured, sketched and took rubbings of a number of duralumin girder fittings lying openly in a railway truck. This information was eagerly absorbed by the Admiralty-Vickers team, working under the general authority of Rear Admiral (then Captain) Sir Murray F. Sueter, the Director of the Naval Air Department, who supervised a small staff which included Hartley Pratt (25) as General Manager, J. E. Temple (24) on general and structural design and Barnes W. Wallis (26) as Head of the Drawing Office in charge of detail design in 1913.

Additional data was gained from a precautionary landing made by the German Army Zeppelin Z.IV (Works No LZ.16) at Luneville in France, on 3rd April 1913, after it had lost its bearings. During the day this Zeppelin airship spent on French soil, remarkably detailed drawings were made of its structure – and copies of them were passed across the English Channel.

A notable fact is that the eleven German Zeppelins built between 1900 and 1912 – up to and including LZ.11, *Victoria Luise,* together with LZ.16 (Z.IV) – were all built of aluminium-zinc alloys. From LZ.12 (Z.III) of 1912, onwards (except for LZ.16), the structural material was changed to duralumin – as was that of the British rigid No. 9, which was as close a copy of the, then, latest German practice as the Admiralty-Vickers team was able to achieve.*

The last four British ships, R.33 to R.38, which had been built up to 1921, were all copies of three German Naval Zeppelins shot down more or less intact on Allied territory in 1916 and 1917; L.33 (LZ.76) at Little Wigborough in Essex, on 24th September 1916, L.48 (LZ.95) at Theberton in Suffolk, on 17th June 1917, and L.49 (LZ.96) at Courbonne-les-Bains in France, on 20th October 1917 There were also three British 'odd-men-out'. Two of them, R.31 and R.32, wooden-framed ships built by Short Brothers at Cardington, were derivatives of the German Schutte-Lanz airships made from drawings smuggled out of Mannheim by a defector, Hermann Müller. The third, the small Vickers R.80 – was an original design by Barnes Wallis, R.80, not flown until January 1921, and then dogged by a series of structural failures.

When British airship flying was closed down, on grounds of economy, in August 1921 after the crash of R.38, there remained intact only R.33 and the uncompleted R.37 at Cardington, and R.36 and R.80 at Pulham.

Though British rigid airships had achieved little during the War, a good deal more success had been gained with a series of British non-rigids operated by the Royal Naval Air Service between 1915 and 1918. In all, the 192 small 'Blimps', of seven different types, flew 87,185 hours between 1st January 1915 and 31st December 1918 – mostly on convoy-escort duties and anti-submarine patrols around the British coasts.

Up to the beginning of 1924, however, by far the most significant

* I am grateful to Mr Peter W. Brooks of British Aerospace for bringing out these important historical facts.

long-range flight by any type of aircraft had been the two-way crossing of the North Atlantic by the Beardmore-built British rigid airship R.34 in July 1919. Under the command of Major G. H. Scott, RAF, R.34 flew from East Fortune, near Edinburgh, to Hazlehurst Field, Minneola, Long Island, New York, in 108 hours and 12 minutes between 2nd and 6th July, the longest duration ever flown by any British aircraft. The airship returned to Pulham, in Norfolk, between 10th and 13th July 1919, in a flying time of 75 hours and three minutes. A crew of 34 was carried westbound and 32 on the homeward journey.

This successful double-crossing of the North Atlantic Ocean in 1919, together with the Zeppelin L.59's flight to Africa and back in 1917, seemed to prove the especial value of the rigid airship for long-distance travel.

There were complications, however; some of them financial and others concerned with the allocation of responsibilities for development.

When the Royal Flying Corps was formed in May 1912, all airships owned by the British Government were placed under the Army's Military Wing. On 1st January 1914 the whole of this Lighter-than-Air section (No 1 Airship, Squadron, RFC) was turned over to the Royal Navy together with its seven non-rigid airships, its 195 personnel and the Airship Station at Kingsnorth, near Hoo, on the Medway. The Royal Flying Corps thereafter concentrated on aeroplanes, together with kite-balloons and spherical balloons for observation work.

Though the Admiralty had taken over the airships, the First Lord, Winston Churchill, was not impressed with their possibilities and went so far as to cancel the Vickers project, No 9, in March 1915 – resuscitated later in that year, after Churchill resigned in May. Thereafter, a modest programme was pursued, concentrated primarily on non-rigids. British rigids had been initiated under Admiralty control; non-rigids by the Army. During the War, both types had been developed and flown under Admiralty control.

In 1917, following increasing losses of German Zeppelin raiders to aeroplane defences over Britain, the German air attack was switched from airships to the new twin-engine Gotha bombers, with dramatic initial results. By an odd stroke of fate, these aeroplane raids led directly to the formation of the Royal Air Force.

Two daylight raids on London, on 13th June and 7th July 1917, precipitated events. Eighteen Gothas in the first and twenty-two in the second caused more casualties in the capital than had all the Zeppelin airship raids combined – 216 killed and 161 injured.

A first result was the recall of No 46 Fighter Squadron, RFC, equipped with Sopwith Pups, from the Western Front and, more significant, an invitation from the Prime Minister, Lloyd George, to a Member of the Cabinet, Lieutenant-General Jan Christian Smuts, to review problems of Home Defence and of future Air Policy, in company with Lieutenant-General Sir David Henderson, Director-General of Military Aeronautics at the War Office.

Two reports were made to the Cabinet, on 19th July and 17th August 1917. The first, signed by Sir David Henderson, recommended that all aircraft of all types (landplanes, seaplanes and airships) should be placed under unified control. The second, from Smuts, recommended the formation of a separate Air Ministry, under an Air Minister, and that the Royal Flying Corps and the Royal Naval Air Service should be amalgamated to form an autonomous Air Service.

These proposals were approved by the Cabinet on 24th August 1917.

Thus, the Royal Air Force came into being on 1st April 1918. Major-General Sir Hugh Trenchard was brought back from France to become the first Chief of the Air Staff, while Major-General Sir Sefton Brancker became Controller-General of Equipment. The first Secretary of State for the Royal Air Force and President of the Air Council was Lord Rothermere – with Sir David Henderson as Vice-President.

Even after the formation of the Royal Air Force, and the Cabinet's acceptance of the recommendation that all types of aircraft, military and civil, should be brought under its authority, the Airship Section of the former Royal Naval Air Service remained with the Admiralty until 5th May 1919. On that date, the 6,000 airshipmen were transferred to the Air Ministry, although the 103 airships (five rigids and 98 non-rigids) remained with the Admiralty until 22nd October 1919. By then, however, most of the personnel had been demobilised and the Cabinet had decided that all but two of the non-rigids (NS.7 and NS.8) should be taken out of commission.

With peace, the importance of India as the brightest jewel in the British Imperial Crown, once more came into the forefront of official thought and public regard.

In September 1919, just before it took over the surviving rigid airships based at Cardington, Howden and Pulham, the Air Ministry put forward a scheme for the operation of a commercial air service to India, using R.33, R.36 and R.80, to be flown under RAF guidance. The proposition came to nothing after it had been turned down by the Treasury on 22nd June 1920, on grounds that the estimated cost of £844,000 spread over two years could not be found or justified. This rejection was followed on 29th July by a Cabinet decision that the Air Ministry should close down all its Lighter-than-Air work – 'HMG in the fullness of its futility', wrote C. G. Grey in *The Aeroplane.*

Although the Air Ministry order to cease all airship work was issued on 20th September 1920, the process of closure dragged on for almost another year, while the new R.38 was completed at Cardington for sale, for £500,000, to the United States Navy. In the event, the crash of R.38 on 24th August 1921, and the loss with it of most of the leading British airshipmen of the day, together with fifteen American colleagues, brought to an end the first phase of British Lighter-than-Air activity.

A review of airship work up to that date revealed disquieting facts. By the end of 1923, out of 154 rigid airships which had been completed and flown by four nations (138 German, 14 British, one French and one American), 104 (68 per cent) had been lost, with a toll of 584 lives. Although 39 of the losses – all German – had been through enemy action in the War, at the same time, 28 airships had been destroyed in crashes in which 301 crew members had been killed. One life had been lost for every 65 airship flying hours.

By contrast, the enthusiasts pointed to the pre-War and post-War record of safety of rigid airships engaged in commercial operations – 3,708 hours and 138,975 miles flown by DELAG (at an average of 37 mph) without a fatality. The disaster to R.38 was traced back to the wartime requirements for an exceptionally light and fragile structure to make possible the attainment of great heights to evade interception.

A case was argued that, without such military needs, new civil airships could be built which would be safer than any flying-machines of past or present. Aeroplanes were still relatively small and of short range, whereas airships, their protagonists pointed out, had shown their capability of range and endurance and could be made to provide a standard and a combination of spaciousness, comfort, quietness and load-carrying ability, not to be matched by any heavier-than-air machine then in prospect. British advocates of Lighter-than-Air watched, with growing envy, the work going forward in Germany, in

France and in the United States on rigid airships – and deplored lost opportunities at home.

In this period of financial stringency and of Government disenchantment with airships, on 27th March 1922, the energetic Commander Dennistoun Burney,* Conservative MP for Uxbridge, the inventor of the mine-sweeping paravane, now retained by Vickers Ltd as a consultant, wrote to the Secretary of State for Air (with a copy to Sir Hugh Trenchard) from Vickers House at No 6, Broadway, Westminster. In this letter, he set out a proposal for the setting up of an Imperial Commercial Airship Service to be run by a private company subsidised by the Government.†

Burney proposed the formation of a subsidiary of Vickers Ltd to build and operate airships. The company would have a nominal capital of £4,000,000, to which Vickers and Shell would each subscribe £100,000; the rest to be raised through 6 per cent Debentures guaranteed by the Government over ten years, in return for the benefits of an airship service in peace-time and the use by the Admiralty of the airships for naval reconnaissance in time of war. He estimated the net cost to the taxpayer at £400,000 a year for seven years – a total of £2,800,000.

To make this possible, Burney proposed that existing airships should be handed over to the company for experimental work and trial flights, together with the airship bases at Cardington and Pulham.

The new company would then build, as a first stage, a new airship of 5,00,000 cubic feet capacity (760 feet long and 110 feet in diameter), which (Burney optimistically estimated) would be able to fly for 3,000 miles at 80 miles an hour, with 44.5 tons available for mails, passengers and cargo; including accommodation for 200 passengers and ten tons of mail out of a disposable load of 82 tons. For naval reconnaissance in wartime:

> the whole of the carrying capacity would be put into extra fuel and wireless equipment. The result would be to increase the range of flight to 12,000 miles at 80 mph, using half-a-ton of fuel an hour, or to 24,000 miles at 40 mph – nearly three weeks without alighting.

He added that, for war, such an airship could carry two fighters and

* Sir Charles Dennistoun Burney (1888–1968) was the only son of Admiral of the Fleet, Sir Cecil Burney, the first baronet. When he retired from the Royal Navy in 1919, Dennistoun Burney became a consultant to Vickers Ltd and a Conservative MP for Uxbridge from 1922 to 1929, in which year he succeeded to his father's baronetcy.

† AIR/8/60, 'The Burney Scheme', 1922–24, and CIT 399B.

three torpedo-carrying aircraft as an airborne aircraft carrier, which could do the work of six cruisers, at a substantial saving in both capital and running costs.

India again came to the forefront as an objective.

For commercial services, Burney proposed that the first new airship should make a demonstration flight to India, after which, as a second phase, five additional airships should be built to provide – as a third phase – a regular, twice a week, mail and passenger service to Egypt and India; later to be extended as a weekly service onwards to Rangoon, Singapore and to Perth in Western Australia. When in full operation, he estimated running costs on the Indian route at £880,000 a year for a revenue of £1,830,000 a year – taking a first-class passenger fare at £70 and a second-class fare at £45.

With his Naval background, though he addressed his letter to the Air Ministry, Burney based his whole scheme on Admiralty support and on turning the airships over to the Admiralty in time of war.*

Burney's proposals were exceedingly optimistic in technical terms as well as in their commercial forecasts. There was no realistic basis for the proposition that any airship of 5 million cubic feet capacity would be able to lift 200 passengers and ten tons of mail and cruise continuously at 80 miles an hour for 3,000 miles; nor – at that time – were there prospects that 400 passengers a week would wish to fly to India and back, or that 40 tons of mail and goods would be on offer. At the same time, Burney's proposals fanned into flames again the old rivalry between the Admiralty and the Air Ministry on airship operations.

The Air Ministry was especially concerned because a decision had only just been made to convert Cardington into a permanent RAF station for two new squadrons of Home Defence fighters. The Burney Scheme was, at the same time, financially unacceptable to the Treasury.

Nonetheless, with the backing of Vickers Ltd, the proposals were taken seriously and, on 17th July 1922, the Prime Minister, Mr David Lloyd George, announced in the House of Commons that a special Sub-Committee of the Committee for Imperial Defence would be set up to go into the whole subject of rigid airships for commercial and strategic use and, in particular, to examine Commander Burney's proposals for the England–India air route. The Sub-Committee would

* AIR 5/908; AIR 19/546 and *The Aeroplane,* 12th April 1922, (P.268) and 21st February 1923 (P.135/136).

be chaired by Mr Leo Amery, Parliamentary Under-Secretary of State at the Admiralty.*

When it reported on 1st August 1922, the Amery Committee found in favour of a revival of rigid airships in Britain and, in principle, in favour of Commander Burney's proposals, subject to more detailed financial and operational analyses. They reported that, compared with aeroplanes, the range of airships could be 'relatively enormous', they could offer important advantages for ocean reconnaissance in wartime; that the Air Ministry would be prepared to add £80,000 a year, for ten years, to its Air Estimates to support airships and that the Admiralty would also consider offering financial support.† 'The essential vulnerability of airships does not preclude their being of very great value for war purposes. There are undoubted great possibilities for the airship in future air warfare as an aeroplane carrier, whether in conjunction with naval operations or for purely air expeditions'.

No action was taken by the Government, however, before the General Election of 15th November 1922, which returned Andrew Bonar Law's Conservative Party to power with a majority of 72.

The Prime Minister thereupon recast the Committee with Sir Samuel Hoare, the new Secretary of State for Air, in the chair, and Leo Amery, now First Lord of the Admiralty, as one of its members.‡

At the same time, an Air Ministry Advisory Panel, as a Sub-Committee of the new Airship Committee (itself a sub-committee of the National and Imperial Defence Committee) was set up under Sir

* C.I.T. 342B and 346B, C.I.T. 363Z and SS(AS)8–363B, The Amery Committee on 'Airships and the Burney Scheme' had a membership of:

L. S. Amery (Admiralty), Chairman;
Admiral Sir Ernle Chatfield (Assistant Chief of Naval Staff);
Vice-Admiral R. Bell Davies, VC (Naval Air Staff);
Air Commodore John M. Steel (Air Ministry);
Sir George L. Barstow (Treasury);
C. A. Place;
Oliver Locker-Lampson (Colonial Office);
Air Commodore Sir W. Geoffrey N. Salmond (Air Ministry);
S. H. Leake;
(Secretary) Sir John R. Chancellor;
(Assistant Secretary) H. R. Moor.

The Committee held five meetings between 26th and 31st July, and reported to the Cabinet on 1st August 1922.

† CIT (SS(AS)8–363B.

‡ The full membership of the 1922/23 Airship Committee was: Sir Samuel Hoare (S of S for Air), Chairman; The Right Hon L. S. Amery (First Lord of the Admiralty); His Grace the Duke of Sutherland (US of S for Air); Sir W. F. Nicholson (Secretary to the Air Ministry); Rear Admiral C. T. M. Fuller (Third Sea Lord); Sir Vincent Baddeley (Admiralty); Major A. Boyd-Carpenter (Financial Secretary of the Treasury) (succeeded by Sir William Joynson-Hicks); and Sir George L. Barstow (Controller of Supply Services, Treasury).

Geoffrey Salmond, Air Member for Supply and Research, to go into technical and operational possibilities while the main Committee explored finances.

Although Sir Geoffrey Salmond's Advisory Panel was made up primarily of officials, it included two notable exceptions. One was Commander Burney and the other, on the recommendation of his old friends, Sir Samuel Hoare and Leo Amery, was Brigadier-General C. B. Thomson who was co-opted as a counter-balance to the independent advocacy of Burney.* So Thomson gained his first, close acquaintanceship with airships and with Cardington. At the same time he developed a common interest and friendship with Geoffrey Salmond and Sefton Brancker, which continued throughout their lives.

On 10th July 1923, Sir Samuel Hoare's Sub-Committee of The National and Imperial Defence Committee on 'Financial Aspects of a Scheme for the Development of a Commercial Airship Service', completed its Report from 2, Whitehall Gardens.†

It made seven recommendations:

1. A commercial airship service should be established between the United Kingdom and India;
2. Negotiations should be entered into with Commander Burney, of Vickers Ltd, for the building and operation of a new fleet of airships, to be undertaken by a commercial company·
3. A Government subsidy should be granted to the proposed company, limited to £250,000 a year, for a bi-weekly service with six airships.
4. The Company to agree to repay to the Government 50 per cent of any profit earned in excess of ten per cent after a period of seven years;
5. All existing airships to be transferred to the Company.
6. The airship bases as Cardington and Pulham to be leased to the Company.
7. The agreement to be limited to a period of 15 years.

Thus, Commander Burney achieved support from the IDC Com-

* The Membership of the Air Ministry Advisory Panel on Airships (1923) was: Sir W. G. H. Salmond (AMSR), Chairman; Sir W. S. Brancker (D of CA); Sqdn. Ldr R. B. B Colmore; C. L. C. Bullock; C. G. Caines, W B Foden; Brigadier-General C B Thomson; Commander C D. Burney; F. W. Munson (Secretary); (AIR 8/60).

† AIR 5/591 C.I.T. 436–B, 42B to 410D.

mittee for almost all that he asked and, on 26th July, Sir Samuel Hoare announced in the House of Commons that the Conservative Government would accept the Burney Scheme in principle for the India route – subject to Treasury approval of the details.

What he did not say was that this decision was reached against the advice of his officials within the Air Ministry and against the majority views of Sir Geoffrey Salmond's Advisory Panel. They were in favour of airship development – but not of the Burney Scheme with its overtones of Admiralty involvement.* Furthermore, they considered Burney's estimates to be grossly optimistic.

During the next few months, discussions were continued between the Treasury and Vickers Ltd. After a meeting between Sir George Barstow of the Treasury and Commander Burney, on 4th September 1923, a 'Provisional Heads of Agreement' was drawn up on 25th September, for the establishment of an Imperial Airship Service.

A consequence of this was the formation by Vickers Ltd, on 20th November 1923, of The Airship Guarantee Company Ltd.

Negotiations were still in train when the General Election of 6th December 1923 threw everything into the melting-pot and, on 22nd January 1924, produced the first Labour Government.

Lord Thomson, as the new Secretary of State for Air, presented all this background of airship history to the Cabinet's 'Airship Development Committee at its first meeting on Leap Year Day of 1924.† The latest situation was, he said, a renewed confidence in the potential of airships for long-range services in no way reduced by the loss of the wartime Zeppelin L.72 (LZ.114) nine weeks before – when, operated by France as the *Dixmude*, it had disappeared over the Mediterranean on 21st December 1923, with the loss of fifty lives.‡

The United States had just completed its first rigid airship for the US Navy – the ZR-1 *Shenandoah* of 2,235,000 cubic feet – which had made its first flight on 4th September 1923. In addition, the United States Government had ordered a new and advanced Zeppelin from Germany, the LZ-126 of 2,762,100 cubic feet – later to be designated ZR-3 *Los Angeles.* It would be delivered by air across the Atlantic later in 1924.

* AIR 5/591 and AIR 8/21.

† AIR 5/908. Note by the Secretary of State for Air.

‡ In September 1923. *Dixmude* (ex-L.72), commanded by Lieutenant de Vaisseau Jean du Plessis de Grenedin, had set up an airship endurance record of 118 hours 41 minutes, flying from Cuers-Pierrefeu, near Toulon, to North Africa and back by way of Paris. On a further long-range flight, the airship was lost, with all hands, off Sardinia, probably through a combination of leaking gas-bags and a lightning strike.

Britain must not stand idly by.

Hence, on 7th March, 1924, Lord Thomson's proposals to the Cabinet for a revised British Airship programme. It was contained in his Cabinet Paper, 'Airships-Programme for Government Research, Experiment and Development'.*

So began the British rigid airship programme of 1924–30 and the origins of R.100 and R.101.

* CP285 (24) 7th March 1924.

Reference Page 65 8th May 1924

APPENDIX 2

AIR MINISTRY SPECIFICATIONS

Nos: 22/24 and 24/24

For Rigid Airships R.100 and R.101

1924

In May of 1924 specifications of requirements were drawn up by a small committee in the Air Ministry under Air Vice Marshal Sir Geoffrey Salmond for the two new airships authorised under the 1924 programme. Specification 22/24 was for the ship to be built by Vickers Ltd; Specification 24/24 was for the airship to be built by the Royal Airship Works at Cardington.

The detailed specifications for the two airships were worked out by the Air Member for Supply and Research, Sir Geoffrey Salmond in conjunction with Squadron Leader Reginald Colmore and Lieutenant Colonel Vincent Richmond.

The outline of requirements (AIR.22–62/509445/2481) dated 8th May 1924, was as follows:

Outline of Requirement

The ship to be constructed to comply with the following general requirements:

General

(1) The gas capacity is to be not less than 5,000,000 cubic feet when full.

(2) The ship must attain a speed of not less than 70 miles per hour as measured by an approved speed indicator at 5,000 ft. under standard conditions, and must be capable of maintaining a normal full speed of 90 per cent of this rate, say 63 miles per hour, for a period of 48 hours.

(3) The total fixed weights of the airship as defined below not to exceed 90 tons: viz.

Structure.
Fabric.
Power cars, and control car.

Machinery including liquids in the system and radiators.
W/T equipment.
Electrical system, including all batteries.
Navigation equipment.
Bow mooring, attachments, winches, wire ropes, etc.
Tanks for fuel and ballast.
Fuel and ballast systems.
Handling and mooring ropes and fittings.
Internal gassing hose.
Passenger car and fittings.

(4) The ship to be capable of ascending, with gas bags full, and descending without danger to the structure of the ship at the rate of 2,000 feet per minute.

(5) The airship to be designed so as to be capable of riiding at mooring mast in winds up to 60 miles per hour and on the assumption that it may take up a yaw of at least 10°.

(6) The Air Ministry will be required to be satisfied before the commencement of the flight to India that the fabrics used for the outer and gas bags of the ship may reasonably be expected to withstand tropical conditions.

(7) Passenger car accommodation to be provided for at least 100 passengers. This accommodation to include adequate and comfortable living and sleeping arrangements including provision of heating, lighting, cooking, ventilating and sanitary facilities.

(8) All usual instruments for navigation and flying, including engine controls, telephones, W/T etc. must be of patterns approved by the Air Ministry, and are to be supplied by the Contractors.

(9) Fuel storage for 57 hours flying at the normal full speed of 63 miles per hour to be provided.

(10) The spare gear, tools, fittings which will be required in flight, referred to in Clause I of the Contract, shall consist of the following, viz:

Main control wires, pulleys and fittings	25% spare.
Automatic Gasbag valves	2 spare.
Gasbag and Outer Cover fabric with solution or dope for attachment	100 sq. yards of each.
Any special fabric fittings such as attachment patches or sleeves	10% spare.
Propellers	25% at least of each type fitted.

Fuel system	2½% replacements of all types of pipes valves and fittings used.
Gasbag pressure gauges	25% spare.
All electric light bulbs fitted	10% of each.
Wireless Spares	As usually carried.
A complete equipment of engine tools to be supplied with each engine.	
Any special tools	One of each.

(11) The ship to be provided with necessary fittings to enable it to be operated from the Air Ministry Mooring Masts.

(12) Sufficient ballast tankage to be provided to fulfil the conditions laid down under trials – of which at least 8 tons must be capable of rapid discharge.

(13) The Contractors to submit to the Technical Co-ordinating Committee referred to in Clause 3 of Schedule I a report on the steps they have taken to ensure satisfactory stability and controllability of the Airship.

Trials

(14) The ship to pass the following trials to the satisfaction of the Secretary of State:

In the United Kingdom

(a) *Shed trials*

Gas bag, lift and trim and shed engine tests to be carried out as may be required by the Secretary of State.

(b) *Flying trials*

A preliminary flight of not less than 3 hours during which no tests will be carried out.

A second flight of not less than 12 hours during which the contract speed trials will be carried out.

A third flight of not less than 12 hours during which the controllability and stability tests will be carried out.

A fourth endurance flight of not less than 48 hours during which opportunity will be taken to test heating, ventilating, W/T etc.

(c) *Mooring Mast trials*

The ship to show herself capable of riding satisfactorily at a mast for any continuous period of not less than seven days with the reservation that the ship will not be kept at the mast in a wind exceeding 60 miles per hour. During these mooring mast trials it must be demonstrated that the gassing, refuelling, ballasting, embarkation and disembarkation of passengers and freight arrangements are satisfactory.

In the event of any major modifications being necessary as a result of any of the above trials the Air Ministry reserve the right to recommence the programme of trials outlined above.

Under Tropical Conditions

(d) *Flying trial*

A satisfactory flight to India with one stop in Egypt for refuelling and necessary repairs.

Airworthiness

(15) The ship to comply with the general requirements as to airworthiness which are to be framed by a special panel of the Aeronautical Research Committee, after consideration by such Committee of any representations which may be made to them by the Contractors.

(16) A certificate of airworthiness will be granted to the airship by the Secretary of State upon his being satisfied that the requirements of the panel mentioned in Clause 15 of this schedule have in fact been complied with in the design and construction of the ship.

Engines

(17) The engines to be fitted to the airship shall comply with the usual air tests imposed by the Air Ministry for aeronautical engines set out in the appendix to this schedule.

Note:

The Operational Requirement of 1924 around which Air Ministry Specification A.24/24 was written, thus called for a payload of 100 passengers together with eight tons of mail and cargo to be carried over a still-air-range of 3,135 nautical miles in 57 hours duration to dry tanks, cruising at 55 knots (or 63 miles an hour). The purpose was to achieve safe, comfortable and relatively fast travel for passengers and goods over the longest non-stop distances then contemplated on Empire routes; that was the 2,930 nautical miles North Atlantic flight from Cardington to Montreal.

Although the duration of 57 hours at 55 knots would not give a sufficient margin to complete the longest sector against substantial headwinds, the theory was that, by reducing speed, the duration and the range could be greatly increased to meet all forecast requirements – including a future air service from Cardington to Lakehurst, New Jersey, in the United States and, eventually from Singapore to Sydney, Australia – a Great Circle Distance of 3,400 nautical miles.

In the event, the ability to achieve a range of more than 3,000 nautical miles carrying 100 passengers was first achieved by the Bristol Britannia Series 300 propeller-turbine aeroplane in 1956.

Reference page 62

APPENDIX 3

BRITISH AIRSHIPS – CONSTRUCTION AND FLYING

Between 1915 and 1930 a total of 230 British airships (214 non-rigids and 16 rigids) flew a total of 91,004 airship hours – of which 87,185 hours 32 minutes were by non-rigid and 3,820 hours and 42 minutes by rigids.

The bulk of the flying – and most of it by non-rigids – was during the four years of the First World War, as follows:

1915	1,496 flying hours and	41,675 miles
1916	8,296 flying hours and	229,107 miles
1917	23,389 flying hours and	579,185 miles
1918	56,536 flying hours and	1,395,765 miles

Total to the end of 1918, 88,717 flying hours and 2,245,810 miles at an average of 25.31 mph.*

Up to 11th November 1918, within these totals, British rigid airships flew 1,532 hours 32 minutes, increased to 1,712 hours 44 minutes by the end of 1919.

Up to the final phase of airship flying performed by R.100 and R.101 in 1929 and 1930, six additional British rigids flew a further 1,683 hours 41 minutes. The largest number of hours flown by any British rigid were by the Armstrong-Whitworth built R.33 (G-FAAF) which, between 6th March 1919, and 23rd November 1926, flew a total of 735 hours and 01 minutes.†

R.100 and R.101 together flew 424 hours and 17 minutes.

In all, therefore, 16 British rigid airships flew a total of 3,820 hours 42 minutes.

The details are as follows:

* AIR 11/156.

† AIR 3 and 43 and AIR 3/45.

FLYING HOURS OF BRITISH RIGID AIRSHIPS
1916 – 1930

	Ship	Constructor and Place Built	Volumetric Capacity cu. ft.	Date First Flight and Date Broken Up	Total Hours Flown	Chief Designer
1.	No 9	*Vickers* 2 Walney is: Barrow	890,000	27 NOV 1916 JUNE 1918	198h 16m	H. B. Pratt First to fly
2.	No 23	*Vickers* 3 Walney is: Barrow	940,000	19 SEPT 1917 SEPT 1919	321h 30m	H. B. Pratt (No 23 Class)
3.	No 24	*Beardmore* 1 Inchinnan	940,000	27 OCT 1917 DEC 1919	192h 53m	H. B. Pratt (No 23 Class)
4.	No 25	*Armstong-Whitworth* 3 Barlow, Selby	940,000	14 OCT 1917 SEPT 1919	225h 05m	H. B. Pratt (No 23 Class)
5.	No 26	*Vickers* 4 Walney is: Barrow	940,000	20 MARCH 1918 MARCH 1919	238h 27m	H. B. Pratt (No 23 Class)
6.	No 27	*Beardmore* 2 Inchinnan	990,600	29 JUNE 1918 AUGUST 1918 Burned Howden	89h 40m	C. I. R. Campbell (23X Class)
7.	R.29	*Armstrong-Whitworth* 4 Barlow, Selby	990,600	20 JUNE 1918 OCT 1919	437h 58m	C. I. R. Campbell (23X Class)
8.	R.31	*Short Brothers* 1 Cardington	1,535,000	1 AUGUST 1918 JULY 1919	08h 55m	C. I. R. Campbell (Wood frame)
FIRST PHASE		TWO YEARS 1916-1918 TO END OF FIRST WORLD WAR	Average: 1,020,775	Av. hrs. flown per ship: 214h 05m Average life: 14.6 months	1,712h 44m	
9.	R.32	*Short Brothers* 2 Cardington	1,535,000	3 SEPT 1919 APRIL 1921	212h 45m	C. I. R. Campbell (Wood frame)
10.	R.33	*Armstrong-Whitworth* 6 Barlow, Selby	1,950,000	6 MARCH 1919 APRIL 1928	735h 01m	C. I. R. Campbell (G-FARG)

	Ship	*Constructor and Place Built*	*Volumetric Capacity cu. ft.*	*Date First Flight and Date Broken Up*	*Total Hours Flown*	*Chief Designer*
11.	R.34	*Beardmore* 4 Inchinnan	1,950,000	14 MARCH 1919 JAN 1921	495h 28m	C. I. R. Campbell
12.	R.36	*Beardmore* 5 Inchinnan	2,101,000	1 APRIL 1921 JUNE 1926	97h 05m	C. I. R. Campbell (G-FARF)
13.	R.38	*Short Brothers* R.A.W., 4/1 Cardington	2,724,000	23 JUNE 1921 AUGUST 1921 crashed	70h 10m	C. I. R. Campbell (ZR-2)
14.	R.80	*Vickers* 5 Walney, Barrow	1,200,000	19 JULY 1921 JULY 1925	73h 12m	B. N. Wallis
SECOND PHASE		YEARS 1919-28	Average: 1,910,000	Av. hrs. flown per ship: 280h 56m Average life: 37.7 months	1,683h 41m	
15.	R.100	*Vickers, A.G. Co.* 1 Howden	5,156,000	16 DEC 1929 NOV 1931	296h 57m	B. N. Wallis (G-FAAV)
16.	R.101A&B	*Royal Airship Works* 2	4,893,740	14 OCT 1929		V. C. Richmond
17.	R.101C	RAW 3 Cardington	5,509753	1 OCT 1930 crashed Allonne OCT 1930	102h 56m 24h 24m	(G-FAAW) V. C. Richmond
THIRD PHASE		1924-30 PROGRAMME	Average: 5,186,498	Av. hrs. flown per ship: 212h 19m Average life: 17.5 months	424h 17m	
TOTAL PROGRAMME		FIRST FLIGHT TO LAST 14 YEARS: 1916-1930	Average capacity: 2,069,747	Av. hrs. flown per ship: 239h 04m Average life: 28.6 months	3,820h 42m	

Reference: Chapter 2, Page 67

APPENDIX 4

THE DESIGN AND CONSTRUCTION OF R.101

19th June, 1929

During Lord Thomson's visit of inspection to the Royal Airship Works at Cardington on 19th June 1929, while R.101 was in the final stages of completion, Colonel Vincent C. Richmond, R.101's Chief Designer, and Wing Commander Reginald Colmore, then Deputy Director of Airship Development, gave to the Secretary of State for Air a comprehensive briefing on the design and construction of R.101 as it had progressed during the previous five years.

Detailed notes of this Briefing Meeting are contained in archives held at the Public Record Office,* in notes left by Lord Thomson and by Sir Christopher Bullock and in Colonel Richmond's records now held in the Royal Air Force Museum at Hendon.

In starting his presentation Colonel Richmond reminded the Secretary of State that, in accordance with his initial instructions in 1924 through AMSR (Sir Geoffrey Salmond) the new airship had been laid out in the summer of that year with the object of incorporating the most advanced technology consistent with a policy of 'Safety First'.

Accordingly, work had been started with a comprehensive programme of Research and Development which ranged from wind tunnel tests to full scale aerodynamic experiments with R.33 and the building and testing of structural specimens.

After the bold decision that the airship should depart from previous practice, to be designed to achieve the best possible streamlined form with a fineness ratio of 5.5 to 1, a second cardinal point had been worked out: that the framework of the hull should be built up of fifteen main longitudinal girders evenly spaced around the outside of the circumference of fifteen ring-shaped transverse frames. So as to achieve a relatively smooth form for the outer cover and in order to avoid very large unsupported panels of fabric, an additional fifteen intermediate auxiliary girders, or 'reefing booms' were to be incorporated in the structure of the hull so as to make the outer form of the airship a 30-sided polygon which was a close approximation to a true

* AIR 5-904-06970.

circle. This 30-sided cross section was increased to 36 sides at the aft end in order to simplify the mounting of the four cantilever monoplane tail fins on a cruciform structure.

The form thus having been settled, a series of wind tunnel tests with various models of R.101 were set in hand during the early design stages, together with full-scale tests with R.33 in flight. When worked out in detail these tests had shown that whereas the long narrow form of R.33 had a drag of 5,861 lb at 60 miles an hour (52 knots), the fatter, streamlined, form of R.101 (without external power cars) could be estimated to have a drag of 6,726 lb at the same speed. This represented the remarkably small increase in drag of less than 15 per cent for an airship which would have 2¼ times the volumetric capacity and gross lift, compared with R.33.

Whereas R.33 required 938 bhp to fly at 60 mph; in theory at least, R.101 would require 1,076 bhp to achieve the same speed in a 'clean condition'.

These tests had given confidence, said Richmond, that, when the flying trials began, R.101 would attain the specified cruising speed of 55 knots – or 63 miles an hour – at a cruising power requirement of about 2,000 horse-power from whatever power-plant might be installed, taking into account the added drag of the power cars.

Then, early in 1925, a further important decision had been taken to contract-out to Boulton and Paul Ltd of Norwich the detail design and manufacture of the structural components of the hull. John D. North * of B & P Ltd was acknowledged to be the leading authority in Britain on metal aircraft construction and he had collaborated fully, in a happy relationship, in the task of making R.101's structure the most advanced of its type yet designed anywhere in the world and – unlike R.100 – an almost complete departure from Zeppelin practice.

For the first time in airship construction extensive use had been made of high-tensile stainless steel strip, rolled into tubular form, to build up the main triangular girders and frames. The result was a saving of at least three tons in weight while achieving a substantial increase in strength compared with the alternative of conventional light-alloy (duralumin) structures.

In fact, said Richmond, about 65 per cent of the weight of R.101's structure was stainless steel strip with duralumin bracing pieces,

* John Dudley North, CBE (1893–1968), educated at Bedford School joined Horatio Barber and Claude Grahame White at Hendon in 1912 and became Chief Engineer of Grahame White Aviation Co. In 1917 he joined Boulton & Paul Aircraft Ltd in Norwich and was Chief Engineer and Chief Designer and a pioneer of metal aircraft construction.

together with sheer-wire-bracing and gas-bag wiring also of steel. The success of this combination had been demonstrated when the test bay, weighing seven tons, had well exceeded all the strength requirements demanded of it.

Richmond added: 'I think we can fairly claim that R.101 is the strongest airship ever built.'

'At the outset,' said Richmond, 'R.101 had been designed around seven engines in separate power-cars to provide the airship with a maximum of rather more than 4,000 horse-power. Thus no more than 50 per cent of the full power would be required in order to attain the specified cruising speed.'

'The Secretary of State will remember,' Richmond went on, 'that the fundamental decision was taken, during the preliminary discussions on the airship programme early in 1924, that – as a result of the terrible loss of life through petrol fires on the water after the disaster to R.38 in 1921 – the new British airships must be powered with compression-ignition diesel engines running on the less volatile, heavy-oil fuel.'*

Following this decision, the Air Ministry team who drew up the specifications for the new airships (Sir Geoffrey Salmond, Reginald Colmore, Vincent Richmond and Tom Cave-Browne-Cave) had explored the possibilities of using the Beardmore Typoon C.I. engine from which a maximum output of 600 bhp at 900 rpm had been promised for a dry weight of 2,200 lb. Thus, in the initial specification of 1924, seven Typhoons were proposed for each airship to give a maximum of 4,200 bhp for a then estimated power-plant weight of 16,800 lb (7.5 tons) including propellers, systems and power-cars. The installed weight would thus have worked out at the respectable figure of precisely four pounds per horse-power. But that was very optimistic.

In the event, the Typhoon engine did not materialise as a practical proposition. Instead, by March 1925, a development of the larger, eight-cylinder in-line Beardmore Tornado C.I. engine – which had a swept volume, or cylinder-capacity, of 84.145 litres (5130.8 cubic inches) was adapted for airship work from its original design for use in Canadian rail-cars.

In its aero-engine form, the Tornado was intended to offer a maximum of 700 bhp at 1,000 rpm for a dry weight of 3,200 lb or 4.57 lb per horse-power but an increase in weight of 1,000 lb per engine for an extra 100 horse-power compared with the Typhoon.

* See also Appendix 2 and Appendix 15.

That would mean an increase of 45 per cent in the power-plant weights.

Seven Tornadoes in place of seven Typhoons would thus provide a maximum of 4,900 bhp for an estimated installed weight of 32,900 lb, or 14.6875 tons; just about double the original estimates.

After long discussions, in which Sir Geoffrey Salmond, as Air Member for Supply and Research, was closely involved, a decision was made on Thursday, 19th March 1925, that the number of engines for R.101 should be cut back to five in order to save weight and complexity.* The result would be a reduction in the maximum available power from the original 4,200 bhp to 3,500 bhp. The designated cruising speed of 55 knots could still be attained, according to estimates, on 2,000 bhp.†

The airship's estimated top speed with five such engines would be reduced by five knots from the originally estimated 73 knots (84 miles an hour) on 4,200 bhp to 68 knots (78.5 miles an hour) on 3,500 bhp but the cruising speed of 55 knots (63 mph) could still be achieved on only 400 horse-power from each engine and there would be a saving of some 5½ tons in weight from the elimination of two engines, their power-cars and four attendant crew members. So it was agreed.

Richmond – always forthright – went on to say that, even so, the latest estimates of R.101's fixed weights were coming out to be somewhat disappointing. This was partly because the Beardmore Tornado engines were inherently heavy in themselves and, even more, because some weight-saving measures – such as aluminium crankcases – had so far not proved practical. In addition there were, as the S of S would know, so many new features in R.101 the weight of which could not be forecast ahead of actual experience.

Though the complete airship had not yet been weighed the expectation now was that it would come out at about 105 tons. If so, that would be 15 tons – or 16.7 per cent – up on the original target of 90 tons for the total of fixed weights including structure, bags, cover and installations. That 90 tons had been set in the specification of 1924 when the seven engines then contemplated were expected to weigh a total of 6.875 tons. Now the five engines, to be installed, as developed for flight, were likely to weigh more than double the 1924 forecast.

Richmond turned to a schedule of estimated weights drawn up on a large wall-chart at the end of the room. It compared the detailed

* AIR 2–262–06,990 and AIR 11–23.
† In practice 55 knots was attained on 1,900 bhp.

weights forecast in Design Memorandum No. 1 * of March 1927, with the latest figures assembled two days previously, on 17th June 1929. By March 1927 the engine bare weights had been set at 4,210 each, because by then it had been accepted that steel instead of aluminium would have to be used for the crank cases to give sufficient stiffness. Even so, at that date in 1927 the fixed weights had still been held at the 90 ton figure by weight reductions elsewhere. The 105 tons now expected meant that the estimated Disposable Lift had been reduced from the 63 tons of Design Memorandum No. 1 to 46.8 tons today.

There had been, said Richmond, three main elements in the increased weight now forecast, compared with the hopes of 1927:

1. *The structure weight* of the hull had gone up by an estimated five tons – or 11 per cent – wholly as a result of the decision to use the reefing booms to achieve a satisfactory shape to the outer cover to reduce drag and improve reliability. The structure weight was now estimated at 49.25 tons compared with an estimate of 44.2 tons in March 1927.

2. *The power-plant* weight had gone up by another 1.5 tons since 1927; an increase of 10.4 per cent, partly because of heavier details and partly because the power-cars in which the engines were installed had come out heavier than expected. The total power-plant weight was now expected to be 15.9 tons compared with 14.4 tons in 1927. Each engine was now recoked at 4,520 lb bare and dry.

3. *The weights of the passenger accommodation* and of the systems had gone up by 4.75 tons, or by 59 per cent. This was chiefly because no passenger accommodation of an equivalent level of comfort and spaciousness had been attempted in any aircraft before. Only experience could show what might be considered to be reasonable for the future. In addition, the kitchen and other miscellaneous equipment had not been included in the original estimates. The accommodation weights were now set at an estimated 12.5 tons compared with 10.2 tons in 1927.

All this was unfortunate. But it had to be faced in a largely experimental, prototype, airship of such radically new form.

'There is this to say,' added Richmond. 'Every strength requirement has not only been met, but substantially exceeded at very large factors of safety. In a repeat R.101, we would have every hope of saving about seven tons in the hull weights of the airship, together with a further valuable amount on the weight of the engine installation.'

* AIR 11/23. (See pages 467 and 475.)

R.101 WEIGHT ESTIMATES 1927 and 1929

	Item	*Royal Airship Works Design Memorandum No 1 7th March 1927*	*Royal Airship Works Revised Weight Estimate 17th June 1929*
	Estimated Volumetric Capacity — cu. ft.	5,050,000	5,000,000
	Theoretical Gross Lift — Tons	153.3	151.8
	Estimated Weights	*Tons*	*Tons*
1.	*Structure:* Longitudinals, frames and joints	36.3	36.3
2.	Reefing booms — aux. longitudinals	—	5.0
3.	Fins, rudders and elevators	2.7	2.95
4.	Wiring: shear, frame and gas bag	5.2	5.0
	Total Main Structure	44.2	49.25
5.	*Power Plant* — Engines, bare, dry	9.4	10.09
6.	Power cars	2.4	3.26
7.	Ancillaries: props, rads, etc.	2.6	2.55
	Total Power Plant	14.4	15.90
8.	*Fabrics* — Gas bags and valves and hoses	7.9	8.5
9.	Envelope: outer cover and empennage	5.3	6.1
	Total Gas Bags and Covers	13.2	14.6
10.	Passenger Accommodation and Control Car	10.2	12.5
11.	*Installations*—Fuel and ballast system	2.3	2.4
12.	Control systems	0.5	0.6
13.	Mooring gear	1.7	1.7
14.	Electrical and W/T systems	0.7	1.2
15.	Crew accommodation, etc.	2.0	2.8
16.	Miscellaneous	0.8	4.05
	Total Installations	8.0	12.75
17.	Total Fixed Weights	90.0	105.0
18.	Theoretical Disposable Lift	63.3	46.8
19.	*Disposable Lift* — Water ballast	8.0	8.0
20.	Fuel and oil	27.0	25.0
21.	Crew and victuals	4.1	4.5
22.	Passengers, mail and cargo	24.2	9.3
	Total Disposable Lift	63.3	46.8
23.	Usable Volumetric Capacity at 97 per cent Filled	4,898.500	4,850.000
24.	Practical Lift available at 97 per cent Purity	148.70	147.23
25.	Practical Disposable Lift	58.7	42.23
26.	Practical Payload	19.6	4.73

In practical terms (when the theoretical maximum gross lift was reduced to take account of the space needed for gas expansion at height), the latest weight estimates showed that R.101 could be expected to be able to transport a useful load of twenty-four passengers and their baggage and victuals, plus one-ton of mail, over the route to India with one stop, for refuelling, in Egypt.

Disappointing as this was, compared with the original, optimistic, requirement for the carriage of 100 passengers and eight tons of mail, or goods, over the same route, the prospective performance of R.101 was well ahead of that of any other existing type of aircraft in payload/range and in the comfort of the accommodation. There was every hope that development work would be able to restore some of the original load. The chief lesson was that airships of 5,000,000 cubic feet capacity were still too small to achieve the proposed performance over Empire routes.

Right from the start, the declared intention had been that R.101 should be planned to advance the state of knowledge and of practice of airship design and construction in every reasonable respect. R.101 had been planned, therefore, as essentially an experimental ship, with experimental engines, to bring the safety of heavy oil to airship operations. In this, as in all other respects, the remit was that nothing should be left undone to ensure the highest possible standards of safety.

To these ends, said Richmond, eleven specific departures from earlier standard practice had been embodied in R.101. They were:

1. The shape of the hull, designed with a profile of a continuous curvature to a specific equation * to achieve as near a perfect streamline form as possible, which would result in no more than two per cent of the resistance of a circular plate of the same maximum diameter. This had produced an airship with a length of 732 feet, a maximum diameter of 132 feet and a total enclosed volume of some 5.755 million cubic feet, including passenger and other accommodation and air space.
2. The elimination of petrol as a fuel so as to make possible the safe operation of a commercial vessel in tropical conditions, where the low-flash point of petroleum vapour would introduce hazards which could hardly be contemplated. The use of compression ignition diesel engines, running on heavy-oil fuel, virtually removed this danger at the expense of a penalty in the

* The equation for the profile of R.101 was as follows, where x=distance from the bow in feet and y=the ordinate or radius in feet: $y=(0.05549\ (400-0.00076X^2\ \frac{1}{2}+0.09968)\ (x(20-0.0276X)\frac{1}{2}$.

installed weight of the power plant – a penalty which was wholly offset over long ranges, because of the relatively low fuel consumption of heavy-oil enginnes. The total weight of heavy-oil engines and fuel for a flight of 50 hours would, thus, be less than the equivalent weight of petrol engines and fuel for a similar duration.

The diesel engine was the power plant *par excellence* for airship work, even though the Beardmore Tornadoes had come out with a heavier weight and a lower power than had been expected when the design began.

3. The extensive use of high-tensile stainless steel for the first time in airship construction, almost wholly on grounds of weight economy, because the size of R.101 makes possible the use of steel of practicable thickness which has a higher specific strength than duralumin. In R.101, the completed hull framework (exclusive of fins, rudders and elevators) would account for about 44 per cent of the total fixed weights; that is about 30 per cent of the gross lift. This hull weight was made up of about 68 per cent girders, 22 per cent joints and attachments and ten per cent steel wire bracings.

Apart from the ten per cent weight of bracing wires, the 90 per cent of the hull weight, represented by girders and joints, was made up of approximately 55 per cent steel and 35 per cent duralumin, with the steel content amounting to some 25 tons. The future use of steel of still higher tensile strength, together with new magnesium alloys for special purposes, such as castings, would make possible substantial further reductions in weight.

4. A new form of transverse ring-frame had been developed without radial bracing. These frames were built up of deep triangular shear-braced structures stiff enough to absorb all transverse forces without any transverse bracing. An added advantage was that the triangular cross-section of the frames afforded convenient stowage space for fuel and water-ballast tanks and other equipment, and required only simple trestles for assembly. Moreover, much fitting work which would otherwise have to be done at a great height above the ground, could be done on the ground before the frames were lifted into place.
5. An advanced type of triangular-shaped cantilever fins had been designed after extensive wind-tunnel tests. They were expected to give good stability at a low drag. The fins were built into the

structure on large cruciform girders attached to frames 14 and 15.

6. A new form of joint had been evolved to assemble the various girders and attachments quickly and accurately by means of bolts in place of the traditional riveted gusset-plates. R.101's joints were formed of machined die-stampings to which the attachments could be bolted at finer tolerances than had been used before on airship structures. As a result, R.101 had been assembled at Cardington relatively quickly once the components had been delivered. The bolted joints also made possible the use of closed section tubular booms of higher efficiency than the usual open-channel sections.
7. An ingenious servo-mechanism had been designed by Squadron Leader Rope for the operation of the rudders and elevators through hydraulic variable-speed gears actuated by a three horse-power electric motor. The control linkage did away with external posts and cables and would give a very light and positive means of moving these large control surfaces.
8. An entirely new form of automatic gas valve, located on the sides of each of the 16 gas-bags at mid-height and actuated through a bellows instead of by the conventional springs. These gas valves – whose automatic opening and closing could be overridden by hand controls – were connected to U-tubes (or syphon pipes), so as to carry discharged hydrogen to the top of the ship and through hoods in the outer cover into the atmosphere.

These valves were the subject of a patent taken out by Michael Rope and represented an important advance in airship engineering practice.

9. A pre-doped outer cover which was stretched to the contour of the ship by means of telescopic kingposts and screw jacks, which pushed out the intermediate reefing girders positioned longitudinally between the main girders of the hull.
10. A system of air vents to maintain the air pressure inside the hull of the airship approximately equal to the external pressure during rapid ascents or descents. This pressure-system was made up of a series of holes, with non-return flaps, arranged radically around the nose and the tail, to provide ingress of air when the ship was descending, together with a series of slots circumferentially around the centre of the ship to allow air to escape when the ship was rising.

11. Finally, a new system of gas-bag wiring to transfer the lift of the sixteen gas bags to the structure of the airship and, at the same time, to locate them in position and clear of the structure, while restraining the gas bags from surging in rough air.

This wiring system was the subject of a joint patent taken out by Colonel Richmond and Michael Rope. In effect, it formed a cage to transfer the pressure of the gas within the bags to the ring frames of the airship's structure and also to form a bulkhead of radiating wires around the ends of each gas bag, rather like parachutes positioned sideways.

So, all in all, R.101 was a radical departure from previous airship practice. This was the deliberate intention in directions which would advance the state of what might be described as both the art and the science of Lighter-Than-Air craft, leading on to still bigger airships. On the whole, said Richmond, the novelties could be judged to be successful thanks to the very substantial effort put into research and development. The one disappointment was the fact that the heavy-oil engines had, so far, turned out to be much heavier and less powerful than their original promise. The concept was undoubtedly right and, in due course, the weight could be expected to come down and the power to go up.*

Such were the main features of R.101 so far – no punches pulled. Richmond sat down and the Secretary of State thanked him for a full and frank account. Clearly, the weights were disappointing, but it was early days yet. What about the time-scale? How had progress gone during the past five years and how did it look for the future?

Colmore got up, notes in hand.† The story was this . . .

As Colonel Richmond had stated, outline design work on R.101, to the Air Ministry specification of requirements, had been started at Cardington in July 1924. At the same time, R.33 had been recommissioned and equipped with a system of pressure gauges, so that, as basic research, the aerodynamic forces and the structural stresses on the airship's hull could be measured in full scale tests. These flying tests, when completed, were checked against wind-tunnel tests on models.

The first practical outcome of this work had been the appearance from Norwich in April 1926 of the component girders of a test bay.

* See Appendix 9 (Page 494) for a note upon Power Plants for the 1924 Airship Programme.

† AIR 5/904/06990.

They had been assembled in the Cardington shed in May of that year.

Satisfactory tests were first made with a single transverse ring, in both horizontal and a vertical position, under varying loads. In July 1926 a second ring was assembled, longitudinals put in place to link up with the first ring and the complete bay bolted up for further tests.

The loads which had been predicted by theory were verified in practice and the bay successfully loaded. The appropriate gas-bag was then placed within the bay, together with ballast bags, fuel tanks and a section of the outer cover, to complete a typical section of the future airship.

By December 1926 a comprehensive series of tests had been completed successfully and – as a final check – the bay was dismantled and the individual girders tested to destruction.

Meanwhile, the general work of design and research had continued. By the end of 1925, Boulton and Paul Aircraft Limited were fully in the picture. The procedure was that Colonel Richmond's staff at Cardington informed Boulton and Paul of the general arrangement of the structural components and the stresses estimated for each member. The Company's Drawing Office then designed a suitable girder, or other component, and submitted drawings to Cardingtonn for approval.

The components were then manufactured by Boulton and Paul, under AID inspection, and delivered to Cardington for assembly.

In January 1927 manufacture had begun of those parts of the airship on which design was complete. By December 1927 all the components of the first of the transverse ring frames had been delivered from Norwich to the Royal Airship Works and it had been assembled and lifted into position in the shed.

By July 1928 Boulton & Paul had completed all the detail drawings and they had been checked and approved by the design staff at the RAW, and eight of the frames and the appropriate longitudinal girders (Frames 4 to 11) had been assembled, lifted and bolted together. In August 1928 the control car was completed and joined up to the rest of the structure. By September the manufacture of the sixteen gas bags had been completed and work had started on the attachment of the outer cover to the completed portion of the ship.

By the end of 1928, the bow structure back to Frame 4 was completed and joined up, the fuel and ballast systems were being installed and work had begun on the forward portion of the other cover.

Frames 12 and 13, at the tail-portion of the airship, had been assembled in January 1929, and joined up. By March, the last components had been delivered from Boulton & Paul and work on the erection of the fins had started. By then, the passenger quarters and the attachment of the outer cover were almost finished.

The final frames (14 and 15) and the fins had been lifted into position in April 1929, and work was in hand on construction of the rudders andd elevators. All of these had been completed in May and placed in position, together with the tail cap.

Now, in mid-June of 1929, the ship was almost complete. Three of the Beardmore Tornado engines had been delivered, the fourth was expected during the next day or two, and the fifth early in July.

Inflation of the gas bags was due to begin very shortly afterwards, to be followed by the shed trials as the last step before the first flight, now likely in September.

None of the engines was, however, yet slung on the ship, though the first power-car, with its engine installed, had begun its gantry tests on 17th June. The engine had started up without difficulty – which promised well for the future.

That, he said, brought the story up to date at 19th June 1929.

Reference pages 68, 119 and 336

APPENDIX 5

R.101 WEIGHTS

(Estimates and Actuals 1927–30)

1927–30

A Wide Variation between optimistic estimates in 1927 and final achievement in 1930, spanned the translation of R.101 from preliminary design to developed prototype – and illustrated the acquisition of practical knowledge.

In the 3½ years from Design Memorandum No 1 in March 1927 to the final weighed weights in September 1930, the concept of R.101 moved from an airship with an estimated nominal gross lift of 153.3 tons to a lift of 167.2 tons – an increase of 13.9 tons or nine per cent in lift. At the same time, the total of fixed weights went up from a specification figure of 90 tons to an actual of 117.9 tons – an increase of 27.9 tons or of 31 per cent, much of it in the increase in the weight of the engines compared with early forecasts.

In consequence the disposable load was reduced from the specified and hoped for 63.3 tons to an actual of 49.3 tons, some 78 per cent of the aimed-for useful lift. In practice, because no airship took off with gas-bags completely filled the practical useful lift of R.101C in its final form was some 42.7 tons. Out of this 18 tons was needed for the 'Service Load' – (crew, ballast, movable equipment, etc.) leaving about 24.7 tons for fuel, oil and payload.

How the estimated and actual weights developed over the years is shown in detail, for the first time, in the accompanying tables from Cardington records.

March 1927/June 1929

PRE FLIGHT
R.101 WEIGHT ESTIMATES
(1927–29)

		7 March 1927, Design Memorandum No 1	*17 June 1929, RAW Weight Estimate*	*Percentage Increase or (Decrease)*
		Tons	*Tons*	%
1.	Estimated Gross Lift	153.3	151.8	(1.0)
	Structure			
2.	Main girders	29.6	30.7	3.7
3.	Joints and attachments	9.4	9.7	3.2
4.	Wire bracing (stainless steel)	5.2	4.6	(13)
5.	Auxiliary reefing-boom girders	—	4.0	—
6.	*Total Hull Framework*	44.2	49.0	10.8
7.	Machinery (Engines & Accessories)	14.4	18.3	27.1
8.	Passenger accommodation & furnishings	10.25	12.5	21.9
9.	Outer cover and gas bags	13.25	14.6	10.2
10.	Installations, systems and misc.	7.90	10.6	34.2
11.	*Total Fixed Weights*	90.00	105.0	26.1
12.	Disposable Lift available	63.3	46.8	(26.1)
	Disposable Load			
13.	Water ballast	8.0	8.0	—
14.	Fuel & Oil for Egypt/India route	27.0	25.0	(7.0)
15.	Crew (three watches and victuals for two days)	4.1	4.5	(9.7)
16.	Passengers, mail and cargo	24.2	9.3	(61.6)
17.	*Total Disposable Load*	63.3	46.8	(26.1)
18.	Passengers, baggage, victuals	(100)	(50)	
	@ 350 lb each	15.6	7.8	(50.0)
19.	Lift available for mail and cargo	8.2	1.5	(81.7)

BRITISH AIRSHIP WEIGHT SCHEDULES

	Airship	*R.101A*	*R.100*	*R.101B*	*R.101C*
1.	Date of Weight Analysis	30 Sept.	20 July	26 June	30 Sept.
2.		1929	1930	1930	1930
		Original Condition	As for Canada	Bags let out	Before India flight
3.	Vol. Capacity: cu. ft.	4,893,740	5,156,000	4,998,501	5,509,753
4.	Nom. Gross Lift — tons	148.5600	156.6214	151.7402	167.1604
5.	Gross Lift at 96% Filled	142.6176	150.2606	145.6706	160.5700
	Weights Tons				
6.	Main Structure	37.1371	30.4907	37.2077	40.3007
7.	Shear Brcng. and Fr. Wires	2.7985	5.9230	2.8103	3.0364
8.	Reef Girders (AX.R.100)	5.7340	1.2647	5.0520	5.3920
9.	Fins, Rudders and Elevators	4.7056	7.7767	4.7836	4.7836
A.	Total Prim. Structure	50.3752	45.4551	49.8536	53.5127
10.	Gas Bags, Hoses, Valves, Pads	9.1843	8.9458	9.1723	9.9743
11.	Gas Bag Nets and Wires	4.5758	4.0927	4.5944	4.9633
B.	Total Gas Installation	13.7601	13.0385	13.7667	14.9376
C.	Outer Cover and Tail Surfs.	7.5774	7.8062	7.6224	8.7614
12.	Passenger Accom. Framework	6.3768	9.4249	6.3768	6.3768
13.	Pass. Accom. Fixts. & Furn.	6.3018	5.2670	5.5426	5.5626
14.	Kitchen & Kitchen Equipment	0.6216	0.6102	0.5368	0.5368
D.	Total Passenger Accom.	13.3002	15.3021	12.4562	12.4762
15.	Off. & Crew Accom. Frmwrk.	1.0131	1.0242	1.0131	1.0131
16.	Control Car Structure	0.4713	0.6212	0.4713	0.4713
17.	Off. & Crew Accom. Equipt.	0.1661	0.1720	0.1611	0.1611
18.	Crdrs., Gngwys. & Ladders	1.7749	2.6321	1.4829	1.6389
E.	Total Operating Accomm.	3.4254	4.4495	3.1284	3.2844
19.	Control Systems	0.7975	0.7067	0.4925	0.6235
20.	Mooring & Hndling. Equipt.	1.7009	2.0754	1.7009	1.7549
21.	Electrical & W/T Systems	1.4589	0.9619	1.1779	1.2809
22.	Water Ballast System	0.7393	0.5550	0.7393	0.7393
23.	Fuel Tanks & System	1.7177	2.7720	1.3077	1.8427
24.	Lub. Oil Tanks & System	0.0598	0.1312	0.0598	0.0598
F.	Total Systems	6.4741	7.2022	5.4761	6.3011
25.	Engines — Bare & Dry	10.0920	5.0625	10.0920	10.5640
26.	Engine Ancills.: Props. Rads. Exhausts	2.5420	1.1598	1.9960	1.6970
27.	Pwr. Cars: Stctre. & Eqpt.	2.6215	3.9670	2.4275	2.4275
G.	Total Pwr. Plt. Instlltn.	15.2555	10.1893	14.5155	14.6885
28.	Misc. Navigating Equipt.	0.2002	0.2100	0.2002	0.2002
29.	Misc. Hatchs, Nose Cab, etc.	0.2701	0.4212	0.2701	0.2701
30.	Misc. Res. Stock & Spares	0.4764	0.5456	0.4764	0.4764

BRITISH AIRSHIP WEIGHT SCHEDULES — *continued*

	Airship	*R.101A*	*R.100*	*R.101B*	*R.101C*
H.	Total Misc. Items	0.9467	1.1768	0.9467	0.9467
I.	Basic Empty Eqppd. Weight	111.1146	104.6197	107.7656	114.9086
31.	Unidentified Items	2.4854	0.9003	3.5344	2.9883
J.	Quoted Weighted Fxd. Wt.	113.6000	105.5200	111.3000	117.8969
32.	Service Load (Inc. Bllst.)	18.0000	18.0000	18.0000	18.0000
33.	Airship Prep. for Svce. Wt.	131.6000	123.5200	129.3000	135.8969
34.	Fuel, Oil Stores Payload	11.0176	26.7406	16.3706	24.6731
35.	Gross Lift for 1200′ Pressure Height	142.6176	150.2606	145.6706	160.5700
	Sources:	Air 5/904	A. E. G. & Wallis Files	Air 11/23	Air 11/23 Cardgtn. 9-12-30

*Summary of Equipment Weights on Board R.101C at Lift and Trim Trials on 30th September 1930 and included in Weight Available for Disposable Lift**
(Total: 3.647 tons = 8,169 lb)

		lb.	*Tons*
A.	*Operating Equipment*		
	1. Engine spares and tools	744	
	2. Stability roller wires	166	
	3. Miscellaneous: Bumper bags, handling bars, echo sounding gear, recording, thermometer	1,626	
	4. Table and chairs	34	
	A. Total Operating Equipment	2,570 lb	1.147 tons
B.	*Crews' Items*		
	5. Crew beds and bedding	610	
	6. Crews' kits	1,230	
	7. Crews' furniture	26	
	B. Total Crews Items	1,866 lb	0.833 tons
C.	*Consumable Stores*		
	8. Fresh water (160 Imp. gallons)	1,600	
	9. Food and emergency rations	281	
	C. Total Consumable Stores	1,881 lb	0.897 tons

* AIR 11–23–X/M.08109.

D.	*Kitchen Equipment*		
	10. Kitchen utensils and ship's silver	190	
	11. Crockery	56	
	D. Total Kitchen Equipment	246 lb	0.110 tons
E.	*Dining Room*		
	12. Chairs and tables	285	
	13. Ship's linen	142	
	E. Total Dining Room	427 lb	0.191 tons
F.	*Lounge*		
	14. Chairs	156	
	15. Tables	103	
	16. Cushions	102	
	17. Pictures and frames	58	
	F. Total Lounge	419 lb	0.187 tons
G.	*Smoking Room*		
	18. Chairs and tables	34	
	19. Cushions	36	
	G. Total Smoking Room	70 lb	0.031 tons
H.	*Sleeping Accommodation*		
	20. Passengers' beds, pillows, blankets, mattresses	358	
	21. Curtains, stools, etc.	83	
	22. Mats and carpets	35	
	23. Lampshades and racks	20	
	H. Total Sleeping Accommodation	496 lb	0.221 tons
I.	*Other Fittings*		
	24. Stair carpets and fixings	33	
	25. Other carpets	146	
	26. Red Cross box	15	
	I. Total Other Fittings	194 lb	0.087 tons
Total Equipment Weights in Disposable Lift		8,169 lb	3.657 tons

BAGGAGE WEIGHTS

(Loaded on board R.101C on 4th October 1930. Weights checked by Arthur Finch – Store Keeper, RAW)

1. *Ship's Officers and Petty Officers* (7)

1.	Flight Lieutenant H. C. Irwin	28 lb	(Captain)
2.	Squadron Leader E. L. Johnston	27 lb	(Navigator)
3.	Lieut-Commander N. G. Atherstone	36 lb	(First Officer)
4.	Flying Officer M. H. Steff (not seen) say	31 lb	(Sec. Officer)
5.	M. A. Giblett	30 lb	(Met. Officer)
6.	G. W. Hunt	20 lb	(Chief Cox'n)
7.	W. Gent	28 lb	(First Engineer)
	Total for ship's Officers and Petty Officers	200 lb	(Av. 28.6 lb)

2. *Crew* (35)

Thirty-five crew members at 10 lb each	350 lb	
Cask of ale	70 lb	
Total of crew's baggage	420 lb	(Av. 12 lb)

3. *Officials from the Royal Airship Works* (6)

1.	Wing Commander R. B. B. Colmore (not seen) say	37 lb	(DAD)
2.	Major G. H. Scott	26 lb	(ADADA)
3.	Lieut-Col V. C. Richmond	44 lb	(ADADT)
4.	Squadron Leader F. M. Rope	30 lb	(A to ADAD)
5.	Mr A. Bushfield	27 lb	(AID)
6.	Mr H. J. Leech	36 lb	(Foreman Eng.)
	Total for officials of R.A.W.	200 lb	(Av. 33.3 lb)

4. *Passengers* (*in order of going aboard*) (6)

1.	Sir Sefton Brancker	30 lb	(DCA)
2.	Major P. Bishop	28 lb	(AID)
3.	Squadron Leader W. Palastra	28 lb	(RAAF)
4.	Squadron Leader W. H. L. O'Neill	28 lb	(DDCA India)

5. Lord Thomson of Cardington	....	254 lb	(S of S for Air)
6. James Buck	....	26 lb	
Passengers total for baggage	....	394 lb	(Av. 65.7 lb)

Summary of baggage loaded: 4th October 1930

1. Ship's officers and petty officers	(7)	......	200 lb	
2. Crew	(35)	......	420 lb	
3. Officials of R.A.W.	(6)	......	200 lb	
4. Passengers	(6)	......	394 lb	
Total baggage	(54)	....	1,214 lb	(Av. 22.5 lb)

(N.B. Lord Thomson's baggage included a carpet roll at 129 lb, a sword at 3 lb and two cases of champagne at 52 lb)

The average weight of baggage was 22.5 lb spread over each of the 54 persons on board R.101.

Reference page 118

APPENDIX 6

R.101 VOLUMETRIC CAPACITIES AND LIFT

1929–30

As with the structural and power-plant weights, the capacity of R.101's gas bags was substantially changed during the translation from drawing board to operational service.

The original specification of 1924 called for a volumetric capacity of five million cubic feet which represented a nominal gross lift of 340,000 lb (at 68 lb per 1,000 cubic feet) – or 151.786 tons.

When the initial Lift and Trim trials were completed on 30th September 1929, the 16 gas bags contained 4,893,740 cubic feet of hydrogen which at 96.15 per cent average purity and at 59°F produced an actual lift of 332,783 lb or 148.564 tons – a reduction of just over two per cent.

By letting out the gas-bag wiring so that the bags could expand, the volumetric capacity had increased, by 26th June 1930, to 4,998,501 at the higher gas purity of 98.35 per cent. The result was 347,662 lb of lift – or 155.2 tons; rather better than the original specification.

In its final form, with the extra bay, as measured at the Lift and Trim trials on 30th September 1930 – exactly a year from the first measurement – R.101C was found to have a volumetric capacity of 5,509,753 cubic feet and a gross lift at 95.4 per cent gas purity, of 371,867 lb or 166.01 tons. That was some nine per cent more than the original specification.

The details are set out in the accompanying tables:

HM AIRSHIP R.101
VOLUMETRIC CAPACITIES OF GAS BAGS IN CUBIC FEET — 1929-1930

	R.101A at 30th September 1929 as Originally Completed	*R.101B* at 26th June 1930 With Gas-Bag Wiring Let Out	*R.101C* at 30th September 1930 With Extra Bay Installed
Gas Bag No.	*Volumetric Capacity in Cubic Feet*	*Volumetric Capacity in Cubic Feet*	*Volumetric Capacity in Cubic Feet*
1.	19,780	19,780	19,780
2.	112,800	116,521	116,521
3.	301,230	311,800	311,800
4.	406,920	416,585	416,585
5.	462,120	474,115	474,115
6.	486,710	496,491	476,491
7.	470,900	478,508	478,508
8.	468,400	476,016	476,016
8A.	—	—	(8a) 510,300
9.	499,460	501,315	501,315
10.	479,600	490,188	490,188
11.	429,540	437,458	437,458
12.	349,480	358,698	358,698
13.	226,070	233,206	233,206
14.	112,130	116,317	116,317
15.	45,810	47,748	48,700
16.	22,790	23,755	23,755
Total	4,893,740	4,998,501	5,509,753
Ratios	100	102.14	112.59
Average Gas Purity	96.15 per cent	98.35 per cent	95.4 per cent
Temperature	59°F		56°F
Pressure	1,013.2 mb.		1010 mb.
Air Density	1,226 g/cu.m.		1,230 g/cu.m.

HM AIRSHIP R.101
GROSS LIFT FROM INDIVIDUAL GAS-BAGS

	R.101A at 30th September 1929 as Originally Completed		*R.101B at 26th June 1930 with Gas-Bag Wiring Let Out*		*R.101C at 30th September 1930 with Extra Bay Installed*	
Gas Bag No.	*Lift in lb.*	*Lift in tons*	*Lift in lb.*	*Lift in tons*	*Lift in lb.*	*Lift in tons*
1.	1,298	0.597	1,334	0.595	1,328	0.595
2.	7,558	3.374	8,026	3.583	7,888	3.519
3.	20,548	9.173	22,016	9.823	21,048	9.395
4.	27,240	12.161	28,548	12.745	28,511	12.728
5.	31,297	13.972	32,941	14.706	32,041	14,304
6.	33,126	14.788	34,471	15.389	33,727	15.057
7.	32,050	14.308	33,094	14,774	32,060	14.312
8.	32,076	14.320	33,126	14.788	30,670	13.691
8A.	—	—	—	—	35,251	15.737
9.	34,240	15.286	34,932	15.595	34,560	15.428
10.	32,842	14.662	34,210	15.272	32,774	14.631
11.	29,390	13,120	30,482	13.608	29,655	13.239
12.	23,762	10.608	25,033	11.175	23,957	10.695
13.	14,992	6.929	15,954	7.122	15,725	7.020
14.	7,688	3.432	8,272	3.693	7,802	3.483
15.	3,125	1.395	3,538	1.579	3,277	1.463
16.	1,551	0.692	1,685	0.752	1,593	0.711
Total	332,783	148.564	347,662	155.206	371,867	166.012
Ratios	100	100	104.47	104.47	111.74	111.74
Average Gas Purity	96.15 per cent	96.15 per cent	98.35 per cent	98.35 per cent	95.4 per cent	95.4 per cent
Temp.	59°F	59°F			56°F	56°F
Pressure	1,013.2 mb	1,013.2 mb			1,010 mb	1,010 mb
Density	1,226 g/mm	1,226 g/mm			1,230 g/mm	1,230 g/mm
Lift per 1000 cu. ft.	68.00	68.00	69.55	69.55	67.49	67.49

Reference pages 55 and 210

APPENDIX 7

THE BRITISH AIRSHIP PROGRAMME

(Expenditure from 1st July, 1924, to 5th October, 1930)

1924–31

A TOTAL of £1,400,000 was forecast in March 1924 to be spent on the projected new British airship programme during the 31 months to September 1927, when the hopes were that two airships would be in service, a first flight to India completed, and Phase Two of the programme ready for launch.

By February 1929 the programme was running rather more than two years late and the estimated cost had gone up to £2,051,000, spread over 67 months to September 1929.

In the event the total cost of the programme came to £2,396,948 over a 6½ year period – an increase of 71 per cent over the original estimate. Out of this, R.101 cost £711,595, including research and development and the insertion of the extra bay. R.100 cost £500,113 at an earlier stage of development. The two airships together thus cost £1,211,708 – which was almost exactly half the cost of the whole programme which included the mooring masts, sheds and the overseas bases.

The details of the estimated and the actual costs are set out in the following tables:

The 1924–1929 British Airship Programme
Total Expenditure 1st July 1924, to 12th October 1929
(First Flight of R.101)

	£
R.101 (Design and construction)	527,000
R.100 (Design and construction)	337,000
Research work	264,000
R.33 and R.36—(July, 1924 to Dec., 1926) (Overhauls and R.33 flights)	103,100
No. 1 Shed Cardington (extensions)	105,800
No. 2 Shed Cardington (from Pulham)	213,200
Cardington Mooring Mast	127,000
Ismailia Mooring Mast and base	93,012
Karachi base, shed and mast	216,000
Total to 12th October, 1929	£1,986,112

R.101 to 12th October 1929 (First Flight)

	R.A.W.	*B & P Ltd.*	*Total*
Design Man Hours (500,000 man hours)	43,000	7,000	50,000
Materials	61,000	47,000	108,000
Direct Labour (2,230,000 man hours)	101,000	44,000	145,000
Overheads	112,000	53,000	165,000
Jigs and Tools	—	13,000	13,000
Plant and Equipment	—	4,000	4,000
Deliveries of finished components	—	1,000	1,000
Power Plant (Beardmore)	41,000	—	41,000
	£358,000	£169,000	
Total			£527,000

Cost of The British Airship Programme
(1924 to 1930)

Item	*R.101*	*R.100*
	£	£
Direct Labour	176,105	145,333
Materials	132,960	
Design Labour	66,240	21,800
Overheads	214,060	
Jigs, tools, plant, etc.:	18,500	
Power plant and spares	56,000	
Inflation, trials, etc.:	29,730	
Boulton & Paul profits	18,000	—
Total at 5th October 1930	£711,595	£471,113

R.101—(To first flight: £527,000) R.100—(To first flight: £337,000)

Total Costs

		£
1.	Research: (including staff costs)	305,200
2.	Development and experimental work (R.33 & R.36)	103,100
3.	Cardington airship sheds	205,800
4.	Karachi base and shed	120,000
5.	Mooring masts	160,000
6.	Overheads, administration, etc., for above tasks	290,084
7.	R.101	711,595
8.	R.100	471,113
9.	Flying operations	30,056
	Total: (1st July, 1924 to 5th October, 1930)	£2,396,948

R.101C
Rectification, modification and refit work and addition of extra bay
1st July to 30th September, 1930

		£
1.	Direct labour @ 436,575 man hours	29,105
2.	Materials	21,960
3.	Design labour @ 162,400 man hours	16,240
4.	Overheads	33,060
	Three months Total:	£100,365

During the period of thirteen weeks between 1st July and 30th September an average of 700 men and women worked on R.101 at the Royal Airship Works and at Boulton and Paul Aircraft Ltd. at Norwich in addition to 270 men in the design offices.

*

The work completed on R.101 during the summer of 1930 after the Hendon flights, in addition to the new bay and gas-bag, included complete repair and overhaul of all existing gas-bags, a substantially new outer cover, extensive padding of all structure in the way of the gas-bags and overhaul and reinstallation of all engines, two of them equipped to run in reverse.

Reference page 122

APPENDIX 8

THE HIGH MOORING MAST FOR AIRSHIPS

One of the chief operational problems with every form of Lighter-than-Air craft has always been its docking arrangements outside its shed before and after flights. Very large ground handling parties were required until more sophisticated mooring gear was developed.

Shortly after the First World War, Major G. H. Scott, when in command of the airship station at Pulham, began work on the concept of the high mooring mast which was tried out in successive variants at Pulham and at Croydon during the early 1920s.

The high mooring mast was, thus, intended as a safe 'open harbour' for airships to which they could be moored and unmoored in any weather and at which they could receive gas, fuel, stores and payload.

The Cardington mast, completed in 1926, was an eight sided steel girder structure, 200 feet high, tapering from 70 feet diameter at ground level to 26 feet 6 inches at the passenger platform, 170 feet from the ground. Above the passenger platform was the 30 feet of the conical housing for the mooring gear. A lower platform 142 feet above ground accommodated searchlights and signalling gear in a galley four feet wide.

The major portion of the mast was built by the Cleveland Bridge and Engineering Company, the contractors for the Cardington airship sheds, under the direction of Major General Sir William Liddell, Director of Works and Buildings at the Air Ministry. The total cost of the mast was £127,000 and the Cardington tower formed the prototype for similar structures built at Ismailia, at Karachi and at St Hubert, Montreal.

Up to the level of the searchlight gallery the tower was an open-work structure made up of eight main legs set in concrete bases and cross-braced on all eight faces to horizontal frames arranged at intervals of 28 feet 4 inches. From the searchlight gallery upwards the surmounting tapered circular turret was framed in steel, enclosed with corrugated sheeting and lit by sixteen windows. The top platform, at the height of 170 feet, from which passengers embarked and disembarked to and from the airships, was 40 feet in diameter and encircled by a heavy parapet. The top rail of this parapet formed a track on which a gangway, let down from the airship, ran on wheels to give freedom for the airship to move around the tower as it swung with the wind.

A lift shaft, 9 ft 6 ins square, ran up the centre of the tower in a separate block, itself 16 feet square. Within this block a stairway ran from ground level to the passenger platform.

The electric passenger lift, with capacity for 12 persons at a time, ascended to a height of 156 feet from the ground at which level there was a steel platform with stairs up the remaining 14 feet to the passenger platform. The lift, 8 feet by 6 feet and 7 feet 6 inches high, was operated by a 30 hp 400 volt electric motor to give a speed of 150 feet a minute.

The tower was designed to withstand a horizontal pull from an airship amounting to 30 tons, exerted 200 feet from the ground, together with a wind pressure load taken at a maximum of 30 lb per square foot. To that end each of the eight concrete base-blocks, in which the tower's legs were set, measured 13 feet 6 inches by 11 feet by 6 feet 6 inches deep, each providing an anchoring force of 60 tons.

The upper portion of the tower, from the passenger platform onwards, was designed and built by Babcock and Wilcox Ltd. This section was a circular steel turret 25 feet in diameter and 12 feet 6 inches high surmounted by a truncated cone with its top 23 feet above the passenger platform.

A three-part telescopic arm, mounted on gimbals, projected through an opening at the top, free to swing from the vertical in any direction up to 30 degrees of movement. The top of the arm consisted of a bell-shaped cup mounted to rotate on ball-bearings. A cable extended through the bell-mouth and linked to a cable, dropped from the airship to be moored, enabled the nose of the airship to be drawn down until a cone on the nose locked home into the cup and so secured the airship to the tower. The telescopic arm was then centred, locked in the vertical position and made free to rotate on a vertical axis so that the airship could swing, nose to tower, in any direction of the wind.

At the base of the tower, in a machinery house, three steam-driven winches, made by Babcock and Wilcox Ltd., operated the hauling gear through drums, 5 feet in diameter to give cable hauling speeds of 50 feet a minute – at three rpm of the drums.

The two-cylinder steam engine designed to meet these requirements had cylinders 12 inches diameter by 8 inches stroke, working at a steam pressure of 80 lb per square inch. The steam engine turned the drum through reduction gears and a clutch designed to slip under a load of 16 tons on the winding cable so as to avoid excessive loads. Each winch operated drums carrying 4,000 feet of wire cable with a breaking strength of 25 tons. The normal operating load was about four tons.

Normally, when an airship approached the mast slowly against the wind, a mooring cable was let out from the nose to the ground and linked, by a ground party, to the end of the mooring cable paid out from the mast head.

The cable was then slowly wound in with the airship riding about 600 feet above the mast and down wind, with one engine running astern to maintain a pull on the cable. At this point, two side wires – or 'yaw guys' – were also connected to cables taken from the nose to pulley blocks some hundreds of feet apart on the ground and thence to winches in the base of the mast.

All three cables were then wound in together; the main pull being taken on the mooring cable while the yaw guys steadied the ship. When all the cable had been wound in an articulated mooring cone on the nose of the ship locked home into a cup on the mast head. The mast fitting was free to rotate as the airship swung with the wind with freedom also for pitch and roll.

A gangway, like a drawbridge, which could be drawn up flush with the nose of the airship, was then let down with its free end resting on the platform running round the mast head. Passengers and crew boarded and disembarked from the ship under cover along this gangway. About twelve men were needed to moor an airship to a mast.

The fuel storage and pumping equipment at the base of the tower was supplied and installed by Zwicky Ltd of Slough, Bucks. Two storage tanks, each capable of holding 10,000 gallons of petrol or fuel oil, were sunk in underground concrete placements. The tanks themselves were of steel, 27 feet long and 9 feet in external diameter with dished ends. The tanks supplied a three-inch diameter main which ran up the tower to an outlet which could be connected to a main of the airship at the mast. The total fuel lift from the centre-line of the tank to the outlet was some 204 feet. Fuel delivery was by means of a Zwicky suction pump capable of supplying 2,000 gallons an hour to a height of 400 feet, driven by a six horse-power Crompton motor running at 900 rpm but driving the pump at 35 rpm.

In addition, a 12 inch gas main ran up the tower for regassing of the airships from the hydrogen-generating plant and further pumps supplied water ballast at up to 5,000 gallons an hour.

The equipment was completed by a boiler house containing two Babcock oil-fired boilers of 997 square feet heating surface, working at 200 lb per square inch boiler pressure. This boiler house, the machinery house (50 feet 6 inches x 57 ft 6 ins) and the accommodation house for officers, crew, mast staff and chart-room facilities were built by Braithwaite and Co of London. The mechanical and electrical portion of the work was supervised by Mr H. E. Oakley, Deputy Director of Works and Buildings at the Air Ministry and Lieutenant-Colonel J. D. K. Restler, the Air Ministry's Chief Mechanical and Electrical Engineer.

At Cardington, the high mooring mast was located 1,000 yards south-east of the airship sheds. The sheds remain but the tower was demolished for use as steel scrap in 1941 during the Second World War.

Reference pages 168/169

APPENDIX 9

R.100 AND R.101 – POWERS, SPEEDS AND FUEL CONSUMPTIONS

DURING their brief flying lives, the two airships, R.100 and R.101 made 22 individual flights and were away from the mooring masts for a total of 424 hours and 17 minutes and spent 387 hours 31 minutes flying on course. Each flight, therefore, averaged a duration of 19 hours 17 minutes away from the mast and 17 hours 36 minutes flying on course.

2. R.100 covered 11,134 nautical miles flying on course, and a total of 14,356 air miles, in a total of 276 hours 02 minutes on course. R.100's average ground speed was thus 40.33 knots (46.44 mph) and its mean true air speed 52.01 knots (59.89 mph).

3. R.101 covered 3,665 nautical miles flying on course and a total of 5,328 air miles, in a total of 111 hours 29 minutes on course. R.101's average ground speed was, therefore, 32.88 knots (37.86 mph) and its average true air speed 47.80 knots (55.04 mph).

4. R.101 differed substantially in its final, lengthened, condition from its original form. Before the insertion of the extra bay, R.101 covered a total of 1,895 nautical miles in 89 hours 35 minutes flying time on course at an average ground speed of 21.2 knots and at an average true air speed of 48.36 knots. After lengthening, R.101C covered 770 nautical miles in 21 hours 53 minutes on course at an average ground speed of 35.2 knots and at an average true air speed of 46.0 knots.

5. In its flying life, therefore, R.100 encountered a mean headwind component of 11.68 knots and R.101 a mean headwind component of 14.92 knots.

6. At a mean true air speed of 52.01 knots R.100 was flown on an average of 1,500 bhp – that is at 38.5 per cent of its maximum available power and at 50 per cent of its maximum continuous cruise power (500 bhp from each engine at 1,730 rpm).

7. At a mean true air speed of 48.26 knots, R.101 in its original, unlengthened, form was flown on an average of about 1,300 bhp – that is at 50 per cent of its maximum available power and at 55.5 per cent of its maximum continuous cruise power (585 bhp from each engiine at 890 rpm). The comparable figures for R.101C were

46.0 knots T.A.S., on 1,350 bhp; in that instance at 41.5 per cent of the maximum available power and at 46 per cent of the maximum continuous cruise power.

8. On its final flight R.101C was flown at a mean true air speed of 53 knots on an average of 2,100 bhp (72 per cent of maximum continuous cruise power) – the fastest and the highest power settings at which either of the airships had ever flown continuously before:

9. In summary the measured performances were as follows – all as averages over their flying lives:

Airship	*True Air Speed Knots*	*Cruising Power b.h.p.*	*Fuel Consumption lb. per hr.*	*Ground Speed Knots*	*Headwind Knots*
R.100	52.1	1,500	870	40.33	11.77
R.101A	50.4	1,520	623	30.44	19.96
R.101B	43.0	920	398	36.98	6.02
R.101C	46.0	1,450	600	35.18	10.92
R.101C (Final flight)	53.0	2,100	885	35.2	17.8

10. Except for R.101C's last flight both airships were thus flown normally at very low power settings and, therefore, at low rates of fuel consumption. For comparisons of performance, if the various airships were all flown on 1,500 bhp at 1,000 feet the relevant figures would have been:

Airship	*No. of Engines*	*r.p.m. for 1,500 B.h.p.*	*True Air Speed Knots*	*Fuel Consumption lb/Hour*	*Power per Engine B.h.p.*	*Percentage of Max. Cruise Power*
R.101 A & B	4	760	50.2	640	375	65%
R.101C	5	705	46.6	625	300	52%
R.100	6	1,260	52.01	870	250	50%

11. Thus, on identical cruise power outputs (1,500 bhp) R.100 was about 1.8 knots faster than the original R.101A and R.101B, and 5.4 knots faster than the enlarged R.101C. The increase in size and length of R.101C thus increased its drag to require some 320 additional horse-power (64 extra bhp per engine) to maintain the same speed of 50.2 knots. On the other hand, flying on five engines – instead of the four of R.101A – the lengthened R.101C on the same output of 375 bhp from each engine would have achieved a speed of 54.2 knots; an increase of four knots over R.101A for the same percentage of available power at a consumption of 800 lb an hour.

12. Details of the respective power-plants of the two airships were as follows:

	R.101	*R.100*
Engine manufacturer	Beardmore Aero Engines Ltd.	Rolls Royce Ltd.
Engine type:	Tornado III C.I.	Condor IIIB
Fuel:	Diesel heavy-oil	Petrol
Number of Cylinders	8	12
Cylinder arrangement	Upright in-line	Upright 60° Vee
Bore and Stroke	8.25 x 12.0 ins.	5.5 x 7.5 ins
Swept volume	84.125 litres 5,130.8 cu ins	35.0 litres 2,140 cu ins
Maximum power	650 bhp	650 bhp
R.P.M.	935	1,900
Fuel consumption	279 lb/hr	382 lb/hr
Specific consumption	0.429 lb/bhp/hr	0.588 lb/bhp/hr
Normal cruise power	475 bhp	410 bhp
R.P.M.	825	1,600 rpm
Fuel consumption	200 lb/hr	233 lb/hr
Specific consumption	0.421 lb/bhp/hr	0.568 lb/bhp/hr
Dry weight, bare engine	4,733 lb	1,420 lb
Specific weight (max. output)	7.281 lb/bhp	2.185 lb/bhp
Installed weight per engine with propeller, cooling, exhaust and starting systems and engine car)	8,580 lb	3,920 lb
Installed weight per max. b.h.p.	13.2 lb/bhp	6.031 lb/bhp
With reversing gearbox		7.495 lb/bhp
Maximum b.h.p. per litre	7.725	18.571

13. A summary of typical powers, speeds and consumptions for the airship variiants is as follows :

POWERS, SPEEDS AND FUEL CONSUMPTIONS
R.101A, R.101C and R.100
(With all available engines operating)

	R.101A Original condition	*R.101C* Lengthened	*R.100* Shortened tail
Number of engines	Four	Five	Six
1. *Maximum output*	2,600 bhp	3,250 bhp	3,900 bhp
2. R.P.M.	935	935	1,900
3. Speed (Knots)	61.0	62.0	70.8
4. Consumption: lb/hr	1,115	1,395	2,290
5. lb per Air N.M.	18.3	22.5	32.3
6. *Maximum Continuous*	2,340 bhp	2,925 bhp	3,000 bhp
7. R.P.M.	890	890	1,730
8. Speed (Knots)	59.0	60.0	65.8
9. Consumption: lb/hr	990	1,237	174.0
10. lb per Aid N.M.	16.8	20.6	26.4
11. *Fast Cruise Power*	1,900 bhp	2,375 bhp	2,820 bhp
12. R.P.M.	825	825	1,700
13. Speed (Knots)	54.9	55.5	64.8
14. Consumption lb/hr	800	1,000	1,630
15. lb per Air N.M.	14.6	18.0	25.1
16. *Normal Cruise Power*	1,800 bhp	2,250 bhp	2,460 bhp
17. R.P.M.	810	810	1,600
18. Speed (Knots)	53.8	54.25	62.5
19. Consumption lb/hr	745	932	1,400
20. lb per Air N.M.	13.8	17.2	22.4
21. *Long Range Cruise*	1,440 bhp	1,800 bhp	1,800 bhp
22. R.P.M.	750	750	1,400
23. Speed (Knots)	49.6	50.0	55.6
24. Consumption lb/hr	600	750	1,050
25. lb per Air N.M.	12.1	15.0	18.9

LONG RANGE CRUISE ON THREE ENGINES

For economical long-range cruise:

Beardmore Tornado III engines:	360 bhp each at 740 rpm and a fuel consumption rate of 150 lb per hour.
Rolls Royce Condor IIIB engine:	300 bhp each at 1,400 rpm and a fuel consumption rate of 175 lb per hour.

Three-engine cruise for maximum still air range

	R.101A	R.101B	R.101C	R.100
1. *Power Output* bhp	1,080	1,080	1,080	900
2. *RPM*	750	750	750	1,400
3. *Cruise Speed* (Knots)	45.3	45.3	41.5	43.5
4. *Fuel Consumption* (lb/hour)	450	450	450	525
5. *Fuel Consumption* (Imp Galls/hr)	53.2	53.2	53.2	70.0
6. *Fuel Consumption* (lb per Air Nautical Mile)	9.93	9.93	10.84	12.07
7. *Still Air Range on 15 Tons of Fuel*	3,380 NM	3,380 NM	3,100 NM	2,785 NM

Note: In all cruise regimes R.101A and R.101B had only four engines available for forwards flight. The drag of both those R.101 variants was, therefore, increased by that of one inoperative power-car during cruise. R.101C had five engines and R.100 six engines available for forwards flight. With all engines running to give totals of 1,800 bhp R.100 at 55.6 knots was 1.8 knots faster than R.101A at 53.8 knots and 5.6 knots faster than the, larger, R.101C which consumed 300 lb less per flying hour. Flying at 55.5 knots R.101C consumed 1,000 lb of fuel an hour and R.100 1,050 lb of fuel an hour.

Reference pages 141 to 144

APPENDIX 10

LOG OF R.101's ENDURANCE TEST FLIGHT

(17th/18th November 1929. 30 hours 32 minutes)

BETWEEN 1033 hrs on Sunday, 17th November 1929, and 1714 hrs on Monday, 18th November, R.101 completed an endurance flight of 30 hrs 41 minutes; the seventh and the longest of its career. In 29 hours flying on course the airship covered 977 nautical miles (1,148 statute miles) from Cardington to Scotland up the East Coast, across to Belfast, back by way of the Isle of Man, then to Dublin and return over Chester and Lichfield.

The mean true air speed made good was some 48.6 knots (56 mph) and the average ground speed 38.3 knots (44 mph). In all 2,366 gallons of fuel (8.76 tons) were consumed and R.101 landed with 1,992 gallons remaining (7.5 tons) sufficient for a further 24 hours of flying at the same rate of consumption. The mean rate of consumption had been 80.55 gallons an hour (676.6 lb an hour).

The airship had left the Cardington mast with a gross lift of 145.256 tons and a disposable lift of 31.657 tons of which 16.23 tons (4,328 gallons) was fuel and 10.25 tons (2,296 gallons) water ballast. There were four officers, 33 crew members and eight passengers on board, a total complement of 45 in all.

Details of the route flown are as follows:

APPENDIX 10

R.101A. ENDURANCE TRIALS

(17th–18th November, 1929)

Flight Log (Seventh Flight)

Disposable Lift on leaving ground: 70,911 lb = 31.7 tons including 4,328 gallons fuel oil (36,355 lb) and 10.25 tons (22,456 lb) of ballast.

Time (GMT)
17th November – Sunday

1033 hrs	Slipped from Cardington Tower (Gross lift 145.3 tons. Disposable lift 31.7 tons).
1040	Set course.
1100	800 feet. Air Temp 37°F. Wind 18 knots 016°. TAS 51 knots.
1114	Thapston.
Noon	Nine miles S.S.E. of Grantham. 1000 ft. 51 knots TAS. Air Temp 35.5° 47 miles run.
1235	No. 1 Engine (Fwd. Stbd.) half speed.
1245	Over Winslow. Heading 337°. Four engines cruising.
1300	1000 feet. Pressure height 1,200 ft. Temp 37°F. TAS 50 knots. 77 n miles run.
1400	1,000 feet Pressure height 1,200 ft. Temp 37°F. Wind N 20 knots. TAS 51 knots. Ground speed 31 knots. 132 miles run. Over Wheldake.
1410	Over York.
1423	Romex pump broken down. Half speed on Wing engines.
1500	1,000 feet. Pressure height 1,200 feet. Temp 35°F. Wind 20 knots N. Course 360°. TAS 48 knots. Wing engines to cruising. 162 n miles run.
1505	Over Northallerton.
1525	Over Darlington.
1600	1,200 feet. Pressure height 1,500 ft. Temp 34°F. Wind 20 knots N. TAS 52 knots. Ground speed 39 knots. Over Durham. 192 n miles run.
1620	Newcastle.
1633	Tynemouth.

1700 1,000 feet. Temp 36°. Wind 20 knots 054° TAS 52 knots. Course 360°. 217 n miles run. Ground speed 45 mph.

1735 Farne Islands.

1800 1,000 feet. Wind 30 knots. Course 323°. TAS 55 knots. 280 n miles run.

1812 St. Abs Head.

1830 Barnesrigg.

1835 Aft engine cruising.

1845 Dunbar (Turned ship over town).

1853 North Berwick.

1900 1,500 feet. Temp 33°. Wind 30 knots variable. 303 n miles run.

1906 Edinburgh.

1925 Engines ordered to half speed on account of Romex pump.

1930 Engines reduced to slow.

1934 Engines cruising. Georgetown.

2000 1,500 feet. 31° Temp. Glasgow. 349 n miles run.

2015 Greenock.

2046 Holy Isle to starboard.

2100 1,700 feet. Temp 30°. Wind 10 knots NE. 385 n miles run.

2102 Ailsa Craig one mile starboard.

2120 Over Corswall Point. Stopped No. 1 engine.

2200 1,700 feet. Temp 32°. Wind 5 knots. TAS 48 knots.

2203 Belfast.

2232 Calf of Man.

2245 Started turning trials.

2300 1,700 feet. Temp 34°F. Wind 5 knots. TAS 38 knots.

(Midnight)

2400 1,700 feet. Temp 32.5°F. Wind 5 knots. TAS 48 knots. 438 n miles run.

0100 1,350 feet. Temp 33° F. Turning Trials. TAS 48 knots.

0200 1,350. feet Temp 34.5°F. Turning Trials. TAS 48 knots.

0245 Set course 108°T. TAS 48 knots.

0300 1,350 feet. Temp 34°F. TAS 48 knots.

0342 Morecambe Bay three miles to port.

0400 1,400 feet. Temp 34°F. Morning Watch (Atherston and Steff) 498 n miles run.

0410 Blackpool. Course 262° Mag.

0500 1,400 feet. Temp 34.5°F. Course 250°. TAS 48 knots. Ground speed 40 knots.

0532 Skerries Light abeam one mile Course 282° Mag.

0600 1,400 feeet. Temp 35.5°. Wind SW 12 knots. Course 260°. TAS 48 knots 566 n miles run.
0645 Rockabill Light. Course 210° Mag.
0700 1,100 feet. Temp 37°F. Wind SW 12 knots TAS 48 knots. Ground speed 40 knots. 634 n miles run.
0710 No. 5 (aft) engine to slow.
0725 No. 5 (aft) engine to cruising.
0732 Stopped No. 1 (mid stbd.) engine.
0752 Howth Light astern one mile.
0800 1,400 feet. Air Temp 37°F. Wind 220° 15 knots 651 n miles run.
0900 1,300 feet. Air Temp 39°F. Wind 20 knots. Course 118°. TAS 50 knots.
0905 Greystone Set course to Holyhead.
1000 1,300 feet. Air Temp 38°F. Wind 190°. 30 knots Course 91° TAS 50 knots. 701 n miles run.
1055 Llandudno.
1100 1,250 feet. Air Temp 37.25°F. Course 100°. TAS 48 knots. 743 n miles run.
1116 Hilbre Point, Hoylake.
1155 Over Chester.
Noon
1200 1,400 feet. Air Temp 35°F (Atherston) 787 n miles run.
1205 Course 160° Mag to Cardington.
1300 1,400 feet. Air Temp 35°F. Wind variable.
1315 Seckington.
1331 Over Hinckley.
1345 Rugby one mile to starboard in fog.
1400 1,500 feet. Air Temp 34°F. Wind light. TAS 51 knots.
1540 Cardington. (Duration ship on course: 29 hours). 932 n miles run.
1620 Ship on wire.
1705 Secured to Tower, Cardington.

Left ground:	1033 hrs 17th November, 1929.
Landed:	1705 hrs 18th November, 1929.
Duration:	30 hours 32 minutes.
Distance covered:	939·5 n miles (Corrected to 997 n miles). (1,148 St miles).
Expended for flight:	2,336 gallons fuel oil = 19,763 lb = 8.82 tons.
Fuel remaining:	1,992 gallons fuel oil = 16,852 lb = 7.52 tons.
Water ballast on landing:	5·75 tons = 12,858 lb = 1,286 gallons (4.5 tons expended).
On board:	Crew 37 @ 5,920 lb. Passengers 8 @ 1,280 lb.
Fuel consumption on course:	approx. 2,312 gallons = 19,561 lb @ 79·72 gallons an hour, at an average of 674 lb an hour on course at 51.6 knots av. TAS.

(Ref: AIR 3/63. C.A.B. 29,261).

Reference page 232

APPENDIX 11

R.101 FOR CANADA

AN INTERESTING SPECULATION is whether R.101 could have been successfully substituted for R.100 for the flight to Canada at the end of July 1930, if R.100 had not successfully passed its final trial flight before departure? In June 1930 both AMSR (Sir John Higgins) and DAD (Reginald Colmore) had proposed this possibility to the S of S.

R.101B – that is the airship with the gas-bags let out before the extra bay had been added – had, by then, a disposable lift of 40.44 tons. There were 3.18 tons of additional net lift, compared with R.101A, from the extension of the gas-bag wiring, together with 3.3 tons of weight reduction from the elimination of various details.

R.100 left for Montreal on 29th July with 14 tons of its disposable load of 48.45 tons made up of the combined weights of officers and crew, water ballast, lubricating oil, food, drinking water, spares, stores and seven passengers. That left 34.45 tons available for fuel.

Had R.101B been similarly loaded, it would have had lift available sufficient for 26.44 tons of fuel out of its disposable lift of 40.44 tons.

On its westbound Atlantic flight, R.100 used 29.5 tons of fuel during a flight of 78 hours and 49 minutes. Out of that 29.5 tons of fuel used, some 29.2 tons were consumed in 74 hours of cruising flight at the rate of about 884 lb an hour. By flying on an average cruise power of 1,540 bhp a mean True Air Speed of 52.5 knots was recorded against an average headwind of seven knots.

At this speed and rate of consumption, R.100 would have had a cruising endurance of 87.3 hours to dry tanks.

To achieve the same endurance from an available fuel load of 26.44 tons, R.101B would have had to have held its fuel consumption to 678 lb an hour, which would have equated to a mean output 1,620 bhp and to a cruising speed of 51.4 knots; well within R.101's capabilities.

Assuming the same length of time (two hours and 37 minutes) taken for mooring at Montreal and the same mean headwind component, R.101B could, therefore, have completed the Atlantic flight from Cardington to Montreal in 78 hours 22 minutes and, carrying the same disposable load as R.100, would have had some 3.3 tons of fuel remaining after mooring, equivalent to an additional 11 hours of flight at the same rate of consumption.

R.101B's flights for the RAF Display at Hendon in June 1930, had shown up two serious defects: first, a weakness in the pre-doped outer cover and, secondly, loss of gas from chafing of the bags on the structure after the wiring had been let out. Both these deficiencies were taken seriously at Cardington. Although a new cover could not have been made and fitted in time for the flight to Canada at the end of July, urgent action could have produced a series of reinforcing bands which could have been adequate for the outwards and return flights to Canada in tolerable weather.

By the time that the lengthened R.101C left for India, the staff at Cardington had satisfied themselves that the much improved standards of padding would be adequate to prevent damage to the gas-bags – a view substantiated by the inspection reports after the final trial flight. Again, strenuous efforts could, probably, have provided this improved standard of padding in time for an Atlantic trial flight by R.101B by the end of July 1930.

Subject to that trial flight, there seems no doubt that R.101B could have been substituted for R.100 for the flight to Montreal early in August 1930, carrying the same Service Load and with the same prospects of success.

In that event, the flight to India would have to have been postponed to await the insertion of the extra bay. There would have been time to complete that work by the Christmas Parliamentary recess of 1930.

In the interim, a demonstration flight by R.100, as far as Egypt, might have been substituted during the course of the Imperial Conference and, in similar circumstances of gale force winds and heavy rain might have suffered the same tragic end as that of R.101 at Allonne in the early hours of 5th October 1930.

Reference page 258

APPENDIX 12

THE FLIGHTS OF R.101 AND R.100

1929–30

IN THE TWELVE MONTHS between 14th October 1929 and 5th October 1930 the two British airships of the 1924–30 Programme made 22 flights and were away from the mast for a total of 424 hours and 17 minutes. In the course of these 22 flights the airships covered a total of 14,799 nautical miles (17,041 statute miles) in 387 hrs 31 mins of flying time on course, at an average ground speed of 38.2 knots, or 44 miles an hour.

In the course of their twenty-two trial flights the two airships carried a total of 330 passengers for a total of 190,607 passenger nautical miles. At a total cost of £2,396,948 for the whole airship programme, (inculding research, and the sheds and masts at Cardington, and Karachi) the expenditure upon the whole airship programme worked out at £6,186 a flying hour, at £162 for each nautical mile flown and at £12 a passenger nautical mile – all in terms of the 1924/30 value of the pound sterling.

The flying of the two airships was divided as follows:

Airship	*Number of Flights*	*Air Duration from Mast*	*Flight Duration on Course*	*Distance Flown N. Miles*	*Passengers Carried*	*Passenger N. Miles*
		hrs mins	hrs mins			
R.101	12	127—20	111—29	3,665	186	41,486
R.100	10	296—57	276—02	11,134	144	149,121
Both Ships	22	424—17	387—31	14,799	330	190,607

During their flying careers the two airships spent a total of 36 hrs 45 minutes in mooring operations at an average of almost 100 minutes each. No doubt further experience and more extensive mechanical handling would have reduced this expenditure of time significantly but in the autumn of 1930 mooring to the mast was still a cumbersome performance.

In all the two airships consumed 38,251 gallons of fuel at an average of 2.58 gallons per airship mile and 5.21 passenger-miles per gallon. The composition was:

Airship	*Manoeuvring Time near Mast*	*Average Manoeuvring Time per Flight*	*Average Ground Speed Knots*	*Average True Air Speed*	*Fuel Consumption*	*Fuel Consumption per Flying Hour*	*Pass. Miles per Gallon*
	hrs mins	minutes	knots	knots	tons	lb	
R.101	15—50	79	32.88	47.79	29.0	583	5.60
R.100	20—55	125	40.34	52.01	101.0	820	4.88
Both Ships	36—45	100	38.19	50.79	130.0	751	5.21

In its brief career of 111½ flying hours R.101 encountered headwinds which averaged almost 15 knots, whereas R.100 encountered average headwinds of 11.6 knots. R.101's compression-ignition engines showed well their consumption-efficiency by consuming 30 per cent less fuel than R.100 while cruising, on average, at 4.2 knots lower speed – most of the time with only four engines for forwards flight.

Further data is set out in the accompanying tables.

SUMMARY OF THE FLIGHTS OF R.101 AND R.100 OCTOBER 1929 TO OCTOBER 1930

Date	*Ship*	*Flt. No.*	*Duration Away from Mast*	*Duration on Course*	*Departed From*	*Arrived At*	*Distance Flown*
1929			hrs min	hrs min			N. Miles
14 October	R.101A	1.	05—33	05—06	Cardington	Cardington	94
18 October	R.101A	2.	09—38	06—30	Cardington	Cardington	182
1 November	R.101A	3.	07—15	05—20	Cardington	Cardington	226
2/3 November	R.101A	4.	14—02	13—00	Cardington	Cardington	330
8 November	R.101A	5.	03—04	02—28	Cardington	Cardington	45
14 November	R.101A	6.	03—09	01—56	Cardington	Cardington	54
17/18 November	R.101A	7.	30—41	29—00	Cardington	Cardington	997
First R.101 Commission	R.101A	7.	73—22	63—20	Cardington	Cardington	1,928
16 December	R.100	1.	05—37	04—07	Howden	Cardington	130
17 December	R.100	2.	06—28	04—28	Cardington	Cardington	174
1930							
16 January	R.100	3.	13—20	11—10	Cardington	Cardington	217
20 January	R.100	4.	07—10	05—10	Cardington	Cardington	214
27 January	R.100	5.	54—52	52—32	Cardington	Cardington	1,779
First R.100 Commission	R.100	5.	87—27	77—27	Cardington	Cardington	2,514
21/22 May	R.100	6.	22—50	21—00	Cardington	Cardington	691
26 June	R.101B	8.	04—35	04—08	Cardington	Cardington	188
27 June	R.101B	9.	12—33	11—05	Cardington	Cardington	401
28 June	R.101B	10.	12—26	11—02	Cardington	Cardington	378
Second R.101 Commission	R.101B	3.	29—34	26—15	Cardington	Cardington	967
25 July	R.100	7.	24—18	23—09	Cardington	Cardington	805
29 July 1 August	R.100	8.	78—49	76—12	Cardington	Montreal	3,364
10 August	R.100	9.	25—57	22—12	Montreal	Montreal	805
13/15 August	R.100	10.	57—36	56—02	Montreal	Cardington	2,955
Third R.100 Commission	R.100	5.	186—40	177—35	Howden	Montreal Cardington	7,929
1/2 October	R.101C	11.	16—51	14—54	Cardington	Cardington	553
4/5 October	R.101C	12.	07—33	07—29	Cardington	Allonne	217
Third R.101 Commission	R.101C	2.	24—24	21—53	Cardington	Allonne	770

R.101 FLIGHTS (1929–30)

Flight No.	*Date*		*Local Time Slipped* H	M	*Secured* H	M	*Total* H	M	*Duration in Powered Flight* H	M	*Distance Covered* St. M.
	1929										
	12th October	Brought out of Shed	05	49	07	32	—				—
1.	14th October	London & Return	11	17	16	50	05	38	05	06	108
2.	18th October	Midlands	08	12	17	50	09	38	06	30	210
	21st October	Ship into Shed	17	30	20	15	—		—		—
	1st November	Ship out of Shed	05	40	07	25	—		—		—
3.	1st November	Norfolk	09	40	16	55	07	15	05	20	260
4.	2/3rd Nov.	I.O.W.	20	12	10	14	14	02	13	00	380
5.	8th November	Local	13	47	16	51	03	04	02	28	52
6.	14th November	Local	13	34	16	43	03	09	01	56	62
7.	17/18th Nov.	Endurance U.K.	10	33	17	14	30	41	29	00	1,148
	30th November	Ship in Shed	17	00	18	29	—		—		—
	First Commission 12th October — 30th Nov. 1929						73h	22m	63	20	2,220
	1930										
	23rd June	Ship out of Shed	15	51	20	26	—		—		—
8.	26th June	Refit Test Flight	15	51	20	26	04	35	04	08	217
9.	27th June	RAF Display Rehearsal	07	22	19	55	12	33	11	05	461
10.	28th June	RAF Display	07	25	19	51	12	26	11h	02	435
	29th June	Ship in Shed					—				—
	Second Commission 23-29th June 30						29h	34m	26h	15m	1,113
	1930										
	1st October	Ship out of Shed			06	34	—		05	10	—
11.	1/2nd October	Final Trial Flight	15	36	08	27	16	51	14	24	637
12.	4/5th October	India Flt.	18	36	02	09	07	33	07	30	248
	Last Commission 1/5th Oct. 1930						24h	24m	21h	54	885
	12 Flights						127h	20m	111h	29m	4,218

R.101 OUTLINE OF FLYING AND TIMES AT MAST

	Date	*Flight No.*	*Duration of Flight*		*Itinerary*	*Wind Speed and Direction and Gas Volume, etc.*
	1929 *29 Sept. 1929* *to 5th Oct. 1929*		*h.*	*m.*	*In Cardington Shed*	*Lift & Trim Trials*
					First Commission R.101A	
1.	12th October		—		Launch — out of Shed 56°F	15/20 mph 260° 97% P
2.	14th October	1	05	33	To London & Return 1011 mb	15/20 mph 290°
3.	18th October	2	09	38	Cardington N. Midlands & Return	S of S on Board
4.	21st October		—		Tower to Shed 1030.8 mb	10 mph 97% P
5.	1st November	3	07	15	Out of Shed Midlands	
6.	{2nd November 3rd November	4	14	02	Isle of Wight Channel & Return 1022.6	20/30 mph 250/310° Speed Tests
7.	8th November	5	03	04	Cardington/ Bedford Local 1065.77	20/30 mph 320° W.P.L. & A/M Officials
8.	11th November		—		At Mast	Severe Gale 10/15 mph 310°
9.	14th November	6	03	09	Cardington/ Bedford Local	
10.	16th November		—		At Mast	M.P.s Visit
11.	{17th November 18th November	7	30	41	Cardington, Scotland, Ireland, I.O.M., Cardington 1008 mb	20/30 mph 360/180° Endurance Flight
12.	23rd November					
13.	30th November		—		Tower to Shed	

R.101 OUTLINE OF FLYING AND TIMES AT MAST (*Continued*)

	Date	*Flight No.*	*Duration of Flight*		*Itinerary*	*Wind Speed and Direction and Gas Volume, etc.*
14.	December 1929 to June 1930				Refit in Shed	Gas Bags let out
	1930		*h.*	*m.*	*Second Commission R.101B*	
14.	23rd June		—		Out of Shed	Cover Split
15.	26th June	8	04	35	Cardington/ Bedford Local 1005.8 mb	10 mph 95.79 P
16.	27th June	9	12	33	Hendon & Return 1004.3 mb	15/25 270° 95.79 P RAF Display Rehearsal
17.	28th June	10	12	26	Hendon & Return 1013.5 mb	RAF Display 15/20 270°
18.	29th June		—		Tower to Shed	For New Bay
19.	28th July		—		In Shed	Parted for New Bay
20.	21st August		—		In Shed	Joined with New Bay
21.	30th Sept. 1930				*Third Commission R.101C*	*Lift and Trim Trials*
22.	1st October		—		Out of Shed	
23.	{1st October {2nd October	11	16	51	Cardington, London, Southend, N. Sea & Return	15/20 mph 120° Last Trial Flight
24.	{4th October {5th October	12	07	33	Cardington, London E. of Hastings, Poix, Beauvais	Left for India 30/50 mph Crashed Allonne
	Totals	12	127	20	1039 hours 50 mins at Mast during 43.3 days	Twelve Flights on 16 days

R.101 FLIGHTS AND FUEL CONSUMPTIONS—1929–30

Flight No.	Date	on Time Course h. m.	Extra Engine Running 120 Time at Lb/Hour h. m.	Total Used Fuel lb.	Fuel Used Etc. Mooring lb.	Fuel Used Course on lb.	Fuel Used Per Course Hour on lb.	Av: BHP	True Speed Air Knots
	1929								
1.	14th October	05.06	00.27	2,115	60	2,055	403	960	43.5
2.	18th October	06.30	03.08	3,892	376	3,515	541	1,000	44.2
3.	1st November	05.20	01.55	4,568	231	4,337	814	1,580	51.0
	2/3rd November	13.00	01.02	6,886	124	6,762	520	1,260	47.5
5.	8th November	02.28	00.36	1,946	72	1,874	760	1,820	54.0
6.	14th November	01.56	01.13	1,523	146	1,377	725	1,390	49.0
7.	17/18th November	29.00	01.41	19,763	202	19,561	674	1,640	51.6
1st COMM	1929	63.20	10.02	40,693	1,204	39,489	623	1,520	50.4 *3,192 ANM
	1930								
8.	26th June	04.08	00.27	2,648	60	2,588	686	1,520	50.4
9.	27th June	11.05	01.28	4,221	180	4,041	373	900	42.5
10.	28th June	11.02	01.24	3,985	170	3,815	346	850	41.6
2nd COMM	1930	26.15	03.19	10,854	410	10,444	398	920	43.0 1,129 ANM
11.	1/2nd October	14.24	02.27	6,785	294	6,491	451	1,120	42.0
12.	4/5th October	07.30	00.03	6,650	10	6,640	885	2,100	53.0
3rd COMM	1930	21.54	02.30	13,435	304	13,131	600	1,450	46.0 1,007 ANM
Total	1929/30	111.29	15.51	64,982	1,918	63,064	566	1,400	47.789 5,328 ANM

R.101 FLIGHTS (1929–30)

Flight No.	Date	Fuel Galls. Lifted	Fuel Galls. Used	On Board Crew	On Board Pass.	Ballast Tons	Disposable Lift
		Galls.	Galls.			Tons	Tons
	1929						
	12th October				—		
1.	14th October	3,055	250	39	14	10	27.8
2.	18th October	3,145	460	38	12	8	26.0
	21st October				—		
	1st November				—		
3.	1st November	3,844	540	39	15	12	34.3
4.	2/3rd November	4,172	214	39	12	11.25	33.4
5.	8th November	2,992	230	44	40	8	29.2
6.	14th November	2,944	180	42	32	15	34.2
7.	17/18th November	4,328	2,336	37	8	10.25	32.66
	30th November				—		
	First Commission	Avge.	Total	Avge.	Avge.	Avge.	Avge.
	12 Oct. — 30 Nov. 1929	3,511	4,210	40	19	10.6	31.08
	1930						
	23rd June						
8.	26th June	4,760	313	37	9	14	37.65
9.	27th June	4,592	499	36	5	12.5	36.75
10.	28th June	4,054	471	39	12	13	34.42
	29th June				—		
	Second Commission	Avge.	Total	Avge.	Avge.	Avge.	Avge.
	23—29 June 1930	4,467	1,283	37	9	13.2	36.38
	1930						
	1st October				—		
11.	1/2nd October	6,037	802	42	15	10.0	42.92
12.	4/5th October	6,713	786	42	12	9.25	44.8
	Last Commission	Avge.	Total	Avge.	Avge.	Avge.	Avge.
	1—5th October 1930	6,375	1,588	42	13	9.62	43.86
	12 Flights	4,228	7,081	39	15	11.10	34.51

R.101 Total Hours on Mooring Tower	1,039H 50M during 43.3 days
Total Hours off Mast:	127H 20M in 12 flights on 16 days
Total Hours out of Shed:	1,167H 10M

Reference pages 360, 411

APPENDIX 13

R.101C's INDIA FLIGHT

Flight Times, Distances and Check Points

ON ITS FINAL FLIGHT, en route to Egypt and India, on the night of 4th/5th October 1930, the lengthened and refitted R.101C, flying for the first time on all five engines by then available for forwards flight, was driven harder than any other British airship had been driven before in the face of the worst weather that any British rigid airship had encountered in free flight.

2. From the start all engines were set at 825 rpm to give 475 bhp each at a consumption of 200 lb of fuel an hour from each engine. With five engines running (2,375 bhp) the true air speed came out at 55.2 knots (63.5 mph); on four (1,900 bhp) – for three hours 20 minutes of the flight – the cruse speed came back to 51 knots (59 miiles an hour).

3. R.101 on its final flight of 7 hours 33 minutes thus flew the first ten minutes at a mean True Air Speed of some 40 knots, 4 hours 03 minutes at 55 knots and 3 hours 20 minutes at 51 knots to achieve a total of about 397 air nautical miles.

4. The ground distance covered was 218 nautical miles (251 statute miles). The mean True Air Speed was thus 52.8 knots and the ground speed 28.9 knots. The mean headwind component thus worked out at 23.9 knots over the duration of the flight.

5. The route taken by R.101 between Cardington at 1836 GMT on 4th October and Allonne at 0209 hrs GMT on 5th October can be set down with precision from wireless reports from the airship and a plotting of more than 500 ground reports of the passage of the airship. After circling Bedford and flying almost to Cranfield before turning on to course for London, at Clophill, R.101C passed over Hatfield, Potters Bar, the Alexandra Palace and Islington (at 2020 hrs GMT) to cross the River Thames at the Isle of Dogs and over the Royal Naval College at Greenwich. The airship was skilfully navi-

gated through the Shoreham, Kent, gap in the North Downs, over Kemsing and past Ightham and Borough Green to fly, at about 1,200 feet just east of Tonbridge and to pass over Hawkhurst and Bodiam Castle to cross the English coast at Haddock Gap, a little east of Hastings at 2135 hrs GMT.

6. The 50 nautical mile Channel crossing brought R.101C over the French Coast at Pointe de St Quentin, on the north-east bank of the Somme estuary exactly five hours after leaving Cardington.

7. The airship flew on towards Orly, Paris, passing Poix at 0105 hrs and arriving overhead Beauvais in storm-force wind and heavy rain at 0200 hrs GMT.

8. At 0209 hrs, three miles south-east of Beauvais, R.101 crashed into the Bois de Coutumes at a height of 200 feet above sea level.

9. The distances flown in each of the seven hours of the flight were as follows:

Time	*Duration*	*Distance Covered During past Hour*	*Position*	*Total Distance Covered*
18.36 hrs	= 0 hour	0 n. miles	Cardington Mast	0 n. miles
19.36 hrs	= 1 hour	36.6 n. miles	Bendish	36.6 n. miles
30.36 hrs	= 2 hours	32.0 n. miles	Beavers Wood, Sidcup	68.6 n. miles
21.36 hrs	= 3 hours	42.4 n. miles	Offshore Hastings	111.0 n. miles
22.36 hrs	= 4 hours	23.6 n. miles	Mid Channel	134.6 n. miles
23.36 hrs	= 5 hours	24.6 n. miles	Pointe de St. Quentin	159.2 n. miles
00.36 hrs	= 6 hours	23.8 n. miles	Verigiés	183.0 n. miles
01.36 hrs	= 7 hours	23.1 n. miles	N. of Blicourt	206.1 n. miles
02.09 hrs	= 7H 33M	11.9 n. miles (in 33 mins.)	Bois de Coutumes	218.4 n. miles

10. The log of the flight was as follows:

R.101's INDIA FLIGHT

4th and 5th OCTOBER 1930 — CARDINGTON/ALLONNE

	Actual Time GMT	*Duration Elapsed hrs mins*	*Distance Flown Ground St Miles*	*Check Points*	*True Air Speed Knots*	*Cumulative Ground Speed Statute MPH*
1.	**18.36**	00.00	—	Cardington Tower	—	—
2.	18.42	00.06	2.0	Shortstown	37.0	20.0
3.	18.45	00.09	3.0	Elstow Manor (A6)	43.0	20.0
4.	18.46	00.10	3.6	South End	47.0	21.6
5.	18.47	00.11	4.4	Bedford, River Ouse	52.0	24.0
6.	18.47½	00.11½	5.0	Bedford School	55.0	26.1
7.	18.48	00.12	7.0	Mowsbury Hill	55.0	35.0
8.	18.49	00.13	8.8	Wood End	55.1	40.6
9.	18.51	00.15	10.0	Twin Wood Farm	55.2	40.0
10.	18.54	00.18	11.5	Clapham High Street	55.2	38.3
11.	18.56	00.20	12.5	River Ouse	55.2	37.5
12.	18.57	00.21	13.0	Biddenham Turn	55.2	37.1
13.	**19.00**	00.24	15.5	Kempston Church End	55.2	38.7
14.	19.03	00.27	17.0	Vicarage Farm	55.2	37.7
15.	19.06	00.30	19.0	W. of Wootton, Thrift Wood	55.2	38.0
16.	19.08	00.32	20.5	Roxhill Manor Farm	55.2	38.4
17.	19.09	00.33	21.5	Marston, Thrift Wood	55.2	39.0
18.	10.15	00.39	24.5	S. of Millbrook Station	55.2	37.7
19.	19.16	00.40	25.5	Luton/Bedford Railway (No 5 engine shut down)	55.2	38.2
20.	19.19	00.43	29.5	Clophill Crossroads	51.0	41.2
21.	19.21	00.45	32.0	Lower Graveshurst	51.0	42.7
22.	19.23	00.47	33.0	Hanscombe End	51.0	42.1
23.	19.29	00.53	37.5	Great Offley	51.0	42.4
24.	19.34	00.58	40.0	Kings Walden	51.0	41.4
25.	**19.36**	01.00	41.2	Bendish	51.0	41.2
26.	19.38	01.01	42.5	West of Hoo End	51.0	41.1
27.	19.39	01.03	43.5	Kimpton	51.0	41.4
28.	19.41	01.05	44.5	West of Shaws Corner	51.0	41.1
29.	19.44	01.08	46.5	River Lea, Water End	51.0	41.0
30.	19.49	01.13	49.5	Hatfield—D.H.	51.0	40.7
31.	19.50	01.14	50.5	West of Hatfield House	51.0	40.9
32.	19.56	01.20	53.25	Brookmans Park (No 5 engine restarted)	51.0	39.9
33.	19.58	01.22	54.75	Potters Bar Station	55.0	40.1
34.	**20.00**	01.24	55.6	Potters Bar	55.2	39.7
35.	20.01	01.25	56.0	Ganwick Corner	55.2	39.5
36.	20.03	01.27	57.25	Hadley Wood	55.2	39.4
37.	20.05	01.29	58.00	Hadley Common	55.2	39.1
38.	20.05	01.29	58.25	New Barnet	55.2	39.0
39.	20.07	01.31	59.00	East Barent (Oakleigh Park)	55.2	38.9
40.	20.09	01.33	60.25	Brunswick Park	55.2	38.9
41.	20.10	01.34	61.00	Friern Hosp. (Colney Hatch)	55.2	38.9
42.	20.13	01.37	62.25	Alexandra Palace Band Stand	55.2	38.5

	Actual Time GMT	*Duration Elapsed hrs mins*	*Distance Flown Ground St Miles*	*Check Points*	*True Air Speed Knots*	*Cumulative Ground Speed Statute MPH*
43.	20.16	01.40	64.00	Crouch End (No 5 engine shut down)	55.0	38.4
44.	20.18	01.42	64.75	Crouch End Station	51.0	38.1
45.	20.19	01.43	56.25	Hornsey Road Station	51.0	38.0
47.	20.20	01.44	65.75	E. of Metro. Cattle Market	51.0	37.9
48.	20.20	01.44	66.00	Bridge St/Westbourne Rd.	51.0	38.0
49.	20.20	01.44	66.25	Offord Road	51.0	38.0
50.	20.22	01.46	67.00	Old Street Station	51.0	38.5
51.	20.23	01.47	68.75	Spitalfields Market	51.0	38.3
52.	20.24	01.48	69.8	Shadwell	51.0	38.8
53.	20.25	01.49	70.0	River Thames—Lower Pool	51.0	38.5
54.	20.27	01.51	71.5	Millwall, Isle of Dogs	51.0	38.6
55.	20.28	01.52	72.5	Royal Naval College, G'wich	51.0	38.8
56.	20.28	01.52	73.0	Royal Observatory	51.0	39.0
57.	20.30	01.54	74.75	Blackheath Park (Kidbrooke)	51.0	39.3
58.	20.34	01.58	77.25	New Eltham Station	51.0	39.2
59.	20.36	**02.00**	79.00	Beavers Wood, Sidcup	51.0	39.5
60.	20.40	02.04	82.75	Crockenhill	51.0	40.0
61.	20.42	02.06	84.00	Hulberry	51.0	40.0
62.	20.50	02.10	86.75	Shoreham Station	51.0	40.0
63.	20.52	02.12	88.25	Otford	51.0	40.1
64.	20.51	02.15	91.25	Kemsing Station	51.0	40.5
65.	20.54	02.18	93.50	Ightham	51.0	40.6
66.	20.56	02.20	95.00	Plaxtol	51.0	40.7
67.	**21.00**	02.24	98.00	Higham Wood	51.0	40.8
68.	21.02	02.26	99.00	Hadlow Stair	51.0	40.7
69.	21.04	02.28	100.5	Tudeley Railway	51.0	40.7
70.	21.05	02.29	101.75	Bouncers Bank	51.0	41.0
71.	21.08	02.32	103.75	Matfield	51.0	40.7
72.	21.13	02.37	108.25	Twysden	51.0	41.4
73.	21.14	02.38	109.5	Bedgebury Cross	51.0	41.6
74.	21.16	02.42	112.5	HighStreet, Hawkhurst	51.0	41.6
75.	21.17	02.43	113.25	The Moor, Hawkhurst	51.0	41.7
76.	21.20	02.46	116.25	Bodiam Castle	51.0	42.0
77.	21.21	02.47	117.0	West of Ewhurst	51.0	42.0
78.	21.24	02.50	119.0	Watts Hill	51.0	42.0
79.	21.28	02.52	120.5	Broad Oak Crossroads	51.0	42.0
80.	21.30	02.54	122.25	River Brede	51.0	42.1
81.	21.31	02.55	123.5	East of Guestling Thorn	51.0	42.3
82.	21.33	02.57	125.25	Pett	51.0	42.4
83.	21.34	02.58	126.25	Marsham Farm	51.0	42.5
84.	21.35	02.59	126.50	Coast at Haddocks Gap (between Fairlight and Cliff End)	51.0	42.4
85.	21.36	**03.00**	127.0	Offshore Hastings	51.0	42.3
86.	**22.00**	03.24	138.0	Over the Channel	51.0	40.5
87.	22.36	**04.00**	155.0	Mid-Channel	51.0	38.75
88.	22.56	04.20	165.0	No. 5 Engine Restarted	52.0	38.10
89.	**23.00**	04.24	167.0	Approaching French Coast	54.0	37.9

	Actual Time GMT	*Duration Elapsed hrs mins*	*Distance Flown Ground St Miles*	*Check Points*	*True Air Speed Knots*	*Cumulative Ground Speed Statute MPH*
90.	23.36	**05.00**	**183.6**	Pointe de St. Quentin	54.0	36.7
91.	23.46	05.10	188.5	Off: Le Crotoy	54.0	36.5
92.	23.50	05.14	190.0	St. Valery-sur-Somme	54.0	36.3
93.	**24.00**	05.24	194.0	Draycourt	54.0	35.9
94.	24.04	05.28	196.5	Quesnoy	54.0	35.9
95.	24.13	05.37	200.1	Toeufles	54.0	35.6
96.	24.21	05.45	203.4	Huppy	54.0	35.4
97.	24.22	05.47	204.5	Doudevainville	54.0	35.3
98.	24.36	06.00	210.7	Vergies	54.0	35.1
99.	24.40	06.04	212.0	Epaumesnil	54.0	34.9
100.	24.42	06.06	213.1	Avesnes	54.0	34.9
101.	24.50	06.14	217.1	Hornoy	54.0	34.8
102.	**01.00**	06.24	222.0	Eplessier	54.0	34.7
103.	01.06	06.30	224.0	Poix	54.0	34.5
104.	01.16	06.40	228.5	Dargies	54.0	34.3
105.	01.26	06.50	234.0	Hetomesnil	54.0	34.2
106.	01.32	06.56	236.5	N.30 West of Crevecour	54.0	34.1
107.	01.36	**07.00**	238.0	East of La Neuville	54.0	34.0
108.	01.40	07.04	239.5	Blicourt	53.5	33.9
109.	01.44	07.08	241.0	Pisseleu	53.5	33.7
110.	01.47	07.11	242.0	Sauquese St. Lucien	53.5	33.7
111.	01.52	07.16	244.5	Houssoy Le Fancy	53.5	33.6
112.	01.52	07.16	244.5	Le Plouy St. Lucien	53.5	33.6
113.	**02.00**	07.24	248.0	Beauvais St. Andre	53.0	33.5
114.	02.05	07.29	249.2	Voismlieu	53.0	33.3
115.	02.06	07.30	249.6	St Lazare Farm	53.0	33.3
116.	02.07	07.31	250.5	West of Allonne	53.0	33.3
117.	02.09	07.33	251.5	Bois de Coutumes	—	33.3

R.101c's final flight: 18.36 hrs on 4th October to 02.09 hrs on 5th October, 1930. Duration: 07 hrs 33 minutes. Distance flown: 251.5 statute miles (218.4 nautical miles). Total air miles flown: 450.2 statute miles (391 nautical miles) at mean true air speed of 59.6 m.p.h. (51.8 knots). Mean ground speed made good: 33.3 m.p.h. (28.9 knots). Average head-wind component: 26.3 m.p.h. (22.8 knots).

Reference pages 362 and 411

APPENDIX 14

R.101's FINAL FLIGHT

Personnel Arrangements

WHEN R.101 left Cardington for India on the evening of 4th October 1930 there were fifty-four on board. Five of them were the ship's officers under Flight Lieutenant Carmichael Irwin; six were officials of the Royal Airship Works led by the Director of Airship Development, Wing Commander Reginald Colmore; six were passengers headed by the Secretary of State for Air, Lord Thomson of Cardington – and thirty-seven were crew members under, respectively, the Chief Coxswain George Hunt, the Chief Engineer, W. R. Gent and the Chief W/T Operator, S. Keeley.

2. R.101's Captain, Flight Lieutenant 'Bird' Irwin, had divided the crew into two Flying Watches of fourteen men each; set out in a note of 23rd September 1930. Six others, the Chief Navigator, Squadron Leader E. L. Johnston, the Meteorological Officer, M. A. Giblett, the Chief Coxswain, George Hunt, the Chief Engineer, W. R. Gent, the Chief W/T Officer, S. Keeley and the Chief Steward, A. H. Savidge, did not keep watch but were detailed to be on duty as necessary.

3. On the long flight to India the day watches were set at four hours each but the night watches shortened to three hours, to make seven watch periods in each 24 hours, as follows – all to be maintained in GMT:—

1. *Forenoon Watch* .... from 0800 to 1200 hrs
2. *Afternoon Watch* .... from 1200 to 1600 hrs
3. *Evening ('Dog') Watch* .... from 1600 to 2000 hrs
4. *First Night Watch* .... from 2000 to 2300 hrs
5. *Middle Watch* .... from 2300 to 0200 hrs
6. *First Morning Watch* .... from 0200 to 0500 hrs
7. *Second Morning Watch* ...... from 0500 to 0800 hrs

4. Four crew meals would be served on the flight each day:—
Breakfast: 0720 to 0830 hrs (For Forenoon and Second Morning Watches)

Lunch:	1120 to 1230 hrs	(For Afternoon and Forenoon Watches)
Tea:	1530 to 1630 hrs	(For Evening and Afternoon Watches)
Supper:	1930 to 2030 hrs	(For First Night and Evening Watches)

Officers and passengers would take meals in the Dining Saloon between the following times:

Breakfast:	0730 to 0930 GMT
Lunch:	1130 to 1300
Tea:	1530 to 1630
Dinner:	1930 to 2030

and 15 minutes after the end of these meal times, the Chief Coxswain, Chief Engineer, Chief W/T Operator, Chief Steward and Mr H. J. Leech would feed in the Dining Saloon.

5. Those detailed to stand watch were as follows:

		No 1 Watch	*No 2 Watch*
1.	*Duty Coxswain*	Flight Sergeant W. A. Potter	C. H. Mason
2.	*Height Coxswain*	M. G. Rampton	L. F. Oughton
3.	*Steering Coxswain*	H. E. Ford	P. A. Foster
4.	*Riggers*: 1	C. E. Taylor	S. Church
5.	2	A. W. J. Norcott	W. G. Radcliffe
6.	3	A. J. Richardson	E. G. Rudd
7.	*Charge Hand Engineer*	G. W. Short	S. E. Scott
	Engineers:		
8.	No 1 Car (Stbd. Fwd)	C. J. Fergusson	M. F. Littlekitt
9.	No 2 Car (Port Fwd)	W. H. King	W. Moule
10.	No 3 Car (Stbd. Mid)	A. C. Hastings	V. Savory*
11.	No 4 Car (Port Mid)	R. Blake	A. J. Cook*
12.	No 5 Car (Aft)	A. V. Bell*	J. H. Binks*
13.	*W/T Operator*	A. Disley*	G. H. Atkins

6. The Officers of the Watch would alternate between Irwin, Atherstone and Steff as circumstances permitted with Johnston to take a duty when appropriate. In the event Irwin and Steff shared the

* Survivors.

first (Evening) Watch on 4th October 1930, followed by Atherstone as Officer of the Watch from 2000 to 2300 hrs (First Night Watch), by Irwin from 2300 to 0200 hrs (Middle Watch) and Steff from 0200 (First Morning Watch) on 5th October.

7. In addition to these arrangements there were, as supernumaries to take over as relief Watch Keepers, Charge-Hand Engineer T. A. Key, Assistant W/T Operator F. Elliott and Engineers C. H. Burton and A. H. Watkins.

Under the Chief Steward, A. H. Savidge, there were Assistant Steward F. Hodnett, E. A. Graham (Cook) and J. W. Megginson (Galley Boy) who would be on duty as required from 0630 to 2130 hrs each day.

8. From survivors' reports and from the position of identified bodies one can reconstruct the location of almost all those on board at the time of the disaster at 0209 hrs on 5th October 1930.

9. Of the eight initial survivors, two of whom died shortly afterwards:

1. *H. J. Leech* (Foreman Engineer) was at 0209 hrs in the smoking room on the lower deck, amidships;
2. *S. Church* (Rigger) was walking forward to undertake an order from G. W. Hunt (Chief Coxswain) to release a half-ton bag of water ballast right forward. (Samuel Church died three days after the crash.)
3. *A. Disley* (W/T Operator/Electrician) was asleep in his bunk on the port side of the lower-deck, alongside the Wireless Room.
4. *W. G. Radcliffe* (Rigger) was on duty in the Chart Room on the lower deck. (W. G. Radcliffe died on 6th October at Beauvais).
5. *V. Savory* (Engineer) was on duty in the No 3 (Starboard Mid) engine car.
6. *A. J. Cook* (Engineer) was on duty in the No 4 (Port Mid) engine car.

7. & 8. *A. V. Bell* and *J. H. Binks* were both on duty in the No 5 (Aft) engine car.

10. Thus, because the airship hit the ground at an angle of about 18 degrees nose down, those in the mid and aft engine cars were able

to scramble clear before the flames, originating from the calcium flares in the control car forward, engulfed the remainder of the ship. Those in the control car and in the forward engine-cars were undoubtedly killed either by the impact or by the collapse of the forward structure above them.

11. Five members of the crew who were killed but not badly burned were identified on 5th October at Allonne. They were:

1. Charge Hand Engineer Sydney Ernest Scott,
2. Rigger E. G. Rudd and
3. W. G. Radcliffe, all of No 2 Watch on duty at the time of the accident. They were all in the Chart Room aft of the control car and almost got clear of the airship.

12. In addition, Flight Sergeant W. A. Potter (Assistant Chief Coxswain) and Engineer Richard Blake of the No 1 Watch, who had just come off duty and had not yet turned in, were pulled out of the wreck, dead, but later identified.

13. Back in London, an additional twenty-one from among those on board R.101 were identified by personal belongings and dental evidence – among them both Flight Lieutenant Irwin and Lieutenant-Commander Atherstone who had retired to rest and were in pyjamas. Those thus identified were:

Passengers

1. Air Vice-Marshal Sir Sefton Brancker (DCA)
2. Squadron Leader W. H. L. O'Neill
3. Mr A. Bushfield (AID)
4. Mr James Buck (PA to S of S)

R.A.W. Officers

1. Wing Commander Reginald Colmore
2. Lieutenant-Colonel V. C. Richmond
3. Squadron Leader F. M. Rope.

And in addition, from among the officers and crew:

1. Flight Lieutenant H. C. Irwin (Captain)
2. Lieutenant-Commander Noël Atherstone (First Officer)
3. M. A. Giblett (Chief Met Officer)
4. W. R. Gent (Chief Engineer)

5. *Engineers*: C. A. Burton
6. Richard Blake
7. Wilfred Moule (No 2 Car)
8. W. H. King
9. *Assistant Coxswains*: L. F. Oughton
10. M. G. Rampton
11. *Wireless Operators*: G. K. Atkins
12. Frank Elliot
13. *Riggers*: C. E. Taylor
14. A. J. Richardson.

14. Finally, twenty-two bodies remained unidentified – all of them either from the Control Car, the Chart Room or from passenger and crew sleeping accommodation, engulfed in the fiercest part of the flames:

1. The Right Hon The Lord Thomson of Cardington (S of S) (Stbd Forward Cabin)
2. Major G. Herbert Scott (ADAD–FT)
3. Squadron Leader E. L. Johnston (Navigator) (Chart Room)
4. Flying Officer M. H. Steff (Second Officer) (Control Room)
5. Major Percy Bishop (AID)
6. Squadron Leader W. Palstra (RAAF)
7. G. W. Hunt (Chief Coxswain) (Chart Room)
8. C. H. Mason (Assistant Coxswain) (Control Car)
9. P. A. Foster (Assistant Coxswain) (Control Car)
10. G. W. Short (Charge-Hand Engineer)
11. *Wireless Operators*: G. H. Atkins
12. C. J. Fergusson (Engineer)
13. A. C. Hastings (Engineer)
14. A. H. Watkins (Engineer)
15. M. F. Littlekitt (Engineer) (No 1 Engine Car)
16. H. E. Ford (Rigger)
17. A. W. J. Norcott (Rigger)
18. S. T. Keeley (Chief Wireless Operator) } sharing a cabin
19. A. H. Savidge Chief Steward) } amidships
20. F. Hodnett (Assistant Steward)
21. E. A. Graham (Cook)
22. J. W. Megginson (Galley Boy).

15. The fire which originated in the control car thus burned first and most fiercely in the lower centre section of the airship in which

lay the Chart Room and, above it, the passenger accommodation including the sleeping cabins in which were the passengers and, on the lower deck, the crew's sleeping accommodation. All those who could not be identified were in these portions of the ship.

16. The full complement on board R.101 on 4th/5th October 1930, was as follows:

Passengers (six)

1. Brigadier-General The Right Hon The Lord Thomson of Cardington, PC, DBE, DSO, psa (HM Secretary of State for Air)
   – Killed; not identified.
2. Air Vice-Marshal Sir W. Sefton Brancker, KCB, AFC (Director of Civil Aviation)
   – Killed; identified on return to London.
3. Major Percy Bishop, OBE (Chief Inspector, AID)
   – Killed; not identified.
4. Squadron Leader W. Palstra RAAF (Representing the Australian Government)
   – Killed; not identified.
5. Squadron Leader W. H. L. O'Neill (Deputy Director of Civil Aviation, India)
   – Killed; identified on return to London.
6. Mr James Buck, Valet to the Secretary of State for Air
   – Killed; identified on return to London.

Officials of the Royal Airship Works, Cardington (six)

7. Wing Commander R. B. B. Colmore, OBE (Director of Airship Development)
   – Killed; identified on return to London.
8. Major G. H. Scott, CBE, AFC (Assistant Director of Airship (Flying and Training)) – Officer in charge of the flight)
   – Killed; not identified.
9. Lieutenant-Colonel V. C. Richmond, OBE (Assistant Director of Airship Development – Technical)
   Killed; identified on return to London.
10. Squadron Leader F. M. Rope (Assistant to Assistant DAD(T))
   – Killed; identified on return to London.
11. Mr. Alexander Bushfield, AID
   – Killed; identified on return to London.
12. Mr. A. H. Leech (Foreman Engineer, AAW, Cardington)
   – survivor.

Officers of R.101 (five)

13. Flight Lieutenant H. Carmichael Irwin, AFC (Captain)
 – Killed; identified on return to London.
14. Squadron Leader E. L. Johnston, OBE, AFC (Navigator)
 – Killed; not identified.
15. Lieutenant-Commander N. G. Atherstone, AFC (First Officer)
 – Killed; identified on return to London.
16. Flying Officer M. H. Steff (Second Officer)
 – Killed; not identified.
17. Mr. M. A. Giblett, MSc (Chief Meterological Officer)
 – Killed; identified on return to London.

Petty Officers and Charge Hands (seven)

18. G. W. Hunt (Chief Coxswain)
 – Killed; not identified.
19. W. R. Gent (Chief Engineer)
 – Killed; identified on return to London.
20. G. W. Short (Charge-Hand Engineer)
 – Killed; not identified.
21. S. E. Scott (Charge-Hand Engineer)
 – Killed; identified at Allonne.
22. T. Key (Charge-Hand Engineer)
 – Killed; not identified.
23. S. T. Keeley (Chief Wireless Operator)
 – Killed; not identified.
24. A. H. Savidge (Chief Steward)
 – Killed; not identified.

Crew Members

25. Flight-Sergeant W. A. Potter (Assistant Coxswain)
 – Killed; identified at Allonne.
26. L. F. Oughton (Assistant Coxswain)
 – Killed; identified on return to London.
27. C. H. Mason (Assistant Coxswain)
 – Killed; not identified.
28. M. G. Rampton (Assistant Coxswain)
 – Killed; identified on return to London.
29. H. E. Ford (Assistant Coxswain)
 – Killed; not identified.
30. P. A. Foster (Assistant Coxswain)
 – Killed; not identified.

31. E. G. Rudd (Rigger)
 – Killed; identified at Allonne.
32. C. E. Taylor (Rigger)
 – Killed; identified on return to London.
33. A. W. J. Norcott (Rigger)
 – Killed; not identified.
34. A. J. Richardson (Rigger)
 – Killed; identified on return to London.
35. W. G. Radcliffe (Rigger)
 – survived; but died on 6th October at Beauvais.
36. S. Church (Rigger)
 – survived; but died on 8th October at Beauvais.
37. R. Blake (Engineer)
 – Killed; identified at Allonne.
38. C. A. Burton (Engineer)
 – Killed; identified on return to London.
39. C. J. Fergusson (Engineer)
 – Killed; not identified.
40. A. C. Hastings (Engineer)
 – Killed; not identified.
41. W. H. King (Engineer)
 – Killed; identified on return to London.
42. M. F. Littlekitt (Engiineer) No 1 car
 – Killed; not identified.
43. W. Moule (Engineer) No 2 car
 – Killed; identified on return to London.
44. A. H. Watkins (Engineer)
 – Killed; not identified.
45. A. V. Bell (Engineer) No 5 car
 – *Survivor.*
46. J. H. Binks (Engineer) No 5 car
 – *Survivor.*
47. A. J. Cook (Engineer) No 4 car
 – *Survivor.*
48. V. Savory (Engineer) No 3 car
 – *Survivor.*
49. G. H. Atkins (Wireless Operator)
 – Killed; identified on return to London.
50. F. Elliott (Wireless Operator)
 – Killed; identified on return to London.

51. A. Disley (Wireless Operator/Electrician)
    – *Survivor.*
52. F. Hodnett (Assistant Steward)
    – Killed; not identified.
53. E. A. Graham (Cook)
    – Killed; not identified.
54. J. W. Megginson (Galley Boy)
    – Killed; not identified.

Reference page 277

APPENDIX 15

THE NEXT STEP – R.102 AND R.103

(*The Projected 1930–35 Programme for British Airships*)

By the Spring of 1929, after almost five years of work on the new airship programme and before either R.100 or R.101 had been completed or had made their first flights, the fact had become clear both to the Design Office at Cardington and to the Airship Directorate at the Air Ministry that these two new airships – although twice the size of earlier craft – would be too small for their designated tasks. Structure and accommodation weights had been underestimated and the heavy-oil diesel engines insisted upon for tropical operations, had turned out to be very much heavier than had been assumed in earlier, optimistic, forecasts.

The Air Ministry's preoccupation with the use of heavy-oil engines was the result of the horrifying experience of the disastrous fires from petrol, floating on the waters of the River Humber, following the breaking up in the air of R.38 in 1921. Most of the lives were lost in burning pools of gasolene. Unfortunately there were no diesel aero-engines of satisfactory power/weight ratio suitable for installation in the new airships. Railcar engines had to be modified, instead.

The consequential need to increase the lift and the payload/range capabilities of the two airships led to plans to insert extra bays from which some nine tons of additional lift could be gained in both new airships, together with proposals to erect three more mooring masts as intermediate refuelling points; one at Malta and another at Hinaidi, Baghdad, on the route to India and a third mast at Moncton in New Brunswick on the Canadian route.

With its extra bay, the lengthened R.101C, when carrying two tons of payload (ten paggengers and a ton of mail) was finally estimated to have a duration of 55½ hours to dry tanks on 24.8 tons of fuel, cruising at 55 knots (63 mph).

Against an assumed 15 knot headwind and allowing contingency fuel reserves of 23 hours duration at 55 knots, on 2,325 bhp, (1,265 still air nautical miles) R.101C would have a practical range of 1,760 nautical miles. This would be adequate for a regular scheduled service to and from India once the projected new masts at Malta and Hinaidi were complete. But the performance was still substantially short of the original requirement.

There was clearly going to be a need for larger airships if the objective of regular and reliable all-the-year-round Imperial Airship Services was to be achieved.

In February of 1929, detailed project studies by Michael Rope at Cardington showed that an airship of 9,300,000 cubic feet capacity would be necessary to achieve the original performance laid down in specification 24/24. Now that realistic assessments could be made of structure and powerplant weights and of reduced standards of passenger accommodation there was no doubt that an airship of almost double the volumetric capacity of R.101 or R.100 would be necessary to fly a regular service between Cardington and Karachi by way of Ismailia, and between Cardington and Montreal direct, on an all-the-year round basis.

All this was put into a note by Rope which formed the basis of discussion between Richmond, Colmore and Bulman of the Air Ministry at Cardington on 8th March 1929.

One of the problems which became clear at this preliminary design conference was that such a substantial further increase in size of ship could be attained only if a completely new shed were to be built at Cardington in which to assemble such a large airship. In addition a great deal of further research would be needed before a design almost twice the size of R.100 and R.101 could be finalised with confidence.

Accordingly, on 18th March 1929, a further design study was examined for a less ambitious project of 7,000,000 cubic feet volumetric capacity. It would have a gross lift of 212.5 tons and be designed for a still air range of 5,000 nautical miles cruising at 65 knots (75 mph) carrying a payload of 15 tons, including 50 passengers and five tons of mail or goods. The length would be 800 feet and the maximum height 147.5 feet.

Even a more modest development of this sort would require expensive modifications to the construction shed at Cardington. So the file was held over for the time being while more urgent work was tackled on R.101. Meanwhile – on 5th June, 1929 – in order to get a 'second opinion' on a feasibility study, a £5,000 short-term contract was placed with the Design Department of the Airship Guarantee Company at Howden – which had the added advantage of keeping the Howden design team employed.

On 13th September 1929 Richmond returned to the New Project File with another discussion with Colmore and Rope looking ahead to the time when design office capacity would be available to start serious work.

The result was a further design study, 'Project C', this time for a ship of 6,750,000 cubic feet which would have a gross lift of 205 tons.

The latest information on the weights of R.101 and on the power-plant likely to be available a year or two ahead were discussed in detail and, as a result, three further design studies were to be examined:

'Project D' – for 8,300,00 cubic feet capacity
'Project E' – for 8,000,000 cubic feet, and
'Project F' – for 7,500,000 cubic feet.

All these were outlined in what would now be described as cost-benefit studies; interrupted by the work resulting from the first flight of R.101 on 12th October 1929. They were examined in detail at a meeting in Colonel Richmond's office on 7th March 1930, when a revised specification was drawn up for a new generation of British airships in the light of all that had been learned with R.101 and R.100 by that date.

A New Project section under Michael Rope was now set up in the Cardington Design Office, and work went forward steadily with design conferences on 13th, 14th, 17th and 18th March 1930.* The meeting on 17th March was at the Air Ministry, chaired by Sir John Higgins, AMSR.

The upshot was yet another, and updated, Outline Requirement (OR) for a new design and a reversion to the ambitious demands and, hence, the very large size first considered in February, 1929. There is no doubt that the pessimism instilled by the early trials of both R.101 and R.100 had led, at this date, to the belief that only by going to an airship of almost double the size again, could the needs be met.

The new OR (A24/30), drawn up on 13th March 1930, set out the specification as follows:

Requirement 1: To fly the North Atlantic. 3,000 nautical miles against a 17 knot (20 mph) headwind with 30 per cent fuel reserves flying at 65 knots (75 mph). Required duration: 81.25 hours.

Requirement 2: To fly to India. 2,400 nautical miles against a 17 knot (20 mph) headwind with 30 per cent fuel reserves at the

* AIR 5/987 and Colonel Richmond's diary.

worst time of year for lift, flying at 65 knots (75 mph). Required duration: 65 hours.

Estimated gross lift required under standard conditions: 288 tons from a volumetric capacity of 9,500,000 cubic feet. Estimated fixed weights: 183 tons. Useful lift: 104 tons.

Length: 906 feet. Diameter 163 feet.

Fuel load: 62 tons. Crew and stores: 22 tons. Payload: 20 tons. Water ballast at five per cent of gross lift: 14.4 tons=3,225 gallons. At seven passengers to a ton, or 320 lb each with food, water and baggage: 70 passengers=22,400 lb or 10 tons.

Alternatively: For an operating range of 1,500 miles, with additional masts, examination to be made of an airship of 7,250,000 cubic feet. Gross lift 227 tons. Fixed weights 147 tons. Useful load 73 tons.

These two new design studies 'Project G' and 'Project H' – were advanced side by side.

Further experience of R.100 and R.101 now provided much more reliable data on which to base the new design. Early in August 1930 the whole situation was reviewed again at the Air Ministry under Sir John Higgins and a final decision was taken to go forward with two sister ships – R.102 and R.103 – as developed versions of R.101 with just over one-third more capacity. Attractive as the 'Project G' had become as a 9,500,000 cubic feet 'all-can-do' airship, the costs and time-scale of research, of a new shed and of construction were judged to be too great. A more modest design, which would be based on R.101, would be quicker, cheaper and – with three new masts – could do the job.

The finally agreed OR was now for the two new airships to be each of 7,500,000 feet volumetric capacity to give a gross lift of 227.7 tons or 218.5 tons when inflated to 96 per cent of capacity for a pressure height of 1,200 feet under standard conditions.

The new ships would have a length of 822 feet and a maximum diameter of 134 feet to give a 'fineness ratio' of 6.13 to 1 with fuller lines at the bow and stern compared with R.101, in order to provide more lift, fore and aft, for the relatively heavy mooring gear at the nose and for the enlarged tail surfaces. Both ships would be built at the Royal Airship Works, Cardington, by the same, proven, methods of construction in association with Boulton and Paul Aircraft Limited of Norwich.

Following this meeting at the Air Ministry design work was set seriously in hand at Cardington and, on 22nd August, Reginald Colmore summed up the intentions in a Project Paper * which gave detailed estimates of the costs of the first of the new airships – now designated R.102 – to be ready for flight by mid 1933, together with a comparison with the recorded costs of R.101.

The new ship woould be 50 per cent larger than R.101, would be built at Cardington and it would begin trials in the summer of 1933 ready for scheduled services in 1934. R.102 – and its proposed sister ship, R.103 – would have 'rather more elaborate accommodation' than R.101, designed for 50 passengers together with ten tons of cargo and mail – a payload now set at 14.24 tons to be carried for a still air range of 4,000 nautical miles, cruising at 55 knots on 2,520 bhp from seven improved Beardmore Tornado diesel engines.

In other words, R.102 would be designed to meet the optimistic specifications of requirements set out in 1924. By 1934 airship technology would have caught up the hopes of ten years before.

Colmore noted that the cost of R.101 from the start in July 1924 up to its launch in October 1929 – 5¼ years – had come to £581,270 all in. He estimated that R. 102 could be designed and built for £535,000. The savings compared with R.101 – in a non-inflationary age† – would come from completing the work in 2½ years and in doing a much greater proportion of the construction at Cardington with now little sub-contracted to Boulton and Paul at Norwich.

Including trials, and an allowance for contingencies, Colmore rounded up the estimated total expenditure on R.102 to the end of 1934 to £600,000 which compared with the £611,000 spent on R.101 up to the end of June, 1930.

Colmore set out the cost comparisons in tabular form : (see page 531).

Thus, with experience of R.101 and a greatly speeded up programme of work, R.102 was optimistically forecast to cost less than the smaller R.101.

A further Paper in September brought into the projected programme the additional costs of the second – and identical – new airship, R.103.‡ This was set at a further £478,550 including flying trials. When the estimated operating costs of R.100, R.101, R.102 and R.103 were added up to the end of 1934, the total expenditure on four British airships through the four years, 1931–1934, was estimated at £1,902,000 at an average rate of expenditure of £475,000 a year. The sum to be sought in the Air Estimates would be around £500,000 a year to put Britain in the lead in the commercial airship business.

The details were set down thus :

* AIR 5/1407. † Except for gas ! ‡ See page 532.

R.101 and R.102 — COST COMPARISONS
(Air 5/987)

Royal Airship Works (22nd August 1930)	*R.101A Actual Costs*	*R.102 Estimated Costs*	*Variation R.102 to R.101A*
1. Period of Design and Construction	1924—1929	1930—1933	
2. Months elapsed	58	38	– 20mhs
3. Nominal Volumetric Capacity (cu. ft.)	5,000,000	7,500,000	+ 50%
4. Actual capacity (cu. ft.)	4,895,060	7,380,000	+ 50%
5. Number of engines	5	7	+ 40%
COSTS	£	£	
1. Direct Labour			
R.A.W.	102,000	150,000	
B & P	45,000	—	
Total Direct Labour:	147,000	150,000	+ 2%
2. Materials			
R.A.W.	64,000	135,000	
B & P	47,000	—	
Total Materials:	111,000	135,000	– 21.6%
3. Design Labour			
R.A.W.	43,000	35,000	
B & P	7,000	—	
Total Design Labour:	50,000	35,000	– 30%
4. Overheads			
R.A.W.	114,000	135,000	
B & P	53,000	—	
Total Overheads:	167,000	135,000	– 20%
5. Jigs & Tools (B&P)	13,000	14,000	
6. Plant & Equipment (B&P)	4,500	6,000	
7. Component Delivery (B&P)	1,000	—	
8. Ex Gratia Payment (B&P)	14,000	—	
9. Profit—Boulton & Paul (B&P)	18,000	—	
Total Jigs, Tools, Etc.:	50,500	20,000	– 39.6%
10. Main Power Plant	(5) 42,000	(7) 52,500	
11. Spare Engines (Part Value)	(2) 14,000	(1) 7,500	
Total Power Plant:	56,000	60,000	+ 7%
Total Design and Construction	581,500	535,000	– 8%
12. Inflation, Trials, Misc.	29,730	30,000	
13. Contingencies	—	35,000	
Total at 30th June 1930:	£611,230	£600,000	– 2%

ROYAL AIRSHIP WORKS
AIRSHIP DEVELOPMENT PROGRAMME 1931–34

	Four Year Programme 1931—34	R.100 £	R.101 £	R.102 £	R.103 £	Totals £
A.	*Design & Construction*					
1.	Shop Labour	27,500	7,000	122,000	102,000	256,500
		New bay				
2.	Materials	27,500	7,500	133,000	133,000	301,000
3.	Design Labour	8,000	2,500	35,000	250,000	70,500
4.	Overheads	29,000	8,000	135,000	75,000	247,000
5.	Jigs & Tools	—	—	14,000	8,000	22,000
6.	Plant & Equipment	—	—	6,000	2,000	8,000
7.	Power Plants	—	—	55,000	55,000	110,000
8.	Inflation & Trials	10,000	—	30,000	30,000	70,000
9.	Contingencies	7,500	2,500	20,000	20,000	50,000
10.	*Total Design & Construction*	107,000	27,500	550,000	450,000	1,135,000
B.	*Direct Operating Costs*					
B1.	*Spares & Replacements*					
1.	Engines & Auxillaries	42,000	75,000	—	—	117,000
2.	Complete Set of Gas Bags	32,000	32,000	—	—	64,000
3.	Complete Outer Covers	9,000	9,000	—	—	18,000
4.	Minor Replacements	16,000	16,000	10,000	4,000	46,000
5.	*Total Spares & Replacements*	99,000	132,000	10,000	4,000	245,000
B2.	*Flying Staff*					
1.	Officers & Crew on Service	56,000	61,600	25,250	12,400	155,250
2.	Officers & Crew Training	—	—	3,600	4,150	7,750
3.	*Total Flying Staff Costs*	56,000	61,600	28,850	16,550	163,000
B3.	*Operations* (Fuel, oil, gas, victuals, etc.)					
1.	Ex: Cardington Base	37,400	18,800	8,200	5,000	69,400
2.	Ex: Ismailia Base	72,050	59,950	—	—	132,000
3.	Ex: Karachi Base	42,900	36,300	—	—	79,200
4.	Ex: Montreal Base	38,150	31,450	5,800	3,000	78,400
5.	*Total Operations*	190,500	146,500	14,000	8,000	359,000
C.	*Estimated Direct Operating Costs*	345,500	340,100	52,850	28,550	767,000
D.	*Total Costs*	453,000	367,600	602,850	478,550	1,902,000
E.	*Planned Flying Hours (1931—34)*	5,300	5,425	400	225	11,350
F.	*Direct Operating Costs Per Flying Hour*	£ 65.19	£ 62.69	£ 132.12	£ 126.89	£ 67.58

During these four years, while the two new ships were in the course of design and construction, the plan would be for scheduled airship services between Cardington and Karachi and Cardington and Montreal to be maintained with R.101 and R.100; both lengthened with additional bays. The new ships would be introduced on to the Atlantic services in 1934.

By 1935, when all four ships would be available, 6,000 flying hours of commercial services were projected at a Direct Operating Cost estimated at £2,291,300 at an average of £70 a flying hour, equivalent too about 28 shilliings (£1.40p) an aircraft mile in 1930's values.

The annual spread of costs was estimated to be likely to be approximately as follows:

FIVE YEAR AIRSHIP PROGRAMME — 1931–35

(Design, Construction, Modifications, Flying Trials, Maintenance and Operating Costs)

	1931	1932	1933	1934	1935	Totals
Preparations	£	£	£	£	£	£
R.100	82,000	11,500	7,000	7,000	—	107,500
R.101	2,000	11,500	7,000	7,000	—	27,500
R.102	90,000	265,000	150,000	35,000	10,000	550,000
R.103	—	100,000	180,000	150,000	20,000	450,000
Total Design Construction, Mods.	174,000	388,000	344,000	199,000	30,000	1,135,000
Spares and Spares Replacements	33,500	62,500	64,500	84,500	125,000	370,500
Flying Staff Costs	30,000	30,000	41,250	61,750	62,800	225,800
Operating Costs	56,000	93,500	97,500	111,500	201,000	568,000
Total Operations	120,000	186,000	203,250	257,750	388,800	1,156,300
Totals By Year	294,000	574,000	547,250	456,750	419,300	2,291,300

A commercial revenue averaging £63 a flying hour was – somewhat optimistically – forecast. It would thus, if realised, come within a short distance of meeting the direct operating costs though not, of course, of making any contribution to the estimated total investment of £4,380,000 on the eleven-year, four airship, programme – and leaving out of account also the projected third airship shed at Cardington and a second Cardington mooring mast to go with it.

In September 1930 the path into the unexplored technical and operational jungle of airship development still seemed full of promise. From the experience gained so far with R.100 and R.101 and further five-year plan, extending to the end of 1935 seemed to the Cardington team, and to the devoted airship band in Whitehall, to be the logical progression. It would fulfil the plans approved by the Cabinet in 1924 – and looked favourably upon by the previous Administration as well – of a four-airship fleet capable of flying regular, scheduled, services to link the Mother Country with the overseas centres of the British Empire.

The plan was drawn up, therefore, to include, in the 6,000 flying hours a year, ten annual round trips across the North Atlantic to Montreal during the summer months and 22 round trips in each full year between Cardington, Ismailia and Karachi.

From these beginnings more expansive plans for new ships would be developed.* Australia would be brought into the operating programme after 1935.

To these ends a revised provisional operating programme was drawn up later in September 1930, between Henry Self and Louis Reynolds at the Air Ministry (Adastral House) and Reginald Colmore, Vincent Richmond and Michael Rope at the Royal Airship Works, Cardington. The total flying hours over the five year period 1931 to 1935 were now set at 24,285 with R.101 at 11,250 flying hours bearing the brunt, and equated to the expenditure now brought up to a total of £2,803,450 over the five years at an average of £560,690 a year, peaking in 1933 at £663,000.

By 1935 the four airships would offer in the year a monthly return service from Cardington to Montreal, a monthly return service to Karachi and a weekly service to and from Ismalia in Egypt. Australia would come into the pattern in 1936.

The revised estimate of flying hours and costs for 1931 to 1935 were worked out thus:

Year	*Annual Costs of New Construction*	*Annual Operating Costs*	*Estimated Total Costs Per Annum*	*Estimate of Annual Hours to be Flown*
1931	£308,000	£120,600	£428,600	3,060
1932	£456,000	£186,600	£642,000	4,730
1933	£456,000	£207,800	£663,800	4,920
1934	£280,000	£247,350	£527,350	5,575
1935	£200,000	£340,100	£540,100	6,000
Totals	£1,700,000	£1,103,450	£2,803,450	24,285
			excluding interest on capital employed	

* AIR 5/988.

The object was to achieve a practical schedule of long-distance airship services by 1935. By then an average revenue of some £80 a flying hour from passengers, mail and cargo was forecast against an operating cost by then set at about £57 a flying hour. Optimistic, but a target to aim for.

By 1935, if all went well, a Third Phase could be contemplated. It would bring into construction the still bigger new airships of the, by then, *'R.104 Class'* of 9,500,000 cubic feet capacity already sketched out on the Cardington drawing boards.

It was all long-term-planning on an imaginative scale, worked out by Colmore, Reynolds, Richmond, Rope and Self with the knowledge that in this they had the enthusiastic support of the Secretary of State for Air.

There was, however, less enthusiasm from the new Permanent succession to Sir Walter Nicholson.

On 13th September 1930 Bullock wrote from Gwydyr House in Whitehall to Hugh Dowding, the new Air Member for Supply and Research at Adastral House, Kingsway. He attached a draft Memorandum to the Secretary of State for Air * on Airship Policy. He sought Dowding's support for the Memorandum whiich – while supporting the long-term plans in principle as a comprehensive view of the airship situation, sought a cautious approach before further commitments were entered into.

> I am quite clear [he wrote] that I myself, if I were Secretary of State, would not consent to further construction at this stage. My primary object has been to ensure that, if the Secretary of State insists on a gamble, he does so with his eyes open and realises that he is making a radical departure from the wisely cautious policy we have hitherto pursued.

The note – with typical civil service caution – suggested that the proposed programme for R.102 and R.103 should be postponed and that a new, curtailed, programme should be substituted as a next stop. It should concentrate upon demonstrating the practicability of an intensive regular operation with R.100 and R.101 for about a year, wholly on the England to Egypt route, flying with one extra mast at Malta. Then, at the end of another year – that is, early in 1932 – the two new

* AIR 5–987–6357 and AIR 8/21–07444.

airships could be put in hand against a background of much more operating experience.

> It seems to me [wrote Bullock] that a slight setting back of the clock should be faced and really does not matter. In developing a new form of transport (which should last until we let atomic energy loose or achieve some form of auto-projection of the individual in the Halanesque style) a year's delay, if that should prove inevitable is, after all, a bagatelle.
>
> I suggest, therefore, that we should concentrate – as the next stage – on what I have called the real crux of the matter; the demonstration of the practicability of regular operations with the enlarged R.100 and R.101.

By this time – 13th September 1930 – authorisation had been given for preliminary design work to be set seriously in hand on R.102 and R.103 pending a full review of the forward programme after the flight of R.101 to India. A decision would then be taken on the pace of the programme.

Meanwhile, at the Air Ministry, a totting up of the total expenditure of time and money upon R.101 was being made as a background to the projected Phase Two of the airship programme. This analysis was completed up to the departure of R.101 for India on 4th October.

For the six years to 30th June 1930 the expenditure of R.101 was found to have added up to a total of £611,830. Between 1st July and 4th October a further £100,365 had been spent primarily on the extra work on the design and installation of the new bay and the allocation to the airship of its share of the Cardington's overheads.

The additional costs of R.101 between 1st July and 4th October 1930 were:

		£
Direct Labour		29,105
Materials		21,960
Design Labour		16,240
Overheads		33,060
	Total	£100,365

Thus the total cost of R.101 from the start of design in November 1924, to the end of October 1930, amounted to £711,595 (at an average of £122,000 per annum) made up as follows:

	£	Percentage of Total
Direct Labour	176,105	24.7%
Materials	132,960	18.7%
Design Labour	66,240	9.3%
Overheads	214,060	30.1%
Jigs, tools, plant, etc.	18,500	2.6%
Power plant and spares	56,000	7.9%
Inflation, trials, miscellaneous	29,730	4.2%
B & P Profit (at 10%)	18,000	2.5%
Total	£711,595	100%

From these costs and from the wage rates which were paid at that time at the Royal Airship Works and at Boulton and Paul at Norwich (16 old pence an hour in the shops and 24 old pence an hour in the design offices) one can deduce that a total of 500,000 direct design man-hours was expended on R.101 up to its first flight, together with 2,205,000 direct shop man-hours of construction and assembly.

In total from first to last R.101 absorbed 662,400 direct design man-hours and 2,641,575 direct shop man-hours of construction; that is a total of 68,462 man-weeks of design and construction at the average of 48.2 man-hours a week with overtime, being worked at the RAW at that time. The average wage then was £4 16s 4d a week in the design office and £3 4s 3d a week on the shop floor.

In the end, R.100 and R.101, together with their sheds, mooring masts, the research programme and the general administration of the airship programme over a period of 76 months between July, 1924, and October, 1930, cost a total of just on £2.4 million. It was divided thus:

The British Rigid Airship Programme 1924—1930

		£
1.	Experiments with R.33 and R.36 (1924—26)	103,100
2.	Cardington sheds (one transferred from Pulham)	205,800
3.	Karachi shed and facilities	120,000
4.	Mooring masts	160,000
5.	Research staff and programmes	305,200
6.	R.100	471,113
7.	R.101	711,595
8.	Flying operations—fuel, oil, gas, maintenance	30,056
9.	Overheads, administration and miscellaneous (Air Ministry, Cardington, Howden and Pulham)	290,084
10.	*Total British Airship Programme (1924—30)*	£2,396,948

For this expenditure a total of 424 flying hours was achieved; 297 hours from R.100 and 127 from R.101. The total cost works out, therefore, at £5,653 a flying hour – plus 48 lives.

In such stark terms, the airship programme was an economic and a human disaster. Yet in technical terms and in those of human endeavour it was, undoubtedly, justified as one of those brave expeditions into the unknown, the potentially rewarding results from which can be proved, or disproved, only by exploration.

Had the flight to India succeeded in October 1930, as it might well have done despite the odds, still the full bright expectations of the bold endeavour inherent in the airship programme were, inevitably doomed to future.

Even if R. 102 and R.103 had been constructed there remained the fundamental deficiencies of Lighter-than-Air operations: first the need to fly low to conserve gas and hence having to fly always in low-level turbulence, often in cloud and rain; secondly the inevitable slow speed of the airship so that quite normal winds would often almost double the time taken on an already lengthy journey. Reasonable punctuality would not be possible. Thirdly, the problem of the 'super-heating' and the 'supercooling' of gas in tropical conditions and in cold weather together with the serious problems of ice formation on the fragile outer covers.

Add to these difficulties the vulnerability of vast structures for Lighter-than Air, including the mooring problems in other than light winds, together with the low productivity of the airship in terms of revenue-ton-miles per annum, and the practical use of airships as transport vehicles evaporates.

Even the evolution of the gas turbine engine – so helpful to the development of the aeroplane – is of little use to the airship because of the poor specific fuel consumption of the gas turbine when used at low levels.

Today, the design and development costs of any new large rigid airship would be immense – in hundreds of millions of pounds.

Delightful and evocative as a new rigid airship would be, it could not survive in the modern, competitive, transport environment nor overcome the millstone of its disastrous record in service in earlier days.

At 1600 hrs on Monday, 31st August 1931, the Cabinet, meeting at 10 Downing Street under the Prime Minister, the Right Hon James Ramsay Macdonald, decided that British airship development should be abandoned and Cardington reduced to a Care and Maintenance basis though keeping a watching brief on overseas airship work. Just

$7\frac{1}{2}$ years had passed since Christopher Thomson had come to the Air Ministry and had presided over a meeting to launch the Rigid Airship programme.

And so closed a gallant chapter in the history of technical development and of air transport.

SOURCES AND ACKNOWLEDGEMENTS

Sources and Acknowledgements

TEN main sources of historic information have been researched in detail over the years in the preparation of this book. They are:

1. *The Airship Files* in the Public Record Office (AIR 2 to AIR 19).
2. *The Cabinet Papers* of 1924 to 1931 likewise held in the P.R.O.
3. *Lord Thomson's* surviving writings, notably portions of his Journal together with letters and articles; many now in the University of Texas at Austin, Texas, USA. Also Lord Thomson's books, *Smaranda* (Jonathan Cape, 1926) and *Air Facts and Problems* (John Murray, 1927).
4. *The Papers* of the Right Hon. J. Ramsay MacDonald, now in the P.R.O. and Mr MacDonald's introduction to *Smaranda.*
5. *Letters and writings* of SA The Princess Marthe Bibesco including her books: *Le Destin de Lord Thomson of Cardington* (Flamourin, 1932); *La Vie d'une Amitie* (Librarie Plon Paris, 1960) and *Egalitie* (Bernard Grasset, Paris, 1935) together with her diaries of 1915 to 1931.
6. *Letters* from Lord Thomson's mother, Mrs Emily Thomson, in 1930 on to her death in 1933.
7. *The diary* of Noël Grabowsky Atherstone, First Officer of R.101.
8. *The Office Day Book* of Colonel Vincent C. Richmond, Chief Designer of R.101 now at the RAF Museum, Hendon.
9. *Letters and papers* of Wing Commander T. R. Cave-Browne-Cave and of Mr J. M. Ward Dyer.
10. *Private papers* and letters to the author of Sir Christopher Bullock, Sir Maurice Dean, Sir Henry Self and Lord Templewood.

I am very much indebted to the Director of the Public Record Office and to his staff for much help and for permission to quote from official documents. I am equally grateful to the Directors and staffs of the Imperial War Museum, of the Royal Air Force Museum, of the Science Museum, and of HM Meteorological Office.

In addition I must acknowledge with much gratitude the generous help of a large number of people who have given me unstintingly of their time and knowledge and of their memories and records of the airship years. Especially I must mention:

Dr. Richard and Miss Rebecca Atherstone, the Right Hon the Lord

Balfour of Inchrye, the Right Hon the Lord Baker, Mr Leslie Barber, the Right Hon the Lord Brabazon of Tara, Mr John Brancker, Mr D. C. Brech, Mr Peter W. Brooks, Mr J. M. Bruce, Sir Christopher Bullock, Mrs Sue Burgess, the Lady Maie Casey, Wing Commander and Mrs T. R. Cave-Browne-Cave, Mr Jonathan Chadwick, Mr Richard M. Clarkson, Mr T. S. D. Collins, Mr Barry Countryman, Mr Robert Crawford, Monsieur André and Monsieur Georges Crucifix, Sir Maurice Dean, Mr F. W. Ward Dyer, Mr J. R. Ede, Sir John Elliot, Mr Samuel Evans, Mrs Joan Faircloth, Mr Henry Farrer, Dr Noble Frankland, the Right Hon the Viscount Gage, Mrs Ian H. Grabowsky, Brigadier John Ghika, Prince and Princess Demetrie Ghika-Comanesti, Air Marshal Sir Victor Goddard, Air Commodore E. B. Haslam, Mrs Olivia Hayes, Mr E. P. Herriz-Smith, Mr Frank Kiernan, the Right Hon the Lord Kings-Norton, Mr R. J. Lewenden, Sir Basil Liddell Hart, the Right Hon Malcolm MacDonald, Captain George Meager, Mr David Marquand, Mr Richard Masefield, Mrs F. E. M. Mattey, Professor Henry Cord Meyer, Mr Robert B. Meyer, Mr Nevil Shute Norway, Sir Alfred Pugsley, Major Louis Reynolds, Mrs Margaret Richardson-Bunbury, Mr W. Roach, Mr Crispin Rope, Professor R. S. Scorer, Sir Henry Self, Mr Daniel F. Sowa, Mr L. A. Speed, Mr John Stroud, Mr R. W. Swales, Mrs F. Szczuka, Mr and Mrs Edwin Taylor, Mr J. E. Temple, Commander David Thomson, Colonel Roger Thomson, Mr H. G. Tibbett, the Rev Peter Tubbs, Sir Frederick Tymms, the Right Hon the Lord Ventry, Mademoiselle Claudia Verhesen, Dr P. B. Walker, Sir Barnes Wallis, Mrs Irene Wagner, Mr Gordon Wansbrough Whiite, Mademoiselle Yvonne Wetten, Mr John W. B. Wimble and Mr Humphrey Wynn.

Among them I am especially grateful for the most valuable assistance which I have received from the Lord Baker, Dr Richard Atherstone, Mr Peter Brooks, Dr Noble Frankland, Brigadier John Ghika and Mr Crispin Rope all of whom have read the proofs and have given me comments and advice for which I am very much in their debt. In addition I acknowledge with gratitude the help and interest of the staffs of the Imperial War Museum, the Historical Branch of the Air Ministry (Ministry of Defence) and of the Public Record Office where the airship reference files are largely contained in File Numbers AIR 2, 3, 5, 8, 11, 12 and 19.

Detailed references to individual sources are made in footnotes to the text.

My thanks are also due to Messrs Michael Joseph for permission to quote from *Tides and Eddies* by Maie Casey and to Dr Jan Van

Loewen Ltd for permission to quote from Noël Coward's *Bitter Sweet*.

Finally I must express my gratitude for the long, patient and cheerful typing of successive drafts of the manuscript over the years on the part of Mrs Margot Bateman, Mrs Margaret Brazier and Mrs Marion Jones and the help with photostat copies by Miss Anne Hamilton, Miss Mary Edwards, Miss Sue Howell and Miss Jane Young.

Without all of these kindly helpers the book could not have been completed.

Index

Italicised roman numerals refer to photographic plate numbers